ATOMIC PHYSICS

ATOMIC PHYSICS

MAX BORN

EIGHTH EDITION

From the original translation of

JOHN DOUGALL

M.A., D.Sc., F.R.S.E.

revised by

R. J. BLIN-STOYLE *and* **J. M. RADCLIFFE**

M.A., D.Phil.   B.Sc., Ph.D.

University of Sussex

DOVER PUBLICATIONS, INC.
New York

Published in Canada by General Publishing Company, Ltd., 30 Lesmill
Road, Don Mills, Toronto, Ontario.
Published in the United Kingdom by Constable and Company, Ltd.

This Dover edition, first published in 1989, is an unabridged and
unaltered republication of the eighth edition (1969) of the work first
published by Blackie & Son Ltd., Glasgow, in 1935. It is reprinted by
special arrangement with Blackie and Son Ltd., Bishopbriggs, Glasgow
G64 2NZ, Scotland.

Manufactured in the United States of America
Dover Publications, Inc., 31 East 2nd Street, Mineola, N.Y. 11501

Library of Congress Cataloging-in-Publication Data

Born, Max, 1882-1970.
  Atomic physics.

  Reprint. Originally published: Glasgow : Blackie, 1969.
  Translation of Moderne Physik.
  Bibliography: p.
  1. Nuclear physics. I. Blin-Stoyle, R. J. (Roger John) II. Radcliffe, J. M.
III. Title.
QC776. B5713 1989   539.7   89-12033
ISBN 0-486-65984-4

PREFACE

THE German edition of this book, which appeared in 1933 under the title *Moderne Physik*, had its origin in a course of lectures which I gave at the Technical College, Berlin–Charlottenburg, at the instance of the Society of Electrical Engineers, and which were reported by Dr. F. Sauter. The choice of material was not left entirely to myself, the Society making certain suggestions which took into account the previous knowledge and the professional predilections of its members.

A year or two later, the idea of an English edition was mooted. But by this time the title of the book had become inappropriate, for in the interval the growing point of physics had shifted, and interest had become centred in the physics of the nucleus, of which only a bare sketch had been given in the lectures. The change in the title, however, has nothing to do with this, but is due to the fact that the publishers already have a book entitled *Modern Physics* on their list of publications.

As a theoretical physicist, I have naturally placed the theoretical interpretation of phenomena well in the foreground. The text itself, however, contains only comparatively simple discussions of theory. Proofs, short but complete, of the formulæ employed are collected in a series of Appendices.

In the lectures two years ago, I emphasized with a certain pride the successes of the theory—now I have rather to direct attention to the immense progress which has been made on the experimental side. But I have not been able to resist the temptation to mention a few ideas of my own, which I hope may be of some assistance in stirring the theory into renewed activity.

MAX BORN.

CAMBRIDGE, 1935.

PREFACE TO THE EIGHTH EDITION

BORN's *Atomic Physics*, well known for so long to students and teachers of physics throughout the world, was first published in 1935 and has now reached its eighth edition. During the thirty or so years of its life there have been dramatic developments in all branches of physics, particularly in the field of elementary particles. In each successive edition Professor Born (later in collaboration with one of us (RJB-S)) incorporated these developments, treating them at the same level and in the same style as the earlier editions. In this eighth edition, we have attempted to continue in this same spirit. Some changes in the ordering have taken place and, in particular, a new chapter on the quantum theory of solids has been introduced. Apart from this we have revised the various chapters to the best of our ability and in doing so have tried to conform to the tone set by Professor Born.

We do not regard the book as a standard text but rather as a broad treatment of basic physics. It is also something of a historical document. For this reason we have chosen to retain many early references which would not normally appear in modern works. Any lapses or gaps in the new text should be attributed entirely to us and not to Professor Born. He is, however, fully responsible for the Conclusion in Chapter X. This Conclusion has hardly changed over the years and we feel that it stands as a remarkable testimony to the physical insight and vision of a great physicist.

R. J. BLIN-STOYLE.
J. M. RADCLIFFE.

CONTENTS

CONTENTS

CHAPTER V

ATOMIC STRUCTURE AND SPECTRAL LINES

CHAPTER VI

SPIN OF THE ELECTRON AND PAULI'S PRINCIPLE

CHAPTER VII

QUANTUM STATISTICS

CONTENTS

CHAPTER VIII

MOLECULAR STRUCTURE

CHAPTER IX

QUANTUM THEORY OF SOLIDS

CHAPTER X

NUCLEAR PHYSICS

APPENDICES

ATOMIC PHYSICS

CHAPTER I

Kinetic Theory of Gases

1. Atomic Theory in Chemistry.

In present-day physics, the concepts of energy and matter are
connected in the most intimate way with atomic theory. We therefore
begin with a brief discussion of the rise of ideas relating to the atom.
The source of these ideas, we know, is chemistry. They suggest themselves almost inevitably when we try to interpret the simple regularities which are at once disclosed when the masses of the substances
transformed in chemical reactions are determined quantitatively with
the balance. It is established in the first place, that in a reaction the
total weight remains unchanged. Secondly, it is found that substances
combine only in fixed simple proportions by weight, so that a definite
weight of one substance can only enter into reaction with definite
weights of a second substance; and the ratio of these weights is independent of the external conditions, such as, for example, the proportion
by weight in which the two substances may have been mixed. These
regularities are expressed in the language of the chemist in the *laws of
constant and of multiple proportions* (Proust, 1799; Dalton, 1808), e.g.:

1 gm. hydrogen combines with 8 gm. oxygen to form 9 gm. water,
1 gm. hydrogen combines with 35·5 gm. chlorine to form 36·5 gm. hydrogen chloride.

An example of the law of multiple proportions is given by the
compounds of nitrogen and oxygen: 7 gm. nitrogen combine with

1×4 gm. oxygen to form 11 gm. nitrous oxide,
2×4 „ „ „ 15 „ nitric oxide,
3×4 „ „ „ 19 „ nitrous anhydride,
4×4 „ „ „ 23 „ nitrogen dioxide,
5×4 „ „ „ 27 „ nitric anhydride.

In the case of gases, simple laws hold not only for the weights of the reacting substances but also for their *volumes* (Gay-Lussac, 1808). Thus (at constant pressure)

2 vols. hydrogen combine with 1 vol. oxygen to form 2 vols. water vapour,
1 vol. hydrogen combines with 1 vol. chlorine to form 2 vols. hydrogen chloride.

The numbers expressing the ratios by volume are those which appear in the corresponding chemical formulæ. In the preceding examples, for instance, we have

$$2H_2 + O_2 = 2H_2O,$$
$$H_2 + Cl_2 = 2HCl;$$

and

$$2N_2 + O_2 = 2N_2O,$$
$$2N_2 + 2O_2 = 4NO,$$
$$2N_2 + 3O_2 = 2N_2O_3,$$
$$2N_2 + 4O_2 = 4NO_2,$$
$$2N_2 + 5O_2 = 2N_2O_5.$$

These facts may be interpreted as follows, as was done by Avogadro: every gas consists of a great number of particles, its atoms or molecules; and equal volumes of all gases, at the same temperature and pressure, contain the same number of molecules.

The significance of this principle in relation to the laws of chemical reactions may be illustrated from the above examples. The fact that two volumes of hydrogen combine with one volume of oxygen to form two volumes of water vapour is (Avogadro, 1811) equivalent to the statement that two molecules of hydrogen combine with one molecule of oxygen to form two molecules of water. Similarly, the combination of one part by weight of hydrogen with eight parts by weight of oxygen to form nine parts by weight of water means that a molecule of oxygen must be eight times, and two molecules of water nine times, as heavy as two molecules of hydrogen.

We are thus led to the concepts of *molecular weight* and *atomic weight*. These are, respectively, the weights of a molecule and of an atom of the substance in question. They are not measured in grammes, but with reference to a standard (ideal) gas, the atomic weight of which is put equal to 1; and it has been agreed to define this, not so that H = 1, but so that C = 12; this convention has turned out very lucky (because of the existence of the heavy isotope of hydrogen, p. 68). We shall denote the molecular weight measured in this way by μ.

That quantity of a substance, the weight of which is μ gm., is called a *mole* (even when the substance is not capable of existence in the chemical sense). A mole of oxygen atoms therefore weighs 16 gm., but a mole of oxygen molecules weighs 32 gm. From this definition of the mole it follows that the quantity "1 mole" always contains the same number of molecules. This number of molecules per mole plays a great part in the kinetic theory of gases. The number of molecules per cubic centimetre from which this quantity can be derived was first measured by Loschmidt in 1865. However, at the present time the convention is to refer to the number of molecules per cubic centimetre as *Loschmidt's number* and the number per mole as *Avogadro's number*. We denote the latter by N_0; its value is (see p. 23)

$$N_0 = 6 \cdot 025 \times 10^{23} \text{ mole}^{-1}.$$

As a consequence of Avogadro's law, 1 mole of any gas at a given pressure p and a given temperature T always occupies the same volume; for a pressure of 760 mm. of mercury and a temperature of $0°$ C. the volume is 22·4 litres.

We add here an explanation of a number of symbols which will be employed below. If m is the mass of a molecule in grammes, then $\mu = N_0 m$; in particular, for atomic hydrogen (μ almost exactly $= 1$), $N_0 m_{\text{H}} = 1$. If further n is the number of molecules in the unit of volume, N that in the volume V, and if ν is the number of moles in the volume V, then we have $\nu N_0 = n V = N$. Finally, we denote by $\rho = nm$ the density of the gas, and by $v_s = 1/\rho$ its specific volume.

2. Fundamental Assumptions of the Kinetic Theory of Gases.

After these prefatory remarks on the atomic theory in chemistry, we now proceed to the *kinetic theory of gases* (Herapath, 1821; Waterston, 1843; Krönig, 1856). Considering the enormous number of particles in unit volume of a gas, it would of course be a perfectly hopeless undertaking to attempt to describe the state of the gas by specifying the position and velocity of every individual particle. As in all phenomena of matter in bulk, we must here have recourse to *statistics*. But the statistics now to be used is of a somewhat different kind from that which we are acquainted with in ordinary life. There, the statistical method consists in recording a large number of events which have occurred, and in drawing conclusions from the numerical data so obtained. Thus, mortality statistics answers the question of how much more probable it is that a man will die at 60 than at 20 years of age; for this purpose we count the number of cases of deaths of men at these ages over a long period, and take the respective numbers found for the two ages as proportional to the probabilities required.

If we propose to apply the statistical method to the theory of gases, the method of procedure must be essentially different; for an enumeration of the molecules which, for example, occupy a given element of volume at a given moment is utterly impossible. We must therefore proceed indirectly, first introducing assumptions which appear plausible, and then building up the theory on these as foundation. As with every scientific theory, the final warrant for the correctness of the assumptions is the agreement of their logical consequences with experience.

We may wish, for example, to know the probability of finding a gas molecule at a definite spot in the box within which we suppose the gas to have been enclosed. If no external forces act on the molecules, we shall be unable to give any reason why a particle of gas should be at one place in the box rather than at another. Similarly, in this case there is no assignable reason why a particle of the gas should move in one direction rather than in another. We therefore introduce the following hypothesis, the *principle of molecular chaos*: For the molecules of gas in a closed box, in the absence of external forces, *all positions in the box and all directions of velocity are equally probable.*

In the kinetic theory of gases we shall only have to do with mean values, such as time averages, space averages, mean values over all directions, and so on. Individual values entirely elude observation. If n_a denotes the number of molecules per unit volume with a definite property a, e.g. with a velocity of definite magnitude, or with a definite x-component of velocity, then by the mean value of a we understand the quantity $\bar{a}$, where

$$\bar{a} = \frac{\Sigma n_a a}{\Sigma n_a}, \quad \text{or} \quad n\bar{a} = \Sigma n_a a,$$

n standing for Σn_a, the number of molecules per cubic centimetre. If, for example, we suppose the velocity of each molecule to be represented by a vector v with the components ξ, η, ζ, and therefore the magnitude $v = \sqrt{(\xi^2 + \eta^2 + \zeta^2)}$, and we wish to find the mean value $\bar{\xi}$ (for molecules with the velocity v), then by the principle of molecular chaos with respect to directions of motion, exactly as many molecules of the gas will have a velocity component $+\xi$ as a component $-\xi$; the mean value $\bar{\xi}$ must therefore vanish. A value of $\bar{\xi}$

differing from zero would imply a mean motion of the whole gas in the direction concerned, with this mean velocity.

On the other hand, $\overline{\xi^2}$ is not equal to zero. From symmetry, we have

$$\overline{\xi^2} = \overline{\eta^2} = \overline{\zeta^2}.$$

If we take the mean value over all directions, keeping v fixed, then it follows from $v^2 = \xi^2 + \eta^2 + \zeta^2$ by taking the mean values that

$$v^2 = \overline{\xi^2} + \overline{\eta^2} + \overline{\zeta^2} = 3\overline{\xi^2}, \quad \text{or} \quad \overline{\xi^2} = \overline{\eta^2} = \overline{\zeta^2} = \frac{v^2}{3}.$$

3. Calculation of the Pressure of a Gas.

With these ideas before us, we are already in a position to calculate the *gas pressure* p as the force acting on unit area (D. Bernoulli, 1738; Krönig, 1856; Clausius, 1857). According to the kinetic theory of gases, this force is equal to the change of momentum of the molecules striking unit of area of the wall per second. Take the x-axis at right angles

Fig. 1.—Number of collisions between the wall and molecules having the velocity v, in the element of time dt; it is equal to the number of molecules at a definite moment in the oblique cylinder of height $\xi\,dt$ and base the element of area struck.

to the wall. If n_v denotes the number of molecules in a cubic centimetre which possess the velocity v, then in the small time dt a given square centimetre of the wall is struck by all those molecules which at the beginning of the time element dt were within the oblique cylinder with edge $v\,dt$ standing on the square centimetre of the wall (fig. 1). Since the height of this cylinder is ξdt,

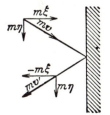

Fig. 2.—Momentum diagram for elastic collision of a molecule with the wall; the energy, the magnitude of the momentum and the component momentum parallel to the wall are not changed by the collision, but the component perpendicular to the wall has its sign reversed; momentum $2m\xi$ is therefore communicated to the wall.

its volume is also ξdt; the number of molecules in it is therefore $n_v\xi dt$. The area of the wall considered is struck per unit time by $n_v\xi$ molecules with the velocity v.

If we think of the molecules as billiard balls, every molecule when it strikes the wall has its momentum changed by $2m\xi$ perpendicular to the wall; the component of momentum parallel to the wall is not altered (fig. 2). The molecules considered therefore contribute to the total pressure p the part $2m\xi^2 n_v$. We sum these values first over all directions of incidence, that is to say, over a hemisphere, keeping v fixed.

The sum in the present case is equal to half the sum over a complete sphere, so that we have

$$2m\Sigma\xi^2 n_v = 2m \cdot \tfrac{1}{2}\overline{\xi^2} n_v = \frac{m}{3} v^2 n_v;$$

where n_v is the number of molecules in a cubic centimetre with a velocity of magnitude v. If we now sum further over all *magnitudes* of the velocity, we find for the total pressure

$$p = \frac{m}{3} \Sigma n_v v^2 = \frac{m}{3} n\overline{v^2}.$$

If V is the total volume of the gas, and N the total number of molecules, it follows from this equation on multiplication by V, since $nV = \nu N_0 = N$ that

$$Vp = N\frac{m}{3}\overline{v^2} = \frac{2}{3}\nu U,$$

where we have put

$$U = N_0 \frac{m}{2}\overline{v^2}.$$

Clearly U denotes the mean kinetic energy per mole, and in monatomic gases is identical with the total energy of the molecules in a mole. In polyatomic molecules the relations are more complicated, on account of the occurrence of rotations of the molecules, and vibrations of the atoms within a molecule. It can be shown, however, that the preceding formula for the gas pressure holds in this case also; U denotes as before the mean kinetic energy of the translational motion of the molecules, per mole, but it is no longer the same as the total energy.

4. Temperature of a Gas.

From the kinetic theory of gases, without a knowledge of the law of distribution of velocities (i.e. of the way in which the number n_v depends on v), we have found that the product of pressure and volume is a function only of the mean kinetic energy of the gas. But we have also an empirical law, the *law of Boyle* (1660) *and Mariotte* (1676), viz.: at constant temperature the product of the pressure and volume of an ideal gas is constant. We must conclude from this that U, the mean kinetic energy per mole, depends only on the temperature of the gas.

In the kinetic theory of gases, the concept of temperature is primarily a foreign element, since in fact the individual molecules are characterized by their velocity alone. But it suggests itself that we should define the absolute gas-temperature T in terms of the mean kinetic energy. This is usually done in accordance with the equation

$$\frac{m}{2}\,\overline{\xi^2} = \frac{k}{2}\,T,$$

where on the left we have the mean kinetic energy of a component of the motion of the centre of inertia of a molecule; k is called *Boltzmann's constant*. For the total motion of the centre of inertia we have therefore also

$$\frac{m}{2}\,\overline{v^2} = \frac{3}{2}\,kT,$$

and, referred to a mole,

$$U = N_0 \frac{m}{2}\,\overline{v^2} = \frac{3}{2}\,RT,$$

where we put

$$N_0 k = R.$$

The justification for this definition of temperature lies in the fact that when we introduce the last expression in the formula for the gas pressure deduced above, we obtain formally the relation which combines the *laws of Boyle-Mariotte and of Gay-Lussac and Charles*:

$$pV = \nu RT.$$

R is called the *absolute gas constant*, and can easily be calculated from measurements of three corresponding values of p, V, and T. Its value is

$$R = 8 \cdot 313 \times 10^7 \text{ erg degree}^{-1} \text{ mole}^{-1} = 1 \cdot 986 \text{ cal degree}^{-1} \text{ mole}^{-1}.$$

We refrain from entering here upon a thoroughgoing discussion of the preceding definition of temperature from the thermodynamic and axiomatic point of view (a complete treatment for the generalized statistics introduced by the quantum theory is given in Appendix XXXV, p. 459), and merely add a brief remark on the units in which temperature is measured.

We use the phrase ideal gas if the product pV is constant at constant temperature; for low pressures this is true for every gas. Deviations from the ideal character of the gas occur when the density of the gas becomes so great that the mean distance between two gas molecules is comparable with the molecular diameter. If we employ such an ideal gas as thermometric substance, the centigrade scale is defined as follows. Let $(pV)_f$ be the value of pV for the gas, when it is brought into contact with melting ice, and $(pV)_b$ its value for contact with boiling water;* then the temperature of the gas, for the general case of any value of pV, is defined according to the centigrade scale by

$$t = 100 \, \frac{pV - (pV)_f}{(pV)_b - (pV)_f}.$$

We see at once that, with this definition, the temperature of melting ice is 0° C. and that of boiling water 100° C.

The change from the centigrade scale to the absolute temperature (Lord Kelvin, 1854) scale, which we have indicated above by the symbol T, is made as follows. It has been established experimentally that at constant pressure an ideal gas expands by 1/273 of its volume at 0° C. when its temperature is raised by 1° C., so that we have, e.g.,

$$\frac{(pV)_b}{(pV)_f} = \frac{T_b}{T_f} = 1 + \frac{100}{273} = \frac{373}{273}.$$

If we retain the unit of the centigrade scale in the absolute scale also, T_b and T_f must differ by 100°. It follows that melting ice ($t = 0°$ C.) has the absolute temperature $T_f = 273°$, and boiling water ($t = 100°$ C.) the absolute temperature $T_b = 373°$. The zero of the absolute temperature scale lies therefore at $-273°$ C.

It may be remarked that the absolute temperature scale is sometimes called the *Kelvin scale*, and to distinguish it from the ordinary centigrade scale is denoted by K.

5. Specific Heat.

The *specific heat* (referred to 1 mole) of a substance is given by the energy which must be supplied to the substance to raise its temperature by 1°. For a monatomic gas, it follows immediately from this definition that the specific heat at constant volume is

$$c_v = \frac{dU}{dT} =: \tfrac{3}{2} R.$$

If heat energy is supplied, but with the pressure kept constant instead of the volume, the gas expands and so does work against the external

---

* f = freezing-point, b = boiling-point.

pressure (which of course in the case of equilibrium must be equal and opposite to the pressure of the gas), the work being

$$p \, \Delta V = R \, \Delta T, \quad \text{i.e.} = R \text{ for } \Delta T = 1°.$$

Thus R is that portion of the specific heat which corresponds to the work of expansion; if we add to this work the specific heat at constant volume $\frac{3}{2}R$, we obtain for the total specific heat at constant pressure

$$c_p = \tfrac{5}{2} R.$$

The ratio c_p/c_v is generally denoted by γ. Hence, for a monatomic gas we have the relation (Clausius, 1857)

$$\gamma = \frac{c_p}{c_v} = \tfrac{5}{3} = 1 \cdot 667,$$

which agrees well with observation, especially for the inert gases.

In polyatomic molecules, besides the three degrees of freedom of the translational motion, other degrees of freedom come in, which correspond to the rotations and vibrations of the molecules and which, when energy is supplied to the gas, can also take up a part of the energy. Now there is a general theorem, the *theorem of equipartition* (Clausius, 1857; Maxwell, 1860), according to which the specific heat is equal to $nk/2$ referred to a molecule, or $nR/2$ referred to 1 mole, where n is the number of kinetic degrees of freedom. For example, a diatomic molecule, considered as rigid (dumb-bell model), possesses essentially two rotational degrees of freedom. In counting the number of degrees of freedom, the degree of freedom corresponding to rotation about the axis joining the two atoms is to be ignored. For idealized atoms concentrated at a point, this is obviously correct; when we take account of the extension of the atoms in space, we encounter a conceptual difficulty, which it requires the quantum theory to clear up (see Chap. VII, § 2, p. 213). In this case we therefore have $c_v = \tfrac{5}{2} R$, $c_p = \tfrac{7}{2} R$, and consequently $\gamma = c_p/c_v = \tfrac{7}{5} = 1 \cdot 4$. These values have in fact been observed, e.g. in molecular oxygen.

6. Law of Distribution of Energy and Velocity.

As the next step in the development of the kinetic theory of gases, we proceed to consider the law of distribution of energy or velocity in a gas, i.e., in particular, the law of dependence of the quantity n_v, employed above, on the velocity. While up to this point a few simple ideas have been sufficient for our purpose, we must now definitely call to our aid the statistical methods of the Calculus of Probabilities.

To begin with a simple example, let us in the first place, without troubling about velocity, consider the question of the number of molecules which on the average are to be found in a definite volume element ω. An exact determination of this number for a given moment, apart from the fact that it is inherently impossible, would be of little use, since the number changes every moment on account of the motion of the molecules; we are therefore concerned only with the average number of molecules. This we determine as follows. Imagine the whole box containing the gas to be divided up into separate cells of sizes $\omega_1, \omega_2, \ldots, \omega_z$ (as an aid to the imagination we may picture everything as two-dimensional, see fig. 3), and "throw" the molecules, which we may think of as little balls, at random into this system of cells. We then note that a definite number n_1 of balls has fallen into the first cell, another number n_2 into the second cell, and so on. At a second trial, we shall perhaps find other values for the number

of balls in the various cells. If we repeat the experiment a great many times, we shall find that a given distribution, defined by the numbers

Fig. 3.—Division into cells for the purpose of finding the most probable density of distribution of the molecules of a gas.

$n_1, n_2, \ldots, n_z$, occurs not merely once, but many times. For a model consisting of a real box divided up into cells, and real balls, the frequency with which a given distribution appears could be determined by a long series of trials. For a gas and its molecules this is not possible; we have to use instead an arithmetical argument, consisting of a numerical part (concerning the numbers $n_1, n_2, \ldots, n_z$) and a geometrical part (concerning the sizes of the cells $\omega_1, \omega_2, \ldots, \omega_z$). In this way we shall find the mathematical probability of a given distribution. And, when we take all possible distributions into account, there is one, for which this probability is greatest, the most probable distribution. Because of the very great number of the molecules in the cubic centimetre, this maximum is overwhelmingly sharp, so that the probability of any distribution deviating essentially from it is negligibly small. It is therefore to be expected that the most probable distribution represents the average state.

At this point, however, a question of principle must be considered. The following difficulty presents itself. If we knew the exact position and velocity of every particle of the gas at a definite moment, the further course of the motion would be completely defined, for the behaviour of the gas is then rigorously determined by the laws of mechanics, and appears from the very outset to have nothing to do with the laws of probability. If we assume that at time $t = 0$ the positions and velocities of the molecules are distributed according to any statistical law, we are not entitled without more ado to expect that at any later time t the state of the gas will be determined by the play of probabilities, independently of the initial state assumed. It would be quite conceivable that, for the initial state chosen, all the molecules would as a consequence of the laws of mechanics be found at the moment t at a definite corner of the box. In order that it may be possible to apply statistics at all, it must be stipulated that there is no coupling of the states at different instants. We must suppose that the collisions, conditioned by the laws of mechanics, by their enormous number completely efface the " memory " of the initial state after only a (macroscopically) short interval of time. Further, it has to be borne in mind that measurements take time; what we determine is not the micro-state at the instant t at all, but its mean value over a considerable period. It is assumed that the time-mean values so determined are independent of the period chosen, and that they agree with the mean values ascertained by considerations of probability alone from the most probable state defined above.

Although this hypothesis, the so-called quasi-ergodic hypothesis, is very plausible, its rigorous proof presents great difficulties. The mathematicians von Neumann and Birkhoff have (1932) proved a theorem which is virtually equivalent to the quasi-ergodic theorem. According to the latest ideas in theoretical physics, it is true, the problem of a rigorous proof of the ergodic hypothesis has become unimportant, inasmuch as it is now meaningless, as will be explained later, to give information about the exact position of the individual molecules. To repeat: the hypothesis asserts that even when the initial state is arbitrary a stationary state is in time attained, owing to the collisions of the molecules with one another and with the wall, and that *this state is the same as the state of greatest probability as defined above*. It is assumed that the walls of the box are " rough ", so that they do not act as perfect reflectors.

We return now to the calculation of the most probable distribution of the molecules in the individual cells of the box. A definite distribution is described by the numbers in the various cells $n_1, n_2, \ldots, n_z$; their sum is of course equal to the number of particles of gas in the box:

$$n_1 + n_2 + \ldots + n_z = n.$$

We denote the ratio of the size of the cell ω_k to the whole volume ω of the box by $g_k = \omega_k/\omega$; we have then

$$g_1 + g_2 + \ldots + g_s = 1.$$

How often will this definite distribution actually occur? It is clear in the first place that we obtain the same distribution, if we permute the individual molecules among themselves; the number of these permutations is $n!$ Here, however, we are including the cases in which the molecules in a given cell are permuted with one another; since these permutations do not represent new possibilities of realization of the distribution considered, we must divide $n!$ by the number $n_1!$ of the permutations within the first cell, and so on, and thus obtain for the number of possibilities of realization

$$\frac{n!}{n_1!\ n_2!\ldots n_s!}.$$

To get the probability of this distribution, we have still to multiply this number by the a priori probability of the distribution, which is $g_1{}^{n_1} g_2{}^{n_2} \ldots g_s{}^{n_s}$, since the a priori probability that a particle should fall into the first cell is g_1, and therefore the same probability for n_1 particles is $g_1{}^{n_1}$, and so on. The probability of our distribution given by the numbers $n_1, n_2, \ldots$ therefore becomes

$$W = \frac{n!}{n_1!\ n_2!\ldots n_z!}\, g_1{}^{n_1} g_2{}^{n_2} \ldots g_s{}^{n_s}.$$

To verify that in our calculation we have actually included all the possibilities of realization, we can find the sum of the probabilities for all possible distributions, which of course must be 1, since it is a certainty that one of the distributions is realized. We therefore form the sum over all distributions $n_1, n_2, \ldots$, for which $n_1 + n_2 + \ldots = n$. This sum is easily found by the polynomial theorem, which gives

$$\sum_{n_1,\, n_2,\ldots} \frac{n!}{n_1!\ n_2!\ldots n_s!}\, g_1{}^{n_1} g_2{}^{n_2} \ldots g_s{}^{n_s} = (g_1 + g_2 + \ldots + g_s)^n = 1.$$

We now transform the above formula for the probability of a definite distribution by means of Stirling's theorem, which for large values of n gives

$$\log n! = n(\log n - 1).$$

We then obtain from the expression for W, by taking logarithms,

$$\log W = \text{const.} + n_1 \log\frac{g_1}{n_1} + n_2 \log\frac{g_2}{n_2} + \ldots.$$

To find the most probable distribution, we must calculate the maximum of $\log W$ for a variation of the numbers $n_1, n_2, \ldots$, subject to the subsidiary condition $n_1 + n_2 + \ldots + n_s = n$. Since $n_1, n_2, \ldots$ are very large we can treat them as continuous variables. By the method of Lagrange's multipliers we obtain as the equations defining this maximum

$$\frac{\partial \log W}{\partial n_1} = \log\frac{g_1}{n_1} - 1 = \lambda, \quad \frac{\partial \log W}{\partial n_2} = \log\frac{g_2}{n_2} - 1 = \lambda, \ldots,$$

where λ is a constant, whose value is determined by the equation $n_1 + n_2 + \ldots = n$. It follows that

$$\frac{g_1}{n_1} = \frac{g_2}{n_2} = \ldots = e^{\lambda+1} = \text{const.};$$

$$n_1 = ng_1 = n\frac{\omega_1}{\omega}, \quad n_2 = ng_2 = n\frac{\omega_2}{\omega}, \ldots.$$

Hence the numbers $n_1, n_2, \ldots$ in the individual cells are proportional to the size of the cells; we therefore have a *uniform distribution of the molecules* over the whole box; the size of the cells does not matter at all.

While the result just deduced for the distribution of the density of the molecules was to be expected from the start, the same method, applied to the distribution of the velocity of the molecules, leads to a new result. The calculations in this case are exactly analogous to those above. We construct a " velocity space " by drawing lines from a fixed point as origin, representing as vectors the velocities of the individual molecules in magnitude and direction. We then investigate the distribution of the ends of these vectors in the velocity space. In this case as before we can make a partition into cells, and consider the question of the number of vectors whose ends fall in a definite cell. There is, however, one essential difference as compared with the former case, in that there are now *two subsidiary conditions*, viz. besides the condition

$$n_1 + n_2 + \ldots + n_s = n$$

for the total number of particles, an additional condition for the total energy E of the gas, viz.

$$n_1\epsilon_1 + n_2\epsilon_2 + \ldots + n_s\epsilon_s = E,$$

where ϵ_l denotes the energy of a molecule, whose velocity vector points into the cell l. Taking account of these two subsidiary conditions, we find for the maximum of the probability (λ and β being the two Lagrangian multipliers)

$$\frac{\partial \log W}{\partial n_l} = \log \frac{g_l}{n_l} - 1 = \lambda + \beta \epsilon_l \quad (l = 1, 2, \ldots, z).$$

This leads to *Boltzmann's law of distribution* (1896)

$$n_l = g_l e^{-1-\lambda-\beta\epsilon_l} = g_l A e^{-\beta\epsilon_l},$$

where A and β are two constants, which have to be found from the two subsidiary conditions. Thus the expression for the number n_l corresponding to the cell l essentially involves the energy belonging to this cell as well as the size g_l of the cell, and that in such a way that

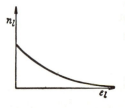

Fig. 4.—Boltzmann's law of distribution: if the cells are of equal size, those with greater energy are more sparsely filled than those with smaller energy.

among cells of equal size one with greater energy is not so well filled as one with smaller energy; the fall in the value of the number n_l with increasing energy obeys an exponential law (fig. 4).

We shall now apply *Boltzmann's law of distribution* to the *special case of a monatomic gas*. Here the energy is given by

$$\epsilon = \frac{m}{2} v^2 = \frac{m}{2} (\xi^2 + \eta^2 + \zeta^2).$$

Position in the velocity space is uniquely defined by the three components ξ, η, ζ. From their meaning the cells g are finite. From the macroscopic standpoint we can consider them as infinitesimal, and denote them by $d\xi \, d\eta \, d\zeta$. In the formation of mean values we can then replace the sums by integrals:

$$\Sigma g_l \ldots \to \int\int\int d\xi \, d\eta \, d\zeta \ldots,$$

the limits of integration being $-\infty$ and $+\infty$ in each case. Since further we are interested in the mean values of v, v^2, ... only ($\bar{\xi}$, $\bar{\eta}$, $\bar{\zeta}$ vanish from symmetry, and $\bar{\xi^2}$, $\bar{\eta^2}$, $\bar{\zeta^2} = \frac{1}{3}\bar{v^2}$), the integrand in each of the integrals to be calculated depends on v only, which suggests that we should introduce polar co-ordinates in the velocity space, with v as radius. Integration with respect to the polar angles can be effected at once, giving the factor 4π, the area of the unit sphere. We have therefore

$$\int \int \int d\xi\, d\eta\, d\zeta \ldots = 4\pi \int v^2 dv \ldots$$

The total number n of the molecules is obtained from Boltzmann's law by calculating the integral

$$n = 4\pi A \int_0^\infty v^2 e^{-\beta \cdot \frac{1}{2}mv^2} dv,$$

while for the total energy we have

$$E = 4\pi A \int^\infty \frac{m}{2}\, v^4 e^{-\beta \cdot \frac{1}{2}mv^2} dv.$$

These two equations determine uniquely the two constants A and β, which up to this point were unknown. The integrals are readily evaluated (see Appendix I, 359); we find the relations

$$n = A\sqrt{\left(\frac{\pi}{\lambda}\right)^3}, \quad \left(\lambda = \frac{\beta m}{2}\right)$$

$$E = \frac{3}{4} mA\sqrt{\left(\frac{\pi^3}{\lambda^5}\right)} = \frac{3}{4}\frac{mn}{\lambda} = \frac{3}{2}\frac{n}{\beta}.$$

But we have seen (p. 7) that on the average a molecule, corresponding to its three translational freedoms, possesses the kinetic energy $\frac{3}{2}kT$. Since the whole kinetic energy of the gas is therefore $\frac{3}{2}nkT$, it follows that

$$\beta = \frac{1}{kT}.$$

The constants in Boltzmann's law are accordingly expressed in terms of the number of molecules in the gas and their absolute temperature.

We inquire next how many mole-

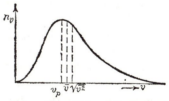

Fig. 5.—Maxwell's law of distribution of velocities, showing the most probable velocity (v_p), the mean velocity ($\overline{v}$), and the root-mean-square velocity $\sqrt{\overline{v^2}}$.

cules have a velocity between v and $v+dv$. This number $n_v dv$ is clearly given by the integrand of the above integral for n:

$$n_v = 4\pi A e^{-\lambda v^2} v^2 = 4\pi n \left(\frac{m}{2\pi kT}\right)^{3/2} e^{-mv^2/2kT} v^2.$$

This relation is known as *Maxwell's law of distribution of velocity* (1860); the graph of n_v as a function of v is shown in fig. 5.

To obtain some idea of the order of magnitude of the velocities of gas molecules, we can calculate from the law of distribution the most probable velocity v_p, or the average velocity v, or other such mean values (see Appendix I, p. 359); we find, e.g., for the most probable velocity the value $v_p = \sqrt{(2RT/\mu)}$, so that, e.g., for molecular hydrogen ($\mu = 2$) at 0° C. ($T = 273°$ K.)

$$v_p = 15 \cdot 06 \times 10^4 \text{ cm./sec.}$$

An experimental test of Maxwell's law of distribution may be carried out as follows. Let a furnace O contain a gas at a definite temperature T. In the wall of the furnace let there be an opening through which the gas molecules can escape into a space which is highly evacuated (fig. 6). The escaping molecules fly on in a straight line with the velocity which they possessed in the furnace at the moment of their exit through the opening. By means of a diaphragm system behind the opening, the stream of molecules issuing in all directions may be cut down to a *molecular ray* (Dunoyer, 1911). The distribution of velocity in the ray, or beam, can be measured directly by various methods, the

Fig. 6.—Diagrammatic representation of the production of a molecular ray. The furnace O, which contains the gas, is heated from outside.

most important of which will be described immediately. When deducing from this distribution the distribution of velocity in the enclosed gas, we must, however, take the fact into account that rapid molecules are relatively more numerous in the beam than in the gas. For the beam consists of all molecules which issue through the opening per unit time; the number of these is (§ 3, p. 5) proportional to $n_v v dv$, whilst $n_v dv$ represents the number of particles having the same range of velocity in the gas. The two distributions therefore differ from each other by the factor v.

A direct method (Stern, 1920) of measuring the velocity in the beam depends on the following principle. A beam, consisting, e.g., of silver atoms, can easily be demonstrated by setting up in the path of the beam a glass plate, on which the silver is deposited. If the apparatus is then rotated about an axis perpendicular to the path of the beam, the molecules are no longer deposited at the same spot on the glass plate as when the tube was at rest, but at a greater or smaller distance from this point according to their velocity, since, during the time of flight of the molecules from the furnace to the glass plate, the tube and with it the glass plate has turned. The distribution of velocity in the beam is therefore found immediately from the plate, by measuring the density of the deposit at various distances from the original point.

A more recent method makes use of the same principle as was used by Fizeau (1849) to measure the velocity of light, viz. that of coupled rotating toothed wheels. Further details of the method need not be gone into here. The measurements, as carried out mainly by Stern and his pupils, showed that the distribution of velocity among the molecules in the furnace does actually satisfy Maxwell's law.

Another method depends on the Doppler effect (1842). If a molecule, which in the state of rest e $\cdot$ its a definite frequency ν_0, moves towards the observer with the relative velocity component v_r, the light appears to him to be displaced in the direction of higher frequencies by the factor $1 + v_r/c$; while for the opposite relative motion the corresponding factor is $1 - v_r/c$. A direct proof of this was given by J. Stark (1905) with the help of *canal rays* (p. 28) consisting of fast-moving positive hydrogen ions. When the light which is emitted by a luminous gas is spectrally decomposed, then besides the frequency ν_0 there occur also all frequencies which are derived from ν_0 in consequence of the Doppler effect due to the motion of the molecules, and that with an intensity' which is given by the number of molecules with a definite component of velocity relative to the observer. Each spectral line has therefore a finite width, and the intensity distribution over its range gives a picture of Maxwell's law.

7. Free Path

We have spoken above of a molecular ray. This consists of molecules which have passed through the diaphragm system and fly on as a beam through the evacuated space. A necessary condition is a *high vacuum*. If, however, there is still a residue of gas in the space (with molecules either of the same or of a different sort), then some of the molecules in the beam, more or fewer according to the pressure, will collide with the gas molecules and be scattered in the process; consequently the molecular beam will be weakened, clearly, in fact, exponentially with the distance travelled, since the number of scattering processes is proportional to the number of molecules present in the beam. If then we denote by $n(s)$ the number of beam molecules which, after traversing a distance s from the opening in the furnace, pass through a plane at right angles to the beam per unit time, we have a law of the form $n(s) = n(0)e^{-s/l}$. Here l is a quantity (to be determined later) of the dimensions of a length; a little consideration shows that this quantity is equal to the distance which on the average a molecule of the beam covers before it comes into collision with a molecule of the gas outside the beam. For, putting $1/l = \beta$, we have for this average

$$\bar{s} = \frac{\int_0^\infty s n(s)\,ds}{\int_0^\infty n(s)\,ds} = \frac{\int_0^\infty s e^{-\beta s}\,ds}{\int_0^\infty e^{-\beta s}\,ds}$$

$$= -\frac{d}{d\beta}\left(\log \int_0^\infty e^{-\beta s}\,ds\right) = -\frac{d}{d\beta}\left(\log \frac{1}{\beta}\right)$$

$$= \frac{d}{d\beta}(\log \beta) = \frac{1}{\beta} = l.$$

We therefore call l " the mean free path " of the ray in the gas (Clausius, 1858). Born, together with E. Bormann, showed (1921) how it can be determined on the basis of its definition by the exponential law, by measuring the attenuation of a beam of silver atoms on passage through the gas which is at rest (air). The more important case is that in which beam and gas consist of molecules of the same kind; the mean free path l is then a property of this gas.

The question of the quantities on which the mean free

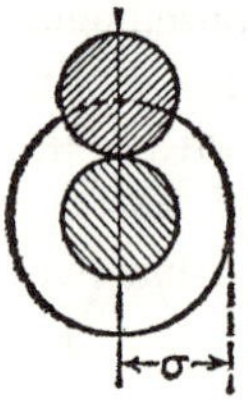

Fig. 7.—Gas-kinetic effective cross-section; in a collision the centres of mass of the two equal molecules cannot come nearer than the distance σ (σ = diameter of a molecule).

path must depend may be examined theoretically. Clearly it is a matter of the number of collisions with the molecules of the gas, in which a definite molecule takes part during its passage through the gas. We may regard the other molecules of the gas as at rest; to take their motion into account does not make any essential difference. We shall regard the molecules as spheres of diameter σ, and have to answer the question of how many collisions such a moving sphere is involved in, during its flight through the gas, which consists of spheres at rest. Since a collision occurs whenever the centre of the moving molecule, in the course of its flight, comes nearer the centre of a molecule at rest than the distance σ, we may also obtain the number of these collisions by considering spheres of radius σ at rest, and a moving point (fig. 7). We have therefore the same problem before us as when a man fires a gun in a wood, and considers the question of the number of trees he has struck. The number is clearly proportional to the thickness of an individual tree, and to the number per unit area; its reciprocal determines the mean range of the bullets. In the case of the gas also, the collision number must be proportional to the number n of the molecules of gas per unit volume and to their (gas-kinetic) cross-section $\pi\sigma^2$. Since the mean free path is inversely proportional to this collision number, it is proportional to

$$\frac{1}{n\pi\sigma^2} = \frac{V}{\pi N_0 \sigma^2},$$

where V is the volume per mole.

Measurement of the mean free path l therefore throws light upon the value of the product $N_0\sigma^2$. A direct method of determining l (for a beam of molecules of a foreign element) has already been given above. As to indirect methods, we must mention first a method, given by Maxwell (1860), depending on the conduction of heat in the gas (Appendix II, p. 361). If the gas molecules did not collide, a rise in temperature anywhere in the gas, that is to say, an increase in the kinetic energy of the particles, would be propagated through the gas with the great velocity of the molecules, say a thousand metres per second; experimentally, however, gases are found to be relatively poor conductors of heat. The reason for this is that a gas molecule can fly for only a relatively short distance, of the order of magnitude of the free path, before it collides with another particle of gas, and thus not only changes its direction of motion, but also gives up a part of its kinetic energy to the particle it has struck. Other methods for the determination of l depend on *viscosity* and on *diffusion* (Appendix II, p. 363). The latter method lends itself well to visual demonstration; if, for example, we let chlorine gas diffuse into air, then, since the colour of the chlorine makes it visible, we can observe directly how slowly the diffusion goes on.

From these experiments, taken together, we find for the order of magnitude of the mean free path at a pressure of 1 atmosphere, $l \sim 10^{-6}$ cm.; at a pressure of 10^{-4} mm. of mercury, which corresponds to the normal X-ray vacuum, $l \sim 10$ cm.

8. Determination of Avogadro's Number.

As has already been remarked above, when the free path is known, so is the product $N_0\sigma^2$, i.e. the product of the square of the molecular diameter and Avogadro's number. To determine σ and N_0 separately, we need a second relation between them. Such a relation is given, at least in respect of order of magnitude, by the *molar volume* of the solid body. In the solid state of aggregation, it is reasonable to suppose that the molecules are in the state of densest packing, so that the volume taken up by 1 mole—up to a factor of the order of magnitude 1—is given by the product of the number of molecules per mole and the volume taken up by one molecule, i.e. by $N_0\sigma^3$. From $N_0\sigma^2$ and $N_0\sigma^3$ we can now determine N_0 and σ, so obtaining

$$N_0 \sim 10^{23}\,\text{mol}^{-1},\ \sigma \sim 10^{-8}\,\text{cm}.$$

Moreover, the influence of the "proper volume" of the molecules is shown not only in the most condensed state (the solid body), but even in the gaseous state by deviations from the law of ideal gases

$$pV = RT.$$

Thus, if the volume V of a definite mass of gas becomes so reduced that the proper volume of the molecules is actually comparable with V, then the free volume at the disposal of the individual molecule is smaller than V, and we obtain the equation of state

$$p(V - b) = RT.$$

The exact calculation gives for b four times the proper volume of the molecules. (For spherical particles of diameter σ,

$$b = 4N_0 \cdot \frac{4\pi}{3}(\sigma/2)^3 = \frac{2\pi}{3}N_0\sigma^3;$$

see Appendix III, p. 365). In dense gases, however, there are also other deviations from the ideal equation of state, which are due to the *cohesion* of the molecules, and have the general effect of making the pressure, for given T and V, smaller than it is according to the formula $pV = RT$. To represent these circumstances, many equations of state have been proposed; the best known is that of van der Waals (1881),

$$\left(p + \frac{a}{V^2}\right)(V - b) = RT.$$

What chiefly interests us here, is that by determination of the constant b we again obtain the product $N_0\sigma^3$. At p. 471 we shall return to the question of the value of a, which measures the cohesion.

    The evaluation of Avogadro's number given above is of course rather rough. We get a more exact method by considering *fluctuation phenomena*. Thus, if we take 1 cm.^{3} of a gas, we shall find exactly as many molecules in it as in another cm.^{3}, viz., at room temperature, about 10^{19}; differences of a few hundred molecules in these huge numbers are of course of no moment. It is another matter if we pass to smaller elements of volume; in a cube of edge 0·1 μ (1 μ = 1 micron = 10^{-3} millimetre) there are on the average only 10^4 molecules, and it is clear that variations of a few hundred molecules are now relatively quite important. If we go on to smaller regions still, we shall finally arrive at volumes which contain only one or two molecules, or none at all. The smaller, therefore, the number of particles considered, the greater will the fluctuations be (Appendix IV, p. 367).

Examples of these fluctuation phenomena are given by the *Brownian movement* (1828), which is observed with microscopic particles (e.g. colloidal solutions, or smoke in air), and manifests itself macroscopically in the oscillations of a mirror suspended by a fine wire; also by the *sedimentation of suspensions*, in which colloidal particles by reason of their weight tend to sink to the bottom of the vessel, but in consequence of collisions with the particles of the solvent are impelled more or less upwards, thus giving rise to a density distribution of the same character as that in the atmosphere, expressed in the barometer formula (Perrin, 1908). A third example is the *scattering of light in the atmosphere* (Lord Rayleigh, 1871), which causes the colour of the sky. If, in fact, the density of air were the same throughout, then just as in an ideal crystal there would be no scattering of light, since the waves scattered by individual molecules would annul each other by interference; the sky would appear black. Scattering is only possible when there are irregularities in the uniformity of distribution, i.e. fluctuations of density; indeed these fluctuations must be so pronounced that they are perceptible within a distance of the order of magnitude of a wave-length. Since the fluctuations in small volumes are greater, short (blue) waves are more strongly scattered than long (red) waves; the sky therefore appears blue (Smoluchowski, 1908).

A much more exact determination of N_0 depends on its connexion with the *elementary electrical charge* e, and the electrolytic unit, the *faraday*. In electrolysis, we have the following fundamental law, discovered by Faraday (1833). In the electrolytic separation of 1 mole of any substance, a quantity of electricity is transported equal to 96,520 coulombs; i.e.

$$eN_0 = F = 96{,}520 \text{ coulombs.}$$

The interpretation is obviously that every ion carries a multiple of the same elementary charge e, and that the total amount of electricity transported is e times the number of atoms. If then we know e, we can calculate from it the value of N_0, or of $m_H = 1/N_0$. A method for determining e was devised by Ehrenhaft (1909) and Millikan (1910), and the latter has developed it to high precision. If electricity actually consists of elementary quanta, the total charge on a body must be an integral multiple of the charge e. Owing to the smallness of the elementary charge, it is of course difficult to test the truth of this by experiments on macroscopic charged bodies. Such experiments offer little

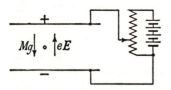

prospect of success unless the whole charge consists of very few quanta; it actually does so in Millikan's oil-drop

Fig. 8.—Condenser in Millikan's method of determining the elementary charge e. The weight Mg is compensated by an electric field E.

method (fig. 8). The charge on an oil droplet may be determined with sufficient exactness, by bringing it into the field of a condenser whose lines of force are directed vertically upwards. There are then two forces acting on the particle in opposite directions, viz. the electric force upwards, and the force of gravity downwards; the droplet will be in equilibrium if the potential difference applied to the condenser is just so chosen that

$$eE = Mg,$$

where e and M are the charge and mass of the oil-drop, E is the electric field strength, and g is the acceleration of gravity. The greatest difficulty is the determination of the mass M. It can be calculated from the density and the radius of the drop, if it is assumed that the density has its normal value. The radius is determined by switching off the electric field, so that the particle falls; and then measuring the velocity of descent, which, in consequence of the viscosity of the surrounding medium, is constant. Inserting this velocity in the formula called Stokes's law, we obtain the radius of the droplet. The experiment, as carried out by Millikan and others, not only proved without ambiguity that the charge on the droplets consists of integral multiples of an elementary quantum, but also allowed a fairly exact determination to be made of this elementary charge e. We may also refer here to various methods for the determination of N_0, which make use of the *radiations from radioactive substances*. We may, e.g., count the number of particles, either by noting the scintillations produced when they strike a screen covered by zinc sulphide, or with the help of the Geiger counter. If we know the mass of our sample and the decay constant, we can calculate N_0.

A more precise method for the determination of N_0 has been developed with the help of X-ray diffraction (p. 30). It is found

possible to measure the wave-length with an ordinary ruled grating (p. 80); the diffraction pattern of a crystal then allows the lattice constant a per molecule to be determined, and N_0 is then found from the known molar volume $N_0 a^3$ (Siegbahn, Compton, 1925). Cohen, DuMond, Layton and Rollett (1955) quote the following values for N_0 and e:

$$N_0 = (6.02486 \pm 0.00016) \times 10^{23} \text{ molecules per mole,}$$
$$e = (4.80286 \pm 0.00009) \times 10^{-10} \text{ electrostatic units.}$$

In conclusion, it may be mentioned in addition that the Boltzmann constant k, which by definition is the quotient of the gas constant R by Avogadro's number, can be also measured directly by determining the *spectral distribution of intensity in the radiation emitted by a black body* (Planck, 1900). The function which expresses the intensity in terms of the frequency and the temperature involves only two universal constants, k and h, the first of which is Boltzmann's constant; the second is called Planck's constant, and is the fundamental constant of the quantum theory (Chap. VII, § 1, p. 210).

CHAPTER II

Elementary Particles

1. Conduction of Electricity in Rarefied Gases.

The development of chemistry and of the kinetic theory of gases has led, as explained in the preceding chapter, to the assumption that matter consists of molecules and atoms. For the *chemist* these particles represent the ultimate constituents of which solids, liquids, and gases are composed, and on which he acts in every reaction produced by purely chemical means.

We now take up this question of the smallest components of matter from the point of view of the *physicist*. The latter has many other phenomena at his command, from the study of which he can collect data upon the structure of matter. Among these phenomena, the processes which occur when electricity passes through rarefied gases have proved to be of special importance.

Under normal conditions, a gas is in general a poor conductor of electricity. If, however, the gas is enclosed in a vessel with two electrodes to which a sufficiently high potential is applied, it is found that at pressures of a few millimetres of mercury, a transport of electricity takes place across the gas, showing itself by the flow of an electric current in the external leads to the electrodes; and that at the same time the gas becomes strongly luminous, a phenomenon which from the theoretical point of view is somewhat complicated, but which in the form of the so-called Geissler tubes is extensively employed in practice, especially for luminous signs.

If the pressure of the enclosed gas is reduced still further, the luminescence disappears (below 0·1 mm. of mercury) almost completely. At very low pressures (below 10^{-3} mm.), however, the presence of rays is observed, proceeding from the cathode and producing a fluorescent appearance on the opposite wall of the glass tube. In certain circumstances these rays can be seen directly in the form of

a bluish thread, stretching from the cathode across the tube.   The rays are called *cathode rays* (fig. 1).

Their properties can be investigated as follows (Plücker, 1858; Hittorf, 1869; J. J. Thomson, 1894). If a body is placed in the path of the rays, it is seen to cast a shadow on the fluorescent part of the glass.   From the geometrical relations it can be inferred that the rays producing the shadow *spread out in*

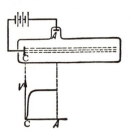

Fig. 1.—How the cathode rays arise. The rays proceed from the cathode C, and can be seen as a bluish thread if the gaseous pressure in the tube is not too small. A is the anode. The propagation of the rays in straight lines from the cathode is due to the fact that the potential between C and A, in consequence of the presence of slight residues of gas (space charges) in the tube, does not rise uniformly from C to A, the rise taking place almost entirely in the immediate neighbourhood of the cathode. [The graph of the potential is shown in the lower part of the figure.] The circuit is closed by a weak transport of positive ions.

straight lines. It is found also that the occurrence of these rays is associated with a transport of electric charge across the tube.   Moreover, the rays can be deflected from their rectilinear path by external electric or magnetic fields, and that in such a way as to permit the inference that the rays consist of rapidly moving, *negatively charged particles*; they are called *electrons*.

The velocity and the specific charge (i.e. the ratio e/m of charge to mass) of these particles can also be determined. If we set up two wire grids in the path of the rays and at right angles to their direction (fig. 2), and apply to them a potential difference V, the electrons in the *longitudinal electric field* between the grids are accelerated or retarded.   The change of speed due to their passage across the field is given by the equation of energy.   If v_0 is their velocity *before*,

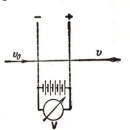

Fig. 2.—Acceleration of electrons by an electric field; the increase of kinetic energy is equal to the potential fall multiplied by the charge $-e$ of the electron.

and v their velocity *after* passing through the field, then for the case where the field is in the opposite direction to the velocity so that the negative electrons are accelerated,

$$\frac{m}{2}\,v_0{}^2 = \frac{m}{2}\,v^2 + eV.$$

When the initial velocity is small, we may put $v_0 \simeq 0$, i.e.

$$\frac{m}{2}\,v^2 = -eV.$$

If V is known, we can therefore determine in this way the value of $(m/e)v^2$.

The same quantity can also be measured by deflection in a *transverse electric field* (fig. 3). If two condenser plates are set up parallel

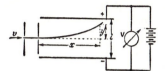

to the path of the rays and at a distance l apart, and a potential difference V is

Fig. 3.—Deflection of an electron in a transverse electric field (between the plates of a condenser). The path is a parabola (path of a projectile).

applied to them, a constant deflecting force eV/l acts on the electrons perpendicular to their original direction of motion. Their path is therefore a parabola, which is defined by the equations

$$x = v_0 t, \quad y = \frac{g}{2} t^2,$$

where v_0 is the initial velocity, and $g = (e/m)(V/l)$ is the acceleration due to the field. Elimination of the time t gives

$$y = \frac{e}{2m} \frac{V}{l} \frac{x^2}{v_0^2}, \quad \text{or} \quad \frac{m}{e} v_0^2 = \frac{V}{l} \frac{x^2}{2y}.$$

By measuring the deflection y of the ray after it has traversed a distance x, we can therefore again find the value of the quantity $(m/e)v_0^2$.

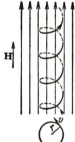

In a *magnetic field* H we know that a moving charge e is acted on by a force which is perpendicular both to the direction of the field and to that of the velocity; the magnitude of this so-called *Lorentz force* is given by $e(v/c)H$, multiplied by the sine of the angle

Fig. 4.—Motion of an electron in a constant magnetic field H; the path is in general a helix, with its axis parallel to the field; in the special case when the electron moves at right angles to H, the path is a circle.

between the directions of velocity and field ($c = 300,000$ km./sec. = the velocity of light). The path of the electron is a helix, whose axis is parallel to the field (fig. 4). In the special case when the component velocity parallel to H vanishes, the helix shrinks into a circle perpendicular to the field. Its radius r is easily calculated. The Lorentz force $e(v/c)H$ is directed towards the centre of the circle. It must obviously be equal to the centripetal force mv^2/r, so that

$$e \frac{v}{c} H = \frac{mv^2}{r}, \quad \text{or} \quad \frac{m}{e} v = \frac{Hr}{c}.$$

Measurement of the radius r and of the magnetic field strength therefore enables us to find the value of $(m/e)v$.

We thus obtain the *result*, that measurements of deflection give in the electric field $(m/e)v^2$, and in the magnetic field $(m/e)v$. Hence the values of e/m and v can be determined. Actual measurements have shown velocities which, with increasing potential difference across the tube, reach near the velocity of light.

With regard to the measurement of e/m, exact experiments have shown that this, the specific charge, is not precisely constant, but depends to some extent on the velocity of the electrons (Kaufmann, 1897). This phenomenon is explained by the *theory of relativity* (Appendix V, p. 370). According to Einstein (1905), the value of the charge e is invariable; the mass, however, is variable, its magnitude in fact depending upon the velocity which it has, relative to the observer who happens to measure it. The electron may have the " rest mass " m_0, i.e. this is its mass for the case when it is at rest relative to the observer; but if it moves relative to the observer with a velocity v, it behaves (e.g. in a field of force) as if it possessed the mass

$$m = \frac{m_0}{\sqrt{\left(1 - \dfrac{v^2}{c^2}\right)}}.$$

This assertion of the theory can be tested by deflection experiments on cathode rays; these experiments confirm it completely (see § 1, p. 55).

The result given by the deflection experiments for the limiting value e/m_0, i.e. for the specific charge of an electron reduced to zero velocity, is

$$\frac{e}{m_0} = 1840\ F,$$

where F denotes Faraday's constant, i.e. represents the quantity of electricity transported in the electrolytic separation of 1 mole. It is given (p. 21) by

$$F = \frac{e}{m_H},$$

where m_H is the mass of a hydrogen atom. For the rest-mass of the electron we have therefore the relation

$$m_0 = \frac{m_H}{1840} = 9{\cdot}1 \times 10^{-28}\ \text{gm.}$$

2. Canal Rays and Anode Rays (Positive Rays).

We have in the cathode rays made the acquaintance of negatively charged particles. Can we not produce positively charged rays also, in the same way as the cathode rays? The answer was given by Goldstein (1886), who succeeded in producing such rays by the following method. If residues of gas are still present in the discharge tube, the electrons on their way from the cathode to the anode will collide with these residual gas molecules and ionize them. The ions thus formed, being positively charged, will be accelerated towards the cathode, in consequence of the potential difference which exists across the discharge tube. They therefore dash against the cathode, and would stick fast in it unless, as Goldstein did, we bored canals through the cathode, which allows the ions

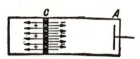

Fig. 5.—Production of canal rays; the positive ions produced in front of the cathode are driven up to the cathode and pass through the canals bored in it.

free passage (fig. 5). The rays so obtained are called *canal rays*. This interpretation of the luminous phenomena observed in the discharge tube was confirmed by Stark's observation of the Doppler effect in the light emitted by the canal rays (p. 17).

Under certain conditions, positively charged rays issue also from the anode, consisting of ions which have been torn out of the atomic fabric of the anode; rays of this sort are called *anode rays*.

The properties of these *positive rays* can be determined by methods analogous to those used for the cathode rays. From the deflection experiments values are found for the specific charge of these particles, of the order of magnitude of Faraday's number F. In these rays, therefore, we are concerned with singly or multiply charged atoms or molecules (ions); and the same values are found for the ratios of the masses of these ions as chemists have found by chemical methods.

For the exact determination of the specific charge, an apparatus was constructed (J. J. Thomson) in which the ions are deflected in an electric field and a magnetic field parallel to it. If a photographic plate is set up perpendicular to the original direction of the rays, the image obtained on the plate is a family of parabolas (fig. 6, Plate I). By a simple calculation (§ 1, p. 26) we can see that the points on a definite parabola arise from a definite set of particles with the same value of e/m; and that the individual points of this parabola correspond to different velocities of the particles, in such a way in fact that the marks due to the particles with smaller velocities, and therefore **more easily deflected**, are farther away from the vertex of the para-

PLATE I

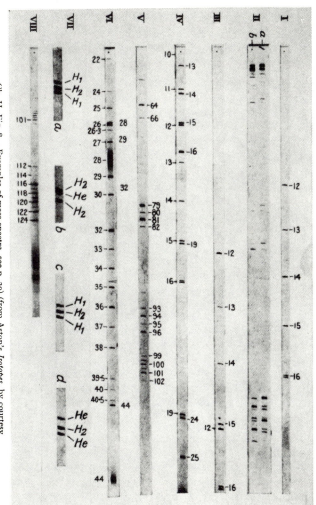

Ch. II, Fig. 8.—Examples of mass-spectra see p. 29) (from Aston's *Isotopes*, by courtesy of the publishers, Edward Arnold)

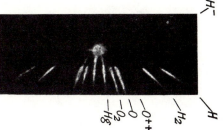

Ch. II, Fig. 6.—Positive ray parabolas (J. J. Thomson). Measurements on these deflection photographs allow the individual parabolas to be assigned to chemical elements and compounds as indicated. The electric and magnetic deflecting fields are parallel to the common axis of the parabolas (see p. 28).

bola. Since a definite parabola corresponds to each value of e/m, we can easily, from the positions of the individual parabolas, determine by measurement the specific charges of the ions contained in the ray, and accordingly their masses also (since we know their charge, which of course is an integral multiple of the elementary charge).

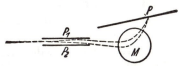

Fig. 7.—Diagrammatic representation of Aston's mass-spectrograph. The positive ray is deflected, first downwards in an electric field (between the condenser plates P_1 and P_2), and then upwards in a magnetic field (indicated by the coil M). By suitable dimensioning and arrangement of the apparatus it can be secured (as was shown by Aston and Fowler) that positive rays with the same e/m but arbitrary velocity are focussed at one and the same point P of the photographic plate. A mass-spectrum is obtained on the photographic plate; examples are shown in fig. 8, Plate I.

Aston's "mass-spectrograph" (for construction, see fig. 7; for mass-spectra, fig. 8) enables us to determine the mass directly. It has the immediate advantage over chemical methods, that the measurement of mass is made on the individual ion, whereas chemists always measure only the mean value of the mass over a very large number of particles. We shall return to this subject later (p. 41).

3. X-rays.

In the year 1895 Röntgen discovered a new kind of ray, distinguished by a penetrating power up to that time unknown. There are still many who remember what a sensation it made when the first photograph of the bones of a living subject was published. The hopes then aroused in the medical profession have been to a large extent fulfilled. But in physics also new paths were opened up by this discovery. The ray-physics characteristic of the present day had begun.

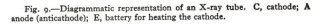

Fig. 9.—Diagrammatic representation of an X-ray tube. C, cathode; A anode (anticathode); E, battery for heating the cathode.

X-rays are produced when a cathode ray impinges on the glass wall of the tube, or on a specially fitted anticathode (fig. 9). Their penetrating power, also called their hardness, increases as the potential exciting the tube is raised. But different substances show different degrees of transparency. The higher the atomic weight, the greater the opacity—hence the possibility of obtaining radiographs of the bones, which contain a much greater proportion of metallic atoms than the surrounding flesh. The question, whether the rays are of corpuscular nature, or waves similar to light, occupied physicists in vain for a long time. Interference and diffraction experiments (Walter and Pohl, 1908), the meaning of which we shall explain later (§ 1, p. 79), gave only one certain result, viz. that, if the rays are waves, their wave-length must be considerably shorter than that of visible or of ultraviolet light. By using a crystal as the diffracting apparatus, von Laue and his co-workers Friedrich and Knipping (1912) succeeded in deciding the question (p. 80): X-rays are light of very short wave-length. This discovery has provided one of the most powerful tools for investigating the structure of matter (molecules and crystals, p. 80) by observing the scattering of X-rays and interpreting the results as interference of secondary waves. But it is remarkable that other experiments with the very same X-rays lead to results completely opposed to the wave view, which have compelled us to interpret light in terms of corpuscles (Chap. IV, § 5, p. 89).

4. Radiations from Radioactive Substances.

We have hitherto been occupied only with radiations produced artificially. As we know, there are also natural radiations, which are emitted by *radioactive substances*, the process involving spontaneous change of the atoms of these substances into other atoms (Becquerel, 1896; P. and M. Curie, 1898; Rutherford, Soddy, 1902). We distinguish three different kinds of radioactive radiations:

1. **α-rays:** The deflection experiments show that in these we have to do with positively charged particles, which are much more difficult to deflect than cathode rays, and must therefore be of much greater mass than electrons. Their e/m ratio corresponds to that of a doubly ionized helium ion, i.e. to a He^{++} particle. That it is actually He^{++} (atomic weight 4) which is in question, and not, say, a singly charged particle of atomic weight 2, which would of course show the same value for the specific charge, is made very probable by the fact (Rutherford, Ramsay and Soddy, 1903) that radioactive substances develop helium, and was proved in most convincing fashion in an experiment due to Rutherford and Royds (1909), who succeeded in capturing α-particles in an evacuated vessel; when the gas composed of α-particles is made luminous, the spectroscope reveals unmistakably the lines of helium.

2. **β-rays:** These, as deflection experiments show, undoubtedly consist of electrons, and differ from the cathode rays only by their higher velocity. While cathode rays can be produced with velocities

within a few hundredth parts of that of light, the velocity of β-rays differs from that of light by a few thousandth parts only.

More exact investigation has shown that two kinds of β-rays are emitted from a given radioactive element. One kind has a continuous "velocity spectrum", i.e. electrons occur with every possible velocity over an extensive range. The other kind has a discontinuous velocity spectrum, i.e. it consists of groups of electrons of definite velocity.

We shall see that the latter sort can be regarded as a secondary effect of slighter importance; they do not originate at all in the nucleus, which is characteristic of the atom (p. 62), but in the external electronic system surrounding the nucleus. The nuclear β-rays proper, which have a continuous spectrum, at first sight present the theoretical physicist with a puzzling problem; for, if particles of all possible energies actually leave the nucleus of the atom, the nucleus cannot remain as a unique structure of definite energy. But all experiments bear evidence that the product of the explosion process—the new atomic nucleus, poorer by an electron—is well defined and unique and has therefore a definite energy. What happens then to the remainder of the energy? We shall see later (p. 67) that the remaining energy is carried away by a most curious particle, the neutrino (ν^0), which has zero charge and probably zero mass and is extremely difficult to detect.

3. **γ-rays:** These are found to be incapable of deflection in the electric or magnetic field. We have to do here with radiation of extremely short wave-length (ultra-X-radiation).

The fundamental law of radioactive transformation (v. Schweidler, 1905) states that the number of atoms disintegrating per unit time $(-dN/dt)$ is proportional to the number of atoms (N) present at the moment:

$$-dN = \lambda N \, dt.$$

The factor of proportionality λ is called the *radioactive decay constant*; it is characteristic of the kind of atom. By integration we find

$$N = N_0 e^{-\lambda t},$$

where N_0 denotes the number of atoms at the time $t = 0$. We can express λ in terms of the *half-value period* T, i.e. the time it takes for half of the atoms to be disintegrated. We have $N_0 e^{-\lambda T} = \frac{1}{2} N_0$,

or
$$T = \frac{\log_e 2}{\lambda} = \frac{0 \cdot 6931}{\lambda}.$$

For radium itself, $T = 1590$ years; but there are radioactive elements with extremely long half-value periods, as thorium with $T = 1\cdot8 \times 10^{10}$ years, and some with extremely short ones, like thorium B' with $T = 10^{-9}$ sec.

The meaning of the law of transformation is, that every atom in a certain measure has the same explosion probability; clearly the law is a purely statistical one. This has been confirmed in two different ways. First, it has been found quite impossible by ordinary physical means (say high temperatures) to accelerate or retard the process of disintegration, or to affect it in any way. Secondly, it has been found possible to determine not only the mean number of particles emitted per second, but also the fluctuations about the mean; and it turns out that these obey the regular statistical laws (Appendix IV, p. 367). Radioactive disintegration is the prototype of an elementary process, which the ideas of classical physics are powerless to explain, but with which modern quantum theory is quite capable of dealing.

We shall now cite a few more experiments, which seem to point unambiguously to the corpuscular nature of the α- and β-rays. We attach special weight to the fact that it seems simply impossible to understand these experiments from any other point of view than that we are actually dealing with discrete particles. In the next chapter, however, we shall discuss a series of experiments on these same rays which seem to indicate just as indubitably that the rays represent a wave process.

We begin with *scintillation phenomena* (Crookes, 1903), which have already been referred to at the end of the last chapter in connexion with methods of determining Avogadro's number. If a fluorescent screen is set up near a radioactive preparation, flashes of light are observed, now here, now there, on the material of the screen. Anyone who has a watch with a " self-luminous " dial can, with the help of a magnifying glass, convince himself of the presence of these flashes of light; the illuminating substance consists of a layer of radioactive material, varnished with zinc blende. When the radiation strikes the zinc blende, the spot is lit up. These phenomena compel us to assume that the α-radiation consists of discrete particles, like a hail of shot, and that the fluorescent screen becomes illuminated wherever it is struck by such a particle. This phenomenon is now used extensively for the detection and measurement of nuclear radiations. In the earliest applications the observer simply counted the number of light flashes in a given interval of time. More recently the light pulse has been changed into a large burst of electrons by means of a photomultiplier. The latter is an instrument consisting of a photocathode from which electrons are emitted when light falls upon it (p. 83) and an electron multiplier. In the electron multiplier electrons emitted by the photocathode are accelerated by a high potential to an anode; here secondary electrons are released and are accelerated to a second anode. This process continues until an avalanche of electrons arrives at the collector anode. The resulting electrical impulse can then be counted as described below.

A second vivid proof of the corpuscular nature of these rays is given by another method of counting the individual particles. The Geiger counter (1913; Geiger and Müller, 1928) consists essentially of a metal

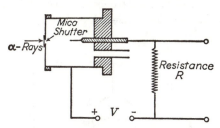

Fig. 10.—Diagrammatic representation of a Geiger counter for recording α- and β-particles. The applied potential is just great enough to prevent spontaneous discharge between the point and the wall of the instrument. An ionizing particle flying in starts the discharge; a resistance R serves to limit the current and thus break off the discharge. The discharges are counted by means of some electronic recording arrangement connected to the circuit.

plate, placed opposite a metallic point (fig. 10); the whole is contained in an air-filled vessel. A potential difference is applied to the plate and point, as great as possible subject to the condition that in spite of the action of the point no discharge passes. If an α-particle now flies past, it will ionize the molecules of air which it strikes. The ions thus produced will be further accelerated in the strong field around the point and produce ions in their turn. This process of multiplication of the primary ion pairs gives a measurable current pulse, producing a voltage drop over the resistance R, which can be detected by an electrometer. After this discharge, the plate and point again become charged, and the process begins afresh. Every particle which flies past causes in this way a momentary discharge which can be recorded by suitable apparatus. For this purpose first standard instruments like sensitive string electrometers were used. After the invention of electronic amplifiers, the discharge could be made audible by a telephone, or could be recorded by a mechanical counter. To-day the technique of counting

fast sequences of electric impulses has been highly developed with the
help of electronic devices, based on the *thermionic valve.*

The simplest form of this instrument (Fleming, 1904) is a highly
evacuated cathode-ray tube like that represented in fig. 1 (p. 25); it
consists of a thin filament which, when made incandescent (by a current),
emits electrons. In a later section we shall deal with this phenomenon of
thermionic emission from the theoretical standpoint (Chap. VII, § 8,
p. 240). Here it suffices to remark that considerable currents can be
transported, whose strength depends mainly on the temperature of the
filament and hardly on the potential applied. Obviously the current in
such a *diode* (tube with two electrodes) can pass only in one direction;
it can therefore be used as an electric valve, or as a rectifier, which

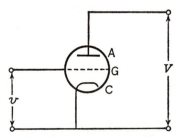

Fig. 11.—Schematic representation of
a triode. A is the anode, C the cathode,
and G the grid. An input potential v
between C and G modulates the anode
current produced by the potential V.

suppresses one phase of an alternating current. Of much wider use is
the *triode* (tube with three electrodes) invented by de Forest (1907) and
much improved by Langmuir (1915); the third electrode is a zigzag
wire or wire netting, called the *grid*, interposed between cathode and
anode (usually in a cylindrical arrangement, the heated wire being the
axis). A potential between cathode and grid will accelerate or retard
the electrons emitted by the cathode according to its sign, and hence
will influence the number of electrons passing through the meshes of
the grid and reaching the anode. By a suitable arrangement of the
outer circuit, considerable amplifications can thus be produced (see
fig. 11) which can still be multiplied by using several valves in succession. The application of this instrument to the production and recep-
tion of wireless waves, and to other technical purposes, is well known,
but does not belong to our subject, the counting of small impulses.

In regard to this problem, a very important step was the invention of *coincidence counters* (Bothe and Kohlhörster, Rossi, 1929) which allowed the study of the correlation of atomic events at different places. In the simplest case two counters are arranged in such a way that only simultaneous impulses in both are recorded. Hence not only the existence of a particle can be ascertained but also its passing two or more definite places (the counters), and thus its path can be roughly determined.

For proving the corpuscular nature of the radiations from radioactive bodies, the most impressive method consists in making the

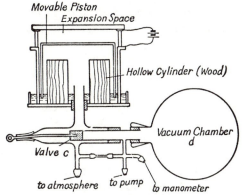

Fig. 12.—Diagrammatic representation of the original Wilson cloud chamber. The movable piston is suddenly lowered by opening the valve *c* and so connecting the vacuum chamber *d* with the part of the apparatus beneath the piston.

individual particles directly visible in the *Wilson cloud chamber* (1912) (fig. 12). If we take pure water vapour in which there are no nuclei on which condensation could take place, in the form of dust particles and so on, and if by sudden *expansion* we cool it sufficiently for the vapour pressure of the water to be exceeded, then in the absence of nuclei of condensation the vapour cannot condense into drops; we obtain supersaturated water vapour. If an α-particle shoots into the gas in this condition, it will ionize the molecules with which it collides. But the ions now act as nuclei of condensation, on which the neighbouring molecules of the supersaturated vapour are deposited as droplets. The path of the α-particle is thus made visible in the form of a series of minute water drops. An ingenious arrangement makes it possible by a single operation to expand the vapour beyond saturation point, to give the radioactive rays free access to the vapour, to expose the whole apparatus to a flash of light, and to take an instantaneous photograph. The plate shows a series of rectilinear or broken tracks. The tracks of the α-particles can be distinguished from those of the β-particles by the greater ionizing action of the former (more intense condensation); they are thicker, straighter, and less broken than the tracks of electrons. By perfecting this technique it was eventually possible to recognize tracks of particles not emitted by natural radioactive substances but produced by artificial disintegration or by accelerating machines like cyclotrons, e.g. protons (Chap. III, § 6, p. 71), and even to discover new particles, e.g. mesons (Chap. II, § 8, p. 49). The greatest difficulty in these investigations was the rarity of events; hundreds or thousands of photographs have to be taken to find the phenomenon. A big advance was the *counter-controlled cloud chamber* (Blackett and Occhialini, 1937); two counters are attached to the chamber, one above and one below it, and so connected that the chamber mechanism is released and a photograph obtained only when a " coincidence " is recorded; i.e. if a particle actually passes through the chamber. The *diffusion cloud chamber* (Lemaitre and Vallarta, 1936; Needels and Nielson, 1950) has more recently come into vogue and is continuously sensitive. Vapour diffuses steadily downwards through a region in which there is a vertical temperature gradient. There is then a zone which is continuously supersaturated and therefore continuously sensitive. Perhaps the most revolutionary type of chamber in recent years is the *bubble chamber* invented by Glaser (1952). Cloud chambers have the basic disadvantage that of necessity the density of material in the chamber is low. To circumvent this difficulty Glaser had the idea of using the instability of superheated liquids against bubble formation. He argued that if ionizing particles passed through such a superheated liquid then condensation nuclei would be created for the formation of bubbles in an analogous fashion to the formation of drops in a conventional cloud chamber. However, the bubble chamber, being filled with liquid, would have a much higher density of material. Several bubble chambers have now been built using liquid hydrogen and have proved extremely useful in the investigation of processes involving protons and other nuclear particles.

Another great advance was made by two Viennese ladies, Misses Blau and Wambacher (1937), who discovered a photographic method of recording tracks of particles. The grains of a photographic emulsion are sensitive not only to light but also to fast particles; if a plate exposed to a beam of particles is developed and fixed the tracks are seen under the microscope as chains of block spots. Their quality depends very much on the size of the grains, and special emulsions with very·small and dense grain have been developed (Ilford, Kodak). The photographic tracks are some thousand times shorter than corresponding tracks in air, because of the higher stopping power of the solid material; they are of the order of some microns. The advantages of this method are its extreme simplicity, the continuity of sensitiveness, and the great number of events recorded on one plate.

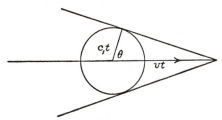

Fig. 13.—Conical headwave

On the other hand high-quality tracks are observed and micro-photographed with oil-immersion objectives which have a narrow depth of focus; hence only a restricted part of a track appears sharp, and several photographs have to be taken with different focus.

A new important method for recording and counting very fast electrons is based on a discovery made by Cerenkov (1934) and theoretically investigated by Frank and Tamm. It is easily understood with the help of an acoustical analogy. A cannon shell moving faster than sound creates a cone-shaped compression wave with the apex in the shell (fig. 13). If c_1 is the velocity of sound and v that of the shell, and if $v > c_1$, it is seen from the figure that the opening angle θ of the cone is given by

$$\cos \theta = \frac{c_1}{v} < 1.$$

Now consider the case of light. While this has the velocity c *in vacuo*, its group velocity in a substance with an index of refraction $n > 1$ is smaller, namely

$$c_1 = \frac{c}{n} < c.$$

If an electron moves with a velocity $v > c_1$ then a conical headwave will develop according to the same law as in the acoustical case. This is the phenomenon experimentally confirmed by Cerenkov. We can use it for counting by letting the conical wave fall on a metal plate where the energy of the photons is transformed into an electric current which is then amplified (photomultiplier, see p. 32). Thus a very sensitive counting device for electrons is obtained.

5. Prout's Hypothesis, Isotopy, the Proton.

Now that we have sufficiently convinced ourselves of the corpuscular character of the α- and β-rays from a radioactive substance, we proceed to consider the question which was raised at the outset, concerning the building up of atoms and molecules from elementary constituents. As far back as the beginning of last century the hypothesis was advanced by Prout that all atoms are ultimately *made up of hydrogen atoms*. It fell into oblivion, however, when chemists became able to determine the atomic weights of the elements more exactly. If Prout's hypothesis were correct, the atomic weight of every atom would of course be a whole number of times that of the hydrogen atom. It is found, however, that for a whole series of atoms this is not the case; a glaring example is chlorine, the atomic weight of which, referred to that of hydrogen as unity, has the value 35·5.

The phenomena accompanying radioactivity led to the conjecture that elements, though chemically absolutely pure, actually represent a mixture of different kinds of atoms, of the same structure indeed, but of different mass. These atoms, perfectly equivalent chemically, but of different mass, are called *isotopes*. To make what follows more easily understood, we should like briefly to remind the reader of the *periodic system* of the elements developed by Newlands (1863), Mendeléef (1870), and Lothar Meyer (1870) (Table I, p. 39). The chemical behaviour of a given element is to a large extent determined by the place which it occupies in the periodic table; thus the alkali metals, with their chemically similar behaviour, occupy the same vertical column, likewise the alkaline earths, the noble metals, the heavy metals, and finally the halogens and the inert or noble gases. The arrangement of the elements in this system was originally carried out on the basis of their atomic weight; to-day we arrange them according to their " atomic number ", as we shall see below (p. 63).

At the end of the periodic system stand the radioactive elements; these too can be arranged according to their chemical behaviour. The method, first applied by the Curies, for separating them and

TABLE I.—SHORT-PERIOD SYSTEM OF THE CHEMICAL ELEMENTS

Period	Group I a	Group I b	Group II a	Group II b	Group III a	Group III b	Group IV a	Group IV b	Group V a	Group V b	Group VI a	Group VI b	Group VII a	Group VII b	Group VIII a	Group VIII b
I														1 H 1·0080		2 He 4·003
II	3 Li 6·940		4 Be 9·02			5 B 10·82		6 C 12·010		7 N 14·008		8 O 16·0000		9 F 19·00		10 Ne 20·183
III	11 Na 22·997		12 Mg 24·32			13 Al 26·97		14 Si 28·06		15 P 30·98		16 S 32·066		17 Cl 35·457		18 A 39·944
IV	19 K 39·096	29 Cu 63·54	20 Ca 40·08	30 Zn 65·38	21 Sc 45·10	31 Ga 69·72	22 Ti 47·90	32 Ge 72·60	23 V 50·95	33 As 74·91	24 Cr 52·01	34 Se 78·96	25 Mn 54·93	35 Br 79·916	26 Fe 55·85 · 27 Co 58·94 ←→ 28 Ni 58·69	36 Kr 83·7
V	37 Rb 85·48	47 Ag 107·880	38 Sr 87·63	48 Cd 112·41	39 Y 88·92	49 In 114·76	40 Zr 91·22	50 Sn 118·70	41 Nb 92·91	51 Sb 121·76	42 Mo 95·95	52 Te 127·61	43 Tc 99 ←→	53 I 126·92	44 Ru 101·7 · 45 Rh 102·91 · 46 Pd 106·7	54 Xe 131·3
VI	55 Cs 132·91	79 Au 197·2	56 Ba 137·36	80 Hg 200·61	57–71 Rare Earths	81 Tl 204·39	72 Hf 178·6	82 Pb 207·21	73 Ta 180·88	83 Bi 209·00	74 W 183·92	84 Po 210	75 Re 186·31	85 At 211	76 Os 190·2 · 77 Ir 193·1 · 78 Pt 195·23	86 Rn 222
VII	87 Fr 223		88 Ra 226·05		89 Ac 227		90 Th 232·12 ←→		91 Pa 231		92 U 238·07 · 93 Np 237	102 No	94 Pu 239 · 95 Am 241		96 Cm 242 · 97 Bk 246 · 98 Cf 249 · 99 E 254s · 100 Fm 256 · 101 Md 256	

RARE EARTHS

VI 57–71	57 La 138·92	58 Ce 140·13	59 Pr 140·92	60 Nd 144·27	61 Pm 147	62 Sa 150·43	63 Eu 152·0	64 Gd 156·9	65 Tb 159·2	66 Dy 162·46	67 Ho 164·90	68 Er 167·2	69 Tm 169·4	70 Yb 173·04	71 Lu 174·99

The numbers in front of the symbols of the elements denote the atomic numbers; the numbers underneath are the atomic weights. The double arrow ←→ indicates the places where the order of atomic weights and that of atomic numbers do not agree. New elements are introduced by their symbols. Four of these fill gaps in older tables, namely, 43 Tc Technetium, 61 Pm Promethium, 85 At Astatine, 87 Fr Francium. The ten others are transuranic elements, 93 Np Neptunium, 94 Pu Plutonium, 95 Am Americium, 96 Cm Curium, 97 Bk Berkelium, 98 Cf Californium, 99 E Einsteinium, 100 Fm Fermium, 101 Mv Mendelevium, 102 No Nobelium.

determining their chemical properties, consists in bringing the radio-active substance with other bodies into solution, and applying various precipitating agents. It is then tested whether the activity, recognizable by the radiating power, is in the precipitate or in the residual solution, or is divided between the two. Each portion is treated in a similar way, until one part of the products is free from radiation, and the other part shows an exponential decay with characteristic half-value period (§ 4, p. 31). Radium itself was isolated thus by the Curies (1898), starting from the mineral pitchblende; barium was left along with the radium to begin with, and could not be separated from it until the end. Radium therefore belongs to the group of alkaline earth metals, and so must be placed in the periodic table in the column under calcium and barium; radon Rn (or Ra emanation), on the other hand, behaves like a noble gas and so ranges under He, Ne, &c. The short-lived succeeding disintegration product RaA ($T = 3$ min.) is found to be chemically analogous to tellurium; the next, RaB ($T = 27$ min.), goes with lead, and so on.

From these and similar results the following important *law of radioactive transformation* has been deduced: emission of an a-particle (loss of charge $+2e$) shifts the residual atom two places to the left in the periodic table, i.e. in the direction towards lower H-valencies; thus radium on giving up an a-particle is transformed into radon (Ra emanation). On the other hand, escape of a β-particle (loss of charge $-e$) displaces the atom one place to the right (Russel, Soddy, Fajans, 1913). If we go through the three radioactive series according to this rule (fig. 14), we find that many places in the table are multiply occupied. To cite a particularly conspicuous example: into the place in the periodic table occupied by ordinary lead, there also come, according to the law of radioactive change, the end-products of the three radioactive series, viz. RaG, AcD and ThD, all three of which have undoubtedly the character of lead. Besides these, RaD, AcB, ThB and RaB also fall into the place mentioned. All these elements, in spite of their like chemical behaviour, have different masses, because in an a-emission the atom gives up the mass 4 of helium, while in a β-disintegration, on account of the small mass of the electron, the mass of the atom remains practically unchanged. The three elements of the radium series mentioned above, RaB, RaD and RaG, must therefore have masses decreasing successively by 4.

It may also happen, we may add (e.g. in every β-disintegration), that two chemically different elements have the same atomic weight; these are called *isobars*.

After the existence of isotopes had been demonstrated in this way for radioactive substances, J. J. Thomson (1913), by deflection experiments on canal rays, succeeded in proving that isotopes occur even among ordinary elements, for instance neon. Ordinary chlorine, whose atomic weight is given by chemists as 35·5, consists of one kind of chlorine of weight 35·0, and another of weight 37·0. These investigations reached their highest development in the mass-spectrograph constructed by Aston (1919), which we have already mentioned (§ 2,

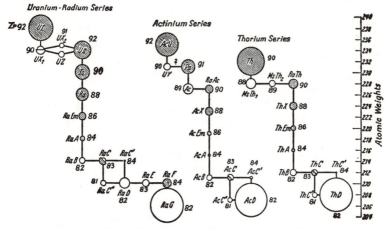

Fig. 14.,—The radioactive transformation series. The changes indicated by vertical lines correspond to α-emission, by horizontal lines to β-emission; in the former the atomic weight falls by 4, the atomic number by 2; in the latter the atomic weight remains approximately constant, and the atomic number increases by 1. Shaded circles indicate α-rays, circles without shading, β-rays. The size of each circle corresponds to the half-value period.

p. 29), and which at present is the most exact method for determining atomic weights (Aston, Dempster, 1918; Bainbridge, 1930; Mattauch, 1937; Nier, 1951).

By magnetic separation of canal rays of lithium, it has even been found possible (Oliphant, Shire and Crowther, 1934) to obtain visibly separated deposits of the two lithium isotopes (6 and 7). This method was used for the separation of the two uranium isotopes (235 and 238) on a large scale (see § 9, p. 353).

Partial separation of isotopes can be obtained by any physical process the rate of which depends directly or indirectly on the mass of the particles: ordinary diffusion and thermal diffusion (Appendix II, p. 361), diffusion assisted by centrifugation, chemical action, fractional distillation, evaporation, ionic migration and electrolysis (Chap. III, § 5, p. 69), photo-chemical action. As an example may be mentioned the partial separation of neon isotopes by Hertz (1933). By means of a suitably devised circulation process, he caused the mixture of isotopes to diffuse several times through a system of clay cylinders. Since the lighter components of the mixture diffuse more rapidly than the heavier, we obtain in this way, as the final products of the circulation process, two mixtures, one of which is richer in the lighter components, the other in the heavier. This method has also been applied successfully for the separation of the heavy isotope of hydrogen (see § 5, p. 68) and for the separation of the uranium isotopes (§ 9, p. 353).

The method of thermal diffusion was successfully applied by Clusius and Dickel (1938) to the cases of neon and chlorine (in the form HCl). With the help of a chemical reaction between HCN gas and a solution of NaCN in water, Hutchinson, Stewart and Urey (1940) were able to increase the fraction of the isotope ^{13}C by a considerable amount.

The result of the detailed investigations on the occurrence of isotopes may be stated in the following form. Every element (in the chemical sense) has either a whole number for its atomic weight, or consists of a mixture of different kinds of atoms whose atomic weights are whole numbers. For practical reasons atomic weights have all along been referred, not to hydrogen equal to 1, but to oxygen equal to 16. This has proved fortunate as an aid to clearness, for the integral character of the isotopic weights stands out much more obviously than with the choice of $H = 1$. Hydrogen indeed, for $C = 12$, has the value $H = 1 \cdot 0080$, differing decidedly from 1, but the atomic weights of all other pure isotopes approximate closely to whole numbers.

To understand how this comes about, we must anticipate so far as to note that the masses of atoms are almost entirely concentrated in the *atomic nuclei*. Every atom consists of a nucleus, which is surrounded by a cloud of electrons (§ 3, p. 62). This nucleus is the essential part of the atom. The result of Aston's investigations must be referred to the nucleus: every nucleus consists of a whole number of hydrogen nuclei or *protons*; its mass (except for trifling divergencies) is an integral multiple of the mass of the proton.

These divergencies, which in spite of their slightness are of the highest importance, will be discussed later (§ 4, p. 64).

Here we may once again summarize the ideas which for something like two decades dominated physics.

There are two primitive atoms, the atoms of electricity, the negative electron and the positive proton; they have equal and opposite charges, but (very remarkably) quite different masses (in the ratio

TABLE II.—TABLE OF ISOTOPES

The isotopes are arranged in each case where known in the order of frequency of occurrence; the radioactive isotopes are indicated by an asterisk. Radioactive isotopes produced artificially are not included.

Element	Z	Isotopes	Element	Z	Isotopes
H	1	1, 2, 3	I	53	127
He	2	4, 3	Xe	54	132, 129, 131, 134, 136,
Li	3	7, 6			130, 128, 124, 126
Be	4	9	Cs	55	133
B	5	11, 10	Ba	56	138, 137, 136, 135, 134,
C	6	12, 13			130, 132
N	7	14, 15	La	57	139
O	8	16, 18, 17	Ce	58	140, 142, 138, 136
F	9	19	Pr	59	141
Ne	10	20, 22, 21	Nd	60	142, 144, 146, 143, 145,
Na	11	23			148, 150
Mg	12	24, 25, 26	Pm	61	147
Al	13	27	Sm	62	152, 154, 147, 149, 148,
Si	14	28, 29, 30			150, 144
P	15	31	Eu	63	153, 151
S	16	32, 34, 33, 36	Gd	64	158, 160, 156, 157, 155,
Cl	17	35, 37			154, 152
A	18	40, 36, 38	Tb	65	159
K	19	39, 41, 40*	Dy	66	164, 162, 163, 161, 160, 158
Ca	20	40, 44, 42, 48, 43, 46	Ho	67	165
Sc	21	45	Er	68	166, 168, 167, 170, 164, 162
Ti	22	48, 46, 47, 49, 50	Tm	69	169
V	23	51	Yb	70	174, 172, 173, 171, 176,
Cr	24	52, 53, 50, 54			170, 168
Mn	25	55	Lu	71	175, 176
Fe	26	56, 54, 57, 58	Hf	72	180, 178, 177, 179, 176, 174
Co	27	59, 57	Ta	73	181
Ni	28	58, 60, 62, 61, 64	W	74	184, 186, 182, 183, 180
Cu	29	63, 65	Re	75	187, 185
Zn	30	64, 66, 68, 67, 70	Os	76	192, 190, 189, 188, 187,
Ga	31	69, 71			186, 184
Ge	32	74, 72, 70, 73, 76	Ir	77	193, 191
As	33	75	Pt	78	195, 194, 196, 198, 192
Se	34	80, 78, 76, 82, 77, 74	Au	79	197
Br	35	79, 81	Hg	80	202, 200, 199, 201, 198,
Kr	36	84, 86, 82, 83, 80, 78			204, 196
Rb	37	85, 87*	Tl	81	205, 203, 207*, 208*, 210*
Sr	38	88, 86, 87, 84	Pb	82	208, 206, 207, 204, 210*,
Y	39	89			211*, 212*, 214*
Zr	40	90, 92, 94, 91, 96	Bi	83	209, 210*, 211*, 212*,
Cb	41	93			214*
Mo	42	98, 96, 95, 92, 97, 94, 100	Po	84	210*, 211*, 212*, 214*,
Tc	43	99			215*, 216*, 218*
Ru	44	102, 104, 101, 99, 100, 96,	At	85	211
		98	Rn	86	222*, 219*, 220*
Rh	45	103	Fr	87	223*
Pd	46	106, 108, 105, 110, 104, 102	Ra	88	226*, 223*, 224*, 228*
Ag	47	109, 107	Ac	89	227*, 228*
Cd	48	114, 112, 111, 110, 113,	Th	90	232*, 227*, 228*, 229*,
		116, 106, 108			230*, 231*, 233*, 234*
In	49	115, 113	Pa	91	231*, 234*
Sn	50	120, 118, 116, 119, 117,	U	92	238*, 235*, 234*
		124, 122, 112, 114, 115	Np	93	—
Sb	51	121, 123	Pu	94	—
Te	52	130, 128, 126, 125, 124,	Am	95	—
		122, 123, 120	Cm	96	—

1 : 1840; see § 1, p. 27). From these all matter is built up, and that in two stages: there is first formed, from protons with some cementing electrons, the very small and compact nucleus; then this is surrounded by a cloud of electrons of relatively loose structure.

But this simple and homogeneous picture, in the light of a series of new discoveries, turned out to be incorrect, as we must now explain.

6. The Neutron.

The helium nucleus of mass 4 and charge 2 will, according to the ideas just developed, consist of 4 protons and 2 electrons; and similarly in higher nuclei some electrons will partly compensate the sum of the proton charges. The question now arises: why is there always a surplus positive charge? May not a nucleus exist with equal numbers of protons and electrons, e.g. as the simplest form, a neutron, consisting of a proton and an electron?

Reflections of this sort have been dwelt upon by many writers, in the course of speculations on the nature of the nucleus and of radioactive disintegration. The actual discovery of neutrons was due to purely experimental results, referring in fact to artificial transformations of atoms. Thus it was observed by Bothe and Becker (1930), when light elements like lithium and beryllium were bombarded by α-rays, that γ-rays were emitted. Then Irène Curie, daughter of the discoverers of radium, and her husband Joliot found (1932) that the radiation from the bombarded beryllium, when passed through substances containing hydrogen, such as paraffin, expelled protons, which γ-rays never do; a new kind of radiation must therefore be present. From an investigation with an ionization chamber it was inferred by Chadwick (1932) that the new radiation consists of uncharged, heavy particles. This was confirmed by Feather (1932) with help of the Wilson chamber. The track of the particle coming from the beryllium remains invisible, showing that it does not ionize the molecules of air, while of course we see the tracks of the nuclei on which it impinges. By comparison of the ionization in nitrogen and hydrogen, Chadwick was able to estimate the mass of the neutron, finding it approximately equal to that of a proton.

With regard to their penetrating power, neutrons behave quite differently from any other kind of radiation, whether light waves or charged particles. With the latter, the process of absorption essentially consists in their giving up energy to the outer electrons of the atoms; as the number of these runs roughly parallel with the mass, so also the absorbing power of different substances runs roughly parallel with their mass. The neutrons, however, pay no attention whatever

PLATE II

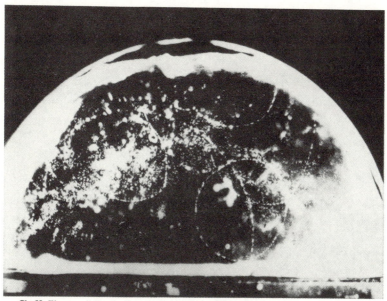

Ch. II, Fig. 15.—" Showers " of electrons and positrons liberated by cosmic rays (see p. 47)

(Rayons Cosmiques, Leprince-Ringuet)

(By permission of the author and publisher, Editions Albin Michel)

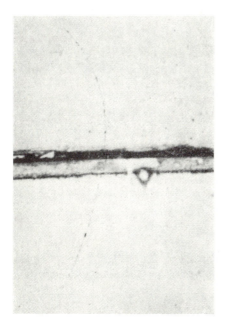

Ch. II, Fig. 16.—A 63-million volt positron passing through a 6-mm. lead plate and emerging as a 23-million volt positron (see p. 47).

(Handbuch der Physik, Julius Springer)

chambers—small, gas-filled vessels, with electrodes—are used. In measuring the strength of the radiation by the ionic current, we must be certain that this current vanishes when there is no radiation. It is found, however, that the current is never completely absent. The residual ionization had been attributed to a weak terrestrial radioactive radiation. Thick lead shielding, however, does not suppress it entirely; a weak ionization always remains (Rutherford, McLennan, 1903), which must arise from a radiation of much higher penetrating power than any known γ-rays. It was observed by Gockel (1909) that this radiation did not diminish, when the apparatus was taken to a height above the ground, in the way it ought to do if it originated in radioactive sources in the earth. Hess then (1912) showed by balloon ascents at heights up to 5 km. that the intensity of the radiation even increases with height; he found also that it is just as great by night as by day (and so cannot come from the sun). The rays appear therefore to come from interstellar space; they are called *cosmic rays*. The experiments were carried to greater and greater heights by Kohlhörster (1914), later by Millikan, Tizard, and Regener; with recording balloons Regener (1935) reached as high as 30 km. The existence of the rays was also demonstrated at the bottom of deep lakes, up to 500 metres under water (Millikan, Regener, 1928).

More exact researches were carried out with the help of the Wilson cloud chamber (first by Skobelzyn, 1929) and a strong magnetic field; the tracks of the particles were seen as circles of slight curvature; their velocity or energy was deduced by measurement of the radius. It was thus found that the rays consist of extremely fast particles; among them it is no rarity to find electrons with velocities to produce which would require a potential of 100 or 1000 million volts.

What we observe here at the earth's surface is certainly not the original cosmic radiation, but a mixture of various secondary particles, amongst them electrons, which are ejected from air molecules by various processes referred to below. Some light appears to be thrown on the question of the nature of the primary radiation by the fact that its intensity depends on the geographical latitude (Clay, 1927); it is weaker at the equator than at the poles. It must therefore be a case of electrically charged particles, which are deflected in the magnetic field of the earth; for a swarm of charged particles, which strikes a magnet, concentrates chiefly round the poles—a fact which of course has long been known, from the explanation of the aurora by Störmer (since 1903). Moreover, the sign of the charge of the incoming particles can also be determined, for positive and negative particles

to the outer electrons—only collisions with a nucleus stop them; since
the size of the nucleus varies very little from one substance to another
(p. 311), it is the number of nuclei per unit volume which now counts.
But 1 gm. of hydrogen contains the same number of nuclei as 16 gm.
of oxygen; the absorbing power per gramme is therefore some 16
times as great for hydrogen as for oxygen. The Wilson photographs
show, moreover, that in some collisions of neutrons and nuclei the
neutron flies on (it is supposed that its diameter is of the same order
of magnitude as that of a nucleus), while in others it is captured by
the nucleus, and causes this to explode, with expulsion of other particles (Chap. III, § 6, p. 70). We thus come to the question of what
part the neutrons take in the structure of nuclei; we shall deal with
this later (p. 66), and only remark here that of course it is now
possible to regard all nuclei, of whatever (positive) charge, as made up
of protons and neutrons alone, without making any use of electrons.

This assumption has been very successful (§§ 4, 5, 6, pp. 64, 68,
70). The fact that electrons can be emitted by nuclei (β-decay) has
then to be explained in a new way which will be dealt with later.

7. Cosmic Rays. Positrons.

To suppose that the positive and negative particles of electricity
have very different masses is an unacceptable asymmetry of nature.
Every theory proposed up to the present is symmetrical in the sign of
the charge. It would therefore long ago have been reasonable to conjecture that there are also particles of both kinds, heavy and light, with
opposite charges to those known: positive electrons and negative
protons. To advance this as a proposition was reserved, however, for
the relativistic form of quantum mechanics proposed by Dirac in 1928
(Chap. VI, § 8, p. 193).

The actual discovery of positive electrons or " positrons " was
achieved independently of this, in connexion with observations on the
so-called " cosmic rays ". These are of great interest for their own sake,
so something may be said here about their discovery and investigation.

As has already been mentioned (§ 1, p. 24), gases in their natural state
are poor conductors of electricity. But they can be made conducting in
various ways, not only, as explained above, by applying potentials at
reduced pressure, but also by irradiation with all kinds of corpuscular
rays or (short-wave) electromagnetic rays. These rays separate electrons
from the atoms or molecules. The electrons, as well as the residues,
the ions, are set in motion by an electric field. To demonstrate the
presence of radiations, and investigate their properties, ionization

are deflected into different directions by the earth's magnetic field. Positive particles are deflected towards the east, negative particles towards the west. From this east-west asymmetry it has been possible to show that at least the majority of the incoming cosmic ray particles are *positively* charged. It is now well established that the majority of these particles are protons. However, in addition, helium nuclei, members of the CNO group and some nuclei as heavy as iron, have been observed. There is still considerable speculation about the origin of cosmic rays although it seems to be fairly well agreed that they obtain their energy either through chaotic scattering by large magnetic clouds (Fermi) or by synchrotron-type acceleration (see p. 76) in, for example, the Crab nebula. In the former case the particles gain energy at the expense of the magnetic clouds causing the scattering; the process is rather akin to the slowing down of neutrons in a low-temperature moderator, except that in this case the neutrons lose rather than gain energy. This Fermi mechanism has the advantage that it gives a natural explanation of the observed energy spectrum of the primary cosmic-ray particles. However, there seems little doubt that the Crab nebula is also a source of cosmic rays, but unfortunately no simple explanation of the energy spectrum is then forthcoming.

In travelling through the atmosphere the primary protons produce secondary particles, amongst them numerous electrons, in a way discussed below.

Anderson was the first to remark that frequently two kinds of track appeared on a photograph, which were of opposite curvature, though otherwise both looked like electron tracks (fig. 15, Plate II, opposite). If then we do not believe in positive electrons, we must assume that the one track is traversed in the opposite sense from the other. This possibility was excluded as follows. A lead plate was set up in the chamber, and tracks were found due to particles which had passed through the plate. But such a particle must necessarily have a smaller velocity, and therefore a greater curvature of path, after it has passed through than before; so that the sense in which the tracks are described can be determined (fig. 16, Plate II, opposite). The existence of positrons was thus proved.

From the theoretical point of view the discovery of the positron is of the highest importance; for it confirms the theory of Dirac already mentioned, which amounts to this, that neither electrons nor positrons are permanent and indestructible, but that by coming into collision they may annihilate each other, with the emission of energy in the form of light waves; and conversely, that a light wave of high energy can in certain circumstances become the source of a " pair " (electron + positron). There is experimental evidence for both processes. The fact that in our actual world negative electrons preponderate is not inconsistent with the theory.

Blackett and Occhialini (1933) found that frequently a whole shower of particles is ejected from the wall of the vessel, their tracks showing, some positive, some negative curvature. These showers consist therefore of positive and negative electrons. Their creation is now fully understood. It takes place as a cascade process (Carlson and Oppenheimer, Bhabha and Heitler, 1937).

Suppose that a very fast electron passes through an atom or molecule, then the electron is deflected by the electrical field of the atomic nucleus. Whenever a fast electron changes its velocity or direction of motion light is emitted. The faster the electron, the shorter is the wave-length of the waves emitted. For the fastest electrons produced by ordinary means in the laboratory the wave-length is that of X-rays. In the case of cosmic ray electrons the wave-length is even shorter than that of γ-rays. Both theoretical and experimental investigations have shown that the intensity of these ultra-γ-rays is so high that the electron, when passing through matter, is stopped very quickly, all its energy being given to these ultra-γ-rays. However, the ultra-γ-rays do not remain inactive. When they pass through matter they are transformed into " electron pairs " as was mentioned above. The positive and negative electrons constituting these pairs have, of course, less energy than the original electron, but if the energy of the latter is sufficiently high, the energy of each positive and negative secondary electron may still be great enough to produce ultra-γ-rays again. And so the process repeats itself, more and more pairs being produced, until the energy is exhausted and the energy of each individual electron too small to lead to any further multiplication. The size of such a shower may range from 2 to several thousand particles, according to the energy of the electron initiating the shower.

The process of shower production is a rather quick one. If, for instance, a cosmic ray electron enters a lead plate, the shower is developed to full size after traversing only 1 to 3 cm. of lead, and soon afterwards all the shower particles are again slowed down to such small energies that they are incapable of emerging from the lead plate if the thickness of the latter exceeds, say, 5 cm. of lead. No electron or its secondaries can therefore penetrate through more than a few centimetres of lead.

However, cosmic ray experiments have revealed the existence of particles with a much greater penetrating power, exceeding 1 m. of lead and even more (Bothe and Kohlhörster, 1928). The nature of these penetrating particles was a puzzle to physicists for quite a long time. It is clear that if the above explanation of the shower process is accepted—and the experiments leave no doubt that it is correct— the penetrating particles cannot be electrons. This dilemma led to the recognition of elementary particles of a new kind, called *mesons*.

8. Mesons and Nuclear Forces.

It is not difficult to see what properties a particle must have in order that it may emit less ultra-γ-ray energy than electrons do, and thus have a greater penetrating power. The deflection which the particle suffers in passing through an atom obviously depends upon its mass. A heavy particle will be deflected less, and therefore will lose less energy, than a light particle. The idea presented itself that the penetrating cosmic ray particles are heavier than electrons. It might be conjectured that they are protons. But then they ought to be *all* positive, which is not the case. It was not until after 1938 (Williams and Pickup, Nishina, and others) that the mass of these particles could be estimated experimentally. (For this purpose the curvature of a track in a magnetic field must be measured, giving the ratio mv/e and, simultaneously, its ionization from which indirectly v can be determined.) The result was this: the mass of the penetrating particles is about 200 times the electron mass, or $\frac{1}{10}$ of the proton mass. For this reason the particles were named mesons; meson (Greek) means "middle".

Probably the meson would not have been discovered until some time later had its existence not been predicted three years before on theoretical grounds. This was achieved by the Japanese physicist Yukawa (1935), whose arguments can only be outlined here, but will be resumed in a later section (Chap. X, § 3, p. 319 ; App. XXXII, p. 448). We have, however, to anticipate a few facts which will now be discussed.

It was already mentioned (at the end of Chap. II, § 6, p. 45), and will be demonstrated later in more detail (Chap. III, § 4, p. 66), that the atomic nucleus consists of two kinds of particles, protons and neutrons. They are very closely packed, for the radius of the nucleus, as determined by different methods (Chap. III, § 3, p. 63), turns out to be only 2 to 9 × 10^{-13} cm.

What, we may ask, keeps the protons and neutrons so close together? There must be some sort of attractive force between them. This cannot be of an electrical nature, because the neutron has no charge and is therefore insensitive to electric fields. Thus Yukawa concluded that a new kind of field must exist, analogous to the electromagnetic field but different in nature, which is responsible for the attraction between a proton and a neutron. Anticipating the later nomenclature we shall call this new field the *meson field*. In close analogy to the ordinary case of a charged particle we assume that both a proton and a neutron carry a " mesonic charge ", f say, which produces such a meson field. A proton and a neutron will then attract each other as two electrically charged particles do, and there will be a mutual potential between them analogous to the Coulomb potential $-e^2/r$ between two opposite electric charges. However, the meson field differs from the electromagnetic fields very appreciably in two respects. As was explained above, the nuclear forces have a very small range (2 to 3 × 10⁻¹³ cm.). Therefore Yukawa assumed that the potential between two nuclear particles is not $-f^2/r$, but

$$-f^2 \frac{e^{-\mu r}}{r},$$

where μ is a new universal constant with the dimensions of a reciprocal length and of the order of magnitude $\mu = 0.3$ to 0.5×10^{13} cm.⁻¹.

Just as electric charges in non-uniform motion emit electromagnetic waves, mesonic charges will, under similar circumstances, emit mesonic waves; but the law of propagation of the latter will be different, as it must depend on the constant μ. Indeed, if ν is the frequency and $\kappa = 1/\lambda$, the wave number of a simple harmonic wave, we have, in the electromagnetic case, $\nu\lambda = c$ or $\nu/c = \kappa$, while it can be shown that the simplest wave equation compatible with Yukawa's static law of force leads to the relation

$$\left(\frac{\nu}{c}\right)^2 = \kappa^2 + \left(\frac{\mu}{2\pi}\right)^2$$

(Appendix XXXII, p. 448). Now we have to anticipate the result of quantum theory that any elementary physical phenomenon has a dual aspect; it can either be described by waves or by particles. The latter carry definite quanta of energy E and momentum p, and the relation of these to the frequency ν and wave number κ of the associated wave is

$$E = h\nu, \quad p = h\kappa$$

where h is Planck's constant (Chap. IV, § 2, p. 82, and Chap. IV, § 5, p. 89). Hence we get from Yukawa's frequency law

$$\left(\frac{E}{c}\right)^2 = p^2 + (\hbar\mu)^2,$$

where we have written $\hbar$ for $h/(2\pi)$ (this notation will be used frequently later on). This equation relates the energy and momentum of the particles.

The final step of Yukawa's argument consisted in interpreting this formula with the help of the theory of relativity (Appendix V, p. 371), according to which energy and momentum of a particle are connected with its rest mass m_0 by

$$\left(\frac{E}{c}\right)^2 = p^2 + m_0{}^2c^2.$$

By comparing this formula with the previous one, Yukawa came to the conclusion that there must exist particles whose rest mass is related to the constant μ of nuclear forces by

$$m_0 = \frac{\mu\hbar}{c}.$$

Inserting here for μ the value given above ($0\cdot3$ to $0\cdot5 \times 10^{13}$ cm.$^{-1}$), we find $m_0 = 130$ to 200 electron masses (p. 27).

In this way Yukawa was led to the prediction of the existence of particles with mass intermediate between electron and proton. This prediction was confirmed in 1939 independently by Neddermeyer and Anderson, and by Blackett and Wilson with the help of cloud-chamber photographs.

Mesons must be considered as elementary particles, i.e. as not composed of other particles. They differ, however, from the electrons and nuclear particles (protons, neutrons) in that they do not occur in the structure of ordinary matter. Why is this? The answer was already given in Yukawa's first paper: mesons are not stable particles. If left alone, a meson decays into other elementary particles. The different possible decay modes are given in Table III. Cosmic rays present us with ample, but rather indirect, evidence for the decay of free mesons.

Since mesons are unstable they cannot be the primary particles of cosmic radiation. They must have been produced somewhere in the earth's atmosphere. By travelling fast enough they might have time to reach sea-level before they decay. We have seen in § 7 that the primary particles coming from interstellar space are probably protons. We can now understand how the mesons are produced by the primary protons. If a fast primary proton enters the atmosphere it may happen that it hits an oxygen or nitrogen nucleus which we know consists also of protons and neutrons. Each of these particles as well as the primary proton is surrounded by a meson field. In a violent collision between the proton and the nucleus, meson waves will be emitted in a similar way to light waves emitted in a violent collision between an electron and another charged particle (which is the very process initiating a shower).

The mesons associated with these waves will decay sooner or later and produce electrons. Most of them indeed decay in the upper atmosphere and only the fast ones reach sea-level. Therefore a large number of electrons arise—as decay products of mesons—in cosmic radiation. Their number will be further increased by cascade multiplication (p. 48). This is the origin of the electrons in cosmic radiation. Yukawa's ideas have thus helped to order and to understand a great many observations. But recent researches have shown that the facts are much more complex than first envisaged.

Some discrepancies between theory and observation, for instance that between calculated and measured lifetime led Marshak and Bethe (1947) to suggest the idea that two different kinds of mesons existed. The first proof of this fact was given by the Bristol school led by Powell (Lattes, Muirhead, Occhialini, and Powell, 1947) and confirmed in the following year by this group and in Berkeley, California (1948). This was achieved by bringing the photographic-track method to a high degree of perfection. It was found that the primary mesons in cosmic rays, called π-mesons, have a mass of about $273m_e$ if positively or negatively charged and $264m_e$ if uncharged. The charged π-mesons ($\pi^\pm$) decay into a μ-meson of the same charge and of mass $207m_e$ and a neutrino, the lifetime of the primary particles for this decay process being about $2\cdot5 \times 10^{-8}$ sec. (see fig. 17 opposite). The negative π-mesons, however, have little opportunity to decay spontaneously and are generally attracted by the positive nuclei of the surrounding atoms and captured by them, producing violent nuclear explosions. The secondary μ-mesons again decay into an electron or positron (according to charge) and two neutrinos; the lifetime is $2\cdot20 \times 10^{-6}$ sec. On the other hand, the neutral π-mesons (π^0) decay into two gamma-ray quanta (γ^0; see p. 82). The fact that nuclear explosions are in general caused by (negative) π-mesons indicates that it is this heavier type of meson which is connected with the nuclear forces as envisaged by

PLATE III

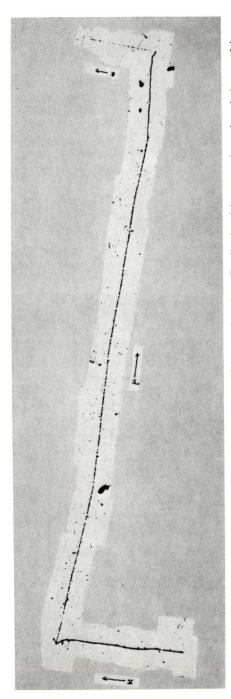

Ch. II, Fig. 18.—A primary π-meson decays into a μ-meson and an invisible (neutral) particle. The direction of the μ-meson is seen from the increase of the number of black grains; the ionizing power of slow particles is greater than that of fast ones. The μ-meson finally disintegrates into an electron and an invisible particle.

The picture is a mosaic of micro-photographs of short sections of the tracks taken with different focus (see pp. 36, 52).

(By courtesy of Prof. C. F. Powell, Bristol

TABLE III. TABLE OF ELEMENTARY PARTICLES
(after A. H. Rosenfeld et al, *Rev. Mod. Phys.* **37**, 633, 1965)

	Particle	Charge	Mass * MeV	Spin $\hbar$	Important Decays	Mean Life sec.
Photon	γ	0	0	1	stable	∞
Leptons	ν_e	0	0	$\frac{1}{2}$	stable	∞
	ν_μ	0	0	$\frac{1}{2}$	stable	∞
	$e^{\mp}$	$\mp$	0·511006 ±0·000002	$\frac{1}{2}$	stable	∞
	$\mu^{\mp}$	$\mp$	105·659 ±0·002	$\frac{1}{2}$	$e+\nu_\mu+\nu_e$	$2\cdot2001 \times 10^{-6}$ ±0·008
Mesons	$\pi^{\pm}$	$\pm$	139·580 ±0·015	0	$\mu+\nu_\mu$ $e+\nu_e$ $\mu+\nu_\mu+\gamma$ $\pi^0+e+\nu_e$	$2\cdot604 \times 10^{-8}$ ±0·007
	π^0	0	134·975 ±0·014	0	$\gamma+\gamma$ $\gamma+e^++e^-$	$0\cdot89 \times 10^{-16}$ ±0·18
	$K^{\pm}$	$\pm$	493·83 ±0·11	0	$\mu+\nu_\mu$ $\pi^{\pm}+\pi^0$ $\pi^{\pm}+\pi^-+\pi^+$	$1\cdot235 \times 10^{-8}$ ±0·005
	K^0_S	0	497·75 ±0·18	0	$\pi^++\pi^-$ $\pi^0+\pi^0$	$0\cdot874 \times 10^{-10}$ ±0·011
	K_L	0	497·75 ±0·18	0	$\pi^0+\pi^0+\pi^0$ $\pi^++\pi^-+\pi^0$ $\pi+\mu+\nu_\mu^0$ $\pi+e+\nu_e$ $\pi^++\pi^-$	$5\cdot30 \times 10^{-8}$ ±0·13
Baryons — Nucleons	p	$+$	938·256 ±0·005	$\frac{1}{2}$	stable	∞
	n	0	939·550 ±0·005	$\frac{1}{2}$	$p+e^-+\nu_e$	$1\cdot01 \times 10^3$ ±0·03
Baryons — Hyperons	Λ	0	1115·50 ±0·08	$\frac{1}{2}$	$p+\pi^-$ $n+\pi^0$	$2\cdot52 \times 10^{-10}$ ±0·04
	Σ^+	$+$	1189·47 ±0·08	$\frac{1}{2}$	$p+\pi^0$ $n+\pi^+$	$0\cdot810 \times 10^{-10}$ ±0·013
	Σ^0	0	1192·54 ±0·10	$\frac{1}{2}$	$\Lambda+\gamma$	$<1\cdot0 \times 10^{-14}$
	Σ^-	$-$	1197·41 ±0·09	$\frac{1}{2}$	$n+\pi^-$	$1\cdot66 \times 10^{-10}$ ±0·03
	Ξ^0	0	1314·9 ±0·8	$\frac{1}{2}$	$\Lambda+\pi^0$	$2\cdot9 \times 10^{-10}$ ±0·4
	Ξ^-	$-$	1321·3 ±0·2	$\frac{1}{2}$	$\Lambda+\pi^-$ $\Lambda+e^-+\nu_e$	$1\cdot73 \times 10^{-10}$ ±0·05
	Ω^-	$-$	1672 ±1	$\frac{3}{2}$	$\Xi+\pi$ $\Lambda+K^-$	$1\cdot1^{+0\cdot6}_{-0\cdot5} \times 10^{-10}$

* Masses are expressed here in terms of the rest energy (Mc^2) of each particle, the energy being measured in " Millions of Electron Volts " (MeV) see p. 65.

Yukawa. However, although his relatively simple theory seems to be qualitatively correct, it does not explain the details of the interactions between nuclear particles.

Experiments carried out at Berkeley confirmed the results of the Bristol school. Gardner and Lattes announced in 1948 that both π- and μ-mesons are produced by bombarding targets of various materials with α-particles of more than 300 MeV from the California cyclotron (p. 76). The masses of π- and μ-mesons have been determined by many investigators and are essentially as follows:

$$m_{\pi\pm} = 273 \cdot 3 m_e, \quad m_{\pi} = 264 \cdot 3 m_e, \quad m_{\mu\pm} = 206 \cdot 9 m_e.$$

Besides the π- and μ-mesons another group (K-mesons) is now well established, whose masses are in the region of $1000 m_e$. They all decay into combinations of charged and neutral π-mesons, or π-mesons and leptons, where *lepton* is used as a collective name for an electron, neutrino or μ-meson. The lifetimes for these decays are of the order 10^{-8} to 10^{-10} sec., and are very slow on a nuclear time scale (a typical nuclear time can be written on dimensional grounds as $t = R/c$, where R is a typical nuclear size, i.e. about 10^{-13} cm. This gives $t \approx 10^{-23}$ sec.). For this reason the basic interaction responsible for these decay processes is said to be *weak*. Another group of particles with masses greater than that of a nucleon has been identified and is now well established. These particles are known as *hyperons*, and they decay into neutrons or protons, and either π-mesons or leptons, with lifetimes of the order 10^{-10} to 10^{-11} sec. (see Table III). Again, the decay processes are very slow, and are attributed to weak interactions. Corresponding to each hyperon is an anti-particle, and in a number of cases these particles have been identified experimentally.

Recently a great many other elementary particle states have been discovered, but with much shorter lifetimes. For this reason they are conventionally referred to as *resonance states* rather than elementary particle states. Considering them in conjunction with the elementary particles listed in Table III, it emerges that there is a very substantial spectrum of elementary particles of different kinds, and attention during the last few years has naturally centred on the symmetries and relationships between these different particles. It has emerged that all these particles and resonance states can be classified into a sort of periodic table, much like the atomic periodic classification. This classification has led to some very profitable speculations about the fundamental nature of the various interactions between elementary particles

CHAPTER III
The Nuclear Atom

1. Lorentz's Electron Theory.

The method of physical science in the investigation of the struc ture of matter has in all times been based on the following principle: laws established for " macroscopic bodies ", i.e. bodies of ordinary size, are applied tentatively to elementary particles; if some disagree ment is then found, an alteration of the laws is taken in hand. In this way, advance essentially depends upon the closest co-operation of observation and theoretical interpretation.

Thus in the preceding pages we have made continual use of the laws of interaction of charged particles, and of their response to the influence of external fields and light rays. The most important result, that the mass of an atom is almost wholly concentrated in a very small nucleus, while its volume, and its physical and chemical proper ties, are determined by a comparatively loose surrounding structure of electrons, we have assumed without proof. This proof we must now supply. First, however, we must look a little more deeply into the laws of the electromagnetic field.

As we know, these are formulated in Maxwell's equations (1855) (App. VI, p. 373). As originally stated, these equations represent the changes in space and time of the electric field strength E, and the magnetic H, in material bodies, and therefore contain certain constants which are characteristic of those bodies (in the simplest case they are the dielectric constant and the magnetic permeability).

The atomic theory of electricity simplifies the field equations, by considering only fields in a vacuum; the units can then be chosen so that the field equations contain no material constants (or at most, the velocity of light $c = 3 \times 10^{10}$ cm./sec., which is retained so that we may be able to use the ordinary units of kinematics, centimetres and seconds). This is the standpoint of Lorentz's electron theory (1897), which domi nated physics about the turn of the century. At that time we were

acquainted with free electrons and were aware that they are constituents of atoms; about the positive atoms of electricity nothing
was known. The electrons were pictured as very small charged bodies,
which generated the field in free space, and conversely were acted on
by forces due to the field. These forces determined the motion of the
electrons in accordance with Newton's law: mass times acceleration
equals force. It was therefore necessary to ascribe to the electrons a
mass m besides their charge e, and we have seen that the fraction
e/m can be determined by deflection experiments. On the basis of these
assumptions the motion of electrons in various fields could be calculated; it was thus found that the size of the electron (its radius, if it
was pictured as a sphere) plays a part, and that for the following
reason. When there is acceleration, the field of the charges in the
electron is modified, and this changed field passes over the electronic
sphere with the velocity of light, and exercises forces on it. For
small accelerations, the forces to a first approximation are proportional to the acceleration and to e^2/a, where e is the charge, and a
the radius (with a numerical factor which depends on the distribution
of the charge); in other words, the effect is the same as if the mass
were increased by a part proportional to e^2/a (J. J. Thomson, 1882;
Heaviside, Searle, 1885). This fact suggested to some physicists
the idea that the electron possesses no " ordinary " mass at all, but
only " electromagnetic " mass; and the smallness of the mass of the
electron seemed to confirm this. Further, the electromagnetic mass
was found to depend on the velocity, and the first observations of this
effect by Kaufmann (1906), were regarded as a brilliant triumph for
the theory. Hasenöhrl (1904) derived the laws of motion of an empty
box with reflecting walls filled with electromagnetic radiation, and
found that this radiation behaved as having a mass E/c^2, where E is
the total electromagnetic energy. This was the first indication of the
general law considered in the following section.

The calculations of the electromagnetic mass of the electron rested
on the assumption (Abraham, 1903) that the electron is rigid, retain-
ing its form throughout the motion; which implies the assumption
of infinitely great internal forces of non-electromagnetic origin. If the
assumption is dropped, we obtain not only other numerical factors,
but also other functions of the velocity. Besides, the assumption of
absolute rigidity is quite incompatible with the theory of relativity
(on account of the Lorentz contraction; see Appendix V, p. 370); if
instead of this we postulate invariability of form in the reference
system in which the electron is instantaneously at rest, we obtain a

formula which agrees, to a numerical factor, with that stated on p. 27 (Lorentz's electron, 1909). The numerical factor, however, must to a large extent remain arbitrary, since it depends on the distribution of the charge, as to which we know nothing. Worse still, even in this relativistic form very great internal forces of cohesion must be assumed, to keep the parts of the electron together, these having charges of the same sign and therefore repelling each other; and this leads to a con- tradiction of a fundamental theorem of the mechanics of radiating systems, of which we have now to speak.

2. The Theorem of the Inertia of Energy.

The theorem of the inertia of energy in its full generality was first stated by Einstein (1905). It asserts that any energy E possesses a mass m, in accordance with the equation $E = mc^2$. Of this relation we shall later continually make use. The relation is very far from being a mere theoretical subtlety; the phenomena connected with it have as a matter of fact the character of large scale effects. Think of a closed box, in the sides of which are fitted two exactly similar instruments (I and II), which are so constructed that they can

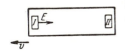

send out a momentary light signal in a definite direction, or completely absorb an incoming light signal (fig. 1). Now let the instrument I, at a

Fig. 1.—Illustration of Einstein's ideal experiment to prove the relation $E = mc^2$. The transmitter I radiates to the receiver II a definite quantity of energy E; in consequence of this the whole box undergoes a recoil.

definite moment, send out a light signal in the direction of the instru- ment II. During this process of emission, the instrument I, and with it the whole box, experiences a recoil. The occurrence of this recoil is due to the radiation pressure. The latter phenomenon was observed experi- mentally by Lebedew (1901), in good agreement with theory; it was in- vestigated later by Nichols and Hull (1903) and others, and more recently very exactly by Gerlach and his collaborators (1923). In consequence of the recoil, during the whole time which the light takes to go from I to II the box will move in the opposite direction, and will not come to rest until the light strikes the instrument II, and the radiation pressure on it again brings the box to a standstill. Now the two instruments may be interchanged; this alters nothing, provided they had originally the same mass. Then let the instrument II send out towards I the same quantity of light as was sent before by I, whereby the whole box is again displaced to the left the same distance as before. Now interchange the instruments once more. By proceeding in this way, it would be possible to displace the box any distance, without any change taking place in the interior of the box or in its neighbourhood—a conclusion obviously at variance with a fundamental property of the centre of inertia.

The contradiction disappears at once, however, if we take account of Einstein's thesis of the equivalence of energy and mass. By the emission of the light signal at the first step of our ideal experiment, the first instrument gives out a definite quantity of energy E; its content of energy, and accordingly its mass also, become smaller. Similarly, the energy content, and therefore the mass, of the instru-ment II is increased on its absorption of the light signal, so that the mass of II is now greater than that of I; to interchange I and II without shifting the centre of inertia therefore requires that the whole box should be displaced a definite distance to the right. If we examine now what the relation between mass and energy must be, in order that the displacement of the box, due to radiation pressure, should be exactly compensated by interchanging the two instruments, a short calculation (Appendix VII, p. 375) leads to Einstein's formula, as quoted above. The most important field of application of Einstein's law is the disintegration of nuclei; we shall see later (p. 73) that this gives an experimental proof of it.

We return now to the problem of electromagnetic mass (p. 56). According to Einstein's theorem, the simplest way of obtaining this mass for small velocities should be, to calculate the internal electric energy of the charge collected in the electron; this energy is in fact proportional to e^2/a, but the numerical factor is in all circumstances different from the one we get by calculating the force of reaction of the electron's own field. The contradiction, as we have already seen, is due to the presence of forces of cohesion, which also should make a contribution to the energy (though this is difficult to reconcile with the postulated rigidity).

In consequence of these unsatisfactory results, the simple classical formulation of electromagnetic mass has gradually been given up. It was found that the theory of relativity suggested in a purely formal way a law of dependence of mass on velocity for every body—a law which has been brilliantly confirmed by experiment (Bucherer, 1909; Neumann-Schäfer, 1914; Guije-Ratnowski-Lavanchy, 1921). As the quantum theory developed, physicists became sceptical about definite models of the elementary particles. They preferred therefore to think of the electron, in regard to all external actions, as a charged point mass, without troubling further about its internal structure. The awkward fact remains, however, that the electron's proper energy, which is proportional to e^2/a, becomes infinite when a is put equal to 0. One way of overcoming this difficulty consists in eliminating the infinite terms from the laws of electrodynamics without changing these other- wise. This has been done by Dirac (1938, 1942) by modifying the expression for the force on a point charge and by Pryce (1938) by changing the definitions of electromagnetic energy and momentum. But this seems to be evading, not solving, a real problem.

For the radius a of the electron has an actual physical significance; by a we understand that length which satisfies Einstein's relation $a \cdot e^2/a = mc^2$, where a is a numerical factor of the order of magnitude 1. We therefore have

$$a = a \cdot \frac{e^2}{mc^2} = a \cdot 2 \cdot 82 \times 10^{-13} \text{ cm.}$$

The simplest phenomenon in which this quantity occurs is the *scatter-ing of light* (or other electromagnetic radiation) by atoms. According to Maxwell's electromagnetic theory, light (as also X-rays) consists of a periodically variable, electromagnetic alternating field. If the light wave strikes a charged particle which can move freely, the latter is set vibrating, and that the more strongly, the lighter the particle is. If the particle is bound to other particles, its induced oscillation will have a greater amplitude the closer the frequency of the incident light is to its proper frequency. Hence, on the one hand, electrons, on account of their trifling mass, will vibrate in sympathy with the light much more strongly than protons, say, or still heavier particles; on the other hand, with visible light, to a very large extent only those particles which occupy places near the surface of the atom, and are therefore relatively loosely bound, will be excited to vibration; while for excitation of the bound electrons farther inside the atom, X-rays will be needed.

Now, as we know, a vibrating charged particle acts like an antenna —it sends out electromagnetic vibrations of its own in the form of spherical waves; and, in fact, the energy drawn per unit time from the primary ray and converted into scattered radiation by a (free or loosely bound) electron is given (Appendix VIII, p. 376) by

$$I = \frac{8\pi}{3} \left(\frac{e^2}{mc^2} \right)^2 I_0,$$

where I_0 is the intensity of the primary radiation. Since I_0 is defined as energy per square centimetre, but I denotes the whole scattered energy, the quotient I/I_0 must have the dimensions of an area. We can put it equal to an " effective cross-section " of the electron πa^2 and have then

$$\frac{I}{I_0} = \frac{8\pi}{3}\left(\frac{e^2}{mc^2}\right)^2 = \pi a^2,$$

so that

$$a = \sqrt{\frac{8}{3}\frac{e^2}{mc^2}}.$$

This quantity is the " radius " of the electron if a is put equal to $\sqrt{(8/3)}$.

The remarkable thing about this method is that it makes no use of any hypothetical extrapolation of electrostatics to the *interior* of the electron, but works with a point electron.

3. Investigation of Atomic Structure by Scattering Experiments.

The most important method for investigating atomic structure consists in causing electromagnetic radiation or a beam of particles of some kind to fall on the atom, and observing the effect. The alteration consists in a weakening (absorption) of the ray which passes on undeflected, and the corresponding production of diffracted or scattered rays. The case of the scattering of light at a free or weakly bound electron has already been considered (§ 2, p. 59). The process can be used for the purpose of counting the loosely bound electrons in an atom. If their number is n, and N is the number of atoms per unit volume, then the energy lost by the primary radiation per centimetre of its path, the absorption constant, is given (J. J. Thomson, 1906) by

$$\frac{nNI}{I_0} = \frac{8\pi}{3}\left(\frac{e^2}{mc^2}\right)^2 nN.$$

Since N is known from kinetic theory data, n can be determined from absorption measurements on X-rays; by making these sufficiently hard, all the electrons, at least in the lighter atoms, can be regarded as practically free, so that n denotes the whole number of electrons, and so also the nuclear charge in terms of the elementary charge as unit.

The measurements showed, for all lighter atoms which could be investigated, that n is approximately equal to half the atomic weight. Let it be remarked at once that to-day, by more refined observations on light and X-rays, we can obtain far fuller information about atomic structure. Thus we have in the first place the phenomenon of dispersion of the light transmitted, which gives information on the binding forces of the electrons; or again, the interference rings which careful observations reveal in the scattered radiation from X-rays (p. 80), and which supply data oh the diameter of the electronic

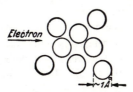

Fig. 2.—Passage of electrons through matter; according to the conceptions of the kinetic theory of gases the molecules are structures o. size ∽1 Å; if they were impenetrable to an electron, the latter could not push its way through thin foil.

envelope of the atom, and on the distance between the centres of the atoms in molecules (Debye, 1929).

We pass now to the attempts to investigate atomic structure by means of scattering experiments on beams of electrons. These were first made by Lenard and his collaborators, who used cathode rays. We have seen in Chapter I that the diameter of an atom is of the order of magnitude 10^{-8} cm. If the atoms were massive spheres, as in the

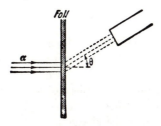

Fig. 3.—Arrangement for observing the scattering of α-particles; a Geiger counter is used to count the number of α-particles deflected from the primary direction through a definite angle θ.

diagram of fig. 2, then collisions with these spheres would necessarily very soon stop a cathode ray particle completely. The systematic investigations of Lenard, however, gave a precisely contrary result

(1903). He found, in fact, that atoms are almost perfectly transparent to swift electrons. His " dynamid " theory of the atom based on these results was, however, not successful.

The title of father of the atomic theory is given to Rutherford, who took up the research with more adequate instrumental resources, and carried it further; to him we owe our concrete, quantitative ideas on atomic structure. For his experiments on scattering Rutherford (1911) used, not, like Lenard, the relatively light electron, but the much more massive α-particle (fig. 3). On account of their greater mass, α-particles are not noticeably deflected by the electrons in the atom, and therefore record only collisions with the more massive particles. The comparative penetrating effect of electrons and α-particles may be illustrated roughly by the difference between a light rifle bullet and a heavy shell.

In the first place, Rutherford's experiments show definitely (1913) that an atom, except for a small massive *nucleus*, is almost perfectly

Fig. 4.—Sca:tering of α-particles by a nucleus of charge Z; the paths of the α-particles are hyperbolas.

empty—a result which had already been found by Lenard. Since the α-particles are perceptibly deflected by the nucleus only, we can deduce the law of deflection from the law of distribution of the α-particles which have been scattered by a piece of foil. The definite result was found that the effective deflecting force is the Coulomb force $2Ze^2/r^2$, where $2e$, as we know, is the charge of the α-particle, and Ze is the charge of the nucleus. The paths of the α-particles are hyperbolas with the nucleus as focus (fig. 4).

If the incident ray contains one α-particle per square centimetre, then according to Rutherford (Appendix IX, p. 377) the number of particles per unit solid angle which suffer a deflection ϕ is given by

$$w(\phi) = Z^2 \left(\frac{m}{M}\right)^2 \left(\frac{e^2}{mc^2}\right)^2 \frac{1}{(v/c)^4 \sin^4 \frac{1}{2}\phi}$$

(m = mass of electron, $M = 4m_H$ = mass of α-particle); here again the effective cross-section of the electron occurs as a factor, but the electronic radius e^2/mc^2 is multiplied by the small factor (characteristic of the dimensions of the nucleus)

$$\frac{m}{M} = \frac{m}{4m_H} = \frac{1}{4 \times 1840} = 1.36 \times 10^{-4}.$$

The experiments showed that the number Z, which gives the nuclear charge, is equal to the number which would be assigned to the element in question in a consecutive enumeration of the elements in the periodic system. If the atom is to be neutral, the number giving the nuclear charge must agree with the number of electrons in the electron cloud surrounding the nucleus, as determined by optical and X-ray scattering experiments. The chemical behaviour of the atom depends of course on the external electrons, so that it is not the mass of the atom, but its atomic number (or number giving the nuclear charge), which determines its chemical properties. Isotopes have the same atomic number.

For collisions which are nearly central, i.e. for scattering through wide angles, deviations occur from the distribution of the scattered α-particles determined by Coulomb's law. From this we must infer that Coulomb's law only holds down to distances of about 10^{-13} cm. The nucleus also has a finite size; it is worthy of remark that the "nuclear radius" is of the same order of magnitude as the "radius of the electron".

According to Rutherford, the nuclear atom may be described as follows. In the centre of the atom there is the nucleus; this was thought of as composed of protons and electrons, but this assumption led to many difficulties, e.g. that of packing the electrons into the nuclear volume, which is of the same order as that of one electron (radius 10^{-13} cm.; see p. 49). Since the discovery of the neutron a more satisfactory model has been developed: viz. a nucleus composed of p protons and n neutrons; the nuclear charge number (atomic number) is then $Z = p$, the atomic mass number is $A = p + n$. The considerations in favour of this model will be presented immediately (§ 4, p. 64).

Round the nucleus, as has been mentioned, there move in the neutral atom Z electrons, which fill a sphere of radius $\sim 10^{-8}$ cm. To get an idea of the dimensions, and the emptiness, of an atom, take the following illustration. If we imagine a drop of water to be expanded to the size of the earth, and all the atoms in it also enlarged in the same proportion, an atom will have a diameter of a few metres. The diameter of the nucleus, however, will be only something like 1/100 mm.

It follows from these relations of magnitude that in the great majority of physical and chemical processes the nuclei act simply as positively charged point masses; only the external electronic system is essential. Research began therefore with the electronic cloud, with such success indeed that we are to-day in possession of a theory which seems to be in complete accord with experiment. Nuclear physics developed later; recently it has made rapid advance, a full account of which would need a separate book. Only the main features of this wide subject will be given here. The following section contains a description of experimental results and their immediate consequences for our theoretical ideas. A more detailed account of nuclear theory, including the meson theory of nuclear forces mentioned already (Chap. II, § 8, p. 49), follows later (Chap. X, § 3, p. 317), when the new mechanics necessary for dealing with atomic systems has been established.

4. Mass Defect and Nuclear Binding Energy. The Neutrino.

We have already mentioned that the mass of a nucleus is not an exact multiple of the mass of the proton. This can be readily understood. In order to remove a proton, neutron or α-particle from the nucleus of a higher element, we must supply energy (except in radioactive substances, in which such disintegrations can occur spontaneously, i.e. without previous supply of energy). This loss of energy when elementary particles become united to a nucleus is, however, according to Einstein (p. 57), equivalent to a loss of mass, so that the final product is lighter than the sum of the weights of the individual components in the separated state. Take a numerical example. The greatest deficit occurs in the formation of a helium atom from four hydrogen atoms. The latter have the total mass 4 × 1·008, and in the combined state (helium) the mass 4·004; the energy freed in the process of combination is therefore, by Einstein's formula,

$$E = (4 \times 1\cdot008 - 4\cdot004)c^2 = 0\cdot028c^2$$
$$= 0\cdot25 \times 10^{20} \text{ ergs/mole.}$$

In order to split up a helium atom into four hydrogen atoms, at least this amount of energy must be supplied. To make it easier to grasp its order of magnitude, convert ergs into calories; we find the value of E to be the enormous one of $6\cdot4 \times 10^8$ kilocalories per mole. For the sake of comparison, it may be noted that heats of combustion are of the order of magnitude of a few hundred kilocalories per mole.

In nuclear physics it is customary to measure binding energies in units of MeV (million electron volts)—1 MeV is the kinetic energy acquired by an electron when accelerated through a potential difference of 1 million volts. Thus in these units the energy associated with the rest mass of an electron $(m_e c^2)$ is approximately 0·51 MeV. Similarly, the binding energy of He is about $B = 27$ MeV. More useful, however, than the binding energy is the binding energy per nucleon (B/A) which for He $(A = 4)$ is about 7 MeV.

The measurement of binding energies made with great accuracy by Aston and others (since 1920) supplies information, therefore, on the " heats of formation " of nuclei, i.e. on the energy relations in nuclear construction. These are found to follow certain perfectly definite rules. It has proved to be convenient to divide up the various nuclei into four groups, according as their mass numbers are divisible by 4, or on division by 4 leave a remainder of 1, 2, or 3; we shall return to this point below (p. 67). The binding energies per particle are shown in fig. 5; it will be seen that, as we ascend in the periodic system, the binding energies (mass defects) increase rapidly at first, then slowly fall off again as we come nearer to the radioactive elements.

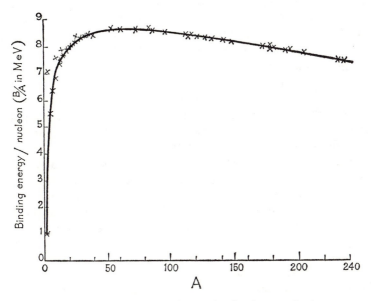

Fig. 5.—Binding energy per nucleon as a function of mass number A

The explanation of these facts depends on the assumption made about the constituents of the nucleus. We have already seen (p. 63) that there is no room for electrons in the nucleus if the radius of the electron is considered as a physical reality. Since this may be doubted, it is important that the impossibility of the existence of individual electrons in the nucleus can be inferred more definitely from measurements of the angular momentum of nuclei, the so-called nuclear spin. We shall return to the matter again (Chap. VIII, § 3, p. 263) ; here let only this much be said: in the quantum theory the angular momentum of a particle is always a multiple of an elementary unit; and if several particles are conjoined, there must be simple whole number relations between the angular momenta of the individual partners and of the whole. These relations, though in the domain of the external electronic envelope they hold without exception, are not fulfilled in nuclei—a specially well-verified case is the nitrogen nucleus N^{14}. If we construct the nucleus from protons and electrons, then the angular momenta behave as if there were only protons present (Heitler, Herzberg, 1929).

Now we have to-day other particles at our disposal, above all the neutron, and it is an obvious suggestion to try the experiment of calling upon these for the elucidation of nuclear structure. All the above difficulties are avoided, if we assume that the nuclei are composed of protons and neutrons (Iwanenko, Heisenberg, 1932). In the first place, with the help of simple assumptions about the interaction forces we can then understand the fundamental fact that in light atoms the atomic weight A is double the atomic number Z, while in heavy atoms it increases somewhat faster than $2Z$. Between two protons there acts the Coulomb repulsion of their positive charges. Between a neutron and a proton there must be a very considerable short-range attraction, to make the formation of the nucleus possible. There are also other forces between a pair of neutrons or a pair of protons; the latter force has been studied with the help of scattering of protons by protons (fig. 6, Plate IV), and it was found to be of the same order as the neutron-proton force. There is reason to assume the same for the neutron-neutron force. Hence we can regard neutron and proton as one and the same particle, called a *nucleon*, which has two states characterized by the charge being zero or one. All experimental evidence shows that the proton-proton force, the proton-neutron force, and the neutron-neutron force are essentially the same when the nucleons are moving in the same states, and when the electrostatic (Coulomb) repulsion between protons has been subtracted out. The nuclear forces between nucleons are independent of their charges, and this has led to the introduction of the phrase " charge independence of nuclear forces ".

PLATE IV

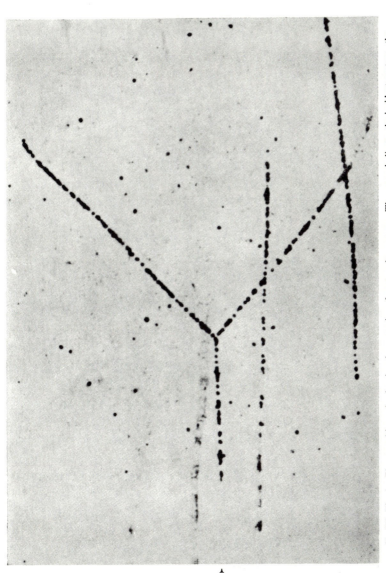

Ch. III, Fig. 6.—Photographic tracks showing the scattering of protons by protons. The arrow indicates the incident proton; the tracks after the collision are orthogonal, which follows from the mechanical laws of collision for equal masses (see pp. 36, 66)

By courtesy of Prof. C. F. Powell, Bristol, and the Oxford University Press

The numbers p of the protons, and n of the neutrons, so adjust themselves that as much energy as possible is set free in the binding process. This means, if we neglect the Coulomb repulsion between the protons, that as many pairs proton-neutron are formed as possible. If now the atomic weight, or more exactly the mass number, $A = p + n$ is given, the most stable state for small numbers is that in which $p = n$; since $p = Z$, the atomic number, it follows that $A = 2Z$. The more protons, however, there are in the combination, the more will the Coulomb repulsion come into account, and that clearly to the disadvantage of the protons; hence the difference $n - p$ becomes > 0, and increases with increasing $p = Z$. It follows that $A = n + p > 2p$, or $A > 2Z$, and that $A - 2Z$ increases as Z increases; and this is actually the case. It is probable that two pairs, proton-neutron, within a nucleus combine so as to form an α-particle, so far as that is possible; for we know this binding to be particularly rigid (the mass defect is great; see p. 64); also, it is α-particles which are emitted from the unstable (radioactive) nuclei; and, finally, the nuclei with atomic weight divisible by 4 form a specially regular series of mass defects (p. 65). Moreover, the stability of the α-particle can be made plausible on the ground of the forces produced by the exchange of charges and the spins of the elementary particles, which have a tendency towards saturation (formation of systems with angular momentum nil, in accordance with Pauli's principle; see § 5, p. 177). In this way the radioactive α-disintegration can be understood. As for β-disintegration, the expulsion of an electron, the new conception has the advantage of reducing this to a single elementary process, viz. the disintegration of the neutron into a proton and an electron. The expulsion of positrons has also been observed from short-lived (artificially produced) nuclei (p. 74); this process would be the disintegration of the proton into neutron + positron. The characteristic of the ordinary β^--emission is, as we have seen (p. 31), a continuous velocity spectrum; the same is true of the β^+-emission also. We have already pointed out the fundamental difficulty arising from this fact; it means that the principle of energy is infringed in these processes (as also, for that matter, the principle of the conservation of angular momentum, or spin). To save it, there is nothing for it but to assume a third particle, which is concerned in the process, but which in consequence of its properties escapes direct observation (Pauli, Fermi, 1934). It must be very light and can have no charge; it is therefore called a " neutrino ". Indirect proof of its existence, quite analogous to the one for the neutron (Chap. II, § 6, p. 44) is available. We see the tracks of atomic nuclei which have been struck; we then infer the existence of the neutron by applying the principle of conservation of momentum. In the case of the neutrino such hits will be too rare; in their stead, however, we can use the recoil at the emission. In ordinary radioactive materials, whose atoms are all very heavy, the track of the recoiling atomic residue at a β-emission is certainly too minute; but to-day we can also manufacture light radioactive atoms artificially (§ 6, p. 70), and with these the recoil would certainly be observable. If then the direction of the tracks of the electron and of the atomic residue are not exactly opposite, the assumption of a third invisible particle would be preferable to giving up the conservation of momentum (Berthe, Peierls, 1934). The first experiments on this effect (Leipunski, 1936; Crane and Halpern, 1938, 1939) gave only weak evidence for the existence of the neutrino. But a new method (Allen 1942) led to convincing results. It uses a radioactive isotope of beryllium which is produced by bombarding normal lithium with deuterons, according to the formula (see Table IV, p. 72) $_3Li^6 + _1D^2 = _4Be^7$. Here, as in the following, the lower suffix is the atomic number Z, the upper suffix the rounded-off atomic weight or mass number A. $_4Be^7$ is less stable than $_3Li^7$ as it contains a proton more (see the foregoing discussion); but the difference is too small for expelling a positron (β^+-decay). Another strange thing happens: the $_4Be^7$ nucleus captures one of the innermost extra-nuclear electrons (K shell) and is transformed in this way into $_3Li^7$ with a decay constant of about 43 days (p. 342). In this case the recoil energy of the resulting atom depends solely on the emitted neutrino. From the mass difference of $_4Be^7$ and $_3Li^7$ (corresponding to 370,000 eV) the recoil energy to be expected is about 58 eV; the experimental results obtained by Allen are in reasonable agreement with this value. A more direct proof of the existence of the neutrino is by an experiment performed by Reines and Cowan in 1956. Near an atomic pile (p. 354) there is an intense flux of neutrinos and the inverse β-process can be used for their detection. In this process a neutrino is captured by a proton, which then decays into a neutron and a positron. Simultaneous detection of the neutron and positron have given convincing proof that the neutrino exists as an elementary particle.

5. Heavy Hydrogen and Heavy Water

Most isotopes play no part in the everyday practice of physicist and chemist, since their differences in mass are relatively slight, and their properties almost identical. It is different with the recently discovered isotope of hydrogen, the nucleus of which has about double the mass of the proton; it has been given the name *deuteron* (the second). In this case essential differences occur in physical and chemical behaviour, so that we can speak of an actually new element, which is called *deuterium*, and is given the symbol D. The importance of this discovery becomes particularly evident when we remember that the most important compound of hydrogen, water (H_2O), which physics from the beginning has taken as a standard substance, now turns out to be a mixture of several kinds of molecule (H_2O, HDO, D_2O), the molecular weights of which differ from each other by 5 and 10 per cent. It is therefore fortunate that, for other reasons of a practical nature, we have long since given up the definition of the kilogram as the mass of 1000 c.c. of water at its maximum density, defining it instead as the mass of a certain piece of platinum-iridium. We may relate the almost romantic story of the discovery, which reminds one of the discovery of argon in the atmosphere by Rayleigh and Ramsay (1894); in both cases the key to the discovery was belief in the reality of minute discrepancies in different measurements of the same quantity. To begin with, two isotopes of oxygen, of masses 17 and 18, were discovered spectroscopically; since the frequencies of vibrations are inversely proportional to the square root of the mass of the vibrating particle, the isotopes betrayed their presence by the occurrence of displaced lines in the molecular spectrum. However, the intensity of the lines was very weak, and the amounts of the isotopes so minute that they could not affect practical measurements. Theoretically, however, the demonstration of their existence, and the measurement of their relative amounts, had important consequences. From these amounts, and the chemical molecular weights, we can in fact calculate the mass of the H-atom, referred to the principal isotope of C as 12. On the other hand, this same quantity has been determined by Aston with the mass-spectrograph. It was found by Birge and Menzel (1931) that a difference of 1/5000 remained over, and they concluded that hydrogen also must contain a small amount, about 1/4000, of a heavy isotope, of mass 2.* Thereupon spectroscopic investigations on hydrogen were undertaken by Urey, Brickwedde and Murphy (1931). We shall see that the position of the atomic lines of an element also depends on the mass of the nucleus (p. 109). Actually, weak satellites of the principal hydrogen lines were found at the right distance. Various methods were then applied to produce enrichment in the heavy isotope D, the most successful of these being ordinary electrolysis (Urey and Washburn, Lewis and Macdonald, 1932)

---

*Later measurements of Aston show, however, that this conclusion which has proved so fruitful was based on incorrect measurements.

It was found that the light H_2 escapes from the cathode 5 or 6 times faster than the heavy D_2. If the original concentration was 1 in 5000, the electrolysis of 6 litres of ordinary water should give us about 1 c.c. of pure heavy water. In point of fact, sufficient quantities of heavy water are now available to allow its properties to be investigated in all directions. Another method of separation, that of Hertz already mentioned (p. 42), depends on the difference in the rates of diffusion of the gases; by this method about 1 c.c. of D_2 was obtained of such purity that in the spectrum the atomic lines of H (the so-called Balmer lines; see p. 106) were no longer visible at all. The exact mass of the D-atom is 2·0147 (Aston, 1936), referred to the principal isotope of O as 16, while that of the H-atom is 1·0081. From the point of view taken by Heisenberg (p. 66), we must look upon the D-nucleus as the combination proton + neutron. This has been directly proved by decomposition into these particles by means of γ-rays (Chadwick and Goldhaber, 1934; p. 75). The properties of heavy water, D_2O, differ very noticeably from those of ordinary water, H_2O; the freezing-point is about 3·8°, the boiling-point about 1·4° higher; the density is as much as about 11 per cent greater than that of H_2O. Similar results hold for other combinations in which H is wholly or partly replaced by D. The velocities of reaction in these show considerable changes. Thus an entirely new branch of chemistry has emerged, which is also important for biology.

For the physicist, however, the importance of heavy hydrogen is mainly due to the fact that we can produce rays of D-atoms just as easily as of H-atoms, and accordingly have now a new means of bombarding other nuclei, which has already given us valuable information about their constitution. Of this we shall speak in the following section.

6. Nuclear Reactions and Radioactive Decay.

As we know, all experience shows that radioactive nuclear disintegration cannot be affected by any ordinary physical means; it takes place spontaneously in accordance with statistical laws. If we had to depend on these processes alone, we would never get direct information as to the structure of the majority of (non-radioactive) nuclei, and would always have to rely on hypotheses.

The greatest importance, therefore, attaches to Rutherford's discovery (1919) that nuclei can be broken up by bombardment with α-rays. The first element with which this was accomplished was

PLATE V

Ch. III, Fig. 7.—Disintegration of a nitrogen nucleus by an α-particle. A proton of very long range is emitted. α-particles of two different ranges are seen, emitted by a mixture of thorium B and C (see p. 71). (From *Proc. Roy. Soc.* A, Vol. 136.)

nitrogen; in the cloud chamber, nuclear hits are occasionally seen in which the track of the incident α-particle suddenly disappears, and the track of the nucleus which was struck, and that of a particle of greater range, begin where the first track ends (figs. 7, 8, 9, Plates V, VI, pp. 71, 73). The same result was afterwards obtained with many other elements, boron, fluorine, neon, sodium, &c. (Rutherford, Chadwick, Kirsch and Petterson, 1920–25), and it was shown by deflection experiments that the light particles expelled are protons. The α-particle is obviously captured by the nucleus; from $_7N^{14}$ is formed the isotope of oxygen of mass 17, already mentioned above, in accordance with the equation

$$_7N^{14} + {}_2He^4 \rightarrow {}_8O^{17} + {}_1H^1.$$

We are here at the starting-point of a sort of nuclear chemistry, the laws of which, as in the example just given, can be written down in a form exactly analogous to chemical formulæ. (We can also add the binding energies, which are analogous to heat evolved or absorbed, and which must exactly compensate the kinetic energies of the particles.) The phenomenon, contrary to what was thought at first and is still frequently said, is not one of destruction of the nucleus, but of a nuclear transformation, which often amounts to nuclear construction. More recently, many other fast particles have been used to bombard the nucleus, and the resulting transformations studied.

Very striking results have been obtained by using the photographic method (p. 36); an example of an explosive disintegration, observed by Powell's school in Bristol, is shown on Plate VII (p. 75).

Cockcroft and Walton found (1932) that artificially produced beams of protons, with energy of 120,000 eV, equal to about $\frac{1}{4}mc^2$, a value quite trifling compared with the energy of natural α-rays (about $16mc^2$), are capable of breaking up the nucleus of the lithium atom, according to the relation

$$_3Li^7 + {}_1H^1 \rightarrow {}_2He^4 + {}_2He^4.$$

In point of fact, it was shown in the Wilson chamber that two helium nuclei, i.e. α-particles, were always shot out simultaneously in opposite directions from the bombarded lithium foil (Kirchner, Dee and Walton, 1934; see Plate VIII, fig. 11, p. 86). Similar transformations were successfully brought about in a whole series of nuclei, and it is remarkable that the measurements of the masses of the nuclei involved with help of the mass-spectrograph and the determination of the kinetic energies

TABLE IV
ARTIFICIAL NUCLEAR TRANSFORMATIONS

Radioactive nuclei emitting $\begin{cases}\text{electrons}\\ \text{positrons}\end{cases}$ are indicated by $\begin{cases}\bar{}\\ +\end{cases}$

Bombarding Particle → / Bombarded Nucleus ↓	γ	$_0n^1$	$_1H^1$	$_1D^2$	$_2He^4$
$_1D^2$	$_1H^1 + _0n^1$			$_1H^3 + _1H^1$ $_2\bar{He}^3 + _0n^1$	
$_3Li^6$		$_2He^4 + _1H^3$	$_2He^4 + _2He^3$	$_3Li^7 + _1H^1$ $_2He^4 + _2He^4$ $_4Be^7 + _0n^1$	
$_3Li^7$		$_3\bar{Li}^8 + \gamma\,(?)$	$_2He^4 + _2He^4$ $_4Be^7 + _0n^1$	$2\,_2He^4 + _0n^1$ $_3\bar{Li}^8 + _1H^1$	$_5B^{10} + _0n^1$
$_4Be^9$	$_4Be^8 + _0n^1$	$_2\bar{He}^6 + _2He^4$	$_4Be^8 + _1D^2$	$_5B^{10} + _0n^1$	$_6C^{12} + _0n^1$
$_5B^{10}$		$_3Li^7 + _2He^4$ $2\,_2He^4 + _1H^3$		$_6\overset{+}{C}^{11} + _0n^1$ $3\,_2He^4$ $_5B^{11} + _1H^1$	$_7\overset{+}{N}^{13} + _0n^1$ $_6C^{13} + _1H^1$
$_5B^{11}$			$3\,_2He^4$	$_6C^{12} + _0n^1$ $_5\bar{B}^{12} + _1H^1$ $3\,_2He^4 + _0n^1$	
$_6C^{12}$		$3\,_2He^4 + _0n^1$ $_4Be^9 + _2He^4$	$_7\overset{+}{N}^{13} + \gamma$	$_7\overset{+}{N}^{13} + _0n^1$	
$_6C^{13}$				$_7N^{14} + _0n^1$ $_5B^{11} + _2He^4$	
$_7N^{14}$		$_5B^{11} + _2He^4$ $_6C^{14} + _1H^1$ $_3Li^7 + 2\,_2He^4$		$_7N^{15} + _1H^1$ $_6C^{12} + _2He^4$	$_8O^{17} + _1H^1$ $_9\overset{+}{F}^{17} + _0n^1$
$_8O^{16}$		$_6C^{13} + _2He^4$			
$_9F^{19}$		$_7\bar{N}^{16} + _2He^4$	$_8O^{16} + _2He^4$	$_9\bar{F}^{20} + _1H^1$	$_{10}Ne^{22} + _1H^1$ $_{11}\overset{+}{Na}^{22} + _0n^1$
$_{10}Ne^{20}$					$_{11}Na^{23} + _1H^1$
$_{11}Na^{23}$		$_{11}\bar{Na}^{24} + \gamma$		$_{11}\bar{Na}^{24} + _1H^1$	$_{12}Mg^{26} + _1H^1$ $_{13}\overset{+}{Al}^{26} + _0n^1$
$_{12}Mg^{24}$		$_{11}\bar{Na}^{24} + _1H^1$			$_{13}Al^{27} + _1H^1$ $_{14}\overset{+}{Si}^{27} + _0n^1$
$_{13}Al^{27}$		$_{11}\bar{Na}^{24} + _2He^4$ $_{12}\bar{Mg}^{27} + _1H^1$		$_{13}\bar{Al}^{28} + _1H^1$	$_{14}Si^{30} + _1H^1$ $_{15}\overset{+}{P}^{30} + _0n^1$

PLATE VI

Ch. III, Fig. 8.—Elastic collision of an α-particle with a nucleus of hydrogen. The tracks of the two particles after the collision are of nearly equal length (see p. 71).

Ch. III, Fig. 9.—Elastic collision of an α-particle with a nucleus of nitrogen. The short track corresponds to the nitrogen (see p. 71). (From *Proc. Roy. Soc.* A, Vol. 134.)

of the particles before and after the disintegration can be performed with such accuracy that one gets a direct experimental verification of Einstein's law $E = mc^2$ (Bainbridge, 1933).

One might think that the nuclear-chemical processes do not take place unless the incident proton approaches the nucleus very closely; but this is not the case. We can calculate how near the nuclei come by equating the potential energy e^2Z/r at the minimum distance to the kinetic energy of a proton eV, where V is the accelerating potential: thus

$$V = \frac{eZ}{r}, \quad r = \frac{eZ}{V}.$$

Hence, with an energy eV of 10^5 electron volts, i.e. 1.59×10^{-7} ergs, so that $V = \frac{1}{3} \times 10^3$ e.s.u., we have, for $Z = 1$,

$$r = 4.80 \times 10^{-10} \times 1 \times 3 \times 10^{-3} = 14 \times 10^{-13} \text{ cm.},$$

which is of the order of magnitude of the nuclear radius (pp. 63 and 307). But the disintegration effect in lithium is demonstrable down to less than 30,000 eV (Rausch von Traubenberg and Döpel, 1933), in which case the proton is still far outside the nucleus. The explanation of this fact will be given later (p. 311).

The heavy hydrogen isotope D has also been used to bring about disintegration (Lewis, Livingstone and Lawrence, 1933) and proved extremely effective. Thus, e.g., we get (according to Cockcroft and Walton, 1934) a transformation of the light lithium isotope into the heavy one,

$$_3\text{Li}^6 + _1\text{D}^2 = _3\text{Li}^7 + _1\text{H}^1,$$

and the protons so obtained have a very long range. There is no need to quote all the transformations so far effected; some of them are collected, with other processes to be discussed immediately, in Table IV (p. 72).

An important discovery is that of new isotopes of hydrogen and helium, viz. $_1\text{H}^3$ (called triton or tritium) and $_2\text{He}^3$. They were first found in disintegration experiments in the Cavendish Laboratory, corresponding to the reactions (Oliphant, Harteck, Rutherford, 1934)

$$_1\text{D}^2 + _1\text{D}^2 \rightarrow _1\text{H}^3 + _1\text{H}^1$$

$$_1\text{D}^2 + _1\text{D}^2 \rightarrow _2\text{He}^3 + _0n^1,$$

and (Oliphant, Kinsey and Rutherford, 1933)

$$_3\text{Li}^6 + _1\text{H}^1 \rightarrow _2\text{He}^4 + _2\text{He}^3.$$

Soon afterwards $_1\text{H}^3$ was directly demonstrated by the mass-spectrograph as a weak admixture with $_1\text{H}^1$ and $_1\text{D}^2$ (Bleakney, Lozier and Smith, 1934). Modern methods which will be indicated below have succeeded in producing these new isotopes $_2\text{H}^3$ and $_2\text{He}^3$ in bulk, though of course in small quantities, but sufficient to study their physical properties.

Neutrons were next employed as bombarding particles (Feather, Harkins, 1932). It is probable on the face of it that these should give a large output, since they are uncharged and so not repelled by the field of the struck nucleus. This surmise has been confirmed, and led to the discovery of a new class of transformations, of which we shall speak presently. In point of fact, these transformations, which may be described as an artificial production of radioactivity, were discovered not with neutrons, but with a-rays, the discoverers being Irène Curie and her husband Joliot (1934). For instance, it was found that aluminium which was bombarded with a-rays and so sent out neutrons continued to radiate after the irradiation was stopped, the emission consisting of positive electrons. Clearly the initial product must, in accordance with the equation

$$_{13}\text{Al}^{27} + {}_2\text{He}^4 \rightarrow {}_{15}\text{P}^{30} + {}_0n^1,$$

be an isotope of phosphorus, which is radioactive, and passes over into an isotope of silicon, with emission of a positron:

$$_{15}\text{P}^{30} \rightarrow {}_{14}\text{Si}^{30} + \beta^+.$$

The half-value period of the " radio-phosphorus " was found to be 3 min. This is long enough to allow the correctness of the interpretation to be tested in these and similar processes by the old method of chemical precipitation (p. 40). The proton and its isotope D also proved useful for the production of radioactive nuclei.

The most abundant output, however, comes from neutrons, with which Fermi and his collaborators (1934) have produced new radioactive nuclei from the majority of known atoms. In spite of the rather short lifetimes of these nuclei (some minutes) it was possible to determine the chemical character of the products, and it was found that the atomic number of the nucleus which captures a neutron changes by 2 or 1 or 0, corresponding to the emission of an a-particle, a proton or of light.

By passing them through substances like water or paraffin wax which contain many hydrogen atoms, neutrons are slowed down by

PLATE VII

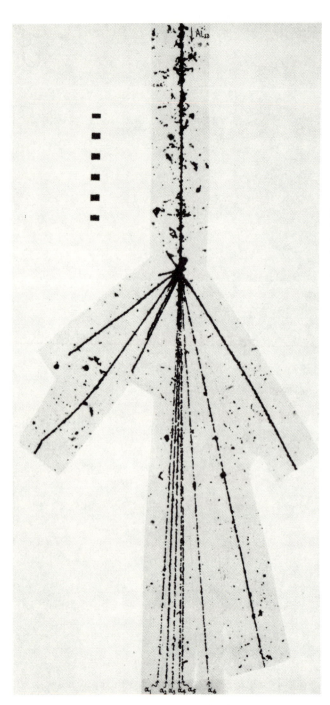

Ch. III, Fig. 10.—An aluminium nucleus ($Z = 13 \pm 1$), moving with a little above half the velocity of light, interacts with a nucleus in the emulsion and decomposes into six α-particles which emerge from the encounter in the form of a jet (see pp. 36, 71).

(By courtesy of Prof. C. F. Powell Bristol)

the impacts with the protons conferring on these particles of equal mass half their energy at each collision. In this way it is possible to have neutrons nearly in thermal equilibrium with the surrounding substance. These slow neutrons are particularly effective in producing new nuclei, and they show many interesting properties, such as selective absorption in special elements. The study of these processes is one of the most powerful methods for revealing the laws of nuclear structure (see Chap. X, § 9, p. 350).

Chadwick and Goldhaber (1934) discovered a new type of disintegration, using γ-rays as bombarding agent. The heavy hydrogen isotope D was shown to break up according to the relation

$$_1D^2 + \gamma \rightarrow {}_1H^1 + {}_0n^1.$$

Thus the binding energy of the D-nucleus ($D = p + n$) was determined and a value for the mass of the neutron could be derived. The binding energy of the deuteron is determined to be 2·237 MeV or 4·38 mc^2, and the mass of the neutron to be 1·008982.

The technical means for producing rays of fast particles have been very much improved in recent years. The simple principle of the well-known electrostatic induction machine which produces a high electric potential by mechanical motion of a charged body has been developed on a gigantic scale by Van der Graaff (1931), who uses a long closed belt as a carrier of the charge. The high potential obtained is fed to the vacuum tube in which charged particles are accelerated in one step (linear accelerators). The problem of accelerating electrons is relativistic since, for example, 1·0-MeV electrons have a velocity 0·94 c (see p. 27 and Appendix V, p. 370). An interesting type of high-energy linear accelerator for electrons uses a circular wave-guide so designed that an electromagnetic wave travels along the guide with continuously increasing speed approaching that of light at the far end. At the injection end the speed is that of the injected electrons, which are then accelerated by the electric field component of the wave and are effectively " carried " along with the wave. A machine of this type at Stanford produces 1-BeV (1000-MeV) electrons.

Other types of instruments are using relatively small potentials, but many accelerating steps, by repeated electromagnetic impulses on a periodic orbit. The earliest of these were the *cyclotron* (Lawrence and Edlefson, 1930), which produces fast ions, and the *betatron* (Kerst and Serber, 1941), which accelerates electrons. In both of them the charged particles are bent into circular orbits with the help of a strong magnetic field (see Chap. II, § 1, p. 26); but the acceleration is produced in different ways. In the cyclotron the two halves of the circular path are kept at different electric potentials (with the help of semicircular metal boxes); the particles crossing the dividing diameter (from one box to the other) are then subject to a strong field. This is fed by a high-frequency source of potential in such a way that each crossing gives an accelerating impulse. In the betatron no electric field is used, but the magnetic field itself is alternating and produces an inductive electromotive force on the circular motion of the electrons, by which they are accelerated. It is possible by proper adjustment to stabilize this motion to such a degree that many hundred thousand circles are performed in spite of occasional collisions with atoms. The energy attainable by an ordinary cyclotron is limited by the relativistic change of mass experienced by the accelerated particle as the energy increases (p. 27). This can be seen as follows. In order that the particle should be continuously accelerated it is essential that the crossings between the two semicircular boxes should be in step with the high-frequency oscillations. For a particle of mass m and charge e moving in a plane perpendicular to a constant magnetic field H in a circle of radius R we have the following force equation

$$Hev = mv^2/R,$$

where v is its velocity. Using $v = \omega R$ where ω is the angular frequency this becomes

$$\omega = He/m,$$

so that if m is constant ω takes the same value for all R. However, if m changes because of relativistic effects, then either ω or H must be changed in order to keep the particle motion in step with the high-frequency oscillations.

In the synchro-cyclotron (Veksler, 1944, McMillan, 1945) the difficulty is overcome by using a periodic time variation of the oscillator frequency. We then obtain bursts of high-energy particles. The 184-inch synchrocyclotron at the University of California, for example, can produce 350-MeV protons at the rate of 120 bursts per second. On the other hand, in the synchrotron, the frequency condition is maintained by varying the magnetic field with time, keeping the oscillator frequency fixed. In an electron synchrotron for energies above about 2 MeV when $v \approx c$, the angular speed is given by $\omega \approx c/R$, so that if ω is constant the orbit radius is approximately constant. This means that if the electrons are injected at sufficiently high energy then the magnet need only fill an annular region, with a consequent considerable saving in expenditure. For a proton synchrotron the condition for an orbit of approximately constant radius is not achieved until the proton energy is in the region of 3 BeV. However, an orbit of constant radius can be obtained if the oscillator frequency is changed whilst the field is expanding in such a way that the relation $v = \omega R$ is satisfied for constant R. A number of synchrotrons have been constructed during the last fifteen years. Typical of these are the proton-synchrotron (Cosmotron) at the Brookhaven National Laboratory which produces 3-GeV protons and the 28-GeV machine at CERN (European Organization for Nuclear Research). Higher-energy machines with energies of the order of hundreds of GeV are being planned, and thought is also being given to the construction of machines with energies in the 1000-GeV region. Using machines of the type described above it has been found possible to produce most of the particles found naturally in cosmic rays (see page 53).

A completely different type of nuclear reaction, called *fission*, where the nucleus breaks into parts of nearly equal size, will be described later (Chap. X, § 9, p. 350) in connexion with a short account of nuclear theory.

At this point we conclude our brief survey of the state of knowledge with regard to nuclei, and now turn to the structure of the external electronic system. We are thus brought into regions whose linear dimensions are several thousand times greater than those of the nucleus. Here, by the co-operation of experimental and theoretical research, the fundamental laws have been made clear in all essential respects. But this was only made possible by the renunciation of ideas to which people's minds had become so accustomed from their experience of events on the large scale that a profound critical analysis had first to be carried out.

CHAPTER IV

Wave-Corpuscles

1. Wave Theory of Light.   Interference and Diffraction.

The ideas which we have arrived at in the preceding chapters with regard to the structure of matter all rest on the possibility of demonstrating the existence of fast-moving particles by direct experiment, and indeed of making their tracks immediately visible, as in the Wilson cloud chamber. These experiments put it beyond doubt that matter is composed of corpuscles. We are now to learn of experiments which just as definitely seem to be only reconcilable with the idea that a molecular or electronic beam is a *wave train*. Before we enter upon this, however, we shall briefly recall the main facts of wave motion in general, using the phenomena of optical diffraction as a concrete example.

While in the eighteenth century physicists almost universally adhered to Newton's emission theory (about 1680), according to which light consists of an aggregate of very small corpuscles, which are sent out by the source of light, and the wave theory of Huygens (1690) could claim only a few supporters (among them the great mathematician Euler), the state of matters changed completely when at the beginning of the nineteenth century Young made the discovery that in certain circumstances two beams of light can enfeeble each other, a phenomenon quite incapable of explanation on the corpuscular theory. The results of the further investigations of Young and Fresnel spoke unequivocally in favour of the wave conception of Huygens, for it is impossible to explain interference phenomena except by a wave theory.

We give here a short discussion of Young's *interference experiment* (fig. 1). The source of monochromatic light Q illuminates the double slit in the diaphragm B with parallel light by means of the lens L. On the screen S behind the diaphragm a system of equidistant bright and dark strips (fringes) appears. How this comes about may be explained as follows. From the two openings in the diaphragm spherical waves spread outwards; these are " coherent ", i.e. they are capable of mutual interference. The two wave motions become superimposed, and reinforce each other at those places where a crest of the one wave coincides with a crest of the other; on the contrary, they destroy each other where a crest of the one wave is superimposed on a hollow of the other. Hence we can tell at once at what places on the screen there will

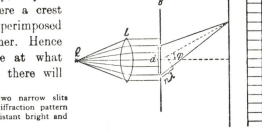

Fig. 1.—Diffraction at two narrow slits close to each other. The diffraction pattern consists of a system of equidistant bright and dark bands (fringes).

be brightness; they are the points whose distances from the two openings in the diaphragm differ exactly by an integral multiple of the wave-length. From fig. 1 we see that the difference of the distances is $d \sin \phi$; there is therefore on the screen

brightness, if $d \sin \phi = n\lambda$ $\quad\quad$ } $(n = 0, \pm 1, \pm 2, \ldots)$,
darkness, $\quad$ if $d \sin \phi = (n + \tfrac{1}{2})\lambda$ }

where d is the distance between the two openings, and ϕ is the angle of deflection.

A similar diffraction pattern is also obtained when light passes through *one slit*. We can picture it roughly as due to the mutual interference of the elementary Huygens waves spreading out from the individual points of the slit. There are two essential differences, however, as compared with the previous case. In the first place, we easily see that the relation

$$d \sin \phi = n\lambda \quad\quad (n = \pm 1, \pm 2, \ldots),$$

where d is the slit-width, does not now give the places at which there is brightness, but those where there is darkness. For in the packet of wave trains which spread out from the slit in the direction given by the equation, all " phases " are in this case represented exactly the same number of times; i.e. we find in the packet exactly as many wave trains which reach the screen with a crest, as trains which arrive at it with a hollow; the trains will therefore extinguish each other. We find, further, that the diffraction maxima are not, as before, almost equally bright; but that their intensity falls off very strongly from the middle maximum outwards, in the way indicated in fig. 2 by the wavy line shown at the side. It should also be specially emphasized that *when the slit-width is reduced the diffraction pattern widens out,* as may easily be deduced either from the above equation defining the position of the diffraction minima, or directly from fig. 2.

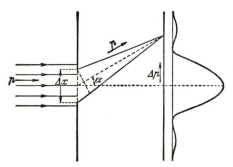

The fact that the form of the diffraction pattern depends essentially on the

Fig. 2.—Diffraction at a slit. The diffraction pattern shows a strong maximum of intensity for the angle of diffraction $a = o$, and also a series of equidistant maxima which become progressively weaker as the angle of diffraction increases.

wave-length of the light makes it possible to carry out spectral investigations by means of interference phenomena (ruled grating, echelon grating, Perot-Fabry plate, Lummer plate). For diffraction patterns to show themselves, it is necessary that the width of the slit employed should be of the order of magnitude of the wave-length of the light. If then we wish to obtain interference phenomena with X-rays, we have to use a grating in which the distance between the rulings is of the order of magnitude of 1 Å. $= 10^{-8}$ cm.

Such gratings are put into our hands by nature, as von Laue (1912) has shown, in the shape of crystals, in which the lattice distances are just of this order of magnitude. If a beam of X-rays is passed through a crystal, we do in fact obtain

Fig. 3.—Diffraction of X-rays at a crystal. As explained by Bragg, the rays are reflected at the lattice planes of the crystal, and thus made to interfere.

interference phenomena. Following Bragg (1913), we can interpret these as due to interference of the rays reflected at different lattice planes of the crystal (fig. 3). Moreover, Compton (1925) and others succeeded in producing X-ray interference in artificial gratings also, this being found possible at grazing incidence of the rays.

Interference of X-rays supplies a powerful weapon for investigating the structure of crystals. For this purpose we do not even require large pieces of the crystal, but can use it in the form of powder (Debye-Scherrer, 1915; Hull, 1917). The interference figures in the latter case are rings round the direction of the incident beam. Indeed, the powder grains may be of molecular size even. What is more, it is found that the interference phenomena due to the individual atoms of the molecule are by no means completely obliterated by the irregular setting and motion of the molecules in liquids and gases. Circular interference rings are observed, from the intensity distribution of which we can draw conclusions with regard to the distances of the atoms in the molecule (Debye, 1929). However, we shall not enter further into these methods of investigating how matter is built up from atoms.

We add here, in the form of a scale, a summary of the wave-lengths of the types of radiation at present known (fig. 4). The scale is

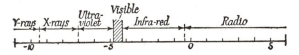

Fig. 4.—Logarithmic scale of wave-lengths; the numbers shown refer to a unit of length $\lambda_0 = $ 1 cm., and represent $\log_{10}(\lambda/\lambda_0)$

logarithmic; the numbers shown are therefore indices of powers of 10. As unit $\lambda_0 = 1$ cm. is taken. Next to the wide range of wave-lengths employed in wireless communication comes the region of infra-red waves, which affect our senses as radiant heat. Then comes the relatively narrow stretch of the visible region (7700 to 3900 Å.), followed by the ultra-violet and Schumann regions, which in turn lead into the domain of X-rays (10 to 0·05 Å.). The radioactive γ-radiation reaches to about 0·001 Å. The cosmic radiation contains γ-rays of very short wave-length (10^{-5} Å.) which would lie beyond the scale of our figure.

2. Light Quanta.

Great as has been the success of classical ideas in the interpretation of interference phenomena, their incapacity to account for the processes of absorption and emission of radiation is no less striking. Here classical electrodynamics and classical mechanics absolutely fail.

To give a few examples of this failure, we recall the experimental fact that a hydrogen atom, for example, emits an infinite series of sharp spectral lines (p. 105). Now the hydrogen atom possesses only a single electron, which revolves round the nucleus. By the rules of electrodynamics, an electron accelerated like this sends out radiation continuously, and so loses energy; in its orbit it would therefore necessarily get nearer and nearer the nucleus, into which it would finally plunge. The electron, which initially revolved with a definite frequency, will radiate light of this frequency; in the case when the frequency of revolution changes (continuously) during the radiation process, it will also emit frequencies close to the ground frequency; but how then the spectrum of hydrogen should consist of a discrete series of sharply separated lines, it is quite impossible to understand.

Further, the stability of the atom is inexplicable. We may think by way of comparison of the system of planets circling round the sun, all, when undisturbed, moving in their fixed orbits. Suppose, however, that the whole solar system arrived in the neighbourhood of Sirius, for example; the mere propinquity would suffice to deflect all the planets out of their courses. If then the solar system moved away again to a distance from Sirius, the planets would now revolve round the sun in new orbits with new velocities and periods of revolution. If the electrons in the atom obeyed the same mechanical laws as the planets in the solar system, the necessary consequence of a collision of the atom with another atom would be that the ground frequencies of all the electrons would be completely changed, so that after the collision the atom would radiate light of wave-lengths also entirely different. In direct opposition to this, we have the experimental fact that an atom of a gas, which by the kinetic theory of gases is subjected to something like 100 million collisions per second, nevertheless, after these as before, sends out the same sharp spectral lines.

Finally, classical mechanics and statistics fail with the explanation of the laws of radiation of heat (or energy). We shall not go into this complex question in detail until later (Chap. VII, p. 204), and here merely quote the result to which Planck (1900) was led by the above considerations. To make the laws of radiation intelligible, he found the following hypothesis to be necessary: *emission and absorption of radiant energy by matter does not take place continuously, but in finite "quanta of energy" $h\nu$* (h = Planck's constant $6 \cdot 62 \times 10^{-27}$ erg sec., ν = frequency). On the other hand, the connexion with the electro-magnetic theory of light is to be maintained to this extent, that the classical laws are to hold for the propagation of the radiation (diffraction, interference).

Einstein (1905), however, went even further than Planck. He not merely postulated quantum properties for the processes of absorption and emission of radiation, but also maintained such properties as in-herent in the nature of radiation itself. According to the *hypothesis of light quanta* (*photons*) which he advanced, light consists of quanta (corpuscles) of energy $h\nu$, which fly through space like a hail of shot, with the velocity of light. Daring as at first sight this hypothesis appears to be, there is nevertheless a whole series of experiments which seem scarcely possible to explain on the wave theory, but which can be understood at once if we accept the hypothesis of the light quantum. Some account of these experiments, which were cited by Einstein himself in proof of his hypothesis, will now be given.

The most direct transformation of light into mechanical energy occurs in the photoelectric effect (Hertz (1887), Hallwachs, Elster and Geitel, Ladenburg). If short-wave (ultra-violet) light falls on a metal surface (alkalies) in a high vacuum, it is observed in the first place that the surface becomes positively charged (fig. 5); it is therefore giving off negative electricity, which issues from it in the form of electrons. We can now on the one hand, by capture of the electrons, measure the total current issuing

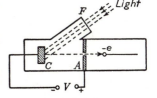

Fig. 5.—Production of photoelectrons (after Lenard). Light entering by the window F strikes the cathode C, and there liberates photoelectrons, which are accelerated (or retarded) in the field between C and A.

from the metal surface, and on the other hand determine the velocity of the electrons by deflection experiments or by a counter field. Exact experiments have shown that the velocity of the emergent electrons does not depend, as one might at first expect, on the intensity of the light; but that only their number increases as the light becomes stronger, the number being in fact proportional to the intensity of the light. The velocity of the photoelectrons depends only on the frequency ν of the light; for the energy E of the electrons the following relation is found:

$$E = h\nu - A,$$

where A is a constant characteristic of the metal.

From the standpoint of the light quantum hypothesis, both these results can be understood at once. Every light quantum striking the metal and colliding with one of its electrons hands over its whole energy to the electron, and so knocks it out of the metal; before it emerges however, the electron loses a part of this energy equal in amount to the work, A, required to remove it from the metal. The number of electrons expelled is equal to the number of incident light quanta, and this is given by the intensity of the light falling on the metal.

Evidence even more patent for the existence of light quanta is given by the classic experiments of E. Meyer and W. Gerlach (1914) on the photoelectric effect with the small particles of metal dust; by irradiation of these with ultra-violet light photoelectrons are again liberated, so that the metallic particles become positively charged. The advance on the previous case consists in this, that we can now observe the time relations in the process of charging the particles, by causing them to become suspended in an electric field, as in Millikan's droplet method for determining e, the elementary electric charge; a fresh emission of a photoelectron is then shown by the acceleration caused by the increase of charge.

If we start from the hypothesis that the incident light actually represents an electromagnetic alternating field, we can deduce from the size of the particles the time that must elapse before a particle of metal can have taken from this field by absorption the quantity of energy which is required for the release of an electron. These times are of the order of magnitude of some seconds; if the classical theory of light were correct, a photoelectron could in no case be emitted before the expiry of this time after starting the irradiation. But the experiment when carried out proved on the contrary that the emission of photoelectrons set in immediately the irradiation began—a result which is clearly unintelligible except on the basis of the idea that light consists of a hail of light quanta, which can knock out an electron the moment they strike a metal particle.

3. Quantum Theory of the Atom.

Planck's original quantum hypothesis was that to every spectral line there corresponds a harmonic oscillator of definite frequency ν which cannot, as in the classical theory, absorb or emit an arbitrary quantity of energy, but only integral multiples of $h\nu$. Niels Bohr (1913) made the great advance of elucidating the connexion of these " oscillators " with one another and with the structure of the atom. He dropped the idea that the electrons actually behave like oscillators, i.e. that they are quasi-elastically bound. His leading thought was something like this. The atom does not behave like a classical mechanical system, which can absorb energy in portions which are arbitrarily small. From the fact of the existence of sharp emission and absorption lines on the one hand, and from Einstein's light quantum hypothesis on the other, it seems preferable to infer that the atom can exist only in definite *discrete stationary states*, with energies E_0, E_1, E_2, ... Thus only those spectral lines can be absorbed for which $h\nu$ has exactly such a value that it can raise the atom from one stationary state to a higher one; the *absorption lines* are therefore defined by the equations $E_1 - E_0 = h\nu_1$, $E_2 - E_0 = h\nu_2$, ..., where

E_0 is the energy of the lowest state in which the atom exists in the absence of special excitation.

If the atom is excited by any process, i.e. if it is raised to a state with energy $E_n > E_0$, it can give up this energy again in the form of radiation. It can in fact radiate all those quanta whose energy is equal to the difference of the energies of two stationary states. The emission lines are therefore given by the equation

$$E_n - E_m = h\nu_{nm}.$$

A direct confirmation of this theory can be seen in the following fact. If Bohr's hypothesis is correct, an excited atom has open to it various possible ways of falling back to the ground state by giving up energy as radiation. For example, an atom in the third excited state can either give up its excess of energy relative to the ground state in one elementary process by radiating a line of the frequency ν_{30}; or it can begin with a transition into the first excited state with the energy E_1 and surrender of the energy quantum $h\nu_{31}$, and then in a second radiation process (frequency ν_{10}) fall back into the ground state; and so

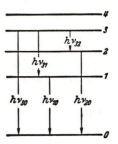

Fig. 6.—Ritz's Combination Principle. An atom in the third excited state can radiate its energy either in the form of a single light quantum of frequency ν_{30}, or as two quanta, the sum of whose frequencies must be exactly ν_{30}.

on (fig. 6). Since the total energy radiated must always be the same, viz. $E_3 - E_0$, the following relation must always exist between the radiated frequencies:

$$\nu_{30} = \nu_{31} + \nu_{10} = \nu_{32} + \nu_{20} = \nu_{32} + \nu_{21} + \nu_{10}.$$

This *combination principle* must of course hold in all cases, and is a deduction from the theory which can easily be put to the test of experiment. Historically, it is true, the order of these two aspects of the matter was reversed; for Ritz, eight years before Bohr's theory was propounded, deduced this combination principle from collected spectroscopic material which had been obtained by experiment. It is by no means the case, however, that all possible " combination lines " do actually occur with perceptible intensity.

Further direct confirmation of Bohr's theory on the existence of discrete energy levels in the atom was given by the experiments of Franck and Hertz (1914). If the atoms are supplied with energy in any way, e.g. by electronic collision, i.e. by bombarding the atom with electrons, then the atoms can only take up such portions of energy as exactly correspond to an excitation energy of the atoms.

Thus, if we bombard the atoms with electrons whose kinetic energy is less than the first excitation energy of the atoms, no communication of energy from the colliding electron to the atom takes place at all (beyond the trifling amount of energy which is transferred in accordance with the laws of elastic collision, and shows itself only in the kinetic energy of the relative motion of the partners in the collision). With respect to collisions of no great strength the atoms are therefore stable in the ground state. Among such slight collisions are those which a gas atom is subjected to in consequence of the thermal motion of the particles of the gas. This is easily verified, roughly, as follows. The mean kinetic energy of a gas particle, by the results of the first chapter, is given by $\bar{E} = \frac{3}{2}kT$, where $k = R/N_0 = 1\cdot37 \times 10^{-16}$ erg/degree; if the whole of this energy were converted at a collision into excitation energy, the energy quantum $h\nu = \frac{3}{2}kT$ would be transferred to the particle struck; where, if we take $T = 300°$ K. (room temperature), ν will be 10^{13} sec^{-1} approximately. On the other hand, frequencies of absorption lines in the visible or even in the infra-red part of the spectrum have values about 10^{14} to 10^{15} sec^{-1}. Higher temperatures will therefore be necessary before " thermal excitation " of the atoms of the gas becomes possible.

We return now to the collision experiments of Franck and Hertz. We see that if the energy E of the electrons is less than the first excitation energy $E_1 - E_0$, the atoms remain in the ground state. If E becomes greater than $E_1 - E_0$, but remains less than $E_2 - E_0$, the atom can be brought by the collision into the first excited state, and consequently when it falls back into the ground state radiates only the line $\nu_1 = \nu_{10}$. If $E + E_0$ lies between E_2 and E_3, the atom which is struck can pass into either the first or the second excited state, and so can radiate the lines ν_{20}, ν_{21}, and ν_{10}; and similarly in other cases (figs. 7a, 7b, Plate VIII opposite). But we can also measure the energies of the electrons after the collision, by causing them to enter an opposing field of known potential difference, and observing the number of electrons which pass through it. In this way also the energy relation was found to be fulfilled exactly to this extent, that the energy loss of the electron due to the collision with an atom was just equal to an excitation energy $E_n - E_0$ of the atom.

PLATE VIII

Ch. III, Fig. 11.—Transmutation of lithium on bombardment with protons. Pairs of α-particles are shot out in opposite directions (see p. 71). (From *Proc. Roy. Soc.* A, Vol. 141.)

Ch. IV. Fig. 9.—Diffraction of electrons by thin silver foil. (After H. Mark and R. Wierl.) The velocity of the electrons (accelerating potential 36 kilovolts) corresponds to a de Broglie wave-length of 0·0645 Å. (Exposure $\frac{1}{10}$ sec.) (See p. 93.)

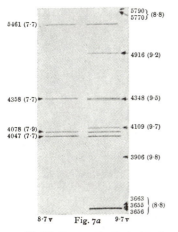

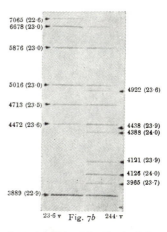

Ch. IV, Figs. 7a, 7b.—Excitation of spectral lines by electronic collision (after Hertz). Only those lines appear in the spectrum whose excitation potential (the number in brackets) is smaller than the energy of the electrons (given under the spectra). Fig. 7a refers to mercury, 7b to helium. The wave-lengths are stated in Å (see p. 86).

4. Compton Effect.

The phenomena described up to this point prove only that energy exchange between light and atoms, or between electrons and atoms, takes place by quanta. The *corpuscular nature* of light itself is proved in the most obvious way by the laws of *frequency change in the scattering of X-rays*. We have in an earlier chapter (Chap. III, p. 59) discussed the classical theory of the scattering of X-rays at comparatively weakly bound (nearly free) electrons, and reached the result that the scattered radiation has always the same frequency ν as the primary radiation; for the electron vibrates in the same rhythm as the electric vector of the incident wave and, like every oscillating dipole, generates a secondary wave of equal frequency.

Compton (1922) investigated the scattering of X-rays by a block of paraffin, and found that the radiation scattered at an angle of less than 90° possesses a greater wave-length than the primary radiation, so that the ν' of the scattered wave, contrary to the prediction of the classical theory, is smaller than the ν of the

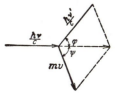

Fig. 8.—Compton Effect. A light quantum on colliding with an electron transfers part of its energy to the latter, and its wavelength becomes greater after the scattering.

incident radiation. On the principles of the wave theory, this phenomenon is unintelligible.

The result, however, can be explained at once (Compton and Debye) if, taking the corpuscular point of view, we regard the process as one of elastic collision of two particles, the electron and the light quantum (fig. 8). For if the light quantum $h\nu$ strikes an electron, it will communicate kinetic energy to the electron, and therefore will itself lose energy. The scattered light quantum will therefore have a smaller energy $h\nu'$. The exact calculation of the energy loss proceeds as in the case of the collision of two elastic spheres; the total momentum must be the same after the collision as before, likewise the total energy. The full calculation is given in Appendix X, p. 380; here we merely quote the result. The *Compton formula* for the change of wave-length of the light quantum due to the scattering process runs:

$$\Delta\lambda = 2\lambda_0 \sin^2 \frac{\phi}{2},$$

$(\lambda_0 = \dfrac{h}{mc} = 0{\cdot}0242 \text{ Å.}$, the Compton wave-length).

The increase in the wave-length is accordingly independent of the wave-length itself, and depends only on ϕ, the angle of scattering. The theory is found to be thoroughly in accord with the facts. In the first place, Compton himself confirmed that the change of wave-length is correctly given by the Compton formula. The recoil electrons, which according to the theory are necessarily produced in the scattering process, and take over the energy loss $h\nu - h\nu'$ of the light quanta in the form of kinetic energy, were successfully demonstrated by Compton; and that not only in scattering by solid bodies, but also in the Wilson chamber, where the tracks of the recoil electrons can be seen directly.

But, as has been shown by Compton and Simon, we can take a further step and test experimentally the relation between the angles of scattering ϕ and ψ of the light quantum and the electron. Certainly a light quantum shows no track in the cloud chamber, but all the same we can determine the direction of the scattered quantum, provided it is scattered a second time and again liberates a recoil electron, the direction of the scattered quantum being found as that of the line joining the initial points of the tracks of the two recoil electrons. Although there is a considerable amount of uncertainty in the interpretation of the experiment, owing to the fact that several tracks may be present, and it is not always possible to determine a pair uniquely as corresponding to each other in the foregoing sense (i.e. produced by one and the same quantum), still Compton and Simon were able to establish agreement between theory and experiment with a fair amount of certainty.

Further confirmation of our ideas about the mechanism of the Compton effect was produced by Bothe and Geiger. They caused X-rays to be scattered in hydrogen, and with a Geiger counter recorded when recoil electrons made their appearance; by means of a second counter they determined the instants at which scattered light quanta appeared. They succeeded in this way in establishing that the emission of the recoil electrons took place at the same moment as the scattering of the light quantum.

Investigations by Bothe, Jacobsen and others (1936) first definitely confirmed the simultaneous appearance of the recoiling light quantum and electron.

In the Compton scattering we have therefore a typical example of a process in which radiation behaves like a corpuscle of well-defined energy and momentum; an explanation by the wave theory of the experimental results which we have described seems absolutely impossible. On the other hand, interference phenomena are quite irreconcilable with the corpuscular view of radiation.

5. Wave Nature of Matter. De Broglie's Theory.

The dilemma was still further intensified when in 1925 de Broglie propounded the hypothesis that the same dualism of wave and corpuscle as is present in light may also occur in matter. A material particle will have a matter wave corresponding to it, just as a light quantum has a light wave; and in fact the connexion between the two opposing aspects must again be given by the relation $E = h\nu$. Further, since from the standpoint of the theory of relativity energy and momentum are entities of the same kind (momentum is the spatial part of a relativistic four-vector, whose time component is energy), it is obviously suggested that for consistency we should write $p = h\kappa$; if ν denotes the number of vibrations per unit time, κ must signify the number of waves per unit length, and therefore be equal to the reciprocal of the wave-length λ of the wave motion; hence

$$p = \frac{h}{\lambda}.$$

The extension of the wave idea from optics to mechanics can be carried through consistently. Before we go into this, however, we should like once more to point out the " irrationality " (as Bohr called it) involved in thus connecting the corpuscular and wave conceptions: E and p refer to a point mass of vanishingly small dimensions; ν and κ, on the contrary, to a wave which is infinitely extended in space and time. The imagination can scarcely conceive two ideas which appear less capable of being united than these two, which the quantum theory proposes to bring into such close connexion. The solution of this paradox will occupy us later.

We shall first develop de Broglie's theory further from a formal point of view. A particle of momentum p in the direction of the x-axis and energy E is to be associated, then, with an infinitely extended wave of the form $u(x, t) = Ae^{2\pi i(\nu t - \kappa x)}$ by means of the two relations

$$E = h\nu, \quad p = h\kappa.$$

This wave advances through space with a definite velocity, the *phase velocity u.* The value of u we can find at once, by considering the (plane) surfaces of constant phase, viz.

$$\phi \equiv \nu t - \kappa x = \text{const.}, \quad \text{or} \quad x = \frac{\nu}{\kappa} t + \text{const.};$$

it follows that

$$u = \left(\frac{dx}{dt}\right)_{\varphi=\text{const.}} = \frac{\nu}{\kappa} = \nu\lambda.$$

Since ν is in general a function of λ, and conversely, this equation embodies the law of *dispersion of the waves.*

But it must be remarked that the phase velocity is a purely artificial concept, inasmuch as it cannot be determined experimentally. In fact, to measure this velocity it would be necessary to affix a mark to the infinitely extended smooth wave, and to measure the velocity of the mark. But the only way in which we can make a mark on the wave train is by superimposing other wave trains upon it, which mutually reinforce each other at a definite place, and so create a hump in the smooth wave function. We have now to determine the velocity with which this hump moves; it is called the *group-velocity.*

A general method is given in Appendix XI (p. 382); here we confine ourselves to a simple special case, which gives the same result, and brings out the difference between phase-velocity and group-velocity with particular clearness. On the primary wave, which we suppose to have the form of $u(x, t)$ above, we superimpose a wave with the same amplitude and a slightly different frequency ν' and wave-length λ'. In this case, as we know, " beat " phenomena occur, and we make use of one of the beat maxima as the mark in our wave train. What we are interested in, then, is the velocity with which the beat maximum moves.

The superposition of the two wave trains gives us mathematically a vibration of the form

$$u(x, t) = e^{2\pi i(\nu t - \kappa x)} + e^{2\pi i(\nu' t - \kappa' x)}.$$

This expression can be written

$$u(x, t) = e^{2\pi i\left(\frac{\nu+\nu'}{2}t - \frac{\kappa+\kappa'}{2}x\right)} \left\{ e^{2\pi i\left(\frac{\nu-\nu'}{2}t - \frac{\kappa-\kappa'}{2}x\right)} + e^{-2\pi i\left(\frac{\nu-\nu'}{2}t - \frac{\kappa-\kappa'}{2}x\right)} \right\}$$

$$= 2 \cos 2\pi \left(\frac{\nu - \nu'}{2}t - \frac{\kappa - \kappa'}{2}x\right) e^{2\pi i\left(\frac{\nu+\nu'}{2}t - \frac{\kappa+\kappa'}{2}x\right)}.$$

It therefore represents a vibration of frequency $(\nu + \nu')/2$ and wave-length $2/(\kappa + \kappa')$, the amplitude of which varies slowly (beats) relative to the vibration itself. The phase, as we deduce at once from the formula, moves with the velocity $(\nu + \nu')/(\kappa + \kappa')$. On the other hand, the maximum of the amplitude moves with the velocity $(\nu - \nu')/(\kappa - \kappa')$. In the limit when ν' tends to ν, and therefore κ' to κ, we find from this the value already found for the phase-velocity

$$u = \frac{\nu}{\kappa} = \nu\lambda,$$

while the group-velocity is given by the limiting value

$$U = \lim_{\nu' \to \nu} \frac{\nu - \nu'}{\kappa - \kappa'}.$$

But this is simply, by definition, the derivative (differential coefficient) of the frequency ν with reference to the wave-number κ, if we regard ν as a function of κ (law of dispersion); hence we have

$$U = \frac{d\nu}{d\kappa} = \frac{d\nu}{d(1/\lambda)}.$$

As is shown in Appendix XI (p. 382), this expression for the group-velocity holds perfectly generally.

We now apply these formulæ to the case of a *free particle* with velocity v. Writing β for v/c, and employing the relativistic formulæ for energy and momentum (see p. 372), we have here

$$\nu = \frac{E}{h} = \frac{m_0 c}{h} \frac{c}{\sqrt{(1 - \beta^2)}}, \qquad \kappa = \frac{p}{h} = \frac{m_0 c}{h} \frac{\beta}{\sqrt{(1 - \beta^2)}}.$$

The phase-velocity is given by

$$u = \frac{\nu}{\kappa} = \frac{c}{\beta} = \frac{c^2}{v},$$

and is therefore greater than the velocity of light c, if the particle's velocity is less than c. The phases of the matter wave are therefore propagated with a velocity exceeding that of light—another indication that the phase-velocity has no physical significance. For the group-velocity we find

$$U = \frac{d\nu}{d\kappa} = \frac{d\nu}{d\beta} \Big/ \frac{d\kappa}{d\beta};$$

it is exactly equal to the particle's velocity, for

$$\frac{d\nu}{d\beta} = \frac{m_0 c}{h} \frac{c\beta}{(1-\beta^2)^{3/2}},$$

$$\frac{d\tau}{d\beta} = \frac{m_0 c}{h} \frac{1}{(1-\beta^2)^{3/2}},$$

and therefore $U = c\beta = v$.

The relationship thus brought out is very attractive; in particular it tempts us to try to interpret a particle of matter as a wave packet due to the superposition of a number of wave trains. This tentative interpretation, however, comes up against insurmountable difficulties, since a wave packet of this kind is in general very soon dissipated. We need only consider the corresponding case in water waves. If we produce a wave crest at any point of an otherwise smooth surface, it is not long before it spreads out and disappears.

6. Experimental Demonstration of Matter Waves.

In view of the boldness of de Broglie's hypothesis, that matter is to be regarded as a wave process, the question of course at once suggested itself, whether and in what way the hypothesis could be put to the test of experiment. The first answer was given by Einstein (1925), who pointed out that the wave idea gives a simple explanation of the *degeneracy of electrons in metals*, which expresses itself in the abnormal behaviour of metals in regard to their specific heat, and as an experimental fact was known to theoretical physicists before de Broglie. The subject will be discussed in detail in Chapter VII (§ 7, p. 237).

Further, it was known from the investigations of Davisson and Germer (1927) that in the reflection of beams of electrons by metals deviations occurred from the result to be expected on classical principles, more electrons being reflected in certain directions than in others, so that at certain angles a sort of selective reflection took place. The conjecture was first propounded by Elsasser (1925) that we have before us here a diffraction effect of electronic waves in the metallic lattice, similar to that which occurs in X-ray interference in crystals (p. 80). The exact investigations which were then undertaken by Davisson and Germer actually gave interference phenomena in precisely the same form as the known Laue interference with X-rays.

Further experiments by G. P. Thomson, Rupp and others showed that when beams of electrons are made to pass through thin foils (metals, mica), diffraction phenomena are obtained, of the same kind as the Debye-Scherrer rings in X-ray interferences (fig. 9, Plate VIII, p. 86). Moreover, when the conditions of interference and the known lattice distances were used as data, it was found that de Broglie's relation between wave-length and the momentum of the electrons was completely confirmed.

The following rough calculation gives an indication of the kind of wave-lengths we have to deal with in beams of electrons. According to de Broglie, we have $\lambda = h/p$ or, if we confine ourselves to electrons of not too high speed, so that we can leave relativistic corrections out of account, $\lambda = h/mv$. On the other hand, the velocity of the electrons is determined by the potential V applied to the cathode tube: $\frac{1}{2}mv^2 = eV$. Hence $\lambda = h/\sqrt{(2meV)}$, or, on inserting the numerical values ($e = 4.80 \times 10^{-10}$ e.s.u., $m = 9.1 \times 10^{-28}$ gm., $h = 6.62 \times 10^{-27}$ erg. sec.), $\lambda = \sqrt{(150/V)}$ Å., when V is expressed in volts. Thus, to an accelerating potential of 10,000 volts there corresponds a wave-length of 0·122 Å.; the wavelengths of the electronic beams employed in practice therefore lie in approximately the same region as those of hard X-rays.

Although it is astonishing that the discovery of the diffraction of electrons was not made earlier, the fact must nevertheless be considered a piece of great good fortune for the development of atomic theory. What confusion there would have been if, soon after the discovery of cathode rays, experiments had been undertaken simultaneously on their charge and capability of deflection, and on their possibilities in regard to interference! Again, Bohr's theory of the atom, which later was to serve as the foundation of the expansion of atomic theory into wave mechanics, was essentially based on the assumption that the electron is an electrically charged corpuscle.

To-day the technique of electronic diffraction is so far advanced that it is employed in industry instead of the earlier methods with X-rays for the purposes of research on materials. One advantage of using electrons is that decidedly higher intensities are available than there are with X-rays. Thus, for example, an interference photograph which may require an exposure of many hours with X-rays can be produced by means of electrons, with the same working data, in something like one second. Another advantage is that the wave-length of the beams of electrons can be varied at will by changing the tube potential; if the setting of the potentiometer is changed, it can be seen at once on the screen how the whole diffraction image contracts or expands according as the wave-length is made shorter or longer. The third and most important advantage of electronic rays is their deflectability by electric and magnetic fields. There is no known method of constructing lenses for X-rays; but by proper arrangement of condensers and magnetic coils one can focus electronic beams (Busch, 1927), and construct lenses and microscopes. Owing to the short wave-length the resolving power is much higher than for optical instruments; and the theoretical limit is much higher still.

Similarly the wave nature of matter can be demonstrated for the case of slow neutrons and the diffraction patterns resulting from the scattering of neutrons can give considerable information about the crystalline structure of solids. Neutrons have a magnetic moment, so that the scattering is sensitive to the magnetic structure of the scattering material. For thermal neutrons (i.e. neutrons with an energy corresponding to a temperature of 300° K.), the de Broglie wavelength is $\lambda = 1 \cdot 81$ Å.

Very important and impressive was the discovery of Stern and his collaborators (1932) that *molecular rays* (of H_2 and He) also show diffraction phenomena when they are reflected at the surfaces of crystals. It was even possible to separate a beam of molecules of nearly uniform velocity with the help of a device similar to the arrangement for measuring the velocity of light: two toothed wheels rotating on the same axis. De Broglie's equation was confirmed for these particles with an accuracy of about 1 per cent. Here, surely, we are dealing with material particles, which must be regarded as the elementary constituents not only of gases but also of liquids and solids. If we intercept the molecular ray after its diffraction at the crystal lattice, and collect it in a receiving vessel, we find in the vessel a gas which has still the ordinary properties.

These diffraction experiments on whole atoms show that the wave structure is not a property peculiar to beams of electrons, but that there is a general principle in question; classical mechanics is replaced by a new *wave mechanics*. For, in the case of an atom, it is clearly the centroid of all its particles (nucleus and electrons), i.e. an abstract point, which satisfies the same wave laws as the individual free electron. Wave mechanics in its developed form does actually render an account of this.

7. The Contradiction between the Wave Theory and the Corpuscular Theory, and its Removal.

In the preceding sections we have had a series of facts brought before us which seem to indicate unequivocally that not only light, but also electrons and matter, behave in some cases like a wave process, in other cases like pure corpuscles. How are these contradictory aspects to be reconciled?

To begin with, Schrödinger attempted to interpret corpuscles, and particularly electrons, as *wave packets*. Although his formulæ are entirely correct, his interpretation cannot be maintained, since on the one hand, as we have already explained above, the wave packets must in course of time become dissipated, and on the other hand the description of the interaction of two electrons as a collision of two wave packets in ordinary three-dimensional space lands us in grave difficulties. The interpretation generally accepted at present was put forward by Born. According to this view, the whole course of events is determined by the laws of probability; to a state in space there corresponds a definite probability, which is given by the de Broglie wave associated with the state. A mechanical process is therefore accompanied by a wave process, the guiding wave, described by Schrödinger's equation, the significance of which is that it gives the probability of a definite course of the mechanical process. If, for example, the amplitude of the guiding wave is zero at a certain point in space, this means that the probability of finding the electron at this point is vanishingly small.

The physical justification for this hypothesis is derived from the consideration of scattering processes from the two points of view, the corpuscular and the undulatory. The problem of the scattering of light by small particles of dust or by molecules, from the standpoint of the classical wave theory, was worked out long ago. If the idea of light quanta is to be applied, we see at once that the number of incident light quanta must be put proportional to the intensity of the light at the place concerned, as calculated by the wave theory. This suggests that we should attempt (Born, 1926) to calculate the scattering of electrons by atoms, by means of wave mechanics. We think of an incident beam of electrons as having a de Broglie wave associated with it. When it passes over the atom this wave generates a secondary spherical wave; and analogy with optics suggests that a certain quadratic expression formed from the wave amplitude should be interpreted as the current strength, or as the number of scattered electrons. On carrying out the calculation (Wentzel, Gordon) it has been found that for scattering by a nucleus we get exactly Rutherford's formula (p. 62 ; Appendix IX, p. 377, and XX, p. 406). Many other scattering processes were afterwards subjected to calculation in this way, and the results found in good agreement with observation (Born, Bethe, Mott, Massey). These are the grounds for the conviction of the correctness of the principle of associating wave amplitude with number of particles (or probability).

In this picture the particles are regarded as independent of one another. If we take their mutual action into account, the pictorial view is to some extent lost again. We have then two possibilities. Either we use waves in spaces of more than three dimensions (with two interacting particles we would have $2 \times 3 = 6$ co-ordinates), or we remain in three-dimensional space, but give up the simple picture of the wave amplitude as an ordinary physical magnitude, and replace it by a purely abstract mathematical concept (the second quantisation of Dirac, Jordan) into which we cannot enter. Neither can we discuss the extensive formalism of the quantum theory which has arisen from this theory of scattering processes, and has been developed so far that every problem with physical meaning can in principle be solved by the theory (Appendix XXV, p. 426). What, then, is a problem with physical meaning? This is for us the really important question, for clearly enough the corpuscular and wave ideas cannot be fitted together in a homogeneous theoretical formalism, without giving up some fundamental principles of the classical theory. The unifying concept is that of probability; this is here much more closely interwoven with physical principles than in the older physics (e.g. the kinetic theory of gases, § 2, p. 3; § 6, p. 9). The elucidation of these relationships we owe to Heisenberg and Bohr (1927). According to them we must ask ourselves what after all it means when we speak of the description of a process in terms of corpuscles or in terms of waves. Hitherto we have always spoken of waves and corpuscles as given facts, without giving any consideration at all to the question whether we are justified in assuming that such things actually exist. The position has some similarity to that which existed at the time the theory of relativity was brought forward. Before Einstein, no one ever hesitated to speak of the *simultaneous* occurrence of two events, or ever stopped to consider whether the assertion of the simultaneity of two events at different places can be established physically, or whether the concept of simultaneity has any meaning at all. In point of fact Einstein proved that this concept must be " relativized ", since two events may be simultaneous in one system of reference, but take place at different times in another. In a similar way, according to Heisenberg, the concepts corpuscle and wave must also be subjected to close scrutiny. With the concept of corpuscle, the idea is necessarily bound up that the thing in question possesses a perfectly definite momentum, and that it is at a definite place at the time considered. But the question arises: can we actually determine exactly both the position and the velocity of the " particle " at a given moment? If

we cannot do so—and as a matter of fact we cannot—i.e. if we can
never actually determine more than one of the two properties (possession of a definite position and of a definite momentum), and if
when one is determined we can make no assertion at all about the
other property for the same moment, so far as our experiment goes,
then we are not justified in concluding that the "thing" under
examination can actually be described as a particle in the usual
sense of the term. We are equally unjustified in drawing this conclusion even if we can determine both properties simultaneously, if
neither can then be determined exactly, that is to say, if from our
experiment we can only infer that this "thing" is somewhere within
a certain definite volume and is moving in some way with a velocity
which lies within a certain definite interval. We shall show later by
means of examples that the simultaneous determination of position
and velocity is actually impossible, being inconsistent with quantum
laws securely founded on experiment.

The ultimate origin of the difficulty lies in the fact (or philosophical
principle) that we are compelled to use the words of common language
when we wish to describe a phenomenon, not by logical or mathematical analysis, but by a picture appealing to the imagination. Common
language has grown by everyday experience and can never surpass
these limits. Classical physics has restricted itself to the use of concepts of this kind; by analysing visible motions it has developed two
ways of representing them by elementary processes: moving particles
and waves. There is no other way of giving a pictorial description of
motions—we have to apply it even in the region of atomic processes,
where classical physics breaks down.

Every process can be interpreted either in terms of corpuscles or in
terms of waves, but on the other hand it is beyond our power to produce
proof that it is actually corpuscles or waves with which we are dealing,
for we cannot simultaneously determine all the other properties which
are distinctive of a corpuscle or of a wave, as the case may be. We
can therefore say that the wave and corpuscular descriptions are
only to be regarded as complementary ways of viewing one and the
same objective process, a process which only in definite limiting
cases admits of complete pictorial interpretation. It is just the
limited feasibility of measurements that defines the boundaries
between our concepts of a particle and a wave. The corpuscular
description means at bottom that we carry out the measurements
with the object of getting exact information about momentum and
energy relations (e.g. in the Compton effect), while experiments which

amount to determinations of place and time we can always picture to ourselves in terms of the wave representation (e.g. passage of electrons through thin foils and observations of the deflected beam).

We shall now give the proof of the assertion that position and momentum (of an electron, for instance) cannot be exactly determined simultaneously. We illustrate this by the example of diffraction through a slit (fig. 10). If we propose to regard the passage of an electron through a slit and the observation of the diffraction pattern as simultaneous measurement of position and momentum from the standpoint of the corpuscle concept, then the breadth of the slit gives the " uncertainty " Δx, in the specification of position perpendicular to the direction of flight. For the fact that a diffraction pattern appears merely allows us to assert that the electron has passed through the slit; at what place in the slit the passage took place remains quite indefinite. Again, from the standpoint of the corpuscular theory, the occurrence of the diffraction pattern on the screen must be understood in the sense that the individual electron suffers deflection at the

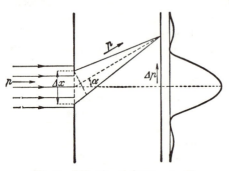

Fig. 10.—Diffraction of electrons at a slit

slit, upwards or downwards. It acquires component momentum perpendicular to its original direction of flight, of amount Δp (the resultant momentum p remaining constant). The mean value of Δp, by fig. 10, is given by $\Delta p \sim p \sin a$, if a is the mean angle of deflection. We know that the experimental results can be explained satisfactorily on the basis of the wave representation, according to which a is connected with the slit-width Δx and the wave-length $\lambda = h/p$ by the equation $\Delta x \sin a \sim \lambda = h/p$. Thus the mean added momentum in the direction parallel to the slit is given by the relation $\Delta p \sim p \, \lambda/\Delta x = h/\Delta x$, or

$$\Delta x \, \Delta p \sim h.$$

This relation, for which a more rigorous derivation will be given in Appendix XII, p. 383, is called *Heisenberg's uncertainty relation*. In our example, therefore, it signifies that, as the result of the definition of the electron's position by means of the slit, which involves the uncertainty (or possible error) Δx, the particle acquires momentum parallel to the slit of the order of magnitude stated (i.e. with the indicated degree of uncertainty). Only subject to this uncertainty is its momentum known from the diffraction pattern. According to the uncertainty relation, therefore, *h represents an absolute limit to the simultaneous measurement of co-ordinate and momentum*, a limit which in the most favourable case we may get down to, but which we can never get beneath. In quantum mechanics, moreover, the uncertainty relation holds generally for any arbitrary pair of " conjugated variables " (p. 385).

A second example of the uncertainty relation is the definition of position by a microscope (fig. 11). Here the order of ideas is as follows. If we wish to determine the position of an electron in the optical way by illuminating it and observing the scattered light, then it is clear, and known as a general rule in optics, that the wave-length of the light employed forms a lower limit to the resolution and accordingly to the exactness of the determination of position. If we wish to define the position as accurately as possible, we will employ light of the shortest possible wave-length (γ-rays). The employment of short-wave radiation implies, however, the occurrence of a Compton scattering process when the electron is irradiated, so that the electron experiences a recoil, which to a certain extent is indeterminate. We may investigate the circumstances mathematically. Let the electron under the miscroscope be irradiated in any direction with

Fig. 11.—Determination of the position of an electron by means of the γ-ray microscope.

light of frequency ν. Then by the rules of optics (resolving power of the microscope) its position can only be determined subject to the possible error

$$\Delta x \sim \frac{\lambda}{\sin a},$$

where a is the angular aperture. Now, according to the corpuscular view, the particle in the radiation process suffers a Compton recoil of the order of magnitude $h\nu/c$, the direction of which is undetermined to the same extent as is the direction in which the light quantum flies off after the process. Since the light quantum is actually observed in the microscope, this indeterminateness of direction is given by the angular aperture a. The component momentum of the electron perpendicular to the axis of the microscope is therefore after the process undetermined to the extent Δp, where approximately

$$\Delta p \sim \frac{h\nu}{c} \sin a.$$

Thus the order of magnitude relation

$$\Delta x \, \Delta p \sim h$$

holds good here also.

Just as every determination of position carries with it an un-certainty in the momentum, and every determination of time an uncertainty in the energy (although we have not yet proved the latter statement), so the converse is also true. The more accurately we determine momentum and energy, the more latitude we introduce into the position of the particle and the time of an event. We give an example of this also, viz. the so-called *resonance fluorescence*. We have seen above (p. 85) that the atoms of a gas which is irradiated with light of frequency ν_{10}, corresponding to the energy difference between the ground state and the first excited state, are raised to the latter state. They then fall back again to the lower state, at the same time emitting the frequency ν_{10}; and if the pressure of the gas is sufficiently low, so that the number of gas-kinetic collisions which occur while the atom remains in the excited state is negligible, then the whole energy which was absorbed will again be emitted. Thus the atom behaves like a classical resonator which is in resonance with the incident light wave, and we speak of resonance fluorescence.

But the energy of excitation of the atoms can also be utilized, not for re-emission of light, but for other actions, by introducing another gas as an indicator. If the latter consists, say, of not too rigidly bound diatomic molecules, the energy transferred in collisions with the excited atoms of the first gas can be utilized for dissociation (Franck, 1922). Again, if the added gas is monatomic, and has a lower excitation level than the first gas, it is itself caused to radiate by the collisions; this is called sensitized fluorescence (Franck). In any case we see that a fraction of the atoms of the first gas is certainly thrown into the excited state by the incident light. We may take the following view of the matter. Excitation by monochromatic light means communication of exact quanta $h\nu_{10}$ to the atom. We there-fore know the energy of the excited atoms exactly. Consequently, by Heisenberg's relation $\Delta E \Delta t \sim h$, the time at which the absorp-tion takes place must be absolutely indeterminate. We can satisfy ourselves that this is so, by considering that any experiment to determine the moment in question would necessarily require a mark in the original wave train—an interruption of the train, for example. But that means disturbing the monochromatic character of the light wave, and so contradicts the hypothesis. A rigorous discussion of the circumstances shows that, if the light is kept monochromatic, the moment at which the elementary act happens does actually altogether elude observation.

The uncertainty relation can also be deduced from the following *general idea*. If we propose to build up a wave packet, extending for a finite distance in the x-direction, from separate wave trains, we need for the purpose a definite finite frequency-range in the monochromatic waves, i.e., since $\lambda = h/p$, a finite momentum-range in the particles. But it can be proved generally (Appendix XII) that the length of the wave packet is connected with the requisite range of momenta by the relation

$$\Delta p \Delta x \sim h.$$

The analogous relation

$$\Delta E \Delta t \sim h$$

can be derived in a similar way.

Bohr is in the habit of saying: the wave and corpuscular views are *complementary*. By this he means: if we prove the corpuscular character of an experiment, then it is impossible at the same time to prove its wave character, and conversely. Let us illustrate this further by an example.

Consider, say, *Young's interference experiment* with the two slits (p. 78) ; then we have on the screen a system of interference fringes. By replacing the screen by a photoelectric cell, we can demonstrate the corpuscular character of the light even in the fringes. It therefore appears as if we had here an experiment in which waves and particles are demonstrated simultaneously. Really, however, it is not so; for, to speak of a particle means nothing unless at least two points of its path can be specified experimentally; and similarly with a wave, unless at least two interference maxima are observed. If then we propose to carry out the " demonstration of a corpuscle ", we must settle the question whether its path has gone through the upper or the lower of the two slits to the receiver. We therefore repeat the experiment, not only setting up a photoelectrically sensitive instrument as receiver, but also providing some contrivance which shows whether the light has passed through the upper slit (say a thin photographic film or the like). This contrivance in the slit, however, necessarily throws the light quantum out of its undisturbed path; the probability of getting it in the receiver (the screen) is therefore not the same as it was originally, i.e. the preliminary calculation by wave theory of the interference phenomenon is illusory. Thus, if pure interference is to be observed, we are necessarily precluded from making an observation of any point of the path of the light quantum before it strikes the screen.

We add in conclusion a few general remarks on the philosophical side of the question. In the first place it is clear that the dualism, wave-corpuscle, and the indeterminateness essentially involved therein, compel us to abandon any attempt to set up a *deterministic theory*. The *law of causality*, according to which the course of events in an isolated system is completely determined by the state of the system at time $t = 0$, loses its validity, at any rate in the sense of classical physics. In reply to the question whether a law of causation still holds good in the new theory, two standpoints are possible. Either we may look upon processes from the pictorial side, holding fast to the wave-corpuscle picture—in this case the law of causality certainly ceases to hold; or, as is done in the further development of the theory, we describe the instantaneous state of the system by a (complex) quantity ψ, which satisfies a differential equation, and therefore changes with the time in a way which is completely determined by its form at time $t = 0$, so that its behaviour is rigorously causal. Since, however, physical significance is confined to the quantity $|\psi|^2$ (the square of the amplitude), and to other similarly constructed quadratic expressions (matrix elements), which only partially define ψ, it follows that, even when the physically determinable quantities are completely known at time $t = 0$, the initial value of the ψ-function is necessarily not completely definable. This view of the matter is equivalent to the assertion that events happen indeed in a strictly causal way, but that we do not know the initial state exactly. In this sense the law of causality is therefore empty; physics is in the nature of the case indeterminate, and therefore the affair of statistics.

CHAPTER V

Atomic Structure and Spectral Lines

1. The Bohr Atom: Stationary Orbits for Simply Periodic Motions.

We have already had before us (Chap. IV, § 2, p. 81) a series of arguments to prove that the classical laws of motion cease to hold good in the interior of atoms. We recall in particular the existence of sharp spectral lines, and the great stability of atoms, phenomena which from the classical standpoint are perfectly unintelligible.

Bohr's explanation of spectra, which we expounded in Chap. IV, § 3, p. 84), points out the road we must follow in setting up a new atomic mechanics. In fact, long even before the discovery of the wave nature of matter, an at least *provisional atomic mechanics* was successfully founded by Bohr and developed by himself and his collaborators, the most prominent of whom was Kramers.

The leading idea (Bohr's *correspondence principle*, 1923) may be stated broadly as follows. Judged by the test of experience, the laws of classical physics have brilliantly justified themselves in all processes of motion, macroscopic and microscopic, down to the motions of atoms as a whole (kinetic theory of matter). It must therefore be laid down, as an unconditionally necessary postulate, that the new mechanics, supposed still unknown, must in all these problems reach the same results as the classical mechanics. In other words, it must be demanded that, for the limiting cases of large masses and of orbits of large dimensions, the new mechanics passes over into classical mechanics. We may obtain a concrete idea of the significance of the correspondence principle from the example of the hydrogen atom, which according to Rutherford consists of a massive nucleus, with an electron revolving round it. By the laws of classical mechanics (Kepler's first law) the orbit of the electron is an ellipse, or in special cases a circle; in the following discussion we shall confine ourselves to this special case. Let the radius of the circular orbit be a, and let it be described

with the angular velocity ω. These two quantities are connected by
the relation

$$a^3\omega^2 = \frac{Ze^2}{m},$$

which corresponds to Kepler's third law; it follows from the equality
of the centrifugal force $ma\omega^2$ and the Coulomb force of attraction
Ze^2/a^2, where Z is the nuclear charge number (1 for H, 2 for He+,
3 for Li++, and so on).

What specially interests us here is the value of the energy of the
revolving electron. By the principle of energy the sum of the kinetic
and the potential energy is constant:

$$\frac{m}{2}\, a^2\omega^2 - \frac{Ze^2}{a} = E.$$

In this equation the energy, which is indeterminate to the extent of
an additive constant, is so normalized that E denotes the work needed
just to release the electron from its connexion to the atom, i.e. to
bring it to rest ($a\omega \to 0$) at an infinite distance ($a \to \infty$) from the
nucleus. By combining the two equations, we find

$$E = -\frac{m}{2}\, a^2\omega^2 = -\frac{Ze^2}{2a} = -\left(\frac{Z^2e^4m\omega^2}{8}\right)^{\frac{1}{3}}.$$

It follows from this that

$$\frac{|E|^3}{\omega^2} = \frac{Z^2e^4m}{8}$$

is constant. We may add that this equation also holds in the case
when the orbit is an ellipse (with semi-major axis a) if we understand
that ω means $2\pi/T$, where T is the period of a revolution (see also
Appendix XIV, p. 387).

So much for the hydrogen atom, according to classical ideas and
fundamental laws; to any orbital radius a, or to any angular velocity
ω, there corresponds a definite value E of the energy, while a, or ω,
can assume any value we choose.

In contrast with this we have the hypothesis of Bohr (Chap. IV,
§ 3, p. 84), according to which the atom can exist only in definite discrete states, and, in a transition from a state with the energy E_1 to
a state with the smaller energy E_2, emits the spectral line for which
$h\nu = E_1 - E_2$. From the frequencies of the emission or absorption
lines we can find the energies of the individual Bohr states.

Now the relations observed in the line spectrum of the hydrogen atom are known very exactly (fig. 1). It was Balmer (1885) who first showed that the lines situated in the visible region—all that were then known—can be represented by the formula

$$\kappa = R\left(\frac{1}{4} - \frac{1}{m^2}\right),$$

with a spectral line corresponding to every integral value of $m > 2$; R being a constant, the so-called Rydberg constant, the value of which was afterwards determined very exactly by Paschen (1916):

$$R = 109678 \text{ cm.}^{-1}.$$

Here κ, in accordance with the usage of spectroscopists, stands for the *wave number*, i.e. the number of wave-lengths per cm.: $\kappa = 1/\lambda$.

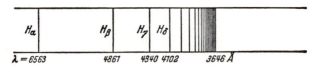

Fig. 1.—Diagram of the Balmer series, showing the notation used for the individual lines

The form of the above equation suggests the idea that the general law for the frequencies may be obtained by taking n^2 instead of 4 ($= 2^2$):

$$\kappa = R\left(\frac{1}{n^2} - \frac{1}{m^2}\right).$$

In fact, on the basis of this formula the lines corresponding to the values $n = 1, 3, 4, \ldots$ could likewise be determined spectroscopically; they were not found in the earlier investigations, since they lie outside the visible spectral region. To-day the following series, called after their discoverers, are known:

$n = 1$: Lyman series (ultra-violet),
$n = 2$: Balmer series (visible),
$n = 3$: Paschen series (infra-red),
$n = 4$: Brackett series (infra-red),

. . . .

It ought further to be particularly remarked that the agreement between the numbers given by the preceding formula and the values found spectroscopically is extraordinarily close. The accuracy of

spectroscopic work is high, and allows the position of the lines to be determined to 5 or 6 figures, yet the values calculated by the formula only differ from the observed values by at most a few units in the last place. The form of Balmer's equation—the difference of two expressions of the same type—suggests, in view of Bohr's theory, that these expressions should be brought into connexion with the energy levels (terms) of the Bohr atomic model:

$$E_n = -\frac{hcR}{n^2} \text{ (Balmer term).}$$

The frequency emitted, when the atom makes a transition from a state n to another state m, is then given, in agreement with the Balmer formula, by the relation

$$h\nu = E_n - E_m.$$

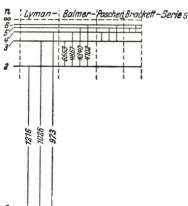

If we know the *term scheme* for an atom, we can at once read off from it the structure of the spectrum. For the hydrogen atom the term scheme has the form

Fig. 2.—Term diagram for the hydrogen atom. The most important lines of the hydrogen spectrum (with some wave-lengths, in Å.) are shown as transitions between two terms.

shown above (fig. 2). The energy scale is normalized in such a way that its zero point corresponds to $n \to \infty$. Hence $-E_n$ is the work needed to remove the electron from the state n to rest at infinity, or the *ionization energy* from the state n. $hcR = -E_1$ is the *ionization energy from the ground state*, a fundamental quantity which is usually expressed in electron volts and referred to as *ionization potential*. Beginning at the lowest term E_1, the terms follow each other more and more closely, and asymptotically approach the limit $E_\infty = 0$.

There seems to be no connexion between this interpretation of the hydrogen lines in terms of transitions between discrete quantum levels and the classical theory of orbital motion. Yet there is such a connexion, a very intimate one indeed, which is revealed by Bohr's *principle of correspondence* and allows the calculation of the so far empirical constant R from the mechanical and electrical properties of the electron.

Applied to the hydrogen atom, this principle asserts that the higher

the Bohr state of the atom is, the more closely the atom obeys the laws of classical mechanics; as n increases, the intervals between the individual energy levels become smaller and smaller, the levels come closer and closer, and the atom approaches asymptotically the state of motion required by classical mechanics.

If then we calculate the emitted frequency by the Balmer formula, for the case when the initial and final states are highly excited states, we find, if $m - n$ is small compared to m and n, the value

$$\nu \sim \frac{2cR}{n^3} \, (m - n).$$

The lowest frequency emitted is got by taking $m - n = 1$, viz.

$$\nu_1 \sim \frac{2cR}{n^3} \, ;$$

for $m - n = 2$ we get a frequency twice as high, for $m - n = 3$ one three times as high, and so on. The spectrum has therefore the same character as that of an electrically charged particle vibrating, on the classical theory, with the proper frequency $\omega = 2\pi\nu_1$, and the associated harmonic frequencies. If we now introduce the ground frequency ν_1 instead of the ordinal number n in the energy formula (Balmer term), we obtain (with $\omega = 2\pi\nu_1$) the expression

$$E \sim - hcR \left(\frac{\nu_1}{2cR}\right)^{\frac{2}{3}} = - \left(\frac{cRh^3\nu_1^2}{4}\right)^{\frac{1}{3}} = - \left(\frac{cRh^3\omega^2}{16\pi^2}\right)^{\frac{1}{3}}.$$

This formula, in regard to the way in which it involves the frequency, has exactly the same structure as the formula which we obtained above (p. 104) for the energy of the classical revolving electron. As there, we have

$$\frac{|E|^3}{\omega^2} = \frac{cRh^3}{16\pi^2} = \text{constant}.$$

But, according to the correspondence principle, in the limiting case of very high values of r, i.e. of very low frequencies, the two formulæ must agree numerically. It follows that

$$R = R_0 Z^2,$$

where

$$R_0 = \frac{2\pi^2 me^4}{h^3 c}.$$

Hence, on the basis of the correspondence principle, and under the assumption that the energy values are actually given by the Balmer terms, we have even at this early stage obtained an unambiguous statement about the value of the Rydberg constant appearing in the Balmer term, and accordingly a possible means of testing the theory.

If we insert the values of e, m and h, as determined by other methods, in the formula for R_0, we find $R_0 \sim 3 \cdot 290 \times 10^{15}$ sec.$^{-1}$, or, when converted into wave numbers, $R_0 \sim 109700$ cm.$^{-1}$; a value which, within the limits of accuracy of our present knowledge of the values of e, m and h (a few thousandth parts), agrees with the experimental value. Moreover, the formula depends in the right way on the ordinal number Z, as is shown by measurements on the ions He$^+$, Li^{++}, ...

Additional evidence for the legitimacy of our argument is given by the fact that by a simple extension of the above formula the *motion of the nucleus* can also be taken into account, and that its effect, as thus found, upon the values of the terms has been completely verified by experiment. In the calculation given above, we proceeded as if the nucleus had infinite mass, and therefore assumed it to be at rest. In reality its mass, though certainly great compared to the mass m of an electron ($\sim 1840\ m$ for the hydrogen nucleus), is still finite. It follows that the nucleus must also move when the electron revolves, and that it describes a circular orbit round the common centre of mass with the same angular velocity ω as the electron, since of course the centre of mass of the whole atom is to be at rest. If we denote the radii of the orbits of electron and nucleus by a_e and a_n, then by the definition of the centre of mass we have $ma_e = Ma_n$, where M is the mass of the nucleus. The equality of centrifugal force and Coulomb attraction gives then

$$ma_e\omega^2 = Ma_n\omega^2 = \frac{Ze^2}{(a_e + a_n)^2}.$$

In the formula for the energy the motion of the nucleus manifests itself, first, in the additional term $\frac{1}{2}Ma_n{}^2\omega^2$ arising from the kinetic energy of the nucleus; and, secondly, in the fact that the potential energy contains, not as before the radius of the circular orbit of the electron, but the sum $a_e + a_n$. By the same process as before (p. 106), we obtain here the formula

$$E = -\left(\frac{Z^2e^4mM\omega^2}{8(m + M)}\right)^{\frac{1}{3}} = -\left(\frac{Z^2e^4m\omega^2}{8(1 + m/M)}\right)^{\frac{1}{3}}.$$

If now, using the correspondence idea, we equate this value of the energy to the value coming from the Balmer term for high quantum numbers, we obtain for the Rydberg constant the expression:

$$R = \frac{R_0 Z^2}{1 + m/M}, \qquad R_0 = \frac{2\pi^2 m e^4}{h^3 c}.$$

The motion of the nucleus manifests itself therefore in a correcting factor, which indeed amounts only to fractions of a thousandth part, but which the high accuracy of spectroscopic methods allows to be exactly tested.

Apart from the correcting factor, every energy term of even order of singly ionized helium coincides, according to the theory, with a hydrogen term, because $Z = 2$ for He$^+$; and therefore the second term of He$^+$ is the same as the first of H, the fourth of He$^+$ the same as the second of H, and so on. If, however, the motion of the nucleus is taken into account, these terms become separated, since the helium nucleus has four times the mass of the hydrogen nucleus, and therefore participates to a slighter degree in the motion of the electron than the latter. In fact, the formula for the energy terms of even order $(2n)$ of He$^+$ gives the expressions

$$(E_{\mathrm{He}^+})_{2n} = -\frac{4R_0 hc}{1 + m/M_{\mathrm{He}}}\,\frac{1}{4n^2} = -\frac{R_0 hc}{(1 + m/M_{\mathrm{He}})n^2},$$

while the hydrogen terms are given by

$$(E_{\mathrm{H}})_n = -\frac{R_0 hc}{(1 + m/M_{\mathrm{H}})n^2}.$$

Every alternate helium line therefore almost, but not quite, coincides with a hydrogen line; the distance between them is easily calculated from the above formulæ; as to order of magnitude, it is approximately in the ratio m/M smaller than the wave-length of the lines themselves and therefore roughly of the order of magnitude of 1 Å., so that it can easily be measured spectroscopically. Experimental tests of this result of the theory have verified it completely.

We may mention further that the value of the Rydberg constant, as measured by Paschen on the lines of hydrogen, which we have given above, is not equal to R_0, but equal to $R_0/(1 + m/M_{\mathrm{H}})$. We thus find the value

$$R_0 = 109737 \text{ cm.}^{-1}.$$

It is at once obvious that the dependence of the frequency of the spectral lines on the mass of the nucleus will hold for other atoms also; it therefore becomes a possibility to discover isotopes spectroscopically. The most important case is that of the heavy hydrogen isotope D itself, for which our formula holds exactly, if we insert the nuclear mass M_D. In point of fact, as has already been mentioned (p 69), the isotope D was first discovered by these spectroscopic methods.

The line of thought followed so far may be summarized as follows. Classical mechanics, on the basis of the picture of the electron revolving round the nucleus, certainly enables us to deduce formulæ for the connexion between orbital radius, frequency of revolution, and energy, but it is incapable of explaining the spectrum emitted by the atom. For the latter purpose we have, following Bohr, to introduce a new hypothesis, viz. that the atom only possesses certain definite energy levels $E_n = -hR/n^2$; and it is the business of the new mechanics to explain these energy levels. From the correspondence principle, i.e. from the single requirement of asymptotic agreement of the new mechanics with the old for large quantum numbers, we have already been able to obtain definite information with regard to the connexion between the experimentally determined Rydberg constant and the atomic constants e, m and h. But even so, there is still no explanation of why the Balmer terms occur at all; up to this point we only know that, if we make the special assumption that the formula $E_n = -Rh/n^2$ is correct, then on account of the correspondence principle the factor of proportionality R must depend in a definite way on e, m and h. It is for the new mechanics to explain these assumptions, or at least to make them intelligible. We may mention in anticipation that this was not accomplished until the introduction of wave mechanics. Nevertheless, we shall in the following pages deal briefly with the leading features and most important results of Bohr's quantum theory of the atom, since these are not only partly required as foundations for the structure of wave mechanics, but are also capable of explaining many experimental results.

In the first place, we collect here, in connexion with the preceding calculations, a few formulæ which are of great importance for the further development of Bohr's theory. If in the Bohr atom we think of the electrons as revolving in definite fixed orbits (ellipses, circles) round the nucleus, then with every energy level of the atom (in the special case of a circular orbit) we can associate a radius, an angular velocity, and so on, on the basis of the formulæ of classical mechanics. It then follows from the formulæ already given (p. 106) that (neglecting the motion of the nucleus) we have for the *radius a* of the circular orbit corresponding to the nth energy state

$$a = -\frac{Ze^2}{2E} = \frac{Ze^2n^2}{2hcR_0Z^2} = a_1\frac{n^2}{Z},$$

where

$$a_1 = \frac{h^2}{4\pi^2me^2} = 0.528 \text{ Å}.$$

represents the radius of the first circular Bohr orbit in the hydrogen atom or, briefly, the "Bohr radius". Similarly, for the *angular velocity*, we find the expression

$$\omega = \frac{4\pi cR_0Z^2}{n^3} = \frac{\omega_1Z^2}{n^3},$$

with

$$\omega_1 = 4\pi cR_0 = 4.13 \times 10^{16} \text{ sec.}^{-1}.$$

Of special importance is the formula for the angular momentum (or, in other words, the moment of momentum) of the electron about the nucleus; from the two preceding formulæ it follows that

$$p = ma^2\omega = ma_1^2\omega_1 n = n\frac{h}{2\pi} = n\hbar.$$

In the Bohr atom the angular momentum is equal to a multiple of $h/2\pi = \hbar$. This fact is called the *quantum condition for angular momentum*.

Conversely, we may postulate this condition, and, by working backwards and using the formulæ of classical atomic mechanics, deduce the Balmer terms. Thus (p. 106), from

$$E = -\frac{Ze^2}{2a}, \quad \text{and} \quad a^3\omega^2 = \frac{Ze^2}{m},$$

it follows that

$$E = -\frac{Z^2e^4}{2ma^4\omega^2} = -\frac{Z^2e^4m}{2p^2},$$

where $p = ma^2\omega$ denotes the angular momentum. If we substitute the value of p from the quantum condition, we find at once

$$E = -\frac{Z^2e^4m}{2}\frac{4\pi^2}{h^2n^2} = -\frac{hcR_0Z^2}{n^2}.$$

It seems now a natural suggestion, that we should regard the quanti sation condition for the angular momentum as an essential feature of the new mechanics. We therefore postulate that it is universally valid. At the same time, we must show by means of examples that the postulate leads to reasonable results. Although from the standpoint of

Bohr's theory the underlying reason for this quantisation rule remains entirely obscure, nevertheless in the further development of the theory it has justified itself by results.

As an example we take the case of the *rotating molecule*, which we treat as a rotator (i.e. as a rigid body which can rotate about a fixed axis). If A is its moment of inertia about this axis, then its kinetic energy when rotating with angular velocity ω is, according to classical mechanics,

$$E = \frac{A}{2}\,\omega^2,$$

and there is no potential energy. The angular momentum about the axis is found from the energy by differentiating with respect to ω:

$$p = A\omega.$$

We now apply the formula for the angular momentum, found for the case of the hydrogen atom:

$$p = n\,\frac{h}{2\pi} = n\hbar,$$

where n denotes a whole number. Eliminating ω, we obtain the energy formula:

$$E_n = \frac{p^2}{2A} = \frac{h^2}{8\pi^2 A}\,n^2.$$

This expression for the value of the energy is called the *Deslandres term*. To this term scheme there corresponds a definite emission spectrum. The transition from the nth state to the $(n-1)$th is bound up with the emission of a line of frequency

$$\nu = \frac{1}{h}\,(E_n - E_{n-1}) = \frac{h}{8\pi^2 A}\,(n^2 - \overline{n-1}^2) = \frac{h}{4\pi^2 A}\left(n - \frac{1}{2}\right).$$

The spectrum consists of a sequence of equidistant lines, giving us the simplest type of *band spectrum*. Other transitions than those to the next lower (and next higher) state cannot occur, as follows from correspondence considerations; for, as we have already shown in the case of the hydrogen atom, and as we also see here at once, transition to the next quantum state but one, or to the next but two, for high quantum numbers involves the emission of double, or three times, the ground frequency, i.e. of harmonics. Now, classically, in the Fourier expansion for circular motion (rotator), only the ground frequency occurs:

$$x = a \cos \omega t, \quad y = a \sin \omega t;$$

and the same of course holds good for the emitted radiation as calculated classically. Therefore, by the correspondence principle, in the quantum theory also no lines can be emitted which correspond to the classical harmonics in the Fourier resolution of the orbital motion; in other words, we have for the rotator the *selection rules*: $n \to n - 1$ for emission transitions, and $n \to n + 1$ for absorption transitions. We have only proved these rules, it is true, for high quantum numbers, since we have made use of the correspondence principle; but it is obviously suggested that we should postulate their validity for all quantum numbers.

2. Quantum Conditions for Simply and Multiply Periodic Motions.

The quantum condition for the angular momentum, which we have deduced from the Balmer term (which is given by experiment), and the application of which to the rotating molecule led us to the Deslandres term (which is in agreement with experiment), gives an indication of the way we have to follow in constructing the new mechanics. It comes to this clearly, that certain quantities can only take values which are whole numbers—they are called quantisable quantities—and the question arises of how in general these quantisable quantities are to be discovered. We shall try to make the matter clear with the aid of simple examples; and in Appendix XIII (p. 384) we give a general formulation of the problem on the basis of the Hamiltonian theory.

An idea of Ehrenfest's (1914) is helpful at this point. Consider a mechanical system, in which such quantisable variables occur, that is to say, quantities capable of integral values only. If we introduce a small disturbance, the new mechanics must hold good in the disturbed system just as in the undisturbed, so that the quantities in question must still have integral values. They must therefore under the action of the disturbance either change by a whole number at a jump, or remain constant. If the actions affecting the system change slowly, the latter must be the case; when that is so, we say that these quantities are *adiabatically invariant*. Hence it is plausible that only adiabatically invariant quantities are quantisable. As a preliminary to the unknown theory it is therefore suggestive to investigate whether there exist such quantities under the classical laws of motion.

We may illustrate this by a simple example. Consider a *simple pendulum* (fig. 3) whose length can be altered, say by drawing the thread over a pulley. If we shorten the thread slowly, we do work,

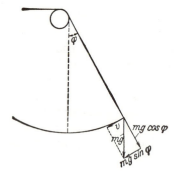

firstly against gravity, secondly against the centrifugal force of the oscillating pendulum. Let the length of the thread be altered slowly from l to $l + \Delta l$; to fix ideas we may suppose Δl to be a negative number, so that the pendulum is shortened. The component of the weight, which stretches the

Fig. 3.—Simple pendulum of variable length. If the length is reduced slowly enough, the ratio of energy to frequency is constant.

thread, is $mg \cos \phi$, and the centrifugal force is $ml\dot\phi^2$, where $\dot\phi$ is the angular velocity. The work done against gravity and centrifugal force is

$$A = \int_{l}^{l+\Delta l} (mg \cos \phi + ml\dot\phi^2) \, (-dl).$$

We now assume that the shortening of the thread takes place extremely slowly, so that during the process the pendulum oscillates to and fro a great many times. We can then disregard the variability of the amplitude of the oscillation with the length of the thread, and integrate over the motion on the supposition that the amplitude is constant. We thus obtain

$$A = -(mg \, \overline{\cos \phi} + m l \overline{\dot\phi^2}) \Delta l,$$

where the bar indicates averaging over the undisturbed motion. If further we confine ourselves to small amplitudes, we can replace $\cos \phi$ by $1 - \phi^2/2$. Thus

$$A = -mg\Delta l + (mg \, \frac{\overline{\phi^2}}{2} - m l \overline{\dot\phi^2}) \Delta l = -mg\Delta l + \Delta W.$$

The first term corresponds to the elevation of the position of equilibrium, which does not interest us. The second part of the expression, i.e. the product of Δl by the expression in brackets, represents the increase ΔW in the energy of the *pendulum motion*. Now the energy of the undisturbed pendulum motion is

$$W = \frac{m}{2} \, l^2\dot\phi^2 + mgl(1 - \cos \phi),$$

where the first term represents the kinetic energy, and the second the potential energy relative to the position of rest. Replacing $1 - \cos\phi$ by the approximation $\phi^2/2$, we get

$$W = \frac{m}{2}\, l^2\dot\phi^2 + mgl\, \frac{\phi^2}{2}.$$

But this is just the energy function for a linear oscillator with the linear amplitude $q = l\phi$. The motion is therefore a simple harmonic vibration $\phi = \phi_0 \cos\omega t$, and we have therefore

$$\overline{\phi^2} = \frac{\phi_0^{\,2}}{2}, \quad \overline{\dot\phi^2} = \frac{\phi_0^{\,2}\omega^2}{2},$$

from which, since $\omega = 2\pi\nu = \sqrt{(g/l)}$, it follows easily that

$$W = \frac{ml^2\phi_0^{\,2}\omega^2}{2} = mgl\, \frac{\phi_0^{\,2}}{2},$$

while, from the second term in the last expression for A,

$$\Delta W = -\frac{ml\omega^2\phi_0^{\,2}}{4}\, \Delta l = -\frac{W}{2l}\Delta l.$$

Hence we have

$$\frac{\Delta W}{W} = -\tfrac{1}{2}\frac{\Delta l}{l}.$$

But on the other hand, since ν varies as $l^{-\frac{1}{2}}$, we have

$$\frac{\Delta\nu}{\nu} = -\tfrac{1}{2}\frac{\Delta l}{l},$$

so that

$$\frac{\Delta W}{W} = \frac{\Delta\nu}{\nu}.$$

This is a differential equation for W as a function of the frequency ν, and its solution is

$$\frac{W}{\nu} = \text{const.} = J.$$

Thus, during the slow (adiabatic) shortening of the pendulum, this quantity J remains constant, so that according to Ehrenfest's principle mentioned above, we can put it equal to an integral multiple of h, i.e.

$$W = nh\nu.$$

We thus obtain the energy levels of the harmonic oscillator, in agreement with Planck's fundamental assumption (p. 84).

The adiabatic invariants for other systems can be determined, theoretically, in a similar way. Still, this direct method is in general extremely troublesome, and one may well ask whether there is no simpler method of discovering invariants. We shall now show how

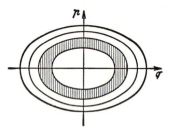

it is possible to do so by means of a geometrical interpretation of the invariant quantity $J = W/\nu$, taking as a special case our previous example of the oscillator (simple pendulum with small oscillations).

Fig. 4.—Phase paths for the linear oscillator; in the pq-plane the phase point describes an ellipse, whose area is an integral multiple of h.

We write down the energy once more, using a somewhat different notation ($q = l\phi$, $p = m\dot{q}$, $f = mg/l$):

$$W = \frac{1}{2m} p^2 + \frac{f}{2} q^2.$$

In a pq-plane (fig. 4) this equation represents an ellipse with the semi-axes

$$a = \sqrt{(2mW)}, \quad b = \sqrt{(2W/f)},$$

as we see at once by writing the equation in the form

$$\frac{p^2}{2mW} + \frac{q^2}{2W/f} = 1.$$

The area of the ellipse is known to be

$$\oint p\,dq = ab\pi,$$

so that, in our case,

$$\oint p\,dq = 2\pi W\sqrt{(m/f)}.$$

(The symbol $\oint$ means that we have to integrate over a whole period. i.e. here, over the whole circumference of the ellipse.) But we have $2\pi\nu = \sqrt{(f/m)}$, so that

$$\oint p\,dq = \frac{W}{\nu} = J.$$

The adiabatic invariant is therefore simply the area of the ellipse, and the quantum postulate states that the area of the closed curve described in the pq-plane (phase plane), in one period of the motion, is an integral multiple of h (Debye, 1913).

The relation thus formulated is capable of immediate generalization. Consider in the first place, as an example with one degree of freedom, the case already treated above (p. 112), that of the *rotator*. Here the co-ordinate is the azimuth $q = \phi$, to which belongs, as canonically conjugated quantity, the angular momentum (or, in other words, the moment of momentum) p. In the free rotation p is constant, i.e. independent of the angle turned through. Thus

$$J = \oint p \, dq = p \oint dq.$$

If we represent the motion in the pq-plane, this integral is to be taken over the straight line $p = $ const., not over a closed curve. But we must observe that in this plane points with the same p, whose q-co-ordinates differ by 2π, represent the same state of the rotator. Properly speaking, therefore, we should consider, not a pq-plane but a pq-cylinder (fig. 5) of circumference 2π, so that the integral has now to be taken over the circumference of the cylinder, and has the

Fig. 5.—Representation of the phase path of a rotator on a cylindrical surface.

value 2π. We therefore obtain $J = 2\pi p$. If we now assume the correctness of our quantisation rule

$$J = \oint p \, dq = nh,$$

it follows that $p = n(h/2\pi)$, the formula which we found before in quite a different way for the hydrogen atom, and successfully applied to the rigid rotator for the interpretation of band spectra.

It is found that the quantisation rule

$$\oint p \, dq = nh$$

can be applied not only in systems with one degree of freedom, but also in complicated systems with many degrees of freedom, and always leads to results which agree with experience. The extension to more than one degree of freedom depends (Sommerfeld, Wilson, 1916) on the fact that in many cases co-ordinates q_1, q_2, ... can be introduced such that the momenta associated with them have the property that p_1 depends only on q_1, p_2 on q_2, and so on. Systems of this kind are said to be *separable*. In general, they are multiply periodic. In that case the motion can be resolved into the superposition of simple periodic vibrations and their harmonics (so-called Lissajous' figures). As an example, consider the case of motion in a plane, where the rectangular co-ordinates x and y oscillate with two frequencies ν_1 and ν_2 (fig. 6). If ν_1 were equal to ν_2, the path would be a circle, an ellipse or a straight line, according to the relation of the phases. If the ratio of ν_1 to ν_2

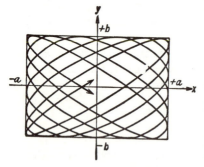

Fig. 6.—Curve representing the motion of a system with two degrees of freedom, in which the two frequencies ν_1 and ν_2 are incommensurable (Lissajous figure).

is a rational number, we again get closed orbits. If ν_1 and ν_2 are incommensurable, i.e. if their ratio is irrational, the curve does not close, but gradually fills the whole rectangle within which the variables range. For multiply periodic motions in general, the orbits are of this type. If, however, an orbital curve closes after a finite number of revolutions, there is in reality only one period, and there will be only one quantum condition of the type

$$\oint p\,dq = nh.$$

If the curve does not close, i.e. if two or more incommensurable periods are present, then there will be as many quantum conditions as there are periods:

$$\oint p_1\,dq_1 = n_1 h,$$
$$\oint p_2\,dq_2 = n_2 h,$$
$$\cdot\ \cdot\ \cdot\ \cdot\ \cdot\ \cdot$$

This is the general case, also referred to as *non-degenerate*, whereas the case of coincident or commensurable periods is called a case of *degeneracy*. If u is the number of periods, v the number of degrees of freedom, then $w = v - u$ is called the *degree of degeneracy*. In Appendix XIII (p. 384) we shall go into these relations a little more deeply.

We shall presently see that this strange combination of classical mechanics with quantum conditions has been replaced by a consistent quantum mechanics in which the integers, $n_1, n_2, \ldots$, appear in a much more natural way. However, the present theory holds even then approximately for high quantum numbers (i.e. if h is small compared with the values of the integrals $\oint p\,dq$), as we should expect from the correspondence principle (see p. 136).

As an example of the above rules we shall now give a discussion of the hydrogen atom, the complete quantisation of which was carried out by Sommerfeld. By Kepler's laws, the orbit of the electron round the nucleus is an ellipse; it is therefore simply periodic. Since the electron has three degrees of freedom, this is a case of double degeneracy. In Appendix XIV (p. 387) we give the quantisation of the Kepler ellipse, which leads to the correct energy levels (Balmer terms).

The degeneracy is partly removed by taking into account the *relativistic variability of mass*, or dependence of the mass of the electron on its velocity. In this case the orbit, according to Sommerfeld, is given by a precessing ellipse (rosette); its major axis revolves in the plane of the ellipse round

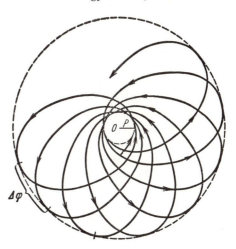

Fig. 7.—Orbit (rosette) of the electron about the nucleus, taking into account the relativistic variability of mass; the motion is doubly periodic, the perihelion being displaced by the angle $\Delta\phi$ per revolution.

the nucleus with constant angular velocity (fig. 7). The orbit is now doubly periodic; besides the original period of revolution, which remains unchanged if the precession is slight, we have now the period of the precessional motion. In accordance with this, we have here *two* quantum conditions:

$$J_1 = nh, \quad J_2 = kh$$

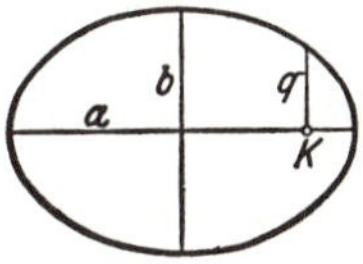

(compare Appendix XIV, p. 387); n determines the semi-major axis a of the approximate ellipse, k its semi-latus rectum q: $a = n^2 a_0$, $q = k^2 a_0$ (fig. 8).

Fig. 8.—Elliptic orbit with the nucleus K as focus.

Further, the calculation shows that the total *angular momentum* is

$$p = k\,\frac{h}{2\pi},$$

and that the *energy* contains a term additional to the Balmer term:

$$E = - \frac{R_0 h Z^2}{n^2} + \epsilon(n, k).$$

From the first formula it follows that for $k = 0$ the angular momentum vanishes; this gives the "pendulum orbits", in which the orbital ellipse degenerates into a straight line. For $k = n$ we obtain the greatest possible angular momentum for a fixed n; this gives the circular orbits. For $k < n$ we get the elliptic orbits. In the spectra, no sign has been found of the energy terms corresponding to the pendulum orbits; the terms not being realized, the theory must delete them from the term scheme. The reason given for their exclusion is that the pendulum orbits pass through the nucleus, so that the electron would collide with the nucleus, which of course is impossible.

The supplementary term $\epsilon(n, k)$, the value of which will be given later, expresses the minuter detail of the hydrogen lines; its effect is to split up every Balmer term into a number of energy terms, corresponding to the quantum number k. The spectral lines themselves are therefore made up of a system of finer lines, which are determined by the transitions between the energy levels of the upper state ($n = n_1$, $k = 1, 2, \ldots, n_1$) and the lower state ($n = n_2, k = 1, 2, \ldots, n_2$).

This is called the *fine structure* of the spectral lines. Its theory was given by Sommerfeld for the case of atoms of the hydrogen type (H, He$^+$, Li^{++}), and was first tested by Fowler and Paschen on the spectrum of singly ionized helium (He$^+$), which was found in complete agreement with the theory. The test is easier with He$^+$ than with H for this reason, that the energy terms of He$^+$ are four times as far apart, on account of the nuclear charge number Z being doubled, whereas the

Fig. 9.—Top-like motion of the normals to the plane of an orbit, in an external field; the normal moves in a cone as shown.

corresponding factor for the fine structure splitting, as the theory shows, is 16; the fine structure of the lines of He$^+$ is therefore more easily demonstrated and measured.

When the relativistic variability of mass is taken into consideration, the degeneracy of the hydrogen atom is certainly removed in part, but the motion is still simply degenerate. This degeneracy is connected with the fact that in the absence of an external field the orbital plane of the electron is fixed in space, and its orientation perfectly arbitrary. The degeneracy is only removed by the introduction of an external field. If the atom is brought into a homogeneous magnetic field **H**, a motion of precession of the orbital plane sets in, round the direction of the field (fig. 9). For the electron revolving in its orbital plane gives the atom angular momentum and also magnetic moment; considered as vectors, these are both at right angles to the orbital plane. The magnetic field takes hold of the magnetic moment and seeks to turn it round into the field direction. The effort to do this is opposed by the rotational inertia due to the revolution of the electron, in a way known, sufficiently well for our purpose, from the theory of the top; the effect, as in the case of the mechanical top, is precessional motion.

This motion gives the third of the periodicities possible in the hydrogen atom. In accordance with our rules, it is to be quantised in the same way as the angular momentum of the orbital motion. This leads to the formula

$$p_\phi = m \frac{h}{2\pi} \qquad (m = -k, -k+1, \ldots, +k).$$

Here p_ϕ is the component of the angular momentum p in the direction of the field, so that this is quantised. Consequently the resultant angular momentum, the magnitude p of which must be an integral multiple of $h/2\pi$, can have only a finite number of possible inclinations to the direction of **H**, in fact $2k+1$ inclinations (fig. 10), corresponding to the $2k+1$ possible values for its component p_ϕ

Fig. 10.—Example of quantisation of direction for the case $k = 3$; the total angular momentum must be inclined at such an angle to the field direction that its projection m on that direction is a whole number (of units $h/2\pi$).

parallel to **H**. We speak of *quantisation of direction*. On the experimental side, this quantisation of direction can be demonstrated by the experiments of Stern and Gerlach, of which we shall give an account later (p. 185).

The precessional motion in the magnetic field contributes an additional quantity of energy $m\nu_L h$, where $\nu_L = eH/(4\pi\mu c)$ is the so-called *Larmor frequency*, μ being the mass of the electron. This value for the energy is obtained as follows. The potential energy of a magnet of moment M in a homogeneous magnetic field **H**, with which it makes the angle θ, is equal to $(-)HM \cos\theta$. In our case the angle θ is defined by the quantisation of direction: $\cos\theta = p_\phi/p = mh/(2\pi p)$.

On the other hand, a revolving charge $(-e)$, which on account of its mass μ gives rise to an orbital angular momentum p, produces by the rules of electron theory a magnetic moment of $(-ep)/(2\mu c)$. By combining these results, we find for the additional energy the value stated.

Every term is thus split up into $2k+1$ terms by the magnetic field. It is therefore to be expected that the spectral lines also possess the property of splitting in the magnetic field. That this is actually so was found experimentally by Zeeman (1896), at a time when nothing was yet known of quantum theory, and attempted explanations of optical processes relied on classical mechanics. Zeeman's earliest observations showed that the lines in question, when observed transversely (magnetic field at right angles to the light path), are split up into three, the middle line oscillating parallel to the field, and the outside lines at right angles to the field; but that, when they are observed longitudinally (field parallel to light path), they split into two, which are circularly polarized in opposite directions. This case is now called the *normal Zeeman effect*. It was shown by Lorentz that a classical oscillator in a magnetic field may actually be expected to show these phenomena. Suppose the linear vibration of the oscillator resolved into two opposite circular motions at right angles to the field, along with a linear motion parallel to the field; then, by the action of the field, according to Lorentz's theory, the frequency of one of the circular motions is increased exactly by the Larmor frequency ν_L, and that of the other circular motion is diminished by the same amount. This agrees with the splitting and polarization observed in one direction or another as described above.

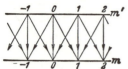

Fig. 11.—Transitions in a magnetic field; on account of the selection rule $\Delta m = 0, \pm 1$, only the transitions indicated by arrows occur; since the terms corresponding to the various values of m are equally far apart in a homogeneous magnetic field, each spectral line splits up in the magnetic field into three lines only (normal Zeeman effect).

It is now easy to see that our quantum theory leads to precisely the same result. In spite of the splitting into $2k+1$ terms, we obtain two, or three, component lines, according as we observe in the direction of the magnetic field, or perpendicular to it. In the light of the correspondence principle, this result can be understood at once. For the classical explanation of the Zeeman effect given by Lorentz implies that the atomic system, by the action of the magnetic field, is given an additional motion of rotation with the Larmor frequency ν_L, that is to say, an additional pure circular motion without harmonics. By the correspondence principle this must also hold asymptotically for the treatment by quantum theory; hence, as in the rotator, we are led to the selection rules $\Delta m = \pm 1$. On the other hand, the component of the motion parallel to the field is not changed in this process; we therefore get the additional selection rule $\Delta m = 0$. For the possible transitions we therefore obtain the scheme of fig. 11. The splitting of the higher and lower terms is shown on horizontal lines, and the arrows correspond to the possible transitions. The height of the energy jump is the same for all the arrows directed obliquely to the left, so that these yield one and the same radiated frequency; similarly with the arrows pointing vertically downwards, or towards the right. We therefore obtain a simple triplet (or doublet). Bohr's theory gives only the *normal Zeeman effect*. In reality (for not too strong fields), the splitting pattern consists in general, not of three lines only, but of a definitely greater number. This *anomalous Zeeman effect* cannot be understood without going beyond the concepts we have developed so far. We shall go into it more deeply in Chap. VI.

We pass now to the *spectra of the alkali atoms* (fig. 12, Plate IX, p. 124). The existence of series of lines in these spectra was discovered by Kayser and Runge (1890) and almost simultaneously and independently by Rydberg (1890). These series are considered to arise in this way: an electron, the so-called radiating electron, moves in the field due to the nucleus and the rest of the electrons, and by itself causes the spectrum. This view is justified in the first place by the fact that in the alkalies, according to experiment, one electron is much more loosely bound than the rest, so that this electron is chiefly responsible for the chemical behaviour of the alkalies; on the other hand, we shall find that the remaining $Z - 1$ electrons form so-called closed shells, round which the odd electron, i.e. the radiating electron, revolves. The field in which it moves is centrally symmetric, so that the potential depends only on the distance from the nucleus; the Coulomb field of the nucleus is "screened" by the remaining $Z - 1$ electrons, and it is just these deviations from the Coulomb field which cause the differences between the *spectra of the alkalies* and that of hydrogen.

Here also the radiating electron moves in a precessing ellipse; the pure ellipse occurs as the form of the orbit in a pure Coulomb field only, any deviation from which, such as that determined by the variability of mass in the hydrogen atom, implies precession. We quantise this motion as before, and so obtain two quantum numbers n and k. By general agreement the terms are denoted by a number and a letter. The number indicates the principal quantum number n. For the

specification of the azimuthal quantum number k the following notation
has established itself:

$$k = 1 \quad 2 \quad 3 \quad 4 \quad \ldots,$$
$$s \quad p \quad d \quad f \quad \ldots \text{ term.}$$

Thus, e.g., $4d$ denotes the term with $n == 4$ and $k = 3$. Since the
precessional motion is purely periodic, the correspondence principle
leads as before to the selection rule $\Delta k = \pm 1$. The only transitions
which occur are therefore those from an s-term to a p-term, from a
p-term to an s or d-term, and so on. For the sake of general distinction, the most important series of spectral lines have been given the
following names:

$$np \to n_0 s \quad \text{principal series,}$$
$$ns \to n_0 p \quad \text{sharp series,}$$
$$nd \to n_0 p \quad \text{diffuse series,}$$
$$nf \to n_0 d \quad \text{fundamental series,}$$

. . . .

In this list we have on the left the terms between which the transition
takes place, n_0 before any letter denoting the lowest value of n that can

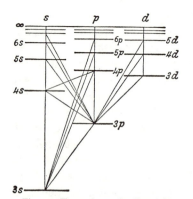

Fig. 13.—Term scheme of sodium (with not
too high resolution). The transitions between
the various levels give the emission lines.

be associated with that letter in the
atom under consideration. On the
right is the name of the series
obtained by giving different values
to n. In the term scheme (fig. 13) it
is to be understood that the energies
increase upwards. For every azimuthal quantum number k there
is a series of energy terms, which
correspond to the various values
of n, and which converge to a
certain value (the *series limit*) when
$n \to \infty$. The oblique lines represent
the possible transitions; those to
the $3s$-term correspond to the
spectral lines of the principal

series, those from s-terms to the $3p$-term to the spectral lines of the
sharp series, and so on. The reason why the lowest value of n is 3,
not 1, will be explained later (p. 181 ff, Table V). By the formation of
such term schemes, founded on theoretical considerations, it has become
possible to introduce order into the chaos of spectral lines as found
experimentally.

PLATE IX

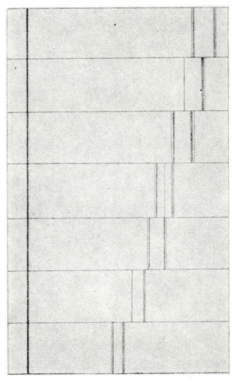

Ch. V, Fig. 14.—Rotating crystal photographs of the K series for elements between arsenic (Z = 33) and rhodium (Z = 47). The figure shows the increase of hardness with increasing atomic number. (The marking on the extreme left is due to the undeflected primary beam.) (See p. 125.)

Ch. V, Fig. 12.— Absorption spectrum of potassium vapour; a sodium line is shown between the two uppermost potassium lines (see p. 123).

As a last application of Bohr's theory we discuss the origin of X-ray spectra (Barkla, 1908). The essential feature distinguishing X-ray spectra from optical spectra is that the former are of the same type in all elements (fig. 14, Plate IX), whereas the latter, though of like structure for elements of like chemical character, are quite different for elements belonging to different columns of the periodic table. This means in the first place that X-ray spectra must have their source in the interior of the atom, whereas for optical spectra, just as for chemical behaviour, it is the outer region of atoms which matters.

The lines found by experiment, which are denoted by K_a, K_β, ..., L_a, L_β, ..., M_a, M_β, ..., can be arranged in a term scheme, by taking advantage of the difference relations

$$K_\beta = K_a + L_a,$$
$$K_\gamma = K_a + L_\beta = K_\beta + M_a, \text{ &c.,}$$

which correspond to Ritz's combination principle; these have been tested experimentally with great exactness, and found to be fulfilled as stated.

The interpretation of X-ray spectra was given by Kossel (1917). In the atom the electrons are arranged in shells, there being a K shell, an L shell, &c. The electrons are most firmly bound in the K shell, less firmly in the L shell, still less firmly in the M shell, and so on. The energy levels indicated by horizontal lines in fig. 15 correspond to the electrons in the individual shells. The excitation of a K-line, according to Kossel, has to be pictured as follows. By some process (collision, absorption of light) an electron of the K shell is ejected from it. For this a definite minimum energy is requisite; all amounts of energy, which are greater than this least amount, can be absorbed. In absorption, we therefore find in the spectrum a sharp edge, the absorption edge; wave-lengths

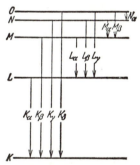

Fig. 15.—Term scheme of X-ray levels (after Kossel), with the transitions corresponding to the X-ray lines.

shorter than this are absorbed. Absorption lines, such as occur in the visible region, are not found in the X-ray region.

If an electron has been ejected in this way from the K shell, an electron of the L or M shell can fall down into the K shell, and a quantum of radiation is emitted; in this case the lines K_a, K_β, ...

appear in the emission. A similar result follows in the case when the excitation causes an electron to be expelled from the L shell.

From his measurements Moseley (1913) was able to deduce the following law for the frequencies of the K_a-lines of the various elements:

$$\nu = \tfrac{3}{4}R_0(Z - a)^2 = R_0(Z - a)^2 \left(\frac{1}{1^2} - \frac{1}{2^2}\right).$$

Here R_0 is the Rydberg constant, which we have already had before us (p. 107); a is called the screening constant, and has approximately the same value 1·0 for all elements. The experimental facts, as stated in Moseley's law, imply therefore that in X-ray spectra we are concerned with terms of the hydrogen type ("hydrogen-like terms"),

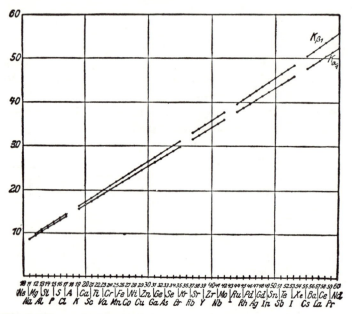

Fig. 16.—Diagram of points with abscissæ equal to the atomic number Z, and ordinates equal to the square roots of the frequencies of K-lines

with the nuclear charge screened off. This result is to be understood in the sense that the innermost electrons, on account of the large nuclear charge, are bound almost solely to the nucleus, and therefore move in a similar way to the electron in the hydrogen atom, without being essentially disturbed in their orbits by the rest of the electrons in the atom. We therefore obtain a term scheme which corresponds to that of the alkali atoms. We must in fact associate with the electrons in the K shell the principal quantum number $n = 1$, with those of the L shell $n = 2$, and so on.

Further, an exact investigation shows that the K-level is single, the L-level triple, and the M-level quintuple, but the explanation of this is postponed to next chapter.

Moseley's law puts at our disposal a simple method of testing the order of succession in the periodic system. If we plot points in a diagram, with abscissa Z, and with ordinate the square root of the frequency of the K_a-lines, the points of the diagram, as determined by experiment, should lie on a straight line (fig. 16). If we gave an element a wrong place in the periodic table, i.e. associated the element with a wrong nuclear charge, then the corresponding point in the diagram would lie off the straight line. In this way it could be proved once again that, for instance, cobalt has a lower nuclear charge number (atomic number) than nickel, in spite of its greater mass; examples of this class are marked with a double arrow in the Table of p. 39. Another result of Moseley's work was the final determination of the gaps in the periodic table, i.e. of the elements which at that time were still unknown; in particular, the existence of hafnium ($Z = 72$) could be predicted in this way before its real discovery (Coster and von Hevesy, 1923).

We may mention further that Bohr in his theory of the structure of the periodic system has substantially used only those simple properties which manifest themselves in the spectra of the alkalies and of X-rays.

Here we conclude our account of Bohr's theory. Although it has led to an enormous advance in our knowledge of the atom, and in particular of the laws of line spectra, it involves many difficulties of principle. At the very outset, the fundamental assumption of the validity of Bohr's frequency condition amounts to a direct and unexplained contradiction of the laws of the classical theory. Again, the purely formal quantisation rule, which stands at the head of the theory, is a foreign element which in the first instance is absolutely unintelligible from the physical point of view. We shall see later how both of these difficulties are removed in a perfectly natural way in wave mechanics.

Bohr's theory leaves some questions unexplained. Why must the pendulum orbits be excluded? The reason given, which points to the collision with the nucleus, is hardly cogent; moreover, it is outside the bounds of Bohr's theory. How does the anomalous Zeeman effect come about? The explanation of this, as we shall see (p. 157), requires us to use the fact that the electron possesses in itself mechanical angular momentum and magnetic moment. Finally, the calculation of the simplest problem of the type involving more than one body, viz. the helium problem, leads to difficulties, and to results contrary to experimental facts.

3. Matrix Mechanics.

The main reason for the break-down of Bohr's theory, according to Heisenberg (1925) is that it deals with quantities which entirely elude observation. Thus, the theory speaks of the orbit and the velocity of an electron round the nucleus, without regard to the consideration that we cannot determine the position of the electron in the atom at all, without immediately breaking up the whole atom. In fact, in order to define its position with any exactness within the atom (whose diameter is of the order of magnitude of a few Ångström units), we must observe the atom with light of definitely smaller wave-length than this, i.e. we must irradiate it with extremely hard X-rays or with γ-rays; in that case, however, the Compton recoil of the electron is so great that its connexion with the atom is immediately severed, and the atom becomes ionized.

Thus, in Heisenberg's view, Bohr's theory fails because the fundamental ideas on which it is based (the orbit picture, the validity of the classical laws of motion, and so on) can never be put to the test. We move, therefore, in a region beyond experience, and ought not to be surprised if the theory, constructed as it is on a foundation of hypotheses which cannot be proved experimentally, partially fails in those deductions from it which can be subjected to the test of experiment.

If a logically consistent system of atomic mechanics is to be set up, no entities may be introduced into the theory except such as are physically observable—not, say, the orbit of an electron, but only the observable frequencies and intensities of the light emitted by the atom. Starting from this requirement, Heisenberg was able to lay down the leading principles of a theory, which was then developed by himself, Born and Jordan (1925), the so-called *matrix mechanics*, which is meant to replace the atomic mechanics of Bohr, and which in its applications has been brilliantly successful throughout. Although in form it is entirely different from the wave mechanics to be expounded later, yet in content, as Schrödinger has shown, the two are identical. On this account we do not consider matrix mechanics in any detail,

but content ourselves with some brief indications. If we start from
the frequencies

$$\nu_{nm} = \frac{E_n}{h} - \frac{E_m}{h},$$

as observable quantities, it is a natural suggestion that we arrange
them in square array as follows:

$$\begin{Bmatrix} \nu_{11} = 0 & \nu_{12} & \nu_{13} & \cdots \\ \nu_{21} & \nu_{22} = 0 & \nu_{23} & \cdots \\ \nu_{31} & \nu_{32} & \nu_{33} = 0 \cdots \\ \cdots \cdots \cdots \cdots \cdots \cdots \cdots \end{Bmatrix}.$$

If now we adhere to the arrangement in this scheme once for all,
so that, say, the place in the fourth row and the second column is
always associated with the transition from the fourth to the second
quantum state, then we can also set out in similar square array
the amplitudes a_{nm} of the " virtual resonators " associated with the
various frequencies emitted, where a_{nm}^2 then denotes the intensity
of the frequency emitted:

$$\begin{Bmatrix} a_{11} & a_{12} & a_{13} & \cdots \\ a_{21} & a_{22} & a_{23} & \cdots \\ a_{31} & a_{32} & a_{33} & \cdots \\ \cdots \cdots \cdots \cdots \cdots \cdots \end{Bmatrix}.$$

Similarly we can insert in an array of this kind other quantities connected with the transition $n \rightarrow m$.

The question now is: how do we calculate with these arrays? Here
the following remark of Heisenberg's is useful: if we multiply two
vibrational factors $a_{nk} = e^{2\pi i \nu_{nk} t}$ and $a_{km} = e^{2\pi i \nu_{km} t}$, then by Ritz's
combination principle we get

$$e^{2\pi i \nu_{nk} t} e^{2\pi i \nu_{km} t} = e^{2\pi i (\nu_{nk} + \nu_{km}) t} = e^{2\pi i \nu_{nm} t},$$

which is a vibrational factor belonging to the same array, so that
by the method specified for forming a product, we merely pass to
another place in the square array, in conformity with the rule assigning places. We can proceed to define the product of two such arrays,
in such a way that this product is again a square array of the same
type. The multiplication rule, which Heisenberg deduced solely from
experimental facts, runs:

$$(a_{nm})(b_{nm}) = (\Sigma_k a_{nk} b_{km}).$$

It was remarked by Born and Jordan that this rule for multiplication is identical with one which has long been known in mathematics as the rule for forming the product of two "matrices", such as occur in the theory of linear transformations and the theory of determinants. We may therefore regard Heisenberg's square arrays as infinite matrices, and calculate with them by the known rules of the theory of matrices.

We now come to the central feature of matrix mechanics, which is this, that a representative matrix of the above type is associated with every physical magnitude. We can form a co-ordinate matrix, a momentum matrix, and so on, and then calculate with these matrices in practically the very same way as we are accustomed to do with co-ordinates, momenta, and so on, in classical mechanics.

There is one essential distinction, however, between matrix and classical mechanics, viz. that when matrices are introduced as co-ordinates q_k and momenta p_k, their product is not commutative; it is no longer true, as it was in classical mechanics, that

$$p_k q_k - q_k p_k = 0.$$

The non-commutativity which presents itself here is not, however, of the most general kind, as the theory shows; for the left-hand expression, with a pair (p_k, q_k) of canonically conjugated variables, can take only the definite value

$$p_k q_k - q_k p_k = \frac{\hbar/i}{2\pi i}.$$

These *commutation laws* (Born and Jordan, 1925) take here the place of the quantum conditions in Bohr's theory. The considerations by which their adoption is justified, as also the further development of matrix mechanics as a formal calculus, are for brevity omitted here. In the next section, however (§ 4, p. 134), it will be found that the analogous commutation laws in wave mechanics are mere matters of course. In Appendix XV (p. 392), taking the harmonic oscillator as an example, we show how and why they lead to the right result.

It may be mentioned in conclusion that the fundamental idea underlying Heisenberg's work was worked out by Dirac (1925) in a very original way, and that in 1964 he put forward views to the effect that although Heisenberg's and Schrödinger's approaches are perfectly equivalent in ordinary (non-relativistic) quantum mechanics, this is not the case in quantum field theory. Here Heisenberg's method turns out to be more fundamental.

4. Wave Mechanics.

Quite independently of the line of thought just explained, the problem of atomic structure has been attacked with the help of the ideas developed in the preceding chapter. According to the hypothesis of de Broglie (p. 89), to every corpuscle there corresponds a wave, the wave-length of which, in the case of rectilinear motion of the corpuscle, is connected with the momentum by the relation

$$\lambda = \frac{h}{p}.$$

It is only consistent to try to extend the theory by applying this wave idea to the atom, that is to say, to the electron revolving round the nucleus; in that case we have to picture the atom as a wave motion round a particular point, the nucleus. What the theory has to do is to deduce the law of this motion.

As the first step, following de Broglie, we shall show that the quantum conditions of Bohr's theory can be interpreted at once on the basis of the wave picture. For this purpose we consider the simple case of a circular motion of the electron round a fixed point (fig. 17).

Fig. 17.—Motion in a circle from the corpuscular point of view; a particle moves in a definite orbit with the momentum p.

On Bohr's theory we have for this motion of revolution the quantum condition of angular momentum

$$p_\phi = rp = n\,\frac{\hbar}{2\pi},$$

where p is the linear momentum mv of the electron. Now imagine a revolving wave instead of the revolving electron. If the radius of the circular orbit is very great, the same relation will hold for the revolving wave as for the plane wave, viz.

$$p = \frac{h}{\lambda}.$$

If we insert this value of the momentum in the preceding quantum condition of angular momentum, we obtain the equation

$$n\lambda = 2\pi r.$$

Here the term on the right represents the circumference of the circle: the formula states that this must be equal to a whole number of wave-lengths.

What then is the meaning of the quantum condition of angular momentum? If we try to construct

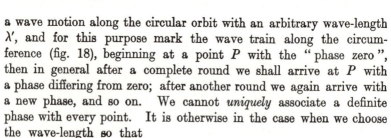

Fig. 18.—Motion of a wave in a circle; a single definite wave form is only possible when the circumference of the circle is a whole number of times the wave-length.

a wave motion along the circular orbit with an arbitrary wave-length λ', and for this purpose mark the wave train along the circumference (fig. 18), beginning at a point P with the " phase zero ", then in general after a complete round we shall arrive at P with a phase differing from zero; after another round we again arrive with a new phase, and so on. We cannot *uniquely* associate a definite phase with every point. It is otherwise in the case when we choose the wave-length so that

$$n\lambda = 2\pi r.$$

Here, in our construction, after a complete round we reach P again with the same phase as we started with. In this case the wave picture, or wave motion, is unique; a complete circuit changes nothing.

The quantum condition of angular momentum is therefore in this example identical with the requirement that the wave function of the corresponding vibratory process be one-valued. We therefore, as a general method, replace the hitherto unintelligible quantum condition of Bohr's theory by the obvious stipulation of *uniqueness (and finiteness) of the wave function* for the whole domain of the independent variables.

We pass now to the considerations which have led to the setting up of a differential equation—the *wave equation*—as the expression of the law of the wave motion in an atom. That the fundamental law takes the form of a differential equation is of course, by analogy with other vibratory processes, only to be expected. There is, naturally, no way of deducing the wave equation by strict logic; the formal steps which lead to it are merely matters of clever guessing.

We begin with the motion of a *free particle*; we describe the associated wave by a wave function

$$\psi = e^{2\pi i(\kappa x - \nu t)} = e^{(2\pi i/h)(px - Et)}.$$

Here ν and κ are the frequency and wave number which, according to de Broglie, are connected with the energy and momentum by the equations

$$\nu = \frac{E}{h}, \quad \kappa = \frac{1}{\lambda} = \frac{p}{h}.$$

By partial differentiation with respect to x and t we find

$$\frac{h}{2\pi i} \frac{\partial \psi}{\partial x} = p\psi, \quad -\frac{h}{2\pi i} \frac{\partial \psi}{\partial t} = E\psi.$$

We can now also read these equations in the converse way: the differential equations being given, it is required to find their solution. In the case when the particle moves in a *straight line*, and all values of x between $-\infty$ and $+\infty$ are permissible, the solution is represented by the function given above. If the particle moves along a circle of circumference l, and we denote by x the co-ordinate of a point, viz. its distance from a fixed point on the circumference measured along the arc, then only values from 0 to $l = 2\pi r$ are essential for x; for an increase of x by l brings us back to the same point. Since ψ on the circle must be a single-valued quantity, an increase of x by $l = 2\pi r$ must make no change in the function. Now the general solution of the first of the two equations above is $\psi = Ae^{(2\pi i/h)px}$, which, when x is increased by l, is multiplied by $e^{(2\pi i/h)pl}$; hence, if ψ is to be a " proper function ", this factor must be equal to unity, i.e.

$$e^{(2\pi i/h)pl} = 1 = e^{2\pi i n}$$

or

$$\frac{pl}{h} = n; \quad p = p_n = \frac{nh}{l} = \frac{nh}{2\pi r}.$$

This signifies that the preceding equation, in the case of circular motion, does not possess a permissible solution for all values of p, but only for the discrete " proper values " $1h/l$, $2h/l$, $3h/l$, ...

The foregoing equations can also be interpreted as follows. When the wave function ψ is known, we obtain the corresponding momentum, or its x-component p_x, by differentiating the wave function partially as to x: $(h/2\pi i)\partial\psi/\partial x = p_x\psi$. To the x-component of the momentum there belongs, as we say, the differential operator

$$p_x = \frac{h}{2\pi i} \frac{\partial}{\partial x};$$

similarly for y and z. The operator belonging to the energy, on the other hand, is

$$E = - \frac{h}{2\pi i} \frac{\partial}{\partial t}.$$

Operators, or entities which operate on any function, that is, which when applied to this function, generate another function, can be represented in the most diverse ways. Heisenberg's matrices are simply one definite kind of representation of such operators; another kind is the set of differential coefficients corresponding to the momentum components and the energy. In the latter kind of representation the Born-Jordan commutation laws admit of a simple interpretation; here, by what we have just seen, $pq - qp$ simply means the application of the differential operator

$$\frac{h}{2\pi i} \frac{\partial}{\partial q} q - q \frac{h}{2\pi i} \frac{\partial}{\partial q}$$

to the wave function ψ. But

$$\frac{h}{2\pi i} \left(\frac{\partial q\psi}{\partial q} - q \frac{\partial \psi}{\partial q} \right) = \frac{h}{2\pi i} \psi.$$

Application of the operator $pq - qp$ is therefore identical with multiplication of ψ by $h/2\pi i$, or, in symbolical form, $pq - qp = h/2\pi i$.

The formalism, which Schrödinger (1926) found suitable for the treatment of the wave theory of the atom, consists of the following rule. Write down the energy function $H(p, q)$ of the Hamiltonian theory as an operator, by replacing p in it everywhere by $(h/2\pi i)\partial/\partial q$; the operator corresponding to terms in p^2 is obtained by repetition, viz.

$$p^2 = \frac{h}{2\pi i} \frac{\partial}{\partial q} \cdot \frac{h}{2\pi i} \frac{\partial}{\partial q} = - \frac{h^2}{4\pi^2} \frac{\partial^2}{\partial q^2}.$$

The energy operator $H\left(\dfrac{h}{2\pi i} \dfrac{\partial}{\partial q}, q \right)$ is to be applied to a wave function ψ. Instead of the energy equation $H(p, q) - E = 0$, we obtain the differential equation

$$\left\{ H\left(\frac{h}{2\pi i} \frac{\partial}{\partial q}, q \right) + \frac{h}{2\pi i} \frac{\partial}{\partial t} \right\} \psi = 0.$$

This is called *Schrödinger's equation* for the problem.

We have thus found the formalism, according to which any mechanical problem can be treated. What we have to do is to find the one-valued and finite solutions of the wave equation for the problem. If in particular we wish to find the *stationary solutions*, i.e. those in which the wave function consists of an amplitude function independent of the time and a factor periodic in the time (standing vibrations), we make the assumption that ψ involves the time only in the form of the factor $e^{-(2\pi i/h)Et}$. If we use this in Schrödinger's equation, we find an equation in which the time does not appear, viz.

$$\left\{ H\left(\frac{h}{2\pi i}\frac{\partial}{\partial q}, q\right) - E \right\}\psi = 0.$$

We have now before us a typical "*eigenvalue*" or "*proper value*" *problem*: we have to find those values of the parameter E (the energy) for which this differential equation possesses a *solution which is one-valued and finite* in the whole domain of the variables (see Appendices XVI, XVII, XVIII, pp. 396, 398, 399). If there is only one solution (eigenfunction) apart from a constant factor, the eigenvalue is called *simple* or *non-degenerate*; if there are several solutions, it is called *degenerate*.

As an example of the method of forming the wave equation we consider a particle moving along a line (co-ordinate q) under the action of a force with the potential energy $V(q)$; then

$$H = p^2/2m + V(q).$$

From the preceding rule we obtain as the Schrödinger equation

$$\left(-\frac{1}{2m}\frac{h^2}{4\pi^2}\frac{\partial^2}{\partial q^2} + V(q) + \frac{h}{2\pi i}\frac{\partial}{\partial t}\right)\psi = 0,$$

and as the time-free equation of the stationary problem

$$(h^2/8\pi^2 m)\partial^2\psi/\partial q^2 + [E - V(q)]\psi = 0.$$

An important example is the linear oscillator, $V(q) = \frac{1}{2}fq^2$, which we shall presently discuss (p. 138).

But before attempting to find exact solutions of wave equations we shall explain why the method of quantized classical orbits described in the preceding sections leads to satisfactory results. We have, of course, to do with a special case of the correspondence principle.

If we consider the classical energy equation $H(p, q) = E$ and substitute the special expression of H for our case, we can solve this equation with respect to p and obtain

$$p(q) = \sqrt{\{2m[E - V(q)]\}}.$$

The wave equation can be written

$$\left(\frac{h}{2\pi}\right)^2 \frac{d^2\psi}{dq^2} + p^2(q)\psi = 0.$$

Let us try to solve it (Jeffreys, 1924; Wentzel, Kramers, Brillouin, 1926) by an expression of the form

$$\psi(q) = A e^{(2\pi i/h)\phi(q)}$$

where the " phase " $\phi(q)$ is a new unknown function; then

$$\left(\frac{h}{2\pi}\right)^2 \frac{d^2\psi}{dq^2} = A e^{(2\pi i/h)\phi}\left(\frac{ih}{2\pi}\frac{d^2\phi}{dq^2} - \left(\frac{d\phi}{dq}\right)^2\right).$$

If, according to the idea of correspondence, h is considered as small and the term with the factor h on the right-hand side neglected, the wave equation reduces to

$$\left(\frac{d\phi}{dq}\right)^2 = p^2(q),$$

with the solution $$\phi(q) = \int p(q)\,dq.$$

If this is substituted in the exponent of the expression for ψ, one sees that the condition for ψ being unique in the domain of the variable q consists in assuming that

$$\frac{1}{h}\oint p(q)\,dq = n$$

is an integer; for in this case the exponent increases by $2\pi in$ every time q completes a cycle of values (pp. 116, 117). But this is just Bohr's quantum condition. We have here a more precise formulation of de Broglie's considerations (p. 131).

It can be shown that this approximate solution is the first term of an asymptotic expansion with respect to powers of h, which is obtained in successive steps by taking not only the phase ϕ, but also the amplitude A as a function of q. This method can also be extended to systems with several degrees of freedom; in this way Sommerfeld's quantum conditions for multiple periodic systems (p. 118) are obtained as a first approximation.

An example of a system with three degrees of freedom is the important case of the *hydrogen atom*. Here the Hamiltonian function is

$$H = \frac{1}{2m}(p_x{}^2 + p_y{}^2 + p_z{}^2) - \frac{e^2 Z}{r},$$

and the differential equation derived from it is

$$\left\{\frac{1}{2m}\left(\frac{h}{2\pi i}\right)^2\left(\frac{\partial^2}{\partial x^2} + \frac{\partial^2}{\partial y^2} + \frac{\partial^2}{\partial z^2}\right) - \frac{e^2 Z}{r} + \frac{h}{2\pi i}\frac{\partial}{\partial t}\right\}\psi = 0.$$

If we now introduce the usual differential symbol ∇^2 for Laplace's operator $\partial^2/\partial x^2 + \partial^2/\partial y^2 + \partial^2/\partial z^2$, and pass to the time-free equation by putting

$$\psi \sim e^{-(2\pi i/h)Et},$$

we obtain the wave equation

$$\left\{\frac{h^2}{8\pi^2 m}\nabla^2 + E + \frac{e^2 Z}{r}\right\}\psi = 0.$$

This is a wave equation in three-dimensional space, whose solutions we shall investigate later. It may make matters easier for the reader if we begin with corresponding problems in one and two dimensions; and for the sake of perspicuity we shall take our examples from classical mechanics (acoustics).

An example of this kind, in one dimension, is that of the *vibrating string*. Its differential equation follows from the theory of elasticity:

$$\frac{\partial^2 \psi}{\partial x^2} - \frac{1}{c^2}\frac{\partial^2 \psi}{\partial t^2} = 0.$$

Here c is a constant depending on the mechanical conditions (thickness of the string, tension); it represents in fact the velocity of a wave running along the string, since any function of $(x \pm ct)$ is a solution. We consider now standing waves periodic in time with $\psi \sim e^{2\pi i \nu t}$; then the equation reduces to

$$\frac{d^2 \psi}{dx^2} + \lambda\psi = 0, \quad \lambda = \left(\frac{2\pi\nu}{c}\right)^2.$$

The eigenvalue parameter λ is here, as in all classical problems, proportional to the square of the frequency ν, whereas in wave-mechanical problems it represents in general the energy $E = h\nu$, and is therefore proportional to ν itself. The solutions of this differential equation are

$$\psi(x) = a \cdot \begin{cases} \cos\sqrt{\lambda}x, \\ \sin\sqrt{\lambda}x. \end{cases}$$

On account of the boundary conditions $\psi(0) = 0$ and $\psi(l) = 0$ (string of length l with ends fixed) the cosine vibration drops out at once as being inconsistent with the first condition. But even the sine vibration is not a solution of the boundary value problem, unless $l\sqrt{\lambda}$ is an exact integral multiple of π, so that ψ vanishes when $x = l$. It is only for definite values of λ (the proper values) that we obtain possible forms of vibration; these are given in fact by

$$\psi(x) = \sin\left(\frac{n\pi x}{l}\right); \quad \lambda = \left(\frac{n\pi}{l}\right)^2.$$

The vibration with $n = 1$ represents the fundamental, that with $n = 2$ the first harmonic (octave), and so on. In the vibrational process, at definite points on the string there are *nodes*, that is to say, points which remain at rest throughout the vibration

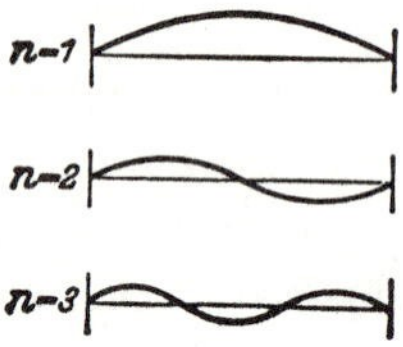

Fig. 19.—Vibrational forms of a string fixed at both ends. The fundamental ($n = 1$), and the first two harmonics ($n = 2$ and $n = 3$).

(fig. 19). The number of nodes is determined by the parameter n, and in fact is clearly equal to $n - 1$.

As an actual example of such a system with one degree of freedom, in the quantum theory, we consider the *harmonic oscillator* the wave equation for which we have already mentioned above (p. 135). Its solution is dealt with in Appendix XVI (p. 396). Instead of Planck's energy levels $E = nh\nu$, wave mechanics, exactly like matrix mechanics (see Appendix XV, p. 392), gives the energy terms

$$E = (n + \tfrac{1}{2})h\nu.$$

The ground state (i.e. as regards energy, the lowest state, $n=0$) accordingly possesses a finite energy $E = h\nu/2$ (zero-point energy). We shall make use of this later (Chap. VIII, § 7, p. 275 ; Appendix XL, p. 471).

We next consider, as a two-dimensional example of a mechanical vibration, the vibrating *circular membrane*, the differential equation of which for standing waves reduces to

$$\left\{\frac{\partial^2}{\partial x^2} + \frac{\partial^2}{\partial y^2} + \lambda\right\}\psi = 0.$$

Here also the proper value parameter λ depends on the nature of the plate and in effect represents the square of the frequency. The differential equation can be easily solved in polar co-ordinates (Appendix XVII, p. 398). In this case again we obtain possible forms of vibration for certain definite values of λ only. Instead of nodes, we have here nodal lines, of which indeed there are two kinds:

1. Lines for which $r =$ const.; these are defined by the radial ordinal number, or " quantum number ", $n = 1, 2, \ldots$.
2. Lines for which $\phi =$ const., corresponding to the azimuthal quantum number $m = 0, 1, 2, \ldots$.

Fig. 20 shows several examples. The signs $+$ and $-$ in the different regions indicate that adjacent regions are always vibrating in opposite phases.

Fig. 20.—Some vibrational forms of a circular membrane fixed at the circumference; the number of radial nodal lines is here (in disagreement with the custom in wave mechanics) called n, that of the azimuthal nodal lines m; n and m are the " quantum numbers " of the state of vibration.

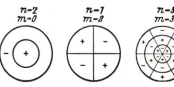

The solution can also be obtained in polar co-ordinates for the *hydrogen atom*, as a three-dimensional quantum problem; this is shown in Appendix XVIII, p. 399). In this problem, it should be added, we cannot speak of ordinary boundary conditions, since the domain over which the independent variables range is the whole of three-dimensional space. Instead of boundary conditions, we have now a rule with regard to the behaviour of the wave function at infinity. The natural condition to impose is that the wave function should vanish at infinity " more strongly " than $1/r$. This follows from the statistical interpretation of the square of the amplitude of the wave function, as the probability of the electron being found at a definite point of space (§ 7, p. 145). The condition is equivalent to this, that the electron must always be at a finite distance.

Taking this " boundary condition " into account, we obtain solutions of the wave equation which correspond to a bound electron (elliptic orbit in the Bohr atom), only for definite discrete values of E. In this way we find as proper values exactly the Balmer terms with the correct Rydberg constant, $E = -Rh/n^2$. Here n is the principal quantum number. Besides this, the azimuthal quantum number l and the magnetic quantum number m also make their appearance. The number of nodal surfaces $r = $ const. is $n - l - 1$; for a given n, l can therefore be any integer from 0 to $n - 1$; as for m, it can take all values from $-l$ to $+l$.

When the relativistic correction for the mass is taken into account, the energy depends on l also. Moreover, as follows from the form of this dependence, $l + 1$ corresponds to the Bohr quantum number k, so that our nomenclature for the terms (see § 2, p. 124) has now to be understood as follows:

$$l = 0 \quad 1 \quad 2 \quad \ldots$$
$$s \quad p \quad d \quad \ldots \text{term.}$$

In a *magnetic field*, E depends on m also, indeed the extra term $m\nu_L h$ occurs as an addition to the energy, exactly as in Bohr's theory. Wave mechanics, so far as we have developed it up to the present, yields only the normal Zeeman effect (as above, § 2, p. 122). To the directional quantisation of Bohr's theory there corresponds here the finite number of values of m, i.e. of energy levels in the magnetic field; there are in fact $2l + 1$ of these, in place of each term which occurs when there is no magnetic field. The splitting of the terms in an *electric field* (Stark effect) is correctly described by wave mechanics, qualitatively and quantitatively (Schrödinger, 1926).

The states of the hydrogen atom, as considered so far, correspond obviously to the elliptic orbits of the old Bohr theory; in both cases the electron remains at a finite distance. But in classical mechanics there are also hyperbolic orbits; what corresponds to these in quantum mechanics? Clearly, solutions of the wave equation which do not disappear at infinity. In order to obtain them we must give up the boundary condition—vanishing of ψ at infinity—and look for solutions which at a great distance from the nucleus behave approximately like plane waves. In point of fact there are such solutions, and that for all positive values of the energy. Physically interpreted, they describe what happens when an electron coming from infinity passes near the nucleus and is deflected by it. It can actually be proved that Rutherford's scattering formula is strictly valid in wave mechanics also; we shall return to this again (§ 7, p. 145; Appendix XX, p. 406). The continuous spectrum of energy reveals itself also in other ways. According to Bohr's principles any transition between two states corresponds to an act of emission or absorption of light. If one or both of the two states belong to a continuous region of energy values, the corresponding frequency spectrum of light will also be continuous. This explains the appearance of continuous X-ray spectra produced by impact of electrons on a metal target (p. 150).

We add a few remarks on the wave mechanics of *many-body problems*. Here of course we are concerned with the solution of a wave equation in many-dimensional space; thus the calculation of the helium spectrum needs as many as six co-ordinates, and that of the lithium spectrum nine. It is clear that in these cases an exact solution cannot be expected, so that we must be content with an approximate solution of the problem. The methods of a highly developed *perturbation theory* enable us to push this approximation as far as we please; the labour involved, however, increases without limit with the order of the approximation. The lowest terms of many atomic configurations have been successfully calculated by this method. The first atoms treated were He, Li$^+$ and Li (Hylleraas, 1930) with results in excellent agreement with experiment.

Deductions which are quite exact can be made from any *properties of symmetry* which the wave function must possess in virtue of the symmetrical character of the problem. The most important of these properties of symmetry is the one which is involved in the complete equivalence of the electrons, and their consequent interchangeability; the wave function must of course be the same, whether, say, the first electron is situated in the K shell and the second in the L shell, or the second in the K shell and the first in the L shell. This leads to general rules for the tabulation of the terms in atoms with several radiating electrons. Still, the results thus obtained are not immediately comparable with experiment, since in wave mechanics, so far as developed above, an essential principle is lacking, which was discovered by Pauli, and which will come before our notice in the next chapter.

5. Angular Momentum in Wave Mechanics.

In Bohr's theory, angular momentum was of special importance for the classification of the spectral lines and the systematic arrangement of the terms; it was found that it corresponded to a quantum number k. This again led to the idea of the quantisation of direction, the experimental confirmation of which by Stern and Gerlach's experiment is perhaps the most impressive evidence we have of the fundamental difference between classical and quantum mechanics. The question is thus raised: how does wave mechanics deal with these things? Is it capable of taking account of quantisation of direction in a natural way?

In wave mechanics, angular momentum, just like linear momentum, has a differential operator corresponding to it, the components of the operator being

$$\mathbf{L}_x = yp_z - zp_y = \frac{\hbar}{i}\left(y\frac{\partial}{\partial z} - z\frac{\partial}{\partial y}\right),$$

$$\mathbf{L}_y = zp_x - xp_z = \frac{\hbar}{i}\left(z\frac{\partial}{\partial x} - x\frac{\partial}{\partial z}\right),$$

$$\mathbf{L}_z = xp_y - yp_x = \frac{\hbar}{i}\left(x\frac{\partial}{\partial y} - y\frac{\partial}{\partial x}\right).$$

What is the significance of these operators? The electronic state is given by the wave function ψ. In order to decide the question whether definite values of the components of angular momentum round the three co-ordinate axes belong to this state, we have to " apply " the above operators, according to the rules of wave mechanics, to the wave function ψ; i.e. we perform the differentiations involved in the operators. There are now two possibilities: either this operation reproduces the wave function except for a constant factor, or it does not. In the first case, the wave function is called a proper function of the equation of angular momentum

$$\mathbf{L}_z\psi = m\hbar\psi,$$

and the state represented by the proper function therefore possesses a definite angular momentum round the z-axis, whose value is given by the " proper value " m, here m is an ordinary number (not, like $\mathbf{L}_z$, an operator). If, however, when we apply the operator $\mathbf{L}_z$ to the wave function we obtain another function, which does not agree with the wave function except for a constant factor, that is to say, if ψ is not a proper function of the equation of angular momentum, this means that the electronic state in question is not associated with a fixed value of the angular momentum about this axis. In the case of the proper functions given in Appendix XVIII (p. 399), by our choice of the polar axis (z-axis) we have specially distinguished this axis from the beginning. According to Bohr's theory, the component angular momentum round this axis must be quantised. Now wave mechanics does in fact show that the proper values of the z-component of the angular momentum are integral multiples of $\hbar$. Thus, on introducing spherical polar co-ordinates

$$\mathbf{L}_z = \frac{\hbar}{i}\left(x\frac{\partial}{\partial y} - y\frac{\partial}{\partial x}\right) = \frac{\hbar}{i}\frac{\partial}{\partial \phi},$$

and on applying this operator to the proper function of the state characterized by the quantum numbers n, l, m, we obtain at once on account of the way in which ϕ is involved ($e^{im\phi}$)

$$\mathbf{L}_z\psi_{nlm} = \frac{\hbar}{i}\frac{\partial}{\partial \phi}\psi_{nlm} = m\hbar\,\psi_{nlm},$$

so that the proper values of the component angular momentum $\mathbf{L}_z$ are actually $m\hbar$. In the other two co-ordinate directions, in the case before us, we may easily satisfy ourselves that we do not get proper values, and therefore do not get definite values of $\mathbf{L}_x$ and $\mathbf{L}_y$. On the other hand, the value of the resultant angular momentum for an atom which can rotate freely (in the absence of an external field) is in all circumstances quantised. For brevity we refer to Appendix XIX (p. 404) for the requisite calculations, and give only the result here: the square of the value of the resultant

$$\mathbf{L}^2 = \mathbf{L}_x{}^2 + \mathbf{L}_y{}^2 + \mathbf{L}_z{}^2$$

is, according to wave mechanics, $l(l+1)\hbar^2$, not, as in Bohr's theory, $l^2\hbar^2$. This special feature is characteristic of the whole wave mechanics of the atom, and in the next chapter we shall often meet it. Moreover, as we shall also see there, the splitting patterns of the terms in the anomalous Zeeman effect give direct confirmation of the fact that the square of the angular momentum is proportional to $l(l+1)$, and not to l^2. If, however, we disregard this difference, we can apply in wave mechanics the representation of the angular momentum by a vector diagram, known from Bohr's theory; we therefore here also represent the total angular momentum by a vector l, with regard to which we must note once for all that it has the absolute value $\sqrt{l(l+1)}$. In the case when the z-direction is specially marked out, say by the fact that an (extremely) weak magnetic field H acts along it, then the component in this direction of this vector angular momentum is capable of integral values m only (in units $\hbar$), and this state of matters continues to hold even in the limiting case $H \to 0$. Here, and especially in the next chapter, for the sake of greater perspicuity we take over the vector representation of the angular momentum precessing round the specially distinguished axis, and having a component in the direction of the axis which can take only integral values.

We expressly emphasize, however, that this idea cannot in this way without more ado be brought into harmony with the conceptual structure of wave mechanics. For the latter purpose, and particularly for the proof that two angular momenta may be compounded vectorially in wave mechanics in the same way as in Bohr's semi-classical theory, higher mathematical methods are required, especially so-called group theory. For this reason we cannot go further into these questions at this point. It may be mentioned, however, that further development has led to the electron being regarded, not as a particle, defined by three space co-ordinates, but as a top-like structure, possessing an angular momentum of its own. This " spin " theory of the electron will be dealt with later (Chap. VI, p. 156).

6. Parity.

We have seen (p. 135) that in dealing with the Schrödinger wave equation, we are interested in those solutions which are one-valued and finite, the so-called eigenfunctions of the equation. Now it is found that these solutions in general have a definite symmetry with respect to the reflection of the co-ordinate axes through the origin provided the Hamiltonian is symmetric under this transformation. Those functions which remain unchanged under this operation are said to have *even parity* and those which change sign are said to have *odd parity*.

That only a possible change of sign results from such a transformation can be seen in a simple case as follows. We will suppose that the Hamiltonian $H\left(\frac{h}{2\pi i}.\frac{\partial}{\partial q}, q\right)$ of a system is symmetric, i.e.

$$H\left(\frac{h}{2\pi i}\frac{\partial}{\partial q}, q\right) = H\left(-\frac{h}{2\pi i}\frac{\partial}{\partial q}, -q\right).$$

Consider a typical eigenfunction $\psi(q)$ of H with eigenvalue E, thus

$$\left\{H\left(\frac{h}{2\pi i}\frac{\partial}{\partial q}, q\right) - E\right\}\psi(q) = 0.$$

Replacing q by $-q$ and using the symmetry of H we have

$$\left\{ H\left(\frac{h}{2\pi i} \cdot \frac{\partial}{\partial q}, q \right) - E \right\} \psi(-q) = 0.$$

Thus $\psi(q)$ and $\psi(-q)$ both satisfy the same differential equation and (see p. 135) provided that we are dealing with a non-degenerate eigen-value E it follows that the two solutions must be related by a constant multiple. Thus

$$\psi(q) = A\psi(-q).$$

Replacing q by $-q$,

$$\psi(-q) = A\psi(q) = A^2\psi(-q).$$

Therefore $A^2 = 1$ and $A = \pm 1$ showing that $\psi(q)$ and $\psi(-q)$ are identical apart from a possible sign difference. For example, surface harmonics $Y_l^{(m)}$ (App. XVIII, p. 399) which are angular momentum eigenfunctions can easily be shown to have even or odd parity according as l is even or odd.

It can furthermore be demonstrated that so long as the Hamiltonian is symmetric under space reflection then the parity of an eigenfunction does not change with time. Parity is said to be *conserved*. We shall see later that although parity is usually conserved there are certain weak processes which have to be described by a Hamiltonian which is not symmetric so that parity is not conserved.

7. The Statistical Interpretation of Wave Mechanics.

In conclusion, we have still to consider the *meaning of the wave function* itself; so far, we have obtained it as a mere by-product, so to speak, in the search for proper values. But in a vibrational process knowledge of the amplitude is at least as important as knowledge of the proper frequency; similarly, it is to be expected that in wave mechanics great physical significance attaches to the wave function ψ, or rather, to the square of its modulus, since of course the instantaneous value of the oscillating function itself cannot play any part, on account of the high frequency. The reason for taking the square of the modulus is that the wave function itself (because of the imaginary coefficient of the time derivative in the differential equation) is a complex quantity, while quantities susceptible of physical interpretation must of course be real.

We have already mentioned the interpretation of the wave function given by Born (p. 95). Let the proper function corresponding to any state of an electron (say) state be ψ_E; then

$$| \psi_E |^2 \, dv$$

is the probability that the electron (regarded as a corpuscle) is in the volume element dv.

This interpretation is almost self-evident, if we consider, not the quantum states proper (with discrete, negative energy-values), but the states of positive energy, which correspond to the hyperbolic orbits of Bohr's theory. We have then to solve a wave equation

$$\left\{ \frac{\hbar^2}{2m} \nabla^2 + E - V(r) \right\} \psi = 0,$$

where, instead of the Coulomb potential

$$- \frac{e^2 Z}{r},$$

$V(r)$ is written somewhat more generally, in order to take into account possible screening of the action of the nucleus by firmly bound electrons. For particles entering the atom with very high speeds, and therefore very large energy E, $V(r)$ will only come into consideration as a small " disturbance ", or " perturbation " ; if we neglect it, we have as a solution of

$$\left\{ \frac{\hbar^2}{2m} \Delta^2 + E \right\} \psi = 0$$

the plane wave $\psi = e^{(i/\hbar)pz}$, where

$$E = \frac{p^2}{2m},$$

and the direction of the wave normal is assumed arbitrarily as parallel to the z-axis. The disturbance can be taken into account to a first approximation by substituting the plane wave expression for ψ in the term $V(r)\psi$ in the original equation; we have then to find a solution (Born, 1926) of

$$\left(\frac{\hbar^2}{2m} \nabla^2 + E \right)\psi = V(r)e^{(i/\hbar)pz},$$

corresponding to a wave receding from the nucleus. It is perfectly clear, especially in the light of the analogy with the scattering of light waves, that the intensity of the secondary wave gives the number of electrons, belonging to a given incident beam, which are deflected in a definite direction; and this in effect implies the statistical interpretation as stated. A more rigorous investigation will be found in Appendix XX (p. 406); it is there shown how the intensity, i.e. the number of particles in a current, per square centimetre and per second, is to be defined. In particular, if $V(r)$ is chosen so as to correspond to a (screened) Coulomb field, the result obtained (Wentzel, 1926) is precisely Rutherford's law of scattering (p. 62). As a matter of fact it is only for fast particles that this proof is valid; but it can be shown (Gordon, Mott, 1928) that the result is strictly correct. The exact solution differs from the approximate one only in terms which have no influence on the current intensity. This is very remarkable, and is analogous to the fact that, in the Coulomb field, the discrete term values, as given by wave mechanics, are in agreement with the values calculated with the help of quantized classical orbits.

If we extend this statistical interpretation to the case of discontinuous states, and if E_n is the energy and ψ_n the proper function of such a state, then $|\psi_n|^2 dv$ is the probability that an electron will be found precisely in the volume element dv; this holds in spite of the fact that the experiment if carried out would destroy the connexion with the atom altogether. The probability of finding the electron somewhere or other in the atom must according to this interpretation be equal to 1; that is to say, the factor in the solution of the (homogeneous) wave equation, which in the first instance is quite indefinite, must be determined so that the equation

$$\int |\psi_n|^2 dv = 1$$

is satisfied. This "normalizing integral", which has no meaning except in the domain of discontinuous energy values, plays a remarkable part in this respect, that it does not vary with the time, even when we do not confine ourselves to stationary states alone, but substitute for ψ any solution at all of the wave equation, in the form containing the time.

We speak frequently of a *density distribution* of the electrons in the atom, or of an *electronic cloud* round the nucleus. By this we mean the distribution of charge which is obtained when we multiply the probability function $|\psi_n|^2$ for a definite state by the charge e of

the electron. From the standpoint of the statistical interpretation its
meaning is clear; it can be represented pictorically in the way shown
in fig. 21, Plate X (opposite). The figures represent the projections
(shadows) of the electronic clouds in various states; the positions of
the nodal surfaces can be recognized at once.

8. Emission and Absorption of Radiation.

From another point of view, the statistical interpretation of wave
functions suggests how the radiation emitted by the atom may be
calculated on wave-mechanical principles. In the classical theory this
radiation is determined by the electric dipole moment p of the atom,
or rather by its time-rate of variation. By the correspondence principle,
this connexion must continue to subsist in the wave mechanics. Now
the dipole moment p is easily calculated by wave mechanics; if we
adhere to the analogy with classical atomic mechanics, it is given by

$$p = e\int |r|\,|\psi_n|^2\,dv = e\int r\psi_n^*\psi_n\,dv,$$

where r stands for the radius vector from the nucleus to the point of
integration, or field point. (As usual, the asterisk denotes replacement
by the conjugate complex quantity.) The integral represents of course
the position of the " electrical centroid of the electronic cloud ". Now,
as is easily proved, this integral vanishes for all states of an atom,
so that the derivative of the dipole moment vanishes, and accordingly
the emitted radiation also; that is, a stationary state does not radiate.
This gives an explanation of the fact—unintelligible from the standpoint of Bohr's theory—that an electron which is revolving about the
nucleus, and according to the classical laws ought to emit radiation
of the same frequency as the revolution, can continue to revolve in
its orbit without radiating. In wave mechanics this absence of emitted
radiation is brought about by the fact that the elements of radiation,
emitted on the classical theory by the individual moving elements of
the electronic cloud, annul each other by interference.

But now, in analogy with the probability function or density
function $\psi_n^*\psi_n$ of a definite state, as defined above (p. 147), we can
form, in the first instance in a purely formal way, the " transition
density " $\psi_m^*\psi_n$ corresponding to a transition from a state n to another
state m; it corresponds physically to the well-known " beat phenomena ", which occur when two vibrations with neighbouring frequencies are superimposed on each other; its rhythm is given by the
time factor

$$e^{-(i/h)(E_n - E_m)t}$$

PLATE X

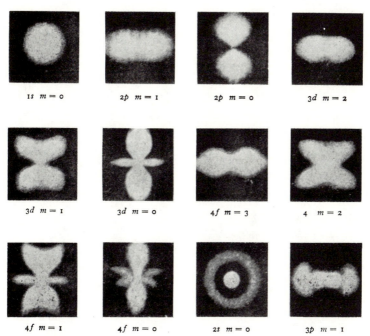

1s $m=0$	2p $m=1$	2p $m=0$	3d $m=2$
3d $m=1$	3d $m=0$	4f $m=3$	4 $m=2$
4f $m=1$	4f $m=0$	2s $m=0$	3p $m=1$

Ch. V, Fig. 21.—The figures show graphically in the form of silhouettes the distribution of charge which, according to Schrödinger's theory, must exist in the electronic cloud. The nodal surfaces corresponding to the quantum numbers stated can be clearly seen, especially in the s-states, and in the states for which m is small (see p. 148).

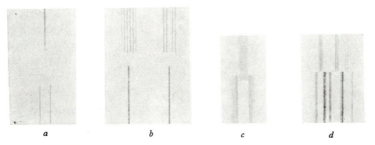

a	b	c	d

Ch. VI, Fig. 2.—Splitting patterns in a magnetic field. (a) Normal Lorentz triplet for the Cd line 6439 (the vibrations in the lower lines are perpendicular to the magnetic lines of force, those in the upper lines parallel to them). (b) Anomalous Zeeman effect for the Na D-lines (below without, above with magnetic field). (c) and (d) show Zeeman effects for Cr lines (above, vibrations parallel to field; below, perpendicular to it). (See p. 157.

of the transition density, and the beat frequency is found from the difference of the energies of the two states:

$$\nu_{nm} = \frac{E_n - E_m}{h}.$$

We now form also, in an analogous way to this, the dipole moment corresponding to the transition from n to m

$$p_{nm} = e \int r \psi_m^* \psi_n \, dv = e r_{nm} e^{-2\pi i \nu_{nm} t}.$$

It oscillates with the beat frequency given above. The quantity r_{nm} is called the *matrix element of the vector co-ordinate* r; as Schrödinger has shown, it is identical with the matrix element which, in Heisenberg's co-ordinate matrix, occupies the nth row and the mth column (Appendix XV, p. 392, see also Appendix XXV, p. 426).

We can now apply the correspondence principle, using the classical expression for the energy emitted by an oscillator of moment $p = es$, namely (Appendix VIII, p. 376)

$$I = \frac{2}{3c^3} \mid \ddot{p} \mid^2.$$

But we have to take into account that each quantum state n has two neighbour states, one above, one below, which, for high quantum numbers, differ by the same amount of energy $h\nu_{nm}$. Hence if we replace p by the matrix element p_{nm}, we must at the same time multiply by the factor 2, so that the radiation emitted per unit time is given by (see also Appendix XXVIII, p. 439).

$$I = \frac{4}{3c^3} \mid \ddot{p}_{nm} \mid^2 = \frac{4}{3} \frac{e^2}{c^3} (2\pi \nu_{nm})^2 \mid r_{nm} \mid^2.$$

Thus we obtain the emitted radiation, in purely correspondence fashion, by the rules of classical electrodynamics. From this it follows automatically that in the spectrum only those lines can occur whose frequency agrees with a beat frequency between two states of the atom. These are exactly the lines to explain which Bohr had to introduce, as a fundamental postulate of his theory, the radiation condition

$$h\nu_{nm} = E_n - E_m,$$

which is perfectly unintelligible from the classical standpoint. It is not to be understood, however, that both states n and m are excited simultaneously when emission occurs; it is rather a matter of their virtual presence. In point of fact, in order that a spectral line may be spontaneously emitted, the upper state must be excited in some way or other; the emission is then a companion process to the jump to the lower quantum state (vibration of an associated virtual resonator).

There is another aspect of the problem of emission of radiation which is dealt with more fully in Appendix XXVIII (p. 439), namely that of stimulated and spontaneous emission. Consider the emitting systems to be immersed in a sea of electromagnetic radiation, such that the energy density of this radiation in the frequency range ν, $\nu + d\nu$ is $u_\nu d\nu$. Then it can be shown that if the system is excited to the state n by some means, the probability of a transition to the state m is enhanced over and above the spontaneous transition probability by an amount which depends on the density of the electromagnetic radiation. This additional effect is known as *induced emission* and turns out to be proportional to $u_{\nu_{nm}}$ that is to the amount of radiation present, having the beat frequency between the two states n and m. Quantitatively, this result can be expressed as follows. If P_{nm} is the probability of a transition from the state n to state m, then

$$P_{nm} = A_{nm} + u_{\nu_{nm}} B_{nm}$$

where A_{nm} and B_{nm} are coefficients (the Einstein A and B coefficients) depending on the atomic states involved. They are related by $B_{nm} = \dfrac{c^3}{8\pi h \nu_{nm}^3} A_m$ (see Appendix XXVIII, p. 439). It should also be noted that if the atomic system is in the state m, then the probability of *absorption* of radiation, so that a transition from state m to n is made, is given by

$$P_{mn} = u_{\nu_{nm}} B_{mn}$$

with $B_{mn} = B_{nm}$.

The intensity of the spectral line is the product of two factors, the number of excited atoms and the radiating strength I of an individual atom, which we have just calculated. Thus, with regard to the conditions of excitation of lines, those ideas in Bohr's theory which are brilliantly verified by experiments are just the ideas which are retained in their entirety in wave mechanics. The latter theory adds a more exact calculation of the intensity I of the individual elementary act, depending on evaluation of the integrals occurring in the matrix elements, while on this question Bohr's theory could only make a few approximate statements, with the help of the correspondence principle. As is shown in Appendix XXI (p. 409), the evaluation of the matrix elements in the case of the hydrogen atom leads to the selection rules

$$\Delta l = \pm 1 \text{ and } \Delta m = 0, \pm 1;$$

that is, all matrix elements vanish which do not correspond to one of the transitions mentioned, and with them vanishes also the radiation of the corresponding frequencies. Application of this to hydrogen-like atoms (such as the halogens) gives the theoretical foundation for the facts that, for instance, transitions occur between s and p terms or between p and d terms, but not between s and d terms or between p and f terms.

Besides the discontinuous states there are also states forming a continuous range (with positive energy); they correspond to the hyperbolic orbits of Bohr's theory (p. 140). The jumps from one hyperbola to another or to a stationary state give rise to the emission of the continuous X-ray spectrum emitted when electrons are scattered or caught by nuclei. The intensity of this spectrum has been calculated by Kramers (1923) from the standpoint of Bohr's theory by a very ingenious application of the correspondence principle. His results can be confirmed and improved by evaluating the matrix elements belonging to two states, one or both of which are in the continuous range (Oppenheimer, 1926).

The method just explained for the calculation of radiation emitted (or absorbed) by atoms can be put on a rigorous basis. It is indeed a special case of a very general theory of transitions between stationary states of a given system induced by external forces, or better, by the coupling to another system with a very dense distribution of stationary states. We obtain for the *transition probability* per unit of time between the states n and m of the given system under the influence of the external system with the density $\rho(\epsilon)$ of states as a function of the energy ϵ

$$P_{mn} = \frac{2\pi}{\hbar} \mid H'_{mn} \mid^2 \rho(\epsilon),$$

where H'_{mn} is the matrix element of the coupling energy corresponding to the two states (Appendix XXVII, p. 435). Dirac has shown how this formula can be applied to the case of the emission, absorption, and scattering of radiation by an atom, by resolving the electromagnetic field into a large number of oscillators of different energies. The result confirms the formula given above for the energy emitted. Many other applications of the transition formula have been made; we shall apply it later (Chap. IX, § 7, p. 334) to the β-decay of atomic nuclei.

In these calculations based on wave mechanics matrix elements appear in a most natural way. In fact the general mathematical formalism of quantum mechanics (Appendix XXV, p. 426) shows that " matrix mechanics ", briefly outlined above, and " wave mechanics " are completely equivalent, two forms of the same content. The first is stressing the similarity of the laws of the new mechanics with those of classical mechanics of particles, while the second operates with the wave picture. Both are compatible only on account of the statistical interpretation, the philosophical import of which has already been discussed in Chap. IV, § 7 (p. 94). It consists in the recognition that the duality of the wave picture and the corpuscular picture leads to mutually exclusive and complementary descriptions of experimental situations, the relation of which is quantitatively defined by the principle of uncertainty. Here we have only one more important point to mention. The uncertainty relations, which we have obtained simply by contrasting with one another the descriptions of a process in the language of waves and in that of corpuscles, may also be rigorously deduced from the formalism of quantum mechanics—as exact inequalities, indeed; for instance, between the co-ordinate q and momentum p we have the relation

$$\Delta q \, \Delta p \geqslant \frac{\hbar}{2},$$

if Δq and Δp are defined as root-mean-squares (see Appendix XXVI, p. 433).

CHAPTER VI

Spin of the Electron and Pauli's Principle

1. Alkali Doublets and the Spinning Electron.

The great success of Bohr's theory and especially of wave mechanics shows that in the interpretation of atomic processes we are on the right road to knowledge. The theory, however, as we have repeatedly emphasized in the preceding chapter, is still incomplete. In particular, explanations cannot yet be given of the anomalous Zeeman effect, the structure of the shells in the atom, &c. The wave mechanics of the atom, in its form as developed up to this point, still needs a far-reaching extension by the introduction of new ideas and hypotheses, which will form the subject of the present chapter.

The starting-point is given by the observation that the lines of the principal series in the alkalies are double. A well-known example is the D-line of sodium, whose doublet nature can be observed even with simple spectroscopic appliances. The splitting of the line is rather considerable—it amounts to 6 Å.; the two components are denoted by D_1 and D_2, their wave-lengths being $\lambda = 5896$ Å. for D_1, and $\lambda = 5890$ Å. for D_2. The term analysis of the alkali spectra, to which the spectrum of sodium belongs, gives the definite information about them that the

$$s\text{-} \qquad \text{terms } (l = 0) \qquad \text{are simple,}$$
$$p\text{-}, d\text{-}, \ldots \text{ terms } (l = 1, 2, \ldots) \text{ are double.}$$

This experimental result cannot be explained either from the standpoint of Bohr's theory, or from that of wave mechanics so far as developed above. We have investigated above (p. 140) the most general motion of an electron in the atom, on the basis of its three degrees of freedom, and have arrived at the conclusion that the motion is completely determined and described by the three quantum numbers n, l, m. Any further splitting of the energy terms than that conditioned by these quantum numbers is therefore unintelligible, so long as we adhere to the idea that the motion of the electron is at most triply periodic.

153

Under the compulsion of experiment, Uhlenbeck and Goudsmit (1925) put forward the following bold hypothesis. If the electron could be regarded as a structure with finite extension, then, like every extended system, it would possess three rotational degrees of freedom besides its three translational ones. It would then have an angular momentum and, because of the rotating charge, also a magnetic moment. Now the idea of a finite extension of the electron was discredited at that time, as we have seen in previous discussions (Chap. III, § 2, p. 57). In spite of this it was suggested by these authors that the experimental facts could be understood by ascribing a mechanical moment (angular momentum) and a magnetic moment to the electron, in the same formal way as a mass m and a charge e are ascribed to it. With regard to the magnitude of the magnetic and the mechanical moments, experiment must of course decide in the first instance; afterwards, we can try to deduce these magnitudes theoretically.

This property of the electron, in virtue of which it has a mechanical and a magnetic moment, is called its *spin*.

The magnitude of the *mechanical moment* follows immediately from known facts about the spectra of the alkalies. The angular momentum of the electron must of course, like every angular momentum, be quantised, and the same holds good for its component in a specially distinguished direction (external magnetic field). If then the value of the mechanical spin-moment is s (in units $h/2\pi$), there must, by the rules for the quantisation of direction, be $2s + 1$ possible " settings " (i.e. orientations or inclinations) with respect to the special direction; the individual components of s, which we call σ, differing from each other by unity. To see this, consider the analogous relations in the Bohr atom, in which the plane of an orbit with angular momentum l has precisely $2l + 1$ possible settings with respect to the special direction, these settings being characterized by the components m of l in that direction (see fig. 10, p. 121). This extension of the concept of orbital angular momentum to spin angular momentum is not only justified by the fact that the consequences deduced from it are found to be in full accord with the facts, but can also be derived from theory. As regards this point, we may refer here once again to what was said in § 5, p. 144, about the applicability of the classical vector-model to the description of atomic states in wave mechanics. In this chapter we shall be concerned almost exclusively with the conceptual scheme, and shall therefore use the pictorial vector-model, postponing the wave-mechanical treatment of the spinning electron until the close of the chapter (§ 8, p. 187).

As has just been brought out, the spin-moment s of the electron must have $2s + 1$ possible settings with respect to a specially distinguished direction.   Now experiment shows that the terms of sodium, excepting the s-terms, are double.   This

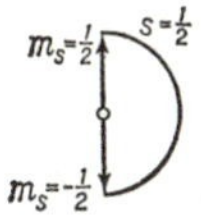

$m_s=\frac{1}{2}$ ⌐ $s=\frac{1}{2}$

$m_s=-\frac{1}{2}$ ⌐

Fig. 1.—Setting of the spin with respect to a specially distinguished direction; there are two possible settings, parallel and antiparallel to this direction.

compels the conclusion that the spin-moment has only two possible settings (fig. 1), unless we are altogether wrong in assuming that this term-splitting is determined by the spin.   Hence we must have $2s + 1 = 2$, or

$$s = \tfrac{1}{2}$$

(in units $\hbar$).   The two possible settings have then the components

$$m_{s1} = +\tfrac{1}{2}, \ m_{s2} = -\tfrac{1}{2}.$$

The occurrence of half-integers here as " quantum numbers " contradicts, at first sight, our ideas regarding the quantisation of angular momenta.   It is to be noted, however, that the idea of a rotating electron, extended in space, possesses merely heuristic value; we must be prepared, on following out these ideas, to encounter difficulties. (For instance, a point at the surface of the electron would have to move with a velocity greater than that of light, if such values as have been determined experimentally for angular momentum and magnetic moment are to agree with those calculated by the classical theory.) The use of half-integral components of angular momentum for the spin consistently leads, however, to results which are in complete agreement with the experimental facts.   On the other hand, the wave mechanics of the spinning electron, in the form given to it by Dirac, leads automatically to this half-integral property, merely as a consequence of the conditions of linearity and relativistic invariance, without any subsidiary assumption.

An electron revolving about the nucleus possesses an orbital moment l; besides this, it has the mechanical spin-moment s.   The question arises: how are these two moments to be combined with each other?   Bohr's theory would reply that they must be combined by the method of vector addition.   This same rule for the composition of l and s holds good according to wave mechanics, although the proof (Wigner, v. Neumann, 1927) requires advanced mathematical methods (group theory). They therefore combine vectorially, giving a resultant (or total) moment j (in units $\hbar$), so that

$$\overrightarrow{j} = \overrightarrow{l} + \overrightarrow{s}.$$

After Sommerfeld, j is sometimes called the " *inner* " *quantum number*; it represents the total mechanical moment of the atom. Admissible values of j must differ by integers. Since $s = \frac{1}{2}$, the only possibilities are

$$j_1 = l + \tfrac{1}{2}, \quad j_2 = l - \tfrac{1}{2};$$

j is therefore half-integral in this case. For each l-value there are accordingly two possible values for the total mechanical moment, so that the corresponding terms are double. The s-terms alone ($l = 0$) form an exception; they are always single, for in this case the only allowable value is $j = s = \frac{1}{2}$, since j, the total angular momentum, must always be positive. The double possibility for the setting of the spin with respect to the orbit is equivalent to a splitting of the energy terms, on account of the magnetic coupling of spin and orbit. The magnitude of the splitting is in fact given directly by the energy which is needed in order to turn the spin round, from one setting relative to l in the magnetic field of the orbit, into the other setting.

We take as an example the case with which we began (p. 153), that of the sodium D-lines. The term analysis shows that the upper state is a p-term, while the lower is an s-term. The former is double, corresponding to the two possible values of the total angular momentum $j = \frac{1}{2}$ and $j = \frac{3}{2}$; the lower term, being an s-term, is single ($j = \frac{1}{2}$). The D_1-line corresponds to the transition from the p-term with the inner quantum number $j = \frac{1}{2}$, the D_2-line to the transition from the term with $j = \frac{3}{2}$.

The rule of vector addition can also be applied to the case of several electrons; in this case the orbital moments l_1, l_2, . . . of the individual electrons, and their spin-moments s_1, s_2 . . . , are combined so as to give the total angular momentum j. Here j is integral or half-integral according as the number of electrons is even or odd. Similarly, the projection m of j on a specially distinguished direction can also be either integral or half-integral.

In conclusion, we may also recall the fact that the system as a whole rotates round the direction of j, with uniform angular velocity; as follows from the meaning of j, the total angular momentum. This implies, as we have fully explained in the preceding chapter (p. 113), that the emitted radiation is subject to the rules $\Delta j = \pm 1$. In addition, however, as the theory agrees with experiment in showing, there are transitions with $\Delta j = 0$; these correspond to changes of state in which the total angular momentum does not change. The fact that these transitions are permitted, while those with $\Delta l = 0$ (or $\Delta k = 0$, see p. 124) are forbidden, is capable of explanation on correspondence principles. We shall not, however, consider the matter more closely.

2. The Anomalous Zeeman Effect.

We shall now show that the electron's own magnetic moment, which is bound up with its mechanical moment, supplies the explanation of the *anomalous Zeeman effect*, i.e. the observed phenomenon that in a (weak) magnetic field a spectral line is split up into a considerable number of lines (fig. 2, Plate X, p. 148); while, according to classical theory, and also according to wave mechanics when spin is not taken into account, we can only have the *normal Zeeman effect*, i.e. the splitting up of every spectral line into a Lorentz triplet.

We may briefly recall the explanation of the normal Zeeman effect. The revolution of the electron produces a *mechanical moment* p_l of the orbital motion, and this is quantised by known rules:

$$p_l = l\hbar.$$

On the other hand, the revolving electron acts like a circular current of strength $I = e(\omega/2\pi)$, where ω is the frequency of revolution, and so generates a magnetic field. But the magnetic field of a circular current I is, as we know, equivalent to that of a magnetic dipole of moment $M = AI/c$, where A is the area enclosed by the circuit, and c is the velocity of light. Hence the revolving electron behaves magnetically like a magnetic dipole of moment $\pi r^2 (e/c)\omega/2\pi$; since, however, the orbital angular momentum is $p_l = \mu r^2 \omega = l\hbar$, the *magnetic moment* M_l of the orbital motion becomes

$$M_l = \frac{eh}{4\pi\mu c} l = \frac{e}{2\mu c} p_l.$$

The value $eh/(4\pi\mu c)$ therefore represents the smallest unit for the magnitude of a magnetic orbital moment in the atom; it is called the *Bohr magneton*.

If a homogeneous magnetic field is applied, the atom is set into precessional motion (fig. 3) about the direction of the field, as has been explained above (p. 121) ; consequently the component m of l in this direction must be a whole number (quantisation of direction). The supplementary energy,

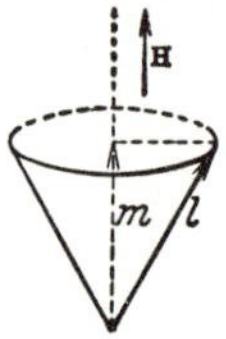

Fig. 3.—Precession of the orbital angular momentum round the direction of the magnetic field (in the absence of the spin it would always lead to the normal Zeeman effect).

by which the energy of the atom is increased owing to the magnetic field, is given by

$$E_{\mathrm{magn}} = -M_l H \cos \theta,$$

where θ is the angle between the magnetic field and the direction of the magnetic moment, i.e. the direction of l. But $\cos \theta$ is obviously equal to m/l, so that

$$E_{\mathrm{magn}} = -\frac{eh}{4\pi\mu c} Hm.$$

The terms therefore split up in the magnetic field, the separation being

$$h\nu_{\mathrm{magn}} = \frac{eh}{4\pi\mu c} H = h\nu_L,$$

where $$\nu_L = \frac{e}{4\pi\mu c} H = 1 \cdot 40 \times 10^6 \, H \, \mathrm{sec.}^{-1},$$

which is the same as the Larmor frequency already introduced (p. 121), i.e. the amount by which the frequency of a vibrating electrical system is changed in a magnetic field, according to the classical theory.

Thus although every term splits up into $2l + 1$ equidistant terms, corresponding to the $2l + 1$ setting possibilities, every line should be split up into only three components, since the precessional motion is purely periodic, and therefore the selection rules $\Delta m = 0, \pm 1$ come into play. In this way, therefore, we get only the *normal Zeeman effect* (see fig. 11, p. 122).

Even taking the spin of the electron into account, nothing in these relations would be changed, if we associated with the electron a magnetic moment, bearing to the mechanical spin-moment $p_s = s(h/2\pi)$ the same ratio as the magnetic orbital moment does to the mechanical, i.e. if

$$\frac{M_s}{p_s} = \frac{M_l}{p_l} = \frac{e}{2\mu c}.$$

For the total angular momentum would then be j, and the total magnetic moment would be $M_j = (eh/4\pi\mu c)j$; thus j and M_j would have the same direction and, setting themselves in the magnetic field in accordance with the quantisation of direction, would precess round the field direction together. The single difference, with spin, would be that now, not $2l + 1$, but $2j + 1$ setting possibilities exist, and that therefore every undisturbed term is split up by the magnetic field into $2j + 1$ terms, but in such a way that the amount of the splitting would be exactly the same as before; in the spectrum there would be no difference at all.

The *anomalous Zeeman effect*, however, can be explained completely by assuming that the magnetic spin-moment is got from the mechanical, not by multiplying by $e/2\mu c$, as with orbital moments, but by multiplying by $e/\mu c$, so that

$$M_s = 2\,\frac{eh}{4\pi\mu c}\,s.$$

Since the mechanical spin-moment is always $s = \frac{1}{2}$, it follows that the magnetic moment of the electron is exactly equal to a Bohr magneton $eh/4\pi\mu c$. This difference in the behaviour of the spin-moments as compared with the orbital moments can be put on a theoretical basis, as has been first shown by Thomas (1926), and later in a much simpler way by Kramers (1935); it is a

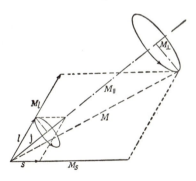

Fig. 4.—Vector model for the anomalous Zeeman effect. The direction of the total angular momentum does not coincide with the direction of the resultant magnetic moment; only the component $M_\parallel$ parallel to j is magnetically effective; the other component $M_\perp$ disappears when averaged, on account of the rotation of the vector figure about j (the total angular momentum).

necessary consequence of the theory of relativity. Moreover, the connexion between mechanical and spin-moment, in the form we are now considering, follows rigorously from the relativistic equation of Dirac.

It is just this difference between the orbital and the spin-moments which is responsible for the anomalous Zeeman effect. The result of it is that the vector sum of the magnetic moments, i.e. the total magnetic moment M, is not in general in the same direction as the total mechanical moment j. In fig. 4 this is shown for the general case of arbitrary vectors l and s in different directions. For clearness, the magnetic orbital moment M_l is shown twice as large as the mechanical orbital moment l ; hence, by the preceding, the magnetic spin-moment M_s must be shown four times as large as the mechanical spin-moment s. The resultants M and j therefore fall in different directions.

In accordance with the meaning of j, the total angular momentum, we must regard the atom, and with it the whole vector figure, as in rotation about the direction given by j; any vector, not in this direction, therefore precesses round it. On account of the high frequency of the precessional motion (it can be shown that the $h\nu$ corresponding to this frequency is of the order of magnitude of the "fine structure splitting" of the terms, which is determined by the coupling of l and s to the vector j; or, in the case, say, of the sodium D-lines, of the order of magnitude of the splitting of the p-terms, i.e. approximately, $\nu \sim 5 \times 10^{11}$ sec.$^{-1}$) only time-averages need be considered for quantities which vary slowly as compared with this frequency. Thus, for example, the atom will behave magnetically, in the presence of an external field, as if the atom had the magnetic moment $\overline{M}$, where the bar indicates a time-average. But the time-average of M is equal to the projection of M on the axis of rotation, i.e. equal to the component $M_{\parallel}$; the component $M_{\perp}$ perpendicular to the axis disappears on averaging.

In the presence of a (weak) external field the atom therefore possesses an effective magnetic moment $M_{\parallel}$ in the direction of j. On account of the angular momentum it precesses about the direction of the field, and the same considerations now hold as those adduced above: j possesses, in consequence of this precessional motion, $2j + 1$ possible settings with respect to the field direction, these being characterized by the component m of j in this direction. The magnetic energies of these settings are given individually by $-M_{\parallel}Hm/j$; the undisturbed term is therefore split up by the magnetic field into $2j + 1$ terms with the separation $M_{\parallel}H/j$.

We thus obtain an "effective" magnetic moment of the atom, which in general is not given by the product of Bohr magneton and total angular momentum, but depends also on the rest of the quantum numbers and in particular on the angles occurring in the vector figure. If we write

$$M_{\parallel} = \frac{eh}{4\pi\mu c} jg,$$

then the factor g represents the differences which occur in our vector model, as compared with the theory of the normal Zeeman effect.

The additional magnetic energy is then given by

$$E_{\text{magn}} = -M_{\parallel}H\frac{m}{j} = -\frac{eh}{4\pi\mu c}Hmg = -h\nu_L mg,$$

where, as above, ν_L is the classical Larmor frequency. The undisturbed term is therefore split up by the magnetic field again into $2j + 1$ equidistant terms, but the amount of the splitting $h\nu_L g$ is not equal to that

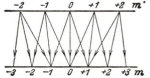

Fig. 5.—Transitions in the anomalous Zeeman effect; since the splitting is different in the various term groups, we get in general just as many separated lines as there are possible transitions altogether.

in the normal Zeeman effect, viz. $h\nu_L$, but differs from it by the factor g, which is called, after its discoverer, the Landé splitting factor (1923). It varies from term to term, and is thus the determining cause of the anomalous behaviour of the atom in the Zeeman effect. The splitting pattern is not the Lorentz triplet; many more lines occur, in accordance with the fact that the energy differences corresponding to the selection rules $\Delta m = -1$, 0, 1 are not now, as they are in the normal Zeeman effect, the same for all values of m (fig. 5).

The splitting up of a line in the anomalous Zeeman effect is therefore essentially determined by the Landé factors for the upper and lower states. These factors, as will now be shown, can be deduced

Fig. 6.—Vector composition of the orbital moment l and the spin-moment s, giving the total angular momentum.

with comparative ease from the vector model. For this purpose we have to calculate the value of the component $M_{\parallel}$ of the magnetic moment. From fig. 6 we can at once read off the relation

$$M_{\parallel} = M_l \cos(l, j) + M_s \cos(s, j).$$

Now for $M_{\parallel}$ we have the expression given above $(eh/4\pi\mu c)jg$, and M_l and M_s can be written

$$M_l = g_l\frac{eh}{4\pi\mu c}l, \ M_s = g_s\frac{eh}{4\pi\mu c}s$$

where g_l and g_s are known as the orbital and spin g-factors. They represent the ratio between the magnetic and mechanical moments for orbital and spin motion respectively. Referring to pages 157 and 159 we see that for an electron $g_l = 1$ and $g_s = 2$. However, we shall later have to deal with this same problem for neutrons and protons where

g_l and g_s have different values, so for generality we shall formulate the problem in terms of g_l and g_s. Thus, the above equation becomes

$$\frac{eh}{4\pi\mu c}\, jg = \frac{eh}{4\pi\mu c}\left\{g_l l \cos(l,j) + g_s s \cos(s,j)\right\}$$

which gives

$$g = g_l \frac{l}{j}\cos(l,j) + g_s \frac{s}{j}\cos(s,j)$$

The values of the cosines appearing in this equation may be written down at once from the triangle of the vectors l, s and j (fig. 6):

$$\cos(l,j) = \frac{j^2 + l^2 - s^2}{2jl}; \quad \cos(s,j) = \frac{j^2 + s^2 - l^2}{2js}.$$

For the Landé splitting factor we thus obtain

$$g = g_l\frac{(j^2 + l^2 - s^2)}{2j^2} + g_s\frac{(j^2 + s^2 - l^2)}{2j^2}$$

According to the correspondence principle this classical formula must hold also approximately in quantum theory if the quantum numbers are large. The quantum-mechanical treatment shows that this is in fact the case. For the quantum number l, belonging to the orbital angular momentum, this is shown in Appendix XIX (p. 404); the result is that the classical l^2 has to be replaced by $l(l+1)$. The same holds for j and s, so that we can adapt the formula for the splitting factor to wave mechanics, by writing everywhere $j(j+1)$, $l(l+1)$, $s(s+1)$, instead of j^2, l^2, s^2:

$$g = \tfrac{1}{2}\{(g_l + g_s) + (g_s - g_l)[s(s+1) - l(l+1)]/j(j+1)\}.$$

For the atomic case where $g_l = 1$ and $g_s = 2$

$$g = \tfrac{3}{2} + [s(s+1) - l(l+1)]/[2j(j+1)].$$

In Appendix XXII (p. 414) the splitting pattern in the anomalous Zeeman effect is calculated for the D-lines of sodium. Even in this simple example—an atom with one radiating electron—it can be seen how complicated, comparatively, is this splitting pattern in the anomalous Zeeman effect. Experiment, however, has fully confirmed the correctness of the calculated pattern, with spectroscopic accuracy. In every case in which the anomalous Zeeman effect has been investigated experimentally, theory and experiment have been found to be in complete agreement. The enormous mass of material which has been collected up to the present time, and which the calculation of Landé's factors has invariably proved capable of explaining, is one of the

strongest pillars on which the quantum theory of the electron is built.
Modern research has provided another more direct piece of evidence
for the correctness of the theoretical idea that the Zeeman effect is
due to precessional motions in the atoms. The "radar" technique
originally developed for radiolocation and detection of distant objects,
has given us the means to produce electromagnetic waves of a few
centimetres wave-lengths which can be used for physical research. Now
$\lambda = 1$ cm. corresponds to a frequency $\nu = c/\lambda \sim 3 \times 10^{10}$ sec.$^{-1}$. If we
compare this with the Larmor frequency (p. 158) $\nu_L = 1 \cdot 4 \times 10^6\ H$ sec.$^{-1}$,
we see that a field of some 10,000 gauss will bring ν_L into the domain
of the radar waves. If the precessions assumed by the theory are real,
an electromagnetic wave passing through a substance, exposed to a
magnetic field, should be strongly affected (scattered, absorbed) if
there is resonance, i.e. equality of frequencies. This effect is applied in
the following way.

The substance under investigation is placed in a static magnetic
field of the order 10,000 gauss and subjected to an oscillating electromagnetic field of known frequency in the microwave region. The magnetic field is then varied until strong absorption of the electromagnetic
radiation implies that the resonance effect has indeed been observed.
It was discovered by the Russian physicist Zavoisky (1946) for
ordinary paramagnetic substances and has been considerably developed
by Bleaney, Halliday and others, in particular for paramagnetic atoms
or ions forming part of a crystal lattice. In this case the atomic electrons
are subject to strong asymmetric crystalline electric fields and the
nature of the resonance is then no longer governed by the simple
formula above. Detailed analysis, however, enables considerable information to be obtained about the form of the crystal and its associated
electric fields. The electrons also interact with the magnetic moment
and electric quadrupole moment (p. 312 et seq.) of the nucleus so giving
a hyperfine structure to the resonance. From the form of this structure
it is then frequently possible to deduce the nuclear moments together
with the magnitude of the nuclear spin.

Ferromagnetic resonance was first studied by Griffiths (1946) and
later by Kittel and others. In a typical experiment the ferromagnetic
material is incorporated in a cavity resonator and a resonance is then
found in the permeability of the ferromagnetic material. A particularly
sharp resonance found by Yager and Bozorth (1947) in the Ni-Fe
alloy Supermalloy is illustrated in fig. 7. This can be interpreted
(after taking account of demagnetizing effects) in terms of a g-factor
for the material of just greater than 2 implying that ferromagnetism

is primarily associated with electron spin (see pp. 161 and 185).
This resonance method is thus a powerful tool for the investigation
of the magnetic properties of atoms and molecules. The same idea
has even been applied to the magnetic moments of the atomic nuclei,
as we shall describe later (Chap. X, § 2, p. 313).

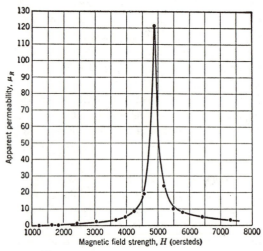

Fig. 7.—Ferromagnetic resonance in Supermalloy

In conclusion, we shall say something about one more phenomenon observed in the Zeeman effect; this is called, after its discoverers, the *Paschen - Back effect* (1921). In the derivation of the Landé factor we have stipulated that the magnetic field is not to be too strong. This assumption was tacitly used when we replaced the magnetic moment by its time-average $\overline{M} = M_{\parallel}$. This is justified as long as the rotation of the whole vector model about the direction of the angular momentum j is much more rapid than its precessional motion about the direction of the magnetic field, the frequency of which is, approximately, $\nu_L = 1.40 \times 10^6 \, H$ sec.$^{-1}$. For the former, in the case of the splitting of the sodium terms, we found (p. 153) $\Delta\lambda \sim 6$ Å., corresponding to $\Delta\nu \sim 5 \times 10^{11}$ sec.$^{-1}$, so that the required condition is certainly fulfilled in this example, for fields up to some thousand gauss.

If, however, we increase the magnetic field strength until the two frequencies are of the same order of magnitude, the foregoing considerations are no longer valid, since it is then not the time-average of M, but M itself, that is in question. We are now in the region of the Paschen-Back effect. We can describe it in another way by saying that the internal energy of precession about j becomes comparable with the external energy of precession of j about H, so that the fine structure splitting depending on the spin is of the same order of magnitude as the splitting of the terms in the magnetic field. If the magnetic field strength is raised further still, so that the energies depending on the setting in the field become much greater than the energy of coupling between the orbital moment l and the spin-moment s, we obtain the normal Zeeman effect; in fact, this coupling is then practically completely annulled, and the orbital and spin-moments precess independently round the direction of H (fig. 8). The total magnetic energy is then given by

$$E_{\text{magn}} = - \frac{eh}{4\pi\mu c} H \Big\{ l \cos(lH) + 2s \cos(sH) \Big\}.$$

If we continue to denote by m_l (a whole number) the projection of l on the field direction, and observe that in this case the spin can only set itself parallel or antiparallel to the field $(m_s = \pm\frac{1}{2})$, it follows from the above equation and the half-integral character of the spin-moment that

$$E_{\text{magn}} = - \frac{eh}{4\pi\mu c} H(m_l \pm 1).$$

We therefore in this case obtain a term-splitting with the term-separation corresponding to the normal Zeeman effect. Thus when the magnetic field is steadily increased, a gradual transition takes place from the anomalous to the normal Zeeman effect; the transitional zone is referred to as that of the Paschen-Back effect. The

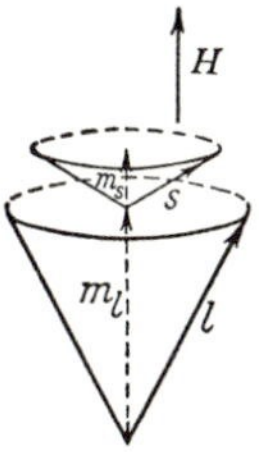

Fig. 8.—Vector model for the Paschen-Back effect (transition to the normal Zeeman effect). Since the energy of the orbital moment and spin-moment in the magnetic field is greater than the magnetic interaction between orbit and spin, the orbital and spin-moments precess separately round the field direction.

nomenclature, it may be noted, is not appropriate—normally (at ordinary field strengths) we get the "anomalous" Zeeman effect, while the "normal" effect is only got at abnormally high field strengths.

3. The Hydrogen Atom and X-ray Terms.

We shall now investigate the question of how the values of the terms are influenced by the existence of spin. We begin with the simple example of the hydrogen atom, and with the schemes of terms which may be called hydrogen-like (alkali terms, X-ray terms). In Chap. V (pp. 124, 140) we discussed the values of these terms, on the basis of Bohr's theory and wave mechanics, without taking into account the spin of the electron. We may state the results once more, in brief summary. The hydrogen-like spectra arise when an electron moves in a Coulomb field (hydrogen terms) or in a Coulomb-like central field (screening by the rest of the electrons; alkali and X-ray terms). The value of the term depends mainly on the principal quantum number: Balmer terms $-Rh/n^2$. If the field deviates from the Coulomb field a correction has to be applied, which depends not only on n but also on the azimuthal quantum number l; it was denoted above (p. 120) by $\epsilon(n, k)$, where $k = l + 1$. A similar correction is required if we take into account the relativistic variability of mass, the influence of which on the hydrogen terms has been mentioned above (p. 119).

The existence of electronic spin leads to a further correction of the terms, since it gives rise to an additional quantum number

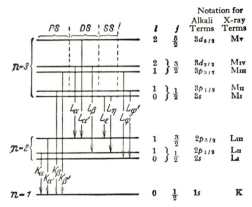

Fig. 9.—Diagrammatic synopsis of the notation for the various terms and lines in alkaline and X-ray spectra.

$j = l \pm \frac{1}{2}$. This causes, as we have explained in § 1, p. 158, a splitting up of the terms into doublets; but the s-terms, and these alone, remain single, for in this case j can have only one value, viz. $\frac{1}{2}$. Instead of the scheme of terms in fig. 13, p. 124, we therefore obtain the scheme of fig. 9. The terms fall in the first place into widely separated groups corresponding to the principal quantum number n ($n = 1, 2, 3, \ldots$). The figure is not at all correct in its proportions; in reality the distance between two groups of terms, with different quantum numbers n, is 10^3 to 10^4 times greater than the splitting within such a group; the transitions between two different groups of terms correspond in the optical spectra to wave-lengths of a few thousand Å., while the fine structure splitting of the lines amounts at most to a few Å.

Within the group of terms with the same principal quantum number, we have in a preceding section (p. 124) discriminated between the individual terms by specifying the azimuthal quantum number k;

instead of k we have now two quantum numbers, viz. the azimuthal quantum number l which corresponds to $k - 1$, and the inner quantum number j. Instead of $l = 0, 1, 2, \ldots$ (or, as formerly, $k = 1, 2, 3, \ldots$), it is customary, as was mentioned on p. 140, to use the letters $s, p, d, \ldots$; the value of j is given as a suffix attached to these letters. Examples of this notation are shown in fig. 9, for the case of the alkali terms. For the X-ray terms a different notation has become established, the terms being distinguished by a capital letter specifying the shell in which they lie (K shell for $n = 1$, L shell for $n = 2$, and so on), and by a roman numeral attached to the letter as a suffix, and corresponding to the order of numbering within the shell.

The case of hydrogen is peculiar in one respect. Experiment gives distinctly fewer terms than are specified in the term scheme of fig. 9; for $n = 2$ only two terms are found, for $n = 3$ only three, and so on. The theoretical calculation shows that here (by a mathematical coincidence, so to speak) two terms sometimes coincide, the reason being that the relativity and spin corrections partly compensate each other. It is found that terms with the same inner quantum number j but different azimuthal quantum numbers l always strictly coincide, for instance, the $ns_\frac{1}{2}$ and the $np_\frac{1}{2}$ term, the $np_\frac{3}{2}$ and the $nd_\frac{3}{2}$ term, and so on; such pairs of terms are drawn close together in fig. 9. For the value of the terms a formula was given by Sommerfeld (1916), even before the introduction of wave mechanics; the same formula is also obtained when the hydrogen atom is calculated by Dirac's relativistic wave mechanics. This formula, which reproduces the values of the hydrogen terms with the greatest exactness, is:

$$E + E_0 = E_0 \left[1 + \frac{a^2 Z^2}{(n_r + \sqrt{n_\phi^2 - a^2 Z^2})^2} \right]^{-\frac{1}{2}}.$$

Here E denotes the energy of the bound electron after deducting the rest energy, and E_0 is the rest energy mc^2; n_r is the radial quantum number; n_ϕ (Sommerfeld) is identical with Bohr's azimuthal quantum number k, and corresponds therefore to the $l + 1$ of wave mechanics. Since, however, as we have just seen, two terms with different l but the same j always coincide when we take the spin into account, discrimination between the terms by means of the quantum number n_ϕ is identical with discrimination by means of j; we therefore have $n_\phi = j + \frac{1}{2}$. The principal quantum number is then found as the sum $n = n_r + n_\phi$. The constant a is given by

$$a = \frac{2\pi e^2}{hc} \sim \frac{1}{137};$$

dimensional considerations show at once that it is a pure number, the only quantity indeed of zero dimensions which can be formed from the three electronic constants e, h and c. Since it gives the amount of the fine structure splitting, it is called (after Sommerfeld) the *fine structure constant*. Z is the atomic number (1 for hydrogen, 2 for He$^+$, and so on). On account of the smallness of a, Sommerfeld's formula can be expanded in ascending powers of a^2Z^2 ; a simple calculation gives

$$E = -\frac{RhZ^2}{n^2}\left\{1 + \frac{a^2Z^2}{n^2}\left(\frac{n}{n_\phi} - \frac{3}{4}\right) + \ldots\right\},$$

where R is written for $E_0a^2/2h$, so that R is simply the well-known Rydberg constant. The Balmer term $-RhZ^2/n^2$ is therefore modified by a correcting factor, which depends on n_ϕ and gives the fine structure; the quantity $\epsilon(n, k)$ of the formula on p. 120 is equal to the additive correction

$$-\frac{Rha^2Z^4}{n^4}\left(\frac{n}{n_\phi} - \frac{3}{4}\right).$$

Sommerfeld's formula was well confirmed not only for the optical spectra (H-atom, He$^+$, Li^{++}, &c.) but also in the X-ray region. As Dirac's relativistic wave equation of the electron leads exactly to the same result, it was for a considerable period regarded as one of the few definite and final achievements of physics. However, this belief was disappointed, as new deviations were discovered.

The tool used in this work was the technique of radar waves already mentioned above in connexion with the magnetic resonance effect. The fine structure splitting corresponds to a precessional motion (see the classical picture, fig. 7, p. 119) and can therefore be detected by its resonance effect on sufficiently short electromagnetic waves. The fine structure splitting is, according to the last formula (with $Z = 1$), of the order $E/h \sim Ra^2/n^2$, where (p. 108) $R \sim 3\cdot3 \times 10^{15}$ sec.$^{-1}$, corresponding to a wave-length

$$\lambda = \frac{c}{\nu} = \frac{ch}{E} = \frac{c}{Ra^2}\,n^2 \sim 0\cdot17n^2 \text{ cm}.$$

For $n = 2$ this is almost 1 cm., and as the actual term displacements are fractions of the frequency used here, the corresponding wave-lengths are in the range of radar waves.

Lamb and Retherford (1948), using an extremely ingenious technique, were able to detect transitions between the very close $2s_{\frac{1}{2}}$ and $2p_{\frac{1}{2}}$ levels induced by interaction with microwaves of a frequency of about 1000 Mc/sec. The observed magnitude of the splitting ("the Lamb shift") between the $2s_{\frac{1}{2}}$ and $2p_{\frac{1}{2}}$ levels agrees to 1 part in 10^4 with the result of perturbation calculations employing a renormalized quantum electrodynamics (see p. 196) which express the shift as a power series in the fine structure constant a.

There seems to be little doubt that the existence of this dimensionless number, the only one which can be formed from e, c and h, indicates a deeper relation between electrodynamics and quantum theory than the current theories provide, and the theoretical determination of its numerical value is a challenge to physics. The solution of this problem seems to be closely connected with a future theory of elementary particles in general. All attempts have so far been in vain. The most notorious of these is that of Eddington, according to whom $1/a$ is the value of

$$\tfrac{1}{2}n^2(n^2+1) \; +1 \text{ for } n=4, \text{ namely exactly } 137.$$

The idea that $1/a$ is an integer is attractive, and seems to be confirmed by the latest experiments, which give $137{\cdot}0388$, very nearly integral. Yet Eddington's theory has failed to predict any new phenomenon (e.g. the different types of mesons, the Lamb-Retherford effect, &c.), and is altogether too fantastic to be acceptable.

We now proceed to consider the spectral lines which are consistent with the above scheme of terms. The selection rules have already been stated (p. 157); they are

$$\Delta l = \pm 1, \quad \Delta j = 0, \quad \pm 1;$$

they do not allow all transitions, but only those indicated by arrows in the scheme of terms (fig. 9, p. 166).

In the hydrogen and alkali spectra, as we have already explained for the simple Bohr theory (fig. 13, p. 124 ; fig. 9, p. 166), the lines which correspond to transitions from a p-term to the ground state (s-term) are called the lines of the principal series; a principal series of higher order contains the lines which lead to the next s-term (in our diagram to the 2s-term). Transitions from the d-terms to the p-terms give the diffuse series, those from the s-terms to the p-terms the sharp series. The higher series (fundamental series, &c.) lead from higher levels to the terms of the third quantum state ($n = 3$), and

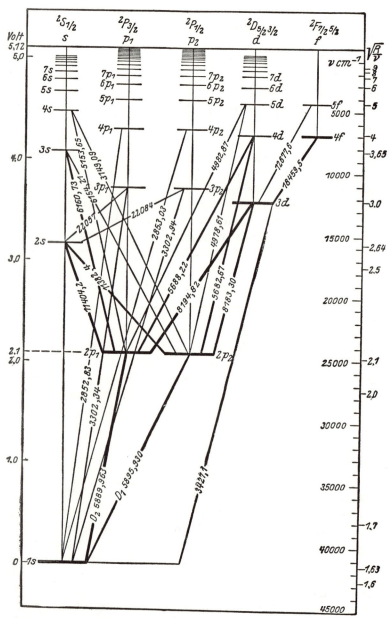

Fig. 10.—Term scheme for sodium, taking account of spin; on the left are shown the energies of the terms in electron volts (measured from the ground state); on the right the frequencies (in wave numbers) which are emitted at a transition downwards from the series limit, and also the values of $\sqrt{(R/\nu)}$.

therefore fall outside the scope of our diagram of terms. We thus obtain—for the case of sodium—the diagram of terms and lines shown in fig. 10, p. 170, which replaces the simple diagram of p. 124 (fig. 13).

For X-ray spectra an essentially different notation has secured general acceptance ; this also is indicated in fig. 9, p.166 (see also fig. 15, p. 125). The lines which correspond to transitions to the K-level are called K-lines; lines whose final state is a term of the L shell are called L-lines; and so on. Individual lines within these general classes are distinguished by small Greek letters in accordance with a recognized convention, which was also exemplified in fig. 15, p. 125. The K_a and $K_{a'}$ lines form the doublet which arises in a transition of the electron from the L to the K shell; K_β and $K_{\beta'}$ correspond to the transitions from the M to the K shell; and so on.

4. The Helium Atom.

In the helium atom two electrons revolve about the nucleus (nuclear charge $2e$); we have therefore 6 co-ordinates to deal with instead of 3, with the result that an exact solution is no longer possible. For the purpose of obtaining a general idea of the possible states, an exact solution is, however, not at all necessary; following Bohr, we can in the first place neglect the mutual interaction of the electrons, and for a first approximation treat the problem as if the two electrons moved undisturbed in the field of the nucleus. Afterwards, the interaction can be taken into account by the methods of the theory of perturbations.

We shall therefore (as with the hydrogen atom) associate three quantum numbers with each of the two electrons:

$$n_1, l_1, j_1;\ \ n_2, l_2, j_2,$$

and we shall assume once for all that in the case of different l-values the first electron possesses the higher azimuthal quantum number $(l_1 > l_2)$. The corresponding angular momenta are composed vectorially (neglecting the interaction between the two electrons); we thus obtain a total angular momentum

$$\vec{j} = \vec{j_1} + \vec{j_2},$$

as also total orbital and spin angular momenta given respectively by

$$\vec{l} = \vec{l_1} + \vec{l_2},\ \ \vec{s} = \vec{s_1} + \vec{s_2}.$$

If, however, the interaction of the two electrons is taken into account, the angular momenta of the electrons separately are no longer integrals of the equations of motion, so that the angular momentum vectors are not now fixed in space. The *total* angular momentum j is constant, however, and of course must still be quantised even for an arbitrary system of electrons, since it corresponds to a rotation of the electronic configuration as a whole. The question is now, whether a many-electron (two-electron) problem can still be characterized, at least approximately, by other angular momentum quantum numbers. The important factor here is the interaction of the two electrons, and the coupling relations between the individual angular momentum vectors (orbital and spin). If in fact, in consequence of the interactions

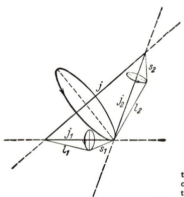

being slight, a precessional period is slow, it will be possible to associate the corresponding angular momenta, approximately, with quantum numbers, viz. those which the angular momenta would have, if there were no coupling. It is entirely a matter of the strength of the various couplings.

An obvious assumption to make

Fig. 11.—Scheme of the so-called (jj)-coupling; to a first approximation the orbital and spin-moments of each electron are compounded into a resultant; the vector sum of these two resultants gives the total angular momentum j.

is the following. For every electron the orbital and spin moments are firmly coupled; but the various electrons influence each other comparatively little. Every electron will then be characterized individually by the quantum number of its angular momentum j (orbital plus spin); that is to say, the vector sum of l_ν and s_ν, for the νth electron, still approximately carries out a motion of precession round the direction of j_ν, in spite of the disturbance by the other electrons; then the j_ν vectors of the various electrons are compounded into a total angular momentum j, and in their turn carry out a precessional motion round the latter (see fig. 11).

Another limiting case of the coupling relations is that in which the spin vectors and the orbital vectors are compounded separately into resultants l and s, so that for two electrons we have the diagram of fig. 12. The vectors l_1 and l_2 rotate round the total orbital moment l, the spin vectors s_1 and s_2 round s, and then these two vectors precess about j. One can understand how this vector model may come into play, by supposing that besides the electrical forces of repulsion other forces of a magnetic nature also act between the electrons; the latter forces being determined by the magnetic moment of the electrons, and producing a strong coupling between the spin vectors of the two electrons.

Experimental results show that as a rule (in helium and the alkaline earths) the second case of coupling is the one which occurs; it is called LS coupling or, after its discoverers, the Russell-Saunders coupling. This is the only case we shall consider here although the case first described (jj coupling) also actually occurs, as well as intermediate stages between these two extreme cases. We may remind the reader of the similar circumstances in the case of the abnormal Zeeman effect, where the

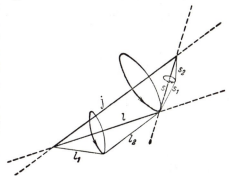

Fig. 12.—Scheme of the Russell-Saunders or (ls)-coupling; to a first approximation the orbital moments of the electrons are compounded into a total orbital moment l, also their spin-moments into a resultant spin-moment s, and finally l and s into the total angular momentum j.

strengthening of the coupling between the external magnetic field and the individual magnetic moments determines a transition to the normal Zeeman effect, via the Paschen-Back effect.

In helium, the coupling relations are normal. The two orbital moments l_1 and l_2 are therefore in the first place combined into a total orbital moment l, which must be a whole number and can therefore take only the values $l = l_1 - l_2, l_1 - l_2 + 1, \ldots, l_1 + l_2$. Similarly, the two spin-moments s_1 and s_2 combine into the total spin-moment which (for two electrons) must likewise be a whole number: since s_1 and s_2 have the value $\frac{1}{2}$, the two values $s = 0$ and $s = 1$ are the only possible ones; in the former case the spins are antiparallel, in the latter parallel. (In fig. 12, for easier visualization we have assumed that the two spins form an obtuse angle with each other: in reality they must always be either parallel or antiparallel to each other.)

The term scheme of helium (see fig. 13) breaks up therefore into two separate groups, according as the total spin is zero or unity; in the former case we speak of *parhelium terms*, in the latter of *ortho-*

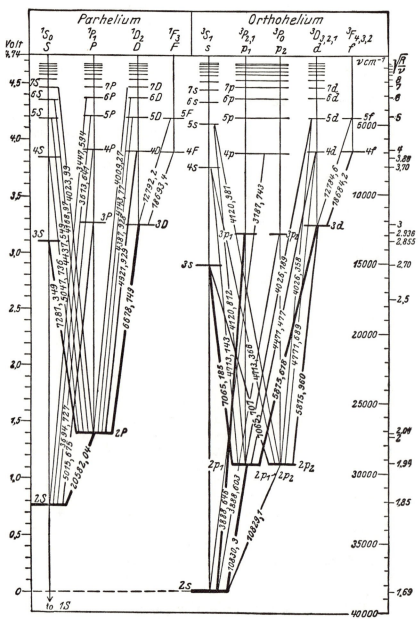

Fig. 13.—Term scheme for helium; there are no inter-combination lines between orthohelium and parhelium; orthohelium (for not too great dispersion) has the character of a doublet system. The 1S term of parhelium is situated much deeper as indicated by the arrow.

helium terms. This separation of the terms into two groups is justified by the fact, which has been proved experimentally and can easily be put on a theoretical footing, that in general there are no transitions between terms of the first group and terms of the second. From the physical standpoint it is accordingly permissible to distinguish sharply between parhelium and orthohelium, especially as it is only in exceptional cases that it is possible to transform a helium atom, which has once shown the parhelium spectrum, in such a way that it emits the lines of orthohelium.

In *parhelium*, then, the total spin moment s is 0; hence the total orbital moment is identical with the total angular momentum: $j = l$. This implies that the whole of the terms of parhelium are singlets, i.e. that to every azimuthal quantum number l there belongs only a single term with the inner quantum number j equal to l.

In *orthohelium*, on the other hand, the total spin moment is equal to 1, and it combines vectorially with the total orbital moment to form the total moment j. Since all three vectors are whole numbers, j in this case can take the three values

$$j = l - 1, \quad l, \quad l + 1.$$

There are therefore three terms for every azimuthal quantum number, and the term scheme of orthohelium is a triplet system. Here also, however, as in the doublet spectra discussed above (p. 153), the s-terms are single; in fact, $l = 0$ for these, so that j can only be equal to s, i.e. to 1. In the alkaline earths (Be, Mg, Ca, Sr, Ba), which like helium have two external electrons, the triplet character can be easily observed in the spectrum. In helium, however, a special feature occurs: it possesses practically a doublet spectrum, since the P_1 and P_2 terms nearly coincide. As an example of this, we take the first line of the sharp series, which corresponds to the transition from $3S$ to $2P$. Exact measurements by Houston have shown that this line consists of three components:

$$
\begin{aligned}
3S \to 2P_0 \ \dots \ 7065{\cdot}707 \ \text{Å,} & \left.\right\} \ \Delta\lambda = 0{\cdot}495 \ \text{Å,}\\
3S \to 2P_1 \ \dots \ 7065{\cdot}212 \ \text{Å,} & \\
3S \to 2P_2 \ \dots \ 7065{\cdot}177 \ \text{Å,} & \left.\right\} \ \Delta\lambda = 0{\cdot}035 \ \text{Å.}
\end{aligned}
$$

We see from these numbers that the splitting between the two lower terms is only about one-fourteenth of that between either and the upper term. The slightness of this splitting is the reason why for a long time the spectrum of helium was referred to as a doublet system.

We add a brief remark on the *notation for the terms*. A few lines above we have used capital letters S, P, D, ... instead of small letters (p. 140). This is customary in the case of several electrons for the purpose of indicating the total orbital moment. It is also customary to define the multiplet character, that is to say, the number of terms belonging to a particular multiplet (all with the same principal and azimuthal quantum numbers), by attaching this number to the term symbol as a left-hand upper index; also, the multiplicity written down is always the one which occurs when l is large, viz. $2s + 1$. In fact, the vector sum $\vec{l} + \vec{s} = \vec{j}$ gives the following values:

for $l \geq s$: $l - s, l - s + 1, \ldots, l + s - 1, l + s$ $(2s + 1$ values$)$,

for $l \leq s$: $s - l, s - l + 1, \ldots, s + l - 1, s + l$ $(2l + 1$ values$)$.

If, for example, $s = 1$, we have a triplet; but this holds only from $l = 1$ (P-term) upwards, while the term $l = 0$ (S-term) is single. This peculiarity is ignored in the use of the notation; the two terms are written ^3S, ^3P, although in reality the former of these is single.

The individual terms of the triplet ^3P are distinguished by right-hand lower indices (suffixes), which indicate the value of j; in helium, for instance,

$$^1P_1 \qquad \text{for parhelium,}$$
$$^3P_0, \,^3P_1, \,^3P_2 \text{ for orthohelium.}$$

Briefly, then, the notation used is as follows. The first number denotes the principal quantum number—it indicates the shell which the electrons occupy; if the electrons are in different shells, this quantum number must be omitted. The letter indicates the total orbital moment, the letters S, P,... corresponding to the quantum numbers $l = 0, 1, \ldots$. The left-hand upper index gives the number of terms (at its maximum) in the term group defined by the principal quantum number and the letter—it is equal to $2s + 1$. The right-hand lower index distinguishes the individual terms of this group by assigning the inner quantum number j for the total angular momentum.

A more detailed theory of the helium spectrum which avoids the crude vector model can be based on the Schrödinger equation for two electrons under the action of the nucleus (Heisenberg, 1927). It leads to the same classification of the terms, and allows moreover approximate numerical calculations of term values. Excellent results have been obtained by Hylleraas (1929) and others. To give one example, the energy necessary to remove one electron from the ground state 1^1S

of parhelium (not represented in the diagram of p. 174, as it lies much lower than the rest of the terms) was found by calculation to be 24·4 eV; this quantity can be measured—its value is given under the heading ionization potential in the last column of the table, p. 181, namely, 24·56 eV. This excellent agreement demonstrates the power of quantum mechanics perhaps better than all qualitative results.

In connexion with this remark about the 1^1S-term the question arises why there is in the diagram no 1^3S-term, i.e. the ground state of orthohelium to be expected. The reason is, that no lines have been found spectroscopically which could correspond to a transition to this term. We must conclude from this that in orthohelium the 1^3S-state is missing. In fact, since we can calculate the position of this term approximately, we can tell approximately the places in the spectrum at which the lines must lie which correspond to transitions to this term. In spite of the most thorough spectroscopic investigations in this region of the spectrum, not a single line has been found which can be brought into connexion with such transitions.

This result is quite inexplicable by the preceding principles as they stand. If we look for its cause in some special property of the term, we find the peculiarity that for this term all the quantum numbers are the same for the two electrons. Thus, both electrons lie in the K shell, and have therefore the same principal quantum number $n_1 = n_2 = 1$, as is also in fact the case for the parhelium term 1^1S. Since the azimuthal quantum number of an electron can never be greater than $n - 1$, it must vanish in this case for both electrons: $l_1 = l_2 = 0$. (For higher S-terms, the two orbital moments could be different from zero, and by antiparallel setting cause the total orbital moment to vanish.) In orthohelium, the two electron spins are parallel, and the components of the spin moments are therefore also equal, $\sigma_1 = \sigma_2 = \frac{1}{2}$, so that the two sets of quantum numbers completely agree. In parhelium, however, the electron spins are antiparallel, so that the quantum numbers of the two electrons are separated in the projection of the two spin components.

5. Pauli's Exclusion Principle.

The state of matters just described in the case of helium, viz. the absence of the 1^3S-term, the expected ground state of the orthohelium term sequence, suggested to Pauli (1925) a general examination of spectra, to see whether, in other elements and under other conditions, definite terms sometimes drop out. It was found that this actually happens; moreover, the term analysis showed in all cases that in these missing terms all the quantum numbers of the electrons agreed. Conversely, it was also found, the terms always drop out when these quantum numbers are the same. This discovery led Pauli to the following principle:

The quantum numbers of two (or more) electrons can never entirely agree; two systems of quantum numbers, which are deducible from each other by interchange of two electrons, represent *one* state.

The second part of the principle enunciates the *indistinguishability* of electrons. It is important for the enumeration of possible states, as required for the theory of the periodic system, and especially for statistics. It is to be remarked also that, in applying the principle, directional degeneracy must be considered to be removed (for instance, by an external magnetic field). The *individual* electron can then be characterized by *four quantum numbers*:

$$n = 1, 2, \ldots,$$
$$l = 0, 1, \ldots, n-1,$$
$$\left. \begin{array}{l} j = l - \tfrac{1}{2}, l + \tfrac{1}{2}, \\ m = -j, -j+1, \ldots, +j \end{array} \right\} \text{ or } \left\{ \begin{array}{l} m_l = -l, -l+1, \ldots, +l, \\ m_s = -\tfrac{1}{2}, +\tfrac{1}{2}. \end{array} \right.$$

Here n denotes the principal quantum number, which can take all values from 1 upwards; l is the azimuthal quantum number, which

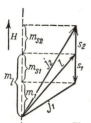

runs from 0 to $n-1$; j is the inner quantum number, and is only capable of the two values stated; m is the projection of j on the specially distinguished direction and, according to the rules for quantisation of direction, runs through the $2j+1$ values between $-j$ and $+j$. Alternatively, instead of j and m we may use the projections of l

Fig. 14.—Orbital moment l, spin moment s, total moment j; with their projections on a specially distinguished direction.

and s on the special direction, namely m_l (p. 165) and m_s (see fig. 14).

Another concept which is frequently employed is this. Two electrons are said to be *equivalent* when they possess the same n and the same l. Two equivalent electrons must therefore, in accordance with Pauli's principle, differ from each other in the direction of l, or in the spin direction; and only certain definite values, not all values, of m_l and m_s are possible for them. It is otherwise with two electrons which are not equivalent, but differ either in the principal or the azimuthal quantum number (or in both); in this case all values of m_l and m_s are possible.

In Appendix XXIII (p. 416) we give an example of the enumeration of the terms for the two cases of non-equivalent and equivalent electrons. We shall not enter here into a detailed discussion—which, though very simple in principle, is somewhat cumbersome to work out—but proceed at once to the most important and concrete application of Pauli's principle, Bohr's theory of the periodic system.

6. The Periodic System. Closed Shells.

We have already frequently referred to the periodic system, or table (see Table I, p. 39; fig. 15 shows a somewhat different form). The periodic table is an arrangement of the elements in a scheme

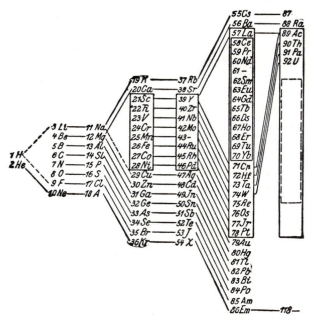

Fig. 15.—Diagram of the periodic system of the elements according to N. Bohr (1921), exhibiting the synthesis of the shells, and the chemical relationship, of the atoms.

which originally was drawn up on the basis of their chemical behaviour and their atomic weights. It has turned out, however, that the real determining factor in the arrangement is not the atomic weight—remember isotopy—but the atomic number Z, i.e. the number of electrons revolving round the nucleus in the neutral atom. Now one of the most important applications of Pauli's principle is to the elucidation of the shell character of atoms, i.e. that property which finds its expression in the periodicity of the atoms with respect to chemical behaviour. Following Bohr (1921) we shall develop the theory of the periodic system by the method of proceeding step by step from a simpler element to the next higher. At each step, therefore, starting from an element with an electronic configuration which we know, we suppose the nuclear charge to be increased by a unit, and at the same time an electron to be inserted in the outer part of the electronic envelope of the known atom; what we want to know is the way in which this addition of an electron takes place.

We begin with the simplest element—hydrogen ($Z = 1$). Its electron in the ground state is in the lowest orbit, and has therefore for its principal quantum number $n = 1$; we say, as has already been remarked (p. 125), that the electron is in the K shell. With helium a second electron appears; thus for both these electrons $n = 1$, and therefore $l = 0$, and also $\mu = 0$. Hence, by Pauli's exclusion principle, they must possess different components of the spin moment. Since there are only two possible settings, viz. $\sigma_1 = +\frac{1}{2}$ and $\sigma_i = -\frac{1}{2}$, $m_5 = \pm\frac{1}{2}$, there are only places for two electrons in the K shell.

If we now add a third electron, there is no place for it in the K shell, and so it must settle in the L shell. Let us count the number of places in the L shell. First, as in the K shell, there are 2 electrons with $l = 0$, distinguishable only by their spin components; next come 6 places for electrons with $l = 1$, since of course μ can take the three values $-1, 0, 1$, and for each of these there are two possible settings for the spin. There are therefore 8 places altogether in the L shell, and these are arranged (Stoner, 1924) in two sub-shells with 2 and 6 electrons. When the atomic number is steadily raised, the L shell becomes gradually filled; the atoms concerned are Li, Be, B, C, N, O, F, Ne. With Ne the L shell is completed; a new electron must settle in the M shell (Table V, pp. 181-3).

We give the enumeration of the places in the M shell in the form of a table:

$l = 0$, $m_l =$	0,	$m_s = \pm\frac{1}{2}$:	2 electrons
$l = 1$, $m_l =$	$-1, 0, 1$,	$m_s = \pm\frac{1}{2}$:	6 ,,
$l = 2$, $m_l = -2, -1, 0, 1, 2$,		$m_s = \pm\frac{1}{2}$:	10 ,,
		in all	18 electrons.

In the M shell there are therefore 18 places, distributed over three sub-groups. The first two groups form together the second short

TABLE V

DISTRIBUTION OF ELECTRONS IN THE ATOMS

Element		K	L		M			N		Ground Term	Ionization Potential (in electron volts)
		1, 0 / 1s	2, 0 / 2s	2, 1 / 2p	3, 0 / 3s	3, 1 / 3p	3, 2 / 3d	4, 0 / 4s	4, 1 / 4p		
H	1	1	—	—	—	—	—	—	—	$^2S_{1/2}$	13·59
He	2	2	—	—	—	—	—	—	—	$^1S_0$	24·56
Li	3	2	1	—	—	—	—	—	—	$^2S_{1/2}$	5·40
Be	4	2	2	—	—	—	—	—	—	$^1S_0$	9·32
B	5	2	2	1	—	—	—	—	—	$^2P_{1/2}$	8·28
C	6	2	2	2	—	—	—	—	—	$^3P_0$	11·27
N	7	2	2	3	—	—	—	—	—	$^4S_{3/2}$	14·55
O	8	2	2	4	—	—	—	—	—	$^3P_2$	13·62
F	9	2	2	5	—	—	—	—	—	$^2P_{3/2}$	17·43
Ne	10	2	2	6	—	—	—	—	—	$^1S_0$	21·56
Na	11		Neon Configuration		1	—	—	—	—	$^2S_{1/2}$	5·14
Mg	12				2	—	—	—	—	$^1S_0$	7·64
Al	13				2	1	—	—	—	$^2P_{1/2}$	5·97
Si	14				2	2	—	—	—	$^3P_0$	8·15
P	15				2	3	—	—	—	$^4S_{3/2}$	10·9
S	16				2	4	—	—	—	$^3P_2$	10·36
Cl	17				2	5	—	—	—	$^2P_{3/2}$	12·90
A	18				2	6	—	—	—	$^1S_0$	15·76
K	19		Argon Configuration				—	1	—	$^2S_{1/2}$	4·34
Ca	20						—	2	—	$^1S_0$	6·11
Sc	21						1	2	—	$^2D_{3/2}$	6·7
Ti	22						2	2	—	$^3F_2$	6·84
V	23						3	2	—	$^4F_{3/2}$	6·71
Cr	24						5	1	—	$^7S_3$	6·74
Mn	25						5	2	—	$^6S_{5/2}$	7·43
Fe	26						6	2	—	$^5D_4$	7·83
Co	27						7	2	—	$^4F_{9/2}$	7·84
Ni	28						8	2	—	$^3F_4$	7·63
Cu	29						10	1	—	$^2S_{1/2}$	7·72
Zn	30						10	2	—	$^1S_0$	9·39
Ga	31						10	2	1	$^2P_{1/2}$	5·97
Ge	32						10	2	2	$^3P_0$	8·13
As	33						10	2	3	$^4S_{3/2}$	10·5
Se	34						10	2	4	$^3P_2$	9·73
Br	35						10	2	5	$^2P_{3/2}$	11·76
Kr	36						10	2	6	$^1S_0$	14·00

TABLE **V** (*continued*)

Element	Configuration of Inner Shells	N 4,2 4d	N 4,3 4f	O 5,0 5s	O 5,1 5p	O 5,2 5d	P 6,0 6s	Ground Term	Ionization Potential (in electron volts)
Rb 37		—	—	1	—	—	—	$^2S_{1/2}$	4·17
Sr 38		—	—	2	—	—	—	$^1S_0$	5·69
Y 39		1	—	2	—	—	—	$^2D_{3/2}$	6·5
Zr 40	Krypton Configuration	2	—	2	—	—	—	$^3F_2$	6·95
Nb 41		4	—	1	—	—	—	$^6D_{1/2}$	6·77
Mo 42		5	—	1	—	—	—	$^7S_3$	7·06
Tc* 43		6	—	1	—	—	—	$^6D_{9/2}$	7·1
Ru 44		7	—	1	—	—	—	$^5F_5$	7·5
Rh 45		8	—	1	—	—	—	$^4F_{9/2}$	7·7
Pd 46		10	—	—	—	—	—	$^1S_0$	8·1
Ag 47			—	1	—	—	—	$^2S_{1/2}$	7·58
Cd 48			—	2	—	—	—	$^1S_0$	8·99
In 49			—	2	1	—	—	$^2P_{1/2}$	5·79
Sn 50	Palladium Configuration		—	2	2	—	—	$^3P_0$	7·30
Sb 51			—	2	3	—	—	$^4S_{3/2}$	8·64
Te 52			—	2	4	—	—	$^3P_2$	8·96
I 53			—	2	5	—	—	$^2P_{3/2}$	10·44
Xe 54			—	2	6	—	—	$^1S_0$	12·13
Cs 55			—			—	1	$^2S_{1/2}$	3·89
Ba 56			—			—	2	$^1S_0$	5·21
La 57			—			1	2	$^2D_{3/2}$	5·61
Ce* 58			2			—	2	$^3H_4$	(6·54)
Pr* 59			3			—	2	$^4I_{11/2}$	(5·76)
Nd* 60			4			—	2	$^5I_4$	(6·31)
Pm* 61			5	The shells 5s to 5p contain 8 electrons		—	2	$^6H_{5/2}$	(6·3)
Sm 62	The shells 1s to 4d contain 46 electrons		6			—	2	$^7F_0$	(5·6)
Eu* 63			7			—	2	$^8S_{7/2}$	5·64
Gd* 64			7			1	2	$^9D_2$	6·7
Tb* 65			8			1	2	^6H	(6·74)
Dy* 66			9			1	2	^5I	(6·82)
Ho* 67			10			1	2	^4I	
Er* 68			11			1	2	^3H	
Tm 69			12			1	2	$^2F_{7/2}$	
Yb 70			13			1	2	$^1S_0$	6·2
Lu* 71			14			1	2	$^2D_{5/2}$	5·0

TABLE V (concluded)

Element	Configuration of Inner Shells	O		P			Q	Ground Term	Ionization Potential (in electron volts)
		5, 2 5d	5, 3 5f	6, 0 6s	6, 1 6p	6, 2 6d	7, 0 7s		
Hf 72		2	—	2	—	—	—	$^3F_2$	5·5
Ta 73	The shells	3	—	2	—	—	—	$^4F_{3/2}$	6·0
W 74	1s to 5p	4	—	2	—	—	—	$^5D_0$	7·94
Re 75	contain	5	—	2	—	—	—	$^6S_{3/2}$	7·87
Os 76	68	6	—	2	—	—	—	$^5D_4$	8·7
Ir 77	electrons	9	—	0	—	—	—	$^4F_{3/2}$	9·2
Pt 78		9	—	1	—	—	—	$^3D_3$	8·96
Au 79			—	1	—	—	—	$^2S_{1/2}$	9·23
Hg 80			—	2	—	—	—	$^1S_0$	10·44
Tl 81			—	2	1	—	—	$^2P_{1/2}$	6·12
Pb 82			—	2	2	—	—	$^3P_0$	7·42
Bi 83			—	2	3	—	—	$^4S_{3/2}$	(8·8)
Po 84			—	2	4	—	—	$^3P_2$	(8·2)
At* 85	The shells		—	2	5	—	—	$^2P_{3/2}$	(9·6)
Rn 86	1s to 5d		—	2	6	—	—	$^1S_0$	10·75
Fr* 87	contain		—	2	6	—	1	$^2S_{1/2}$	(4·0)
Ra* 88	78		—	2	6	—	2	$^1S_0$	5·27
Ac* 89	electrons		—	2	6	1	2	$^2D_{3/2}$	
Th 90			1(—)	2	6	1(2)	2	$^3F_2$	
Pa* 91			2(1)	2	6	1(2)	2	(^4K)	
U* 92			3	2	6	1	2	$^5L_6$	4·0
Np* 93			5(4)	2	6	—(1)	2	(^6M)	
Pu* 94			6(5)	2	6	—(1)	2	(^7K)	
Am* 95			7	2	6	—	2	^8H	
Cm* 96			7	2	6	1	2	^9D	

The table shows not only those distributions of electrons and ground terms which have actually been ascertained from the spectra, but also those which have been determined by considerations of analogy; the latter are indicated by an asterisk attached to the symbol of the element. In some cases (Actinides) two alternative possibilities are given, one in brackets (Seaborg, 1949).

period (of 8 elements); it extends from Na to A. Then, however, a deviation occurs from the apparently natural order of succession in which the electrons settle. This order is always determined by the energy released when a new electron settles (i.e. is bound). The energy relations, however, are not always such that one shell must be completed before an electron settles in the next shell. On the contrary, it may happen that an electron in an s-orbit of a higher shell is, from the energy point of view, more firmly bound than in a d- or f-orbit of the lower, still incomplete, shell. This case occurs in the further development of the M shell; the ten $3d$-terms for the 10 electrons still wanting are, as experiment shows, higher as regards energy than the $4s$-terms which correspond to binding of the electrons in the N shell. The next two electrons to be added (K, Ca) settle therefore in the N shell. Then, but not before, begins the filling up of the M shell, which continues from Sc $(Z = 21)$ to Zn $(Z = 30)$. The formation of the N shell then proceeds up to the element Kr $(Z = 36)$. The process described is repeated in the N shell; the two next electrons (Rb, Sr) settle in the O shell $(n = 5)$; the filling up of the N shell then begins, along with intermittent settlement of electrons in the O shell. The rare earths (also called Lanthanides, i.e. the elements following Lanthanum) correspond to the completion of the N shell, which is accompanied by the binding of 2 in the P shell; this explains the great chemical similarity of these elements. All through, as has already been mentioned, the chemical behaviour of the elements is determined by the electronic configuration in the outermost shell; elements with outermost shells of similar structure possess in large measure equivalent chemical properties. This feature may appear again in a group of heavy elements, the Actinides, Ac 89 to Cm 95. On the other hand, the same principle explains the occurrence of periods in the system of the elements. Thus, for example, the inert gases Ne, A, Kr, Xe and Rn have all closed shells of 8 electrons; the alkalies are characterized by the fact that one electron revolves round the atom outside closed shells; the halogens, on the other hand, lack one electron to make up a closed shell. The problems of chemical binding and the chemical behaviour of the atoms will be taken up again in detail in Chap. VIII (p. 266).

7. Magnetism.

In § 2 (p. 157) we have investigated at length the splitting up of the terms of an atom in a magnetic field; it is determined by the directing force of the magnetic field acting on the magnetic moments of the spin and of the orbital motion of the electrons in the atom.

The magnetic moments within the atoms are also responsible for the magnetic behaviour of the substance made up of these atoms. Two cases have to be distinguished. A substance is called *paramagnetic* if its atoms (or molecules) possess a magnetic moment. The magnetization of the substance, which depends on this, is clearly in the same direction as the field, since of course it arises from the orientation of the magnetic moments in the direction of the field. Further, it is characteristically strongly dependent on the temperature, since the orientation of the elementary magnets is opposed by the smoothing-out action upon direction which is due to the thermal motion. We shall deal with this more fully in Appendix XXXVII (p. 462). On the other hand, a substance is called *diamagnetic* if its atoms do not possess a permanent magnetic moment; in this case the magnetization is in the opposite direction to the field, and is practically independent of the temperature.

A few solid substances, e.g. the elements of the iron group, Fe, Co, Ni, have in a certain temperature interval the property that a magnetization in bulk remains, even if the producing field is removed. They are called *ferromagnetic*. There is a brief discussion of ferro-magnetism in Chap. IX, p. 298.

Atomic theory gives the magnetic moments (compare the results obtained above in connexion with the anomalous Zeeman effect, p. 157). Thus, for atoms with closed outer shells the moment 0 is found (in the inert gases, but also, according to measurements by Stern, in Zn, Cd, Hg). S-terms have no orbital moment ($l = 0$), so that the magnetic moment of the atom is entirely due to the spin; thus the alkalies, with one radiating electron in an s-orbit, have a magnetic (spin) moment of 1 Bohr magneton, and the same is true also of the noble metals (Cu, Ag, Au). Those elements which have an intermediate shell not yet completed possess large magnetic moments (for instance, the elements of the iron group, and the rare earths).

The assertions of the theory about the magnitude of the magnetic moments of atoms can be tested by determining the magnetic sus-ceptibility of the substances in question (Weiss, about 1910). However, this only gives a value averaged over all those directions, and all those values of magnetic moments, which occur; a direct test of the theory cannot therefore be immediately obtained by this method.

The magnitudes of the moments can also, however, be measured in the individual atom, as was shown by Stern and Gerlach (1921). Their method bears approximately the same relation to the macro-scopic method of Weiss, as Aston's method for the determination of atomic masses bears to the macroscopic method of determining atomic weights, which only gives mean values over all the isotopes present. The method of Stern and Gerlach is based upon the deflection of a molecular beam in a non-homogeneous magnetic field. We may regard the atom with the magnetic moment as an elementary magnet of dimen-sions which though small are still finite. If we bring this magnet into a homogeneous magnetic field, it will not be acted on by a resultant force, for the magnetic force acts on its north pole with the same strength as on the south pole, but in exactly the opposite direction; thus the axis of the magnet may possibly execute a pendulum motion or precession about the direction of the field, but the centre of inertia of the magnet either remains at rest or moves in a straight line.

It is otherwise when the field is not homogeneous. In that case the forces acting on the north and south poles are not quite the same, so

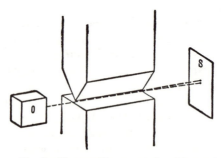

Fig. 16.—Diagrammatic representation of Stern and Gerlach's apparatus. A molecular beam issues from the oven O, and passes between the pole-pieces of the magnet (one of which has the form of a knife-edge) to the receiving screen S.

that, besides the couple which the two magnetic forces exert on the elementary magnet even in the case of a homogeneous field, a resultant force also acts on the magnet as a whole, and imparts to it an acceleration either in the direction of the magnetic field or in the opposite direction. If then an elementary magnet flies through a non-homogeneous magnetic field, it will be deflected from its rectilinear path. The amount of the deflection is determined by the degree of inhomogeneity of the field; in fact, to produce a deflection of sensible amount the inhomogeneity of the field must be so marked that the field changes decidedly even within the small length of the elementary magnet (which in our case is at most of the order of magnitude of atomic linear dimensions, viz. 10^{-8} cm.). Stern succeeded in producing a sufficient degree of inhomogeneity by suitable construction of the pole-pieces of a magnet, one piece being shaped as a knife-edge, while the other piece, which he set opposite it, had a flat face, or was provided with a groove (fig. 16). The magnetic lines of force consequently crowd together at the knife-edge, so that the magnetic field strength is considerably greater there than at the other pole-piece. A fine beam of atoms is now projected from an oven through a diaphragm system so as to pass between the pole-pieces. Each individual atom

PLATE XI

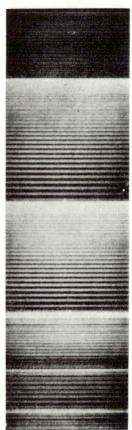

Ch. IX, Fig. 8.—N₂-band system, consisting of a set of rotational bands, each of them corresponding to the same change by 1 of the rotational quantum number, associated with a jump of the vibrational and the electronic quantum number (which displaces the system in the visible region). Each line is a very narrow triplet, produced by electronic-spin transitions (see p. 262).

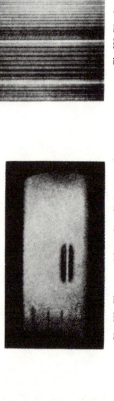

Ch. VI, Fig. 17.—Magnetic splitting up of a lithium beam by the method of Stern and Gerlach (see p. 187).

Ch. IX, Fig. 9.—CO-band near λ = 5610 Å, one of the so-called Ångström bands. The triplets which are clearly seen at the end away from the band head correspond to the three rotational changes by −1, 0, +1 (see p. 262).

is deflected in the non-homogeneous field, according to the magnitude and direction of its moment. The traces of the individual atoms can be made visible on the receiving screen (by intensification if necessary, as in photography).

According to the classical theory, a broadening of the beam must be produced in this way on the screen, since the moments of the atoms flying through the magnetic field can, on that theory, have all directions relative to the field. On the quantum theory, however, on account of the quantising of direction, not all settings are possible, but only a certain discrete number, as we have seen above in detail when considering the anomalous Zeeman effect (p. 157). The beam, as it appears on the screen, will therefore be *split up* into a finite number of discrete beams; in fact, there must appear on the screen exactly $2j + 1$ separate traces, if the atoms are in a state with the inner quantum number j; for in this case there are just $2j + 1$ possible settings, relative to the direction of the field, for the total angular momentum, and accordingly also for the total magnetic moment.

The experiment when carried out did in fact show a separation of the beam into several distinct beams; thus a beam of lithium atoms was split up into two beams (fig. 17, Plate XI), as it should be, since the ground term of the lithium atom is a ^2S-term (one valency electron in an s-orbit $[l = 0]$ with the spin moment $\frac{1}{2}$). The magnitude of the magnetic moment also could be determined from the amount of the separation. In this way Gerlach succeeded in producing a direct proof that the magnetic spin moment is exactly equal to one Bohr magneton. A systematic investigation, upon various elements, yielded results which were throughout in complete agreement with the theory.

Stern and Rabi have succeeded in so refining the accuracy of the measurements as to make it possible to measure the magnetic moments of nuclei. But this belongs to the domain of nuclear physics, and will be dealt with in a later section (p. 313).

8. Wave Theory of the Spin Electron.

In the preceding pages we have dealt with the theory of spin, and the questions connected with it—fine structure, Pauli's exclusion principle, &c.—solely on the basis of the vector model, regarding the angular momentum vectors as given magnitudes, and operating with them according to the rules of the quasi-classical theory of Bohr. We mentioned at the outset (p. 162) that this procedure can be justified by wave mechanics. We cannot give a complete treatment here of the wave mechanics of the spinning electron, but we should like to point out at least in what way electronic spin is actually brought within the ambit of wave mechanics.

This extension of wave mechanics was introduced by Pauli (1925). The leading idea of Pauli's theory is somewhat as follows. For simplicity consider an isolated electron. Its state, according to Schrödinger, is described by a wave function $\psi(x, y, z, t)$, where $|\psi|^2$ gives the probability of the electron being found at the point considered. We might now, keeping in view the idea of the rotating electron, introduce the spin into the wave equation by taking into account the rotational degrees of freedom. This, however, at once proves to be impossible; for in that case two new quantum numbers would appear in the solutions, as in every rotating system (e.g. l and m in the revolution of an electron); so that we would obtain an essentially greater number of states than the number actually found experimentally. The idea of the rotating electron is not to be taken literally.

As has already been proved in § 1 (p. 153), it follows unequivocally from the spectra, especially those of the alkalies, that for a fixed set of three quantum numbers n, l, m the electron can be in two, and only two, different states, which have also different energies. We can take this new degree of freedom into account formally, by introducing besides the ordinary co-ordinates an additional co-ordinate σ, which can take only two values altogether; we shall denote these values by $\sigma = +$ and $\sigma = -$, respectively. We can picture this to ourselves by supposing, say, that one value of this variable characterizes the state in which the spin is parallel to a specially distinguished direction, while the other value of the variable denotes the antiparallel setting. We thus obtain a wave function which now depends on five co-ordinates: $\psi = \psi(x, y, z, t, \sigma)$. It suggests itself, however, to split up this wave function into the two components

$$\psi = \begin{pmatrix} \psi_+(x, y, z, t) \\ \psi_-(x, y, x, t) \end{pmatrix}$$

which represent it respectively for the two possible values of the variable σ. It is evident that $|\psi_+|^2$ specifies the probability of hitting upon the electron at the place considered, and that with a spin direction parallel to the special direction; while $|\psi_-|^2$ is the corresponding probability for the opposite direction of the spin.

The question now meets us: how are we to calculate with these functions? For this Pauli has found the following method. In wave mechanics, there corresponds to every physical magnitude an operator, which is to be applied to the wave function. As such operators we can employ, as in Schrödinger's theory, differential operators [e.g. the component momenta $p_x = (\hbar/i)(\partial/\partial x)$]; or, as in Heisenberg's theory, matrices or other similar mathematical tools. It is therefore reasonable to associate operators also with the components s_x, s_y, s_z of the spin moment $\hbar$; these operators, however, do not act on the co-ordinates x, y, z, t, but on the two-valued variable σ. By this we are to understand that the application of these operators either changes the value of the quantity σ or leaves it unchanged, while at the same time the whole wave function may possibly be multiplied by a factor. It has proved advantageous to regard the operators as " linear transformations " (matrices), and we usually write them in the form $s = \frac{1}{2}\hbar\sigma$ where the components of s are the Pauli matrices

$$\sigma_x = \begin{pmatrix} 0 & 1 \\ 1 & 0 \end{pmatrix}, \quad \sigma_y = \begin{pmatrix} 0 & i \\ -i & 0 \end{pmatrix}, \quad \sigma_z = \begin{pmatrix} 1 & 0 \\ 0 & -1 \end{pmatrix}.$$

If, generally,

$$a = \begin{pmatrix} a_{++} & a_{+-} \\ a_{-+} & a_{--} \end{pmatrix}$$

is such a matrix, then the operation a, applied to the vector of two components $\psi = \begin{pmatrix} \psi_+ \\ \psi_- \end{pmatrix}$, means the production of a new vector in accordance with the rule

$$a\psi = \begin{pmatrix} a_{++} & a_{+-} \\ a_{-+} & a_{--} \end{pmatrix} \begin{pmatrix} \psi_+ \\ \psi_- \end{pmatrix} = \begin{pmatrix} a_{++}\psi_+ + a_{+-}\psi_- \\ a_{-+}\psi_+ + a_{--}\psi_- \end{pmatrix}.$$

Thus we have

$$\sigma_x\psi = \begin{pmatrix} 0 & 1 \\ 1 & 0 \end{pmatrix} \begin{pmatrix} \psi_+ \\ \psi_- \end{pmatrix} = \begin{pmatrix} \psi_- \\ \psi_+ \end{pmatrix},$$

$$\sigma_y\psi = \begin{pmatrix} 0 & i \\ -i & 0 \end{pmatrix} \begin{pmatrix} \psi_+ \\ \psi_- \end{pmatrix} = \begin{pmatrix} i\psi_- \\ -i\psi_+ \end{pmatrix},$$

$$\sigma_z\psi = \begin{pmatrix} 1 & 0 \\ 0 & -1 \end{pmatrix} \begin{pmatrix} \psi_+ \\ \psi_- \end{pmatrix} = \begin{pmatrix} \psi_+ \\ -\psi_- \end{pmatrix}.$$

The matrices are so chosen that they are subject to the same commu-tation rules as the ordinary components of angular momentum (§ 5, p. 141); further, in this case the z-direction is the specially distin-guished direction, in which the spin, represented by the fourth co-ordinate, has set itself. For σ_z is a diagonal matrix with the proper values (diagonal elements) $+1$, -1; the z-component of the angular momentum has therefore one of the fixed values $+1$ or -1. On the other hand, σ_x and σ_y are not diagonal matrices; their values are therefore not measurable simultaneously with σ_z, but are only statistically determinate.

If now a magnetic field exists parallel to the z-axis, then, by § 2 (p. 157), the magnetic energy of the setting of the spin with respect to the field direction is given by

$$E_{\text{mag.}} = -2\,\frac{e}{2\mu c}\,\frac{1}{2}\,\hbar\,\boldsymbol{\sigma} \cdot \boldsymbol{H} = -\frac{e\hbar}{2\mu c}\,H\sigma_z.$$

The factor 2 here arises from the anomaly in the relation between magnetic and mechanical moment in the spin, as compared with the similar relation in the orbit—an anomaly mentioned on p. 159, and explained by Thomas as a consequence of the theory of relativity. In this equation $\boldsymbol{\sigma}$ is the Pauli spin operator and is its component in the direction of the field.

As for the wave equation itself, in the absence of magnetic fields it runs exactly like the Schrödinger equation for the problem in question:

$$(W_0 - E)\psi = 0.$$

Here W_0 denotes the energy operator, consisting of the kinetic and the potential energy; for a one-electron problem we have

$$W_0 = -\frac{\hbar^2}{2m}\,\nabla^2 + V.$$

Since W_0 does not act on the spin variable σ, we can split up the wave equation into the two components

$$(W_0 - E)\psi_+ = 0, \quad (W_0 - E)\psi_- = 0.$$

This signifies that without magnetic fields no transition is possible between the two possible settings of the spin, and that the electrons behave as if they possessed no magnetic moment.

If the atom is in a homogeneous magnetic field, the direction of which we shall take parallel to the z-axis, the usual energy operator W_0 is supplemented by a term due to the magnetic field, viz. $-(e\hbar/2\mu c)H\sigma_z$, so that the wave equation now runs:

$$\left(W_0 - \frac{e\hbar}{2\mu c}\,H\sigma_z - E\right)\psi = 0.$$

Here again the separation into the two partial functions can be effected at once, taking into account the meaning of the operator σ_z.

We thus obtain the two equations

$$\left(W_0 - \frac{e\hbar}{2\mu c} H - E\right)\psi_+ = 0,$$

$$\left(W_0 + \frac{e\hbar}{2\mu c} H - E\right)\psi_- = 0,$$

from which we see that the magnetic energy is simply added to or subtracted from the ordinary energy, so that an unperturbed term E_0 is split up by the magnetic field into the two terms

$$E_0 + \frac{e\hbar}{2\mu c} H, \quad \text{and} \quad E_0 - \frac{e\hbar}{2\mu c} H.$$

This double possibility of setting is the reason for the occurrence of doublets in the spectra of the one-electron atoms.

We shall not proceed further here with the development of Pauli's theory, but only remark that it has proved thoroughly successful, so long at least as the electron velocities are not too great. It must be emphasized, however, that Pauli's theory cannot be regarded as an explanation of the existence of spin, since of course when it was constructed the experimental facts, like the twofold possibility for the spin, and the ratio of the mechanical and magnetic moment, were simply inserted in the theory from the outside.

An essential advance in this direction is made in Dirac's *relativistic wave theory of spin* (1928). The object aimed at was to set up a wave equation which should satisfy Einstein's principle of relativity. According to this principle, the space co-ordinates x, y, z and the time t (the last multiplied by $c\sqrt{-1}$, where c is the velocity of light) stand on a precisely equal footing. But Schrödinger's differental equation (p. 134) is not symmetrical in the four co-ordinates; it is of the second order in the spatial differential coefficients, but of the first order in the time derivative. Now Dirac has set up a wave equation which satisfies the relativistic postulate of symmetry, and, as regards form, is of the first order in all four variables. Although we cannot discuss this theory in detail, we have to emphasize particularly that, from very general principles, and with no special assumption about the spin, it deduces all those properties of the electron which we have summarized under the word spin (mechanical and magnetic moment in the correct ratio). In its results it is to a large extent equivalent to Pauli's theory, but it goes decidedly beyond the latter, especially in regard to fast electrons. The main difference is that Dirac's theory operates with a 4-component wave function instead of Pauli's 2-component function (ψ_+, ψ_-). There must be therefore, apart from the two settings of the spin, some other property with two possible values for the state of the electron. What may this be?

In the theory of relativity the relation between the energy and the component momenta is quadratic in all these quantities, viz. (Appendix V, p. 370)

$$E^2 = p^2c^2 + m_0{}^2c^4.$$

For a given momentum p there are therefore always two values of the energy, $+E$ and $-E$. It is just this double sign of energy which necessitates the use of a 4-component ψ-function in Dirac's wave equation. But what is the meaning of negative energy values for a free particle? The idea is certainly unfamiliar and strange, since in classical (non-quantum) mechanics the energy can always be regarded as positive; this is evident in the non-relativistic case, since the kinetic energy is proportional to the square of the velocity, but holds also in relativistic mechanics. For with the smallest value of p, viz. $p = 0$, we associate the value $E_0 = +m_0c^2$; since p and E vary continuously, we have therefore, corresponding to all values of $p > 0$, the values $E > m_0c^2 > 0$. This is different in quantum theory; here changes of state may take place by a jump; what is then to prevent E from falling by more than $2m_0c^2$ in some process, and accordingly, by leaping over the excluded region from $-m_0c^2$ to $+m_0c^2$, getting into the negative region $E < -m_0c^2$? A simple example of this sort has been pointed out by O. Klein; it is called Klein's paradox. If an electron is caused to fly through two wire grids, set up one behind the other (fig. 18), and a counter field is applied between the grids, the electron in passing through loses part of its kinetic

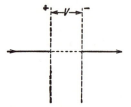

Fig. 18.—Illustration of Klein's paradox. According to Dirac's relativistic wave mechanics, an electron can overcome a counter field, if the potential difference exceeds twice the rest energy m_0c^2 of the electron.

energy. If the retarding potential is increased, the loss of energy by the electron becomes greater and greater, until finally it is unable to pass through the grids, but is reflected. So far, everything is classically in order. But if we go on to increase the counter field still further, Dirac's theory shows that electrons can once more pass through the grids, provided the potential difference between them is greater than twice the rest energy m_0c^2 of the electron. In this case, however, the electrons emerge from the field with negative energy. If now we calculate the probability that the passage through this counter field will take place, we find that it increases with increasing field strength between the two grids, but that for all electrostatic fields which can be produced experimentally it is vanishingly small. Although this is reassuring as it leaves unquestioned the applicability of Dirac's theory to ordinary problems, the possibility of transitions to states of negative energy remains, and there is no doubt that in atomic processes there occur cases of sufficiently extreme potential gradient to produce such transitions. We are therefore faced with the problem as to what these states of negative energy really are in terms of experimental language.

Dirac has answered this question by the following bold line of argument.

In the formal theory, states of negative energy are on a perfectly equal footing with those of positive energy. This equality does not appear, however, to exist in nature; ordinary processes of motion all take place in the domain of positive energy. To give expression to this fact, Dirac assumes that all the states of negative energy are normally occupied by negative electrons, which, however, do nothing to make themselves physically observable. We are therefore to consider that we are continually surrounded on all sides by electrons with all possible (negative) energies, without our noticing it at all. That of course implies an amendment to the ordinary accepted laws of electrodynamics: only electrons with positive energy can generate a field. From this sea of electrons, which extends from $(-\infty)$ to the level $-m_0 c^2$, electromagnetic fields, say in an act of absorption in the atom, can raise an electron, and bring it into the energy region $E > m_0 c^2$; it then takes its place as an ordinary electron and generates electromagnetic fields in accordance with the ordinary laws. There remains behind, however, an empty place in the sea of negative energy states—a hole—and clearly this has entirely the same character as a positive charge. Dirac's formulæ show that this hole in the midst of the ordinary negative electrons with negative energies behaves exactly like a positive electron with positive energy. Klein's paradox can now be solved very simply by applying Pauli's exclusion principle, according to which each state (defined by a full set of quantum numbers) can be occupied by one electron only. The state of negative energy, into which the incident electron would go after passing through the opposing field, is already occupied by another electron; the process is therefore impossible.

Dirac thought at first that these holes corresponded to the protons, though he was conscious of the difficulty of the difference in the masses. But when the first indications of the existence of positrons were found by Anderson (see p. 47), Dirac at once saw in this a confirmation of his theory, and by prediction of phenomena took a hand in the direction of experimental research.

The most important consequence of the theory is that, by absorption of light quanta rich in energy, a pair, electron + positron, can be produced, and that conversely two such particles can unite, and produce a pair of light quanta. The electron and positron can also exist for a short time before annihilation in a metastable bound state known as *positronium* similar to the hydrogen atom, but in which the positron now replaces the proton. The calculation of the probability for these processes follows the usual methods of wave mechanics, since the first process, according to Dirac, really consists in the raising of an electron out of a state (of negative energy), the second in the fall of an electron from a high state into a hole. Both processes are positively demonstrated to-day, as experimental results. We have the first in the showers which are produced by cosmic rays (see p. 48); further, cases are observed where by irradiation of light elements with γ-rays positive and negative electrons are liberated, which can be attributed to this process of materialization. In the converse process, the neutralization of two electrons of opposite sign, the energy $2m_0c^2 = 10^6$ electron volts becomes free (if the kinetic energy of the electrons is neglected); but momentum must also be conserved, and this excludes the case of emission of a light quantum, which of course has the definite momentum $m_0c^2/c = m_0c$. There are therefore two possible cases: either two light quanta originate in empty space, each equal to $m_0c^2 = \frac{1}{2} \times 10^6$ electron volts; or a particle in the neighbourhood takes up the residual momentum. If this is an electron, it will carry off a third of the energy $2m_0c^2$, so that the light quantum will retain $\frac{4}{3}m_0c^2 = 680,000$ electron volts; if it is a nucleus, it will on account of its great mass take over very little energy, and the light quantum will therefore take the full amount of $2m_0c^2 = 10^6$ electron volts. In point of fact, in scattering of γ-rays by atoms, besides the ordinary Compton scattering monochromatic waves are emitted, one of which has an energy of very exactly half a million electron volts; it arises presumably from the reunion of a positron, produced primarily at the absorption of the γ-rays, with a negative electron (Gray and Tarrant, Joliot). The process has been demonstrated unequivocally by Klemperer (1934). He made boron and carbon artificially radioactive by bombardment with protons and deuterons (see Chap. III, § 5, p. 68) ; these materials then become a powerful source of positrons. The positrons are absorbed by a metal piece which is enveloped on its two sides by two hemispherical counters, adjusted for γ-rays (see p. 31). It is then observed that the two counters give a simultaneous response, with a frequency exceeding that of any possible chance coincidences, so that two light quanta are being emitted at the same time in opposite directions; and it is found by absorption measurements that it is the half-million electron volts radiation which is involved. The other two processes are (in agreement with the theory) extremely rare.

Recently (1955-6), the discoveries of the anti-proton (p^-) and the anti-neutron ($\tilde{n}^0$) by Chamberlain, Segre, Wiegand and Ypsilantis, have added further support to the basic ideas of Dirac's theory. These particles bear the same relation to the proton and neutron respectively as the positron does to the electron.

In spite of its successes, Dirac's hole theory of the positron is still far from satisfactory, since proper account is not taken of the inter-action of the great body of negative-energy electrons with the radiation field. Now we have seen (p. 80) that when dealing with the interaction of electromagnetic radiation with atoms and electrons it is necessary to treat the radiation as being quantized. Formally, it is possible to set up a well-defined scheme for quantizing such a wave field, the procedure being referred to as " second quantization ". Further, it might be expected for reasons of symmetry that the Dirac wave field should also be quantized and indeed such a quantization can be carried out in a similar fashion. The result of this quantization is that we now have a theory which represents a physical system consisting of an assembly of positive-energy electrons and positrons interacting with an assembly of photons, and which provides for the possible creation or annihilation of all members of the system. However, new difficulties now present themselves. One of these is connected with the infinite self-energy of the electron caused by its interaction with the radiation field. Another is associated with a phenomenon usually referred to as polarization of the vacuum. Even if no electron-positron pairs are actually formed, the virtual existence of these states has physical consequences, namely, the appearance of charge density distributions in the vacuum as if this had a dielectric polarizability; the mathematical expressions for these quantities often lead to infinite values.

It has been found possible, however, to remove these infinities by the techniques of mass and charge renormalization (Bethe, 1947; Kramers, 1948; Tati and Tomonaga, 1948; Schwinger, 1948; Dyson,

1949). The effect of the interaction of an electron with the radiation field, for instance, is to give it an additional energy proportional to its kinetic energy, the constant of proportionality having the form of an infinite integral which diverges logarithmically. Now this additional energy could also be produced by an increase in the rest mass of the electron from m to $m + \delta m$ where δm includes the infinite integral. However, the observed rest mass of any electron is $m + \delta m$ since it can always interact with the radiation field, so that the quantity $m + \delta m$ should be used *ab initio* in the theory. We must therefore compensate for it in the present theory by subtracting the additional contribution to the kinetic energy due to δm. This will then just cancel the infinite electron self-energy part. This process is known as *mass renormalization* and removes the infinities occurring in the theory resulting from radiation interaction. The remaining infinities due to polarization of the vacuum can be removed by a similar process known as *charge renormalization*. In this case the infinities arising are equivalent to changing the electronic charge by δe, where δe represents the charge induced in the vacuum and includes all diverging integrals. However, since the vacuum is always present, the observed charge is always $e + \delta e$ so that, as with mass renormalization, to compensate for this a subtraction of the terms due to δe must be made, and this again exactly cancels the troublesome infinities.

Thus, although it is a major unsolved difficulty of the theory that infinite quantities should appear, it is possible to remove them in an unambiguous manner by the process of renormalization.

9. Density of the Electronic Cloud.

We now return to the problems considered in § 6 of this chapter, and discuss the way in which the probability density $|\psi|^2$ of the electron cloud can be investigated experimentally by using intensity measurements of X-rays and electronic rays scattered by the substance.

We have seen (Chap. III, § 2, p. 59; Appendix VIII, p. 376) that ordinary light is scattered by free electrons. It is found that the intensity I_0 of the incident beam produces radiation of intensity I_f at a point at a distance r and on a line making an angle θ with the incident direction given by

$$I_f(\theta) = I_0 \left(\frac{e^2}{mc^2}\right)^2 \frac{1 - \frac{1}{2}\sin^2\theta}{r^2}.$$

If the electrons are not free but bound in an atom, this expression is not valid any more, but has to be multiplied by a factor which depends on the density of the electronic cloud; the calculation (carried out in Appendix XXIV, p. 419) gives

$$I(\theta) = |\, F(\theta) \,|^2\, I_r(\theta),$$

where
$$F(\theta) = 4\pi \int_0^\infty r^2 \,|\, \psi(r) \,|^2 \frac{\sin Kr}{Kr}\, dr$$

is called the *atomic form factor*. The quantity K appearing in it is a function of the angle θ between the direction of the incident light and the direction of observation, namely,

$$K = 2k \sin\tfrac{1}{2}\theta,$$

where $k = 2\pi/\lambda = 2\pi x$ is the wave number per 2π cm.

From an observation of $F(\theta)$ for all angles of deflection θ, we can determine $|\, \psi(r) \,|^2$ by solving the above given integral equation. Though this is mathematically simple, namely, reversing an ordinary Fourier integral, its practical application is extremely complicated; for we do not encounter single atoms, but have to do with agglomerations of great numbers in gases, liquids, or solids, which by their structure produce interference patterns superposed on that expressed by the form factor. In gases and liquids, where the atoms are arranged in a disorderly fashion, we obtain rings of diminishing strength around the incident beam, depending essentially on the mean distance of the atoms. In solid crystals we have the well-known Laue-Bragg pattern of interference. The form factor modifies the intensity distributions of these effects, but it is not the only influence: there are purely geometrical factors, like the widths and distances of the slits used, the temperature effect produced by the heat motion of the atoms, the influence of imperfections of the crystal, &c. Hence it is not easy to separate the form factor from all other factors determining the observed intensity.

In the case of scattering of electrons we obtain for sufficiently large velocities for the number deflected through the angle θ per unit of solid angle (Appendix XXIV, p. 419), the so-called *differential cross-section*,

$$Q(\theta) = |f(\theta)|^2,$$

where
$$f(\theta) = \frac{e^2}{2mv^2} \frac{1}{\sin^2\frac{1}{2}\theta} \{Z - F(\theta)\},$$

which reduces to Rutherford's formula (Chap. III, § 3, p. 62; Appendices IX, p. 377, and XX, p. 406), if $F(\theta)$ is neglected. The *form factor* $F(\theta)$ represents therefore the screening of the nuclear charge by the electronic cloud.

The practical application of this formula encounters the same difficulties as in the case of X-rays, due to the collective action of many atoms in ordered or disordered arrangement, and to all the other effects mentioned above; moreover, the scattering formula holds only for fast electrons, for slow ones more elaborate methods must be used.

These considerations show the desirability of a theoretical determination of the form factor. This problem is equivalent to a calculation of $|\psi|^2$ from the theory.

The simplest method, suggested independently by Thomas (1926) and Fermi (1928), consists in neglecting completely the different states of the single electrons, and replacing the individual wave functions by a statistical average. As we have seen in Chap. V, § 2, p. 116, the original quantum conditions of Bohr and Sommerfeld can be interpreted in the following way:

If a pair of conjugated co-ordinate and momentum, q, p is represented as co-ordinates in a qp-plane, the quantum condition means the partition of this plane by a set of curves of constant energy, $H(p, q) = $ const., in such a way that the areas between neighbour curves are all equal to h. If such an area is called a cell, we can say that there is exactly one state per cell. Quantum mechanics confirms this fact. This is seen from the uncertainty law of Heisenberg, according to which $\Delta q \Delta p \sim h$, which can be interpreted as meaning that no subdivision of the qp-plane in cells smaller than h has a physical significance. This qualitative reasoning can be transformed into an exact statement by considering in detail the possible states of a free particle, as will be shown later (Chap. VII, § 2, p. 216). If it is extended to spatial motion, we have three co-ordinates and three momenta; hence, instead of the qp-plane, we have a 6-dimensional space, and this is to be divided in cells of 6-dimensional volume h^3.

We consider now all the electrons of the atom to be under exactly the same conditions, namely, each electron subject to the law of conservation of energy, with a potential energy $e\phi$, where ϕ is the averaged potential due to the nucleus and all the other electrons.

Then we apply the Pauli principle, which says that in each state, i.e. in each cell of volume h^3, there can be only two electrons (not one, as there are two spin orientations possible). If P is the maximum

momentum which can occur, the momentum space will be filled from its origin $P = 0$ up to the boundary of a sphere with radius P, hence with volume $\frac{4}{3}\pi P^3$; and this represents also the volume of the 6-dimensional qp-space if the volume in co-ordinate space is taken to be unity. If we multiply this volume by 2, and divide it by h^3, we obtain the number of electrons per unit volume with a momentum smaller than P, namely, $\frac{8}{3}\pi P^3/h^3$, and the product of this into the charge e of an electron is the density of the electrons with $p < P$,

$$\rho = -\frac{8\pi e}{3}\left(\frac{P}{h}\right)^3.$$

Now we can express both ρ and P in terms of the mean electrostatic potential ϕ.

For ρ this is very simple; we have Poisson's equation, which, under the assumption that both ρ and ϕ are symmetrical about the nucleus and depend only on the distance r from it, reads

$$\nabla^2\phi = \frac{1}{r}\frac{d^2}{dr^2}(r\phi) = -4\pi\rho.$$

P on the other hand can be obtained from the energy law

$$\frac{p^2}{2m} - e\phi(r) = E$$

by the consideration that only those electrons are bound in the atom for which the work necessary to bring them to the " surface " of the atom is positive. In the case of the hydrogen atom (Chap. V, § 1, pp. 104, 106) there is only one electron, and therefore no such surface; the zero of energy corresponds to infinite distance. It will be seen later that the same holds for all neutral atoms; but in order to include ions, with a resulting charge ez, we have to introduce a surface, namely, a sphere of radius R, outside of which the potential is the Coulomb function ez/r; it is obvious that this has not to be counted in determining the work necessary to remove an electron. Taking $E = 0$ for the electron on this sphere, we have for the maximum momentum

$$P = [2me\phi(r)]^{1/2},$$

and for ϕ the boundary conditions

$$\phi(R) = 0, \quad \left(\frac{d\phi}{dr}\right)_R = \left[\frac{d}{dr}\left(\frac{ez}{r}\right)\right]_R = -\frac{ez}{R^2}.$$

A further boundary condition is obtained by remarking that close to the nucleus the contribution of the electrons to the potential is negligible, so that $\phi(r) \to eZ/r$, or

$$r\phi(r) \to eZ, \quad \text{for} \quad r \to 0.$$

If the value for P is introduced into the formula for ρ, we obtain

$$\rho = -\frac{8\pi}{3} \frac{e}{h^3} [2me\phi(r)]^{3/2}.$$

This multiplied by -4π is the right-hand side of Poisson's equation, which now contains the only unknown function $\phi(r)$. We can get rid of all numerical constants by introducing new units for length and potential, writing

$$x = \frac{r}{a}, \quad \Phi(x) = \frac{1}{Ze} r\phi(r)$$

$$X = \frac{R}{a}, \quad a = a_1 \left(\frac{9\pi^2}{128Z}\right)^{1/3},$$

where $a_1 = h^2/(4\pi^2 me^2)$ is the first radius of Bohr's hydrogen model (Chap. V, § 1, p. 111). If this is substituted into the differential equation, it reduces, after an elementary calculation, to

$$\frac{d^2\Phi}{dx^2} = \sqrt{\frac{\Phi^3}{x}},$$

while the boundary conditions become

$$\Phi(0) = 1, \quad \Phi(X) = 0, \quad X\left[\frac{d\Phi}{dx}\right]_x = -\frac{z}{Z}.$$

Hence the solution depends only on the quotient z/Z of the ionic charge of the total system to the nuclear charge, and different ions with the same z/Z have similar distributions, differing only in the scale of radius and potential.

For the neutral atom ($z = 0$), we have the boundary conditions $\Phi = 0$ and $d\Phi/dx = 0$ for $x = X$; the differential equation then shows that $d^2\Phi/dx^2 = 0$ for $x = X$, and, by repeated differentiation, that all higher derivatives vanish for $x = X$. This is only possible if $X = \infty$, as otherwise $f(x)$ would vanish identically. Hence the radius of the neutral atom, according to this simple theory, turns out to be infinite (as mentioned before).

The corresponding curve is shown in the diagram (fig. 19) as the outermost line; it approaches zero so quickly that the theoretical infinity of the radius does not matter practically. This curve and the others, representing positive ions, have been numerically computed by Fermi, and confirmed using a differential analyser (Bush and Caldwell, 1931). The sections of the abscissæ, denoted by X_1, X_2, correspond to the radii of the positive ions, and the tangents at these points $(X_1, 0)$, $(X_2, 0)$ intersect the ordinate axis in the points $(0, z_1/Z)$, $(0, z_2/Z)$, . . . corresponding to the different stages of ionization.

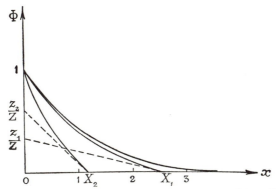

Fig. 19.—Thomas–Fermi curve for the neutral atom (outermost line) and the positive ion, for two degrees of ionization (inner lines); X_1, z_1 correspond to low ionization, X_2, z_2 to high ionization.

From " Atombau and Spektralinien " by A. Sommerfeld, by permission of the publishers, Messrs. Vieweg & Sohn

These curves give, of course, only a crude picture of the electronic distribution; it will be fairly accurate for the interior of heavy atoms (large Z), but quite wrong for light atoms and for the outer layers of all the atoms. There is no shell structure evident from direct inspection. However, Fermi has shown that his simple model accounts for the most important feature of the periodic system connected with the shell structure; we can determine the value of the atomic number Z for which the quantum number l of angular momentum m first reaches one of the values 0, 1, 2, 3, . . . , or in other words, for which element orbits of the type s, p, d, f, . . . first appear. For this purpose we have to calculate that fraction of points in the sphere of radius P in the momentum space which corresponds to a given value of angular momentum; it turns out that for a given l there is no such part of the sphere if Z is smaller than a given value. Thus it has been found that the s, p, d, f-state should first appear for $z = 1, 5, 21, 55,$

while they actually appear (see Table V, p. 181–3)   for H 1, B 5, Sc 21, Ce 58, in fair agreement.

We can obtain an explicit expression for the atomic form factor by introducing $\rho = -e \, |\, \psi \,|^2$ into the general formula above, and using the result for ρ in terms of $\phi(r)$ or $\Phi(x)$, namely,

$$F(\theta) = Z \int_0^\infty x^{1/2} \left[\Phi(x) \right]^{3/2} \frac{\sin \sigma x}{\sigma x} \, dx, \quad \sigma = \frac{4\pi a}{\lambda} \sin \tfrac{1}{2}\theta,$$

where a is the characteristic length introduced above, which is proportional to $Z^{-1/3}$. The universal function $F(\theta)/Z$ can be tabulated

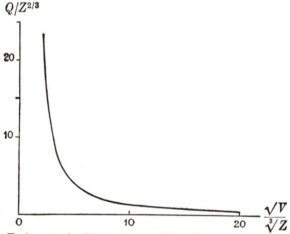

Fig. 20.—Total cross-section Q for an electron colliding with an atom of atomic number Z (without loss of energy through excitation, " elastic "), calculated by use of the Thomas–Fermi field. The abscissa is $V^{1/2}Z^{-1/3}$, where V is the accelerating potential in volts; the ordinate is $QZ^{-2/3}$ in units πa_1^2.

From " Atomic Collisions " by Mott and Massey, by permission of the publishers, Clarendon Press, Oxford

numerically as a function of σ, which depends on the ratio of the wavelength to the atomic " radius " a.

In a similar way we can calculate the differential and total cross-section for electronic collisions. It is found that the total cross-section Q, multiplied by $Z^{-2/3}$, is a function of $V^{1/2}Z^{-1/3}$ alone, where V is the accelerating potential. The resulting curve is represented in fig. 20; it shows in a condensed way the size of the obstacle which atoms of different Z form in the way of a beam of fast electrons.

Here the possibility of excitation of the atom is neglected. The theory of this effect can also be developed on similar lines (Born,

1926), and leads at once to an explanation of the experiments of Franck and Hertz (pp. 85, 86). In this case the incident electrons lose energy which is transferred to the electrons of the atom; we speak, therefore, of *inelastic scattering*. The cross-sections for each of these processes can be expressed in terms of the potential energy of interaction with the help of the same kind of approximation used for the elastic collisions (valid for high velocities). For the actual calculation, the Thomas-Fermi model is of course useless, as it holds only for the ground state. We have, in this case, to use more elaborate methods to obtain approximate wave-functions for the atomic electrons.

A very powerful method has been devised by Hartree (1928), known as the method of the *self-consistent field*. The Schrödinger equation is solved for each electron separately under a potential which is the average of that produced by all the other electrons. The latter depends again on the motion of these electrons, and thus the problem can be solved only by a method of trial and error. We may start, for instance, with the Thomas-Fermi potential of a group of electrons and determine the motion of a further electron produced by this field; doing this for one electron after the other, we can calculate a corrected mean potential, and repeat the procedure until the result is self-consistent, i.e. no further improvement is obtained by repetition. Fock (1934) has shown how the indistinguishability of the electrons and the Pauli principle can be taken into account.

A great amount of numerical work in this line has been done, and tables for the wave functions of electrons, ordered according to shells (K-, L-, . . .) are available for many atoms. These can be used for the calculation of form factors, scattering cross-sections, and other observable quantities.

CHAPTER VII

Quantum Statistics

1. Heat Radiation and Planck's Law.

We have already mentioned several times that the quantum theory is of statistical origin; it was devised by Planck when he was endeavouring to deduce the law of heat radiation (1900).

If, contrary to the order of historical development, we have discussed the quantum theory of the atom before quantum statistics, we have our reasons. In the first place, the failure of the classical theory displays itself in atomic mechanics—for instance, in the explanation of line spectra or the diffraction of electrons—even more immediately than in the attempts to fit the law of radiation into the frame of classical physics. In the second place, it is an advantage to understand the mechanism of the individual particles and the elementary processes before proceeding to set up a system of statistics based upon the quantum idea.

In this chapter we shall supply what is lacking, and begin with the deduction of the *law of heat radiation*, following Planck's method. We think of an enclosure, say a box, whose walls are heated by some contrivance to a definite temperature T. The walls of the enclosure send out energy to each other in the form of heat radiation, so that within the enclosure there is a radiation field. We characterize this electromagnetic field by specifying the energy density u, which in the case of equilibrium is the same for every internal point; if we split up the radiation into its spectral components, we denote by $u_\nu d\nu$ the energy density of all radiation components whose frequency falls in the interval between ν and $\nu + d\nu$. Thus the function u_ν extends over all frequencies from 0 to ∞; it represents a continuous spectrum. Up till now we have been occupied with line spectra, which are emitted by individual atoms in rarefied gases. But even molecules, consisting of only a limited number of atoms, do not send out isolated lines as single atoms do, but narrow " bands ", often not resoluble. And the more numerous and more densely packed the atoms are, the more the lines run into one another so as to form continuous strips.

A solid represents an infinite number of vibrating systems of all frequencies, and therefore emits a continuous spectrum.

Now there is a theorem of Kirchhoff's (1859) which states that the ratio of the emissive and absorptive powers of a body depends only on the temperature of the body, and not on its nature; otherwise radiative equilibrium could not exist within a cavity containing substances of different kinds. (By emissive power is meant the radiant energy emitted by the body per unit time, by absorptive power the fraction which the body absorbs of the radiant energy which falls upon it.) By a *black body* is meant a body with absorptive power equal to unity, i.e. a body which absorbs the whole of the radiant energy falling upon it. The radiation emitted by such a body—called " black radiation " —is therefore a function of the temperature alone, and it is important to know the spectral distribution of the intensity of this radiation. The following pages are devoted to the determination of the law of this intensity.

With regard to the experimental production of black radiation, it has been proved by Kirchhoff that an enclosure (oven) at uniform temperature, in the wall of which there is a small opening, behaves as a black body. In fact, all the radiation which falls on the opening from the outside passes through it into the enclosure, and is there, after repeated reflection at the walls, finally completely absorbed by them. The radiation in the interior, and hence also the radiation which emerges again from the opening, must therefore possess exactly the spectral distribution of intensity which is characteristic of the radiation of a black body.

Without going beyond thermodynamics and the electromagnetic theory of light, we can deduce two laws regarding the way in which black body radiation (or, as it is also called, cavity radiation) depends on the temperature. *Stefan's law* (1879) states that the total emitted radiation is proportional to the fourth power of the temperature of the radiator; the hotter the body, the more it radiates. Proceeding a step further, W. Wien found the *displacement law* (1893) which bears his name, and which states that the spectral distribution of the energy density is given by an equation of the form

$$u_\nu = \nu^3 F\left(\frac{\nu}{T}\right),$$

where F is a function of the ratio of the frequency to the temperature, but cannot be determined more precisely with the methods of thermodynamics. These two theorems can be proved, as is shown in Appendix XXXIII (p. 451), by treating the radiation as a thermo-dynamic engine which—by means of a hypothetical movable mirror—can do work in virtue of the radiation pressure; so that, on account of the Doppler effect, arising from the motion of the mirror, the frequency of the radiation is altered, and therefore its energy content also. We may remark that Wien's law includes Stefan's law; to deduce the latter, we have only to integrate over the whole spectrum:

$$\int u_\nu d\nu = \int \nu^3 F\left(\frac{\nu}{T}\right) d\nu.$$

Putting $x = \nu/T$, and taking x as a new variable of integration, we find

$$\int u_\nu d\nu = T^4 \int x^3 F(x)\, dx,$$

so that the total radiation energy is proportional to the fourth power of T, since the integral with respect to x is independent of T, being a mere constant.

The reason for calling Wien's law the "displacement law" is this. It was found experimentally that the intensity of the radiation from an incandescent body, maintained at a definite temperature, was represented graphically, as a function of the wave-length, by a curve of the form sketched in fig. 1. For extremely short, as also for extremely long waves, the intensity is vanishingly small; hence it must have a maximum value for some definite

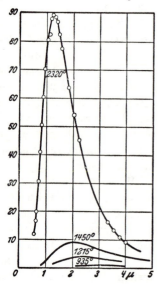

Fig. 1.—Distribution of intensity of heat radiation as a function of the wave-length, according to measurements by Lummer and Pringsheim.

wave-length λ_{max}. If we now change the temperature of the radiating body, the intensity graph also changes; in particular the position of the maximum is shifted. In this way it was found from the measurements that the product of the temperature and the wave-length, for the corresponding maximum of intensity, is constant; or

$$\lambda_{max} T = \text{const.}$$

This relation is explained at once by Wien's law. We have up to now referred to the energy distribution as a function of the frequency ν, u_ν representing the radiation energy in the frequency interval $d\nu$. The displacement law, however, refers to a graph showing the intensity distribution as a function of λ, so that we have now u_λ representing the energy in the wave-length interval $d\lambda$. The conversion from u_ν to u_λ is easy: obviously we must have $u_\nu d\nu = u_\lambda d\lambda$; and, since $\lambda\nu = c$, we have, as the relation between $d\nu$ and $d\lambda$, $|\,d\nu\,|/\nu = |\,d\lambda\,|/\lambda$. Hence, for the spectral distribution of energy expressed as a function of the wave-length, we find

$$u_\lambda = \frac{c^4}{\lambda^5}\, F\!\left(\frac{c}{\lambda T}\right).$$

We can now prove the displacement law at once, by calculating the wave-length for which u_λ is a maximum. The condition for this is $du_\lambda/d\lambda = 0$, or

$$\frac{c^4}{\lambda^5}\left\{-\frac{5}{\lambda}F\!\left(\frac{c}{\lambda T}\right) - \frac{c}{\lambda^2 T}F'\!\left(\frac{c}{\lambda T}\right)\right\} = 0,$$

from which there follows

$$\frac{c}{\lambda T}F'\!\left(\frac{c}{\lambda T}\right) + 5F\!\left(\frac{c}{\lambda T}\right) = 0.$$

This is an equation in the single variable $c/\lambda T$, whose solution, assuming it exists, must of course have the form $\lambda T = \text{const}$. Thus the theorem about the displacement of the intensity maximum with temperature follows immediately from Wien's law. The value of the constant, it is true, cannot be determined until the special form of the function F is known.

About this function, however, thermodynamics by itself has nothing to say; to determine it, we must have recourse to special representations by a model. All the same, it is clear from thermodynamical considerations that the form of the law given by the function F must be independent of the special mechanism. As the simplest model of a radiating body, Planck therefore chose a linear harmonic oscillator of proper frequency ν. For this oscillator, we can on the one hand determine the energy radiated per second—this being the radiation emitted by an oscillating dipole (p. 149); it is given by

$$\delta\epsilon = \frac{2e^2(\overline{\ddot{r}})^2}{3c^3} = \frac{2e^2}{3mc^3}\,(2\pi\nu)^2\,\bar{\epsilon},$$

where ϵ is the energy of the oscillator, and the bars denote mean values over times which, while great in comparison with the period of vibration, are yet sufficiently small to allow us to neglect the radiation emitted during its time of continuance. According to the equation of motion we have $\ddot{r} = -(2\pi\nu)^2 r$, and

$$\bar{\epsilon}_{\text{kin}} = \tfrac{1}{2}m\overline{\dot{r}^2} = \tfrac{1}{2}m\overline{(2\pi\nu r)^2} = \bar{\epsilon}_{\text{pot}} = \tfrac{1}{2}\bar{\epsilon}.$$

On the other hand, the work done on the oscillator per second by a radiation field with the spectral energy density u_ν is

$$\delta W = \frac{\pi e^2}{3m} u_\nu,$$

as follows from the equation of motion of the oscillator under the action of the field, and is proved in Appendix XXXIV (p. 455). In the case of equilibrium, these two amounts of energy must be equal. Hence we have

$$u_\nu = \frac{8\pi\nu^2}{c^3}\bar{\epsilon}.$$

If therefore we know the mean energy of an oscillator, we know also the spectral intensity distribution of the cavity radiation.

The value of $\bar{\epsilon}$, *as determined by the methods of classical statistics* (§ 5, p. 9), would be

$$\bar{\epsilon} = kT,$$

where k is Boltzmann's constant; this is a special case of a very general result of statistical mechanics, the *law of equipartition*, according to which any term in the Hamiltonian which is proportional to the square of a co-ordinate or a momentum contributes the same amount to the mean energy, namely, $\tfrac{1}{2}kT$. For the oscillator there are two such terms, hence the energy kT. We can prove this also directly by a simple calculation; according to Boltzmann's theorem (proved on p. 14), in the state of equilibrium the value ϵ for the energy of the oscillator occurs with the relative probability $e^{-\epsilon/kT}$, so that we obtain ϵ by averaging over all values of ϵ, with this weight factor. Putting for brevity $\beta = 1/kT$, we find

$$\bar{\epsilon} = \frac{\int_0^\infty \epsilon e^{-\beta\epsilon}\,d\epsilon}{\int_0^\infty e^{-\beta\epsilon}\,d\epsilon} = -\frac{d}{d\beta}\log\int_0^\infty e^{-\beta\epsilon}\,d\epsilon$$

$$= -\frac{d}{d\beta}\log\frac{1}{\beta} = \frac{1}{\beta} = kT.$$

If the classical mean value of the energy of the oscillator, as thus determined, is substituted in the radiation formula, the result is

$$u_\nu = \frac{8\pi\nu^2}{c^3}\, kT.$$

This is the *Rayleigh-Jeans radiation law* (1900, 1909). We see in the first place that it agrees, as of course it must, with Wien's displacement law, which, as a deduction from thermodynamics, holds in all cases. For the long-wave components of the radiation, i.e. for small values of ν, it also reproduces the experimental intensity distribution very well; in this region the intensity of the radiation increases with the frequency, in proportion to its square. For high frequencies, however, the formula is wrong; we know from experiment that the intensity function reaches a maximum at a definite frequency and then decreases again. The above formula, however, fails entirely to show this maximum; on the contrary, according to the formula, the spectral intensity distribution increases as the square of the frequency, and for extremely great frequencies, i.e. for extremely short waves, becomes infinite; the same is true of the total energy of radiation $u = \int_0^\infty u_\nu\, d\nu$; the integral diverges. We have here what is called the " ultra-violet catastrophe ".

Attempts have been made to remedy this very conspicuous failure in the theory by the following hypothesis. Suppose that, as in chemistry, a finite reaction time is needed before equilibrium is attained, and let the reaction velocity in the case of cavity radiation be very small, so that a very long time must elapse before equilibrium is reached, during which time the system as a whole, in consequence of external influences, will in general, we may well imagine, have completely changed. But this certainly does not go to the root of the matter; for we can suppose of course, at least theoretically, that the cavity can be maintained unchanged at constant temperature for as long as we please, so that the above abnormal state of equilibrium would necessarily set in some time after all.

In these circumstances, Planck put forward the bold idea that these difficulties can be removed by postulating *discrete, finite quanta of energy* ϵ_0; the energy of the oscillators is to be (besides $\epsilon = 0$) either equal to ϵ_0 or to $2\epsilon_0$ or to $3\epsilon_0$, and so on. In this way we do in fact obtain Planck's radiation law, which is verified experimentally with striking accuracy. The essential point is the determination of the mean energy $\bar{\epsilon}$; in point of form the investigation only differs from the earlier one (p. 208) in replacing integrals by sums; for the individual energy values, now as before, occur with the weight given by the Boltzmann factor, but now only the energy values $n\epsilon_0$ ($n = 0, 1, 2, 3, \ldots$) are possible, instead of all values as before. Hence, we have for the mean value,

$$\bar{\epsilon} = \frac{\sum\limits_{n=0}^{\infty} n\epsilon_0 e^{-\beta n\epsilon_0}}{\sum\limits_{n=0}^{\infty} e^{-\beta n\epsilon_0}} = -\frac{d}{d\beta} \log \sum_{n=0}^{\infty} e^{-\beta n\epsilon_0}$$

$$= -\frac{d}{d\beta} \log \frac{1}{1 - e^{-\beta\epsilon_0}} = \frac{\epsilon_0 e^{-\beta\epsilon_0}}{1 - e^{-\beta\epsilon_0}}$$

$$= \frac{\epsilon_0}{e^{\beta\epsilon_0} - 1}, \qquad (\beta = 1/kT)$$

so that, on substitution of this expression in the radiation formula, we obtain

$$u_\nu = \frac{8\pi\nu^2}{c^3} \frac{\epsilon_0}{e^{\epsilon_0/kT} - 1}.$$

In order that this formula may not be inconsistent with Wien's displacement law, which, being deduced from thermodynamics alone, is certainly valid, we must assume that

$$\epsilon_0 = h\nu,$$

where h is a universal constant (Planck's constant); for the temperature can only appear in the formula in the combination ν/T. This gives *Planck's radiation law*

$$u_\nu = \frac{8\pi h\nu^3}{c^3} \frac{1}{e^{h\nu/kT} - 1}.$$

As we remarked above, this radiation formula is in very good agreement with experimental results. The distribution of intensity for various temperatures is shown graphically in fig. 2. For low frequencies the function increases approximately as the square of ν; for, if $h\nu/kT \ll 1$, we can expand the exponential function in the denominator, and obtain a series in ascending powers of $h\nu/kT$, the first term of which corresponds precisely to the Rayleigh-Jeans radiation formula:

$$u_\nu = \frac{8\pi h\nu^3}{c^3} \frac{1}{(1 + h\nu/kT + \ldots) - 1} = \frac{8\pi\nu^2}{c^3} kT + \ldots;$$

for long-wave radiation Planck's formula therefore agrees with the
classical one.   Not so in the short-wave region; if $h\nu/kT \gg 1$, the
exponential function has a value much greater than unity, so that we

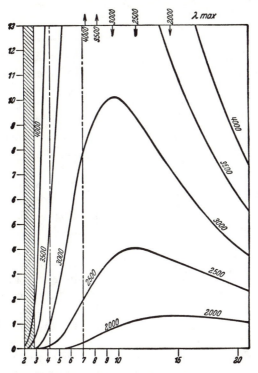

Fig. 2.—Spectral distribution of intensity of heat radiation according to **Planck**,
for temperatures from 2000° to 4000°. The unit for the abscissæ is 0·1μ. The
shaded part represents the ultra-violet region, as far as 0·28μ; the visible region
lies between the two dotted lines. The numbers on the curves are absolute temperatures. Note the agreement with the experimental results shown in fig. 1, p. 206.

can leave out the 1 in the denominator; for short waves we thus
obtain the radiation law in the approximate form

$$u_\nu = \frac{8\pi h \nu^3}{c^3}\, e^{-h\nu/kT},$$

which had previously (1896) been obtained by Wien in an attempt
to explain his measurements in this region of the spectrum (fig. 3).
Between the domains of these two limiting laws there lies a continuous transitional region, which contains the maximum of the
distribution curve.   This maximum, in accordance with Wien's law,
is displaced towards the region of short waves as the temperature

rises. Calculation of the intensity maximum leads in the same way as before to the relation

$$\frac{kT}{h\nu_{max}} = \frac{k}{hc}\,\lambda_{max}T = C,$$

where C is a constant, which is found by solving a transcendental equation, and has the value 0·2014. We have mentioned above (Chap, I, § 8, p. 23) that the atomic constants k and h can be determined by means of the spectral distribution of energy in black body radiation. The first determination of h, in fact, was effected by Planck by means of Stefan's law and Wien's displacement law. According to the former, the total energy radiated, from 1 sq. cm. of the surface of a body heated to temperature T, is $J = \sigma T^4$; σ is called the radiation constant, and has the value $\sigma = 5\cdot67 \times 10^{-5}$ erg/(cm.^{2} sec. degree^4). We obtain it theoretically by integrating Planck's distribution function over the whole spectrum; the result is $\sigma = 2\pi^5 k^4/(15c^2 h^3)$. On the other hand, the measurements give for the constant in Wien's law $[\lambda_{max}T = C(hc/k) =$

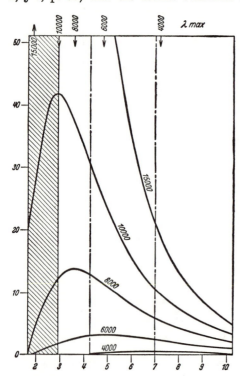

Fig. 3.—The same as fig. 2 for higher temperatures; the unit for the ordinates is smaller than in fig. 2

$0\cdot2014\,hc/k]$ the value 0·290 cm. degree. (To give an idea of its order of magnitude, it is sufficient to remark that the intensity maximum of the radiation from the sun, which radiates like a black body at the temperature $T = 6000°$, lies in the green region of the spectrum, i.e. at a wave-length of about 4500 Å.) From these two empirically determined constants, h and k can be calculated; the values so obtained agree very well with those determined by other methods.

We may add one brief remark. From the general course of the distribution curves we can see how low the efficiency is when glowing heated bodies, such as incandescent lamps, are used for purposes of illumination; for the visible spectral region occupies only a small strip in the figure giving the intensity curves for the heat radiation, and the rest of the radiated energy is lost, so far as illumination is concerned.

To return to the discovery of Planck's formula: it goes without saying that his hypothesis met at first with the most violent opposition. Physicists were unwilling to believe that the law of radiation could only be deduced by introducing the quantum hypothesis; this they regarded as a mathematical artifice, which might in some way or other be interpreted without going outside the circle of classical ideas. All attempts, however, at such an interpretation have been a complete failure.

To Einstein is due the credit of having been the first to point out that, apart from the law of heat radiation, there are other phenomena which can be explained by means of the quantum hypothesis, but which are unintelligible from the standpoint of classical physics. In the year 1905, Einstein put forward the *light quantum hypothesis*, and as experimental evidence for it cited the law of the photoelectric effect; this we have already dealt with (Chap. IV, § 2, p. 82).

2. Specific Heat of Solids and of Polyatomic Gases.

A year or two later (1907) Einstein showed that Planck's formula for the mean energy of an oscillator,

$$\bar{\epsilon} = \frac{h\nu}{e^{h\nu/kT} - 1},$$

finds direct confirmation in the thermal behaviour of solids. It is a fact of experiment that at high temperatures a law holds called *Dulong and Petit's law*, which states that the molecular heat, i.e. the specific heat per mole (p. 8), is approximately 6 cals/degree for all solids. From the classical standpoint this law is at once intelligible. In a solid, every atom can be regarded as a harmonic three-dimensional oscillator, since of course in a solid we think of the atom as quasi-elastically bound to a certain position of equilibrium. The mean total energy falling to them as three-dimensional oscillators, by the rules of classical statistics, is $3kT$; so that a mole of the substance possesses the energy $U = 3N_0 kT = 3RT$, where R is the gas constant, and is approximately equal to 2 cals/degree. The specific heat is found from this as the increase of energy for a temperature increase of 1°, so that

$$c_v = \frac{\partial U}{\partial T} = 3R \sim 6 \text{ cals/degree.}$$

Experiment, however, shows deviations from this rule; and the harder the body, i.e. the more firmly the atoms are bound to their equilibrium positions, the greater are the deviations; thus, for diamond, the specific heat per mole at room temperature is only about 1 cal/degree.

Einstein's explanation of these deviations is that it is not permissible in this case to use the classical expression for the mean energy of the oscillators, but that we must apply the expression obtained by Planck for the mean energy of a quantised oscillator. In that case the mean energy of the oscillators per mole is

$$U = \frac{3N_0 h\nu}{e^{h\nu/kT} - 1} = 3RT \frac{h\nu/kT}{e^{h\nu/kT} - 1}.$$

In this formula $h\nu$ is the elementary quantum of the vibrational energy of the oscillators; it is so much the greater, the more firmly the atoms are bound to their equilibrium positions; loose binding is equivalent to small vibrational energy, and so to low frequency. The important point now is, whether $h\nu$ is less or greater than kT. In general, at room temperature $h\nu/kT \ll 1$, so that we can simplify the formula for the energy of the oscillator by expansion. Hence in this case it passes over into the classical formula

$$U = 3RT \frac{h\nu/kT}{(1 + h\nu/kT + \ldots) - 1} = 3RT + \ldots,$$

and so gives Dulong and Petit's law.

If, however, the atoms are very firmly bound (as in diamond), or if the specific heat is measured at very low temperatures, $h\nu/kT$ becomes comparable with unity, or even greater than unity, and deviations occur from Dulong and Petit's law. We then obtain for the specific heat a curve of the form shown in fig. 4, which for large values of T rises asymptotically to the classical value of 6 cals/degree, but falls away for low temperatures, and at $T = 0$ passes through the origin. Experimental investigations to test the results given by this theory, mainly carried out by Nernst and his collaborators, showed approximate agreement between theory and experiment, particularly as regards the fact that the specific heat approaches zero as the temperature diminishes. Nevertheless they indicated deviations which proved that the theory in the form in which it then existed still required improvement. These improvements were made by Debye, and independently by Born and v. Kármán (1912). They rest on the following considerations. Up to this point we have dealt with the individual atoms of the solid (crystal), as if they performed undisturbed harmonic vibrations inde-

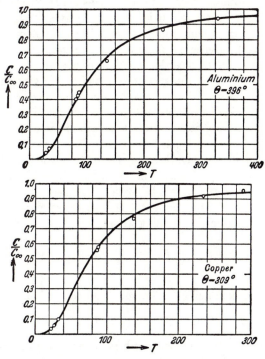

Fig. 4.—Graph of specific heats at low temperatures according to Debye; the small circles show observed points, the continuous curves correspond to Debye's theory. Θ is a temperature characteristic of the substance, such that $C(= C_v)$ is a function of T/Θ.

pendently of one another. This, however, is by no means the case, since of course the atoms in the lattice are very strongly coupled together. We therefore ought not to say that the N_0 atoms in the crystal perform oscillations of the same frequency, but must rather deal with the coupled system of $3N_0$ vibrations (corresponding to the $3L$ degrees of freedom of the N_0 atoms per mole), and accordingly write down the energy in the form

$$U = \sum_{r=1}^{3L} \frac{h\nu_r}{e^{h\nu_r/kT} - 1},$$

where ν_r is the frequency corresponding to an individual vibration.

It would of course be a difficult undertaking to try to evaluate this sum directly on the basis of some assumed model. But we can obtain approximate formulæ; the simplest method—which serves moderately well for monoatomic crystals—was given by Debye, and is as follows. The proper vibrations of the individual atoms of the crystal lattice appear in the ordinary theory of elasticity as elastic proper vibrations of the whole crystal, though, it is true, only waves whose lengths are much greater than atomic distances actually come under observation (as sound waves). Hence, for an approximate evaluation of the above energy sum, we may replace the spectrum of the proper vibrations of the individual atoms by the spectrum of the elastic vibrations of the whole crystal. We need to calculate the number of modes of oscillation with frequencies in the interval ν to $\nu + d\nu$. This number is needed continually in physics, and the result we are about to obtain will be used again in discussing black-body radiation, the states of electrons in crystals and in β-decay theory. The problem is exactly that of counting the number of modes of vibration in an elastic body, or simplest of all, those of an elastic string of length L. It was shown earlier (Chapter V, p. 138) that if the ends of the string are nodes then the eigenfunctions are the standing waves

$$\psi_n(x) = \sin\left(\frac{2\pi nx}{2L}\right),\ n = 1, 2, 3, \ldots$$

with the string containing an integral number of half-wavelengths. We prefer in practice to label the modes by their wave number $\kappa_n = \dfrac{n}{2L}$. Then the difference in the wave numbers of successive modes is

$$\delta\kappa = \kappa_{n+1} - \kappa_n.$$

The number of standing waves with wave numbers in an interval $d\kappa$ is thus

$$dN = \frac{d\kappa}{\delta\kappa} = 2L\, d\kappa$$

Since the frequency of the modes is given by $\nu = c\kappa$, where c is the velocity of a wave on the string, we find that the number of modes in the frequency interval ν to $\nu + d\nu$ is

$$dN = \frac{2L}{c}\, d\nu.$$

Rather than go on directly to the case of three dimensions, we will solve the problem again for a string of length L, using this time travelling waves rather than standing waves, since the former are more useful in many cases. Travelling waves $e^{2\pi i \kappa x}$ are indeed the appropriate solutions of the differential equation for the string (see Chapter V, 4, p. 137) if we use the periodic boundary condition (Born–von Kármán) that

$$\psi(x + L) = \psi(x).$$

Substituting the expression for a travelling wave, the allowed values of κ are seen to satisfy the equation

$$e^{2\pi i \kappa L} = 1.$$

The solutions are $\kappa = 0, \pm \dfrac{1}{L}, \pm \dfrac{2}{L}, \ldots$ and the string now contains an integral number of wavelengths. The difference in the wave numbers of successive modes is $\delta\kappa = \dfrac{1}{L}$ and the number of travelling waves with wave numbers in the interval $d\kappa$ is now

$$dN = \frac{d\kappa}{\delta\kappa} = L\, d\kappa.$$

Remembering that κ can have either sign, and that $\nu = c \mid \kappa \mid$ we see that the number of modes in the frequency interval ν to $\nu + d\nu$ is the sum of the number of modes in the wave-number intervals $\dfrac{\nu}{c}$ to $\dfrac{\nu + d\nu}{c}$ and $-\dfrac{\nu}{c}$ to $-\dfrac{\nu + d\nu}{c}$, so that again

$$dN = \frac{2L}{c}\, d\nu.$$

The number of modes is given correctly by both boundary conditions.

These results have a straightforward generalization to the case of a rectangular body with sides L_x, L_y, and L_z. The eigenfunctions have the form $e^{2\pi i \kappa \cdot r}$ and, when periodic boundary conditions are applied, the allowed values of κ are $\left(\dfrac{n_x}{L_x}, \dfrac{n_y}{L_y}, \dfrac{n_z}{L_z}\right)$ with $n_x = 0, \pm 1, \pm 2, \ldots$, $n_y = 0, \pm 1, \ldots$ etc. These values form a regular dense lattice of points. The number of modes having vector wave numbers with components in the intervals κ_x to $\kappa_x + d\kappa_x$, κ_y to $\kappa_y + d\kappa_y$, κ_z to $\kappa_z + d\kappa_z$ is then

$$dN = L_x L_y L_z \, d\kappa_x d\kappa_y d\kappa_z$$

which can be written

$$dN = V d^3 \kappa$$

where V is the volume of the box. This result is in fact true (Weyl) whatever the shape of the box, provided it is sufficiently large. We can use this formula now to compute the number of modes with frequencies in the interval ν to $\nu + d\nu$. These modes will have vector wave numbers in a spherical shell of radius $\kappa = \nu/c$ and thickness $d\kappa = d\nu/c$, where c is the velocity of the wave. The volume of this shell is $4\pi\kappa^2 \, d\kappa$ and we obtain the result

$$dN = \frac{4\pi V}{c^3} \, \nu^2 \, d\nu.$$

However, the crystal can support both transverse and longitudinal waves, the former having two degrees of freedom corresponding to the two possible mutually perpendicular directions of polarization. Taking both types of wave into account, the number of modes in the frequency interval ν to $\nu + d\nu$ is

$$dN = 4\pi V \left[\frac{2}{c_t{}^3} + \frac{1}{c_l{}^3} \right] \nu^2 \, d\nu$$

where c_t and c_l are the velocities of the transverse and longitudinal waves respectively. Defining a mean velocity of sound $\bar{c}$ by the equation

$$\frac{1}{\bar{c}^3} = \frac{2}{c_t{}^3} + \frac{1}{c_l{}^3}$$

this becomes

$$dN = \frac{4\pi V}{\bar{c}^3} \, \nu^2 \, d\nu.$$

We now return to the determination of the mean energy of the atoms in a solid. The formula already deduced (p. 215), viz.

$$U = \sum_{r=1}^{3L} \frac{h\nu_r}{e^{h\nu_r/kT} - 1},$$

can now be transformed by the method used above, i.e. by regarding the frequencies ν_r no longer as proper vibrations of the atoms, but as frequencies of elastic waves in the body. Since we know the number of these vibrations in the frequency interval between ν and $\nu + d\nu$, we can rewrite the sum as an integral, in the integrand of which this number of vibrations appears as the weighting factor for this vibrational state, alongside the original summand; thus

$$U = \int \frac{h\nu}{e^{h\nu/kT} - 1} \frac{4\pi V}{\bar{c}^3} \nu^2 \, d\nu.$$

It must be observed, however, that the total number of proper vibrations of the crystal is finite, viz. equal to $3N_0$; a maximum frequency ν_m therefore exists, which is defined by the equation

$$3N_0 = Z = \frac{4\pi V}{3\bar{c}^3} \nu_m^3, \quad \text{or} \quad \nu_m = \bar{c} \sqrt[3]{\left(\frac{9N_0}{4\pi V}\right)}.$$

The preceding integral is therefore not to be taken to ∞, but only to this limiting frequency as its upper limit. For the energy U we thus obtain the expression

$$U = \frac{4\pi V}{\bar{c}^3} \int_0^{\nu_m} \frac{h\nu^3 \, d\nu}{e^{h\nu/kT} - 1} = \frac{4\pi V}{\bar{c}^3} \left(\frac{kT}{h}\right)^4 h \int_0^{x_m} \frac{x^3 \, dx}{e^x - 1} = 3RT \frac{3}{x_m^3} \int_0^{x_m} \frac{x^3 \, dx}{e^x - 1},$$

where for brevity x_m is written for $h\nu_m/kT = \Theta/T$; the quantity $\Theta = h\nu_m/k$ is called Debye's *characteristic temperature*. *Debye's formula* gives a better approximation than Einstein's model of independent oscillators, all with the same frequency (fig. 4, p. 215). Born and v. Kármán derived independently a still more rigorous formula by taking into account the lattice structure of crystals, to which in the preceding counting process we have paid no heed whatever. They thus obtained separate formulæ for the various types of crystal structure which, however, for not too low temperatures, can be approximately represented by a sum of a few Debye functions with different Θ-values.

Empirically, this theory has been confirmed throughout. E.g. experiment gives a T^3-law for the specific heat at extremely low temperatures, whereas the simple Einstein theory leads to an exponential law for its increase with temperature. Debye's theory, however, leads to the correct law for low temperatures; in fact, for these temperatures, x_m tends to infinity, so that the integral in the formula for U becomes practically constant, while the fourth power of T stands before the integral; the specific heat is found from the energy by differentiation with respect to the temperature, and the experimental T^3-law therefore follows as it ought to do. Investigations by Blackman (1935), it is true, have indicated that sometimes the measured specific heat seems to obey the T^3-law at temperatures not low enough to justify the preceding theoretical explanation. He has shown that a refined application of the theory of Born and v. Kármán which takes account of the lattice structure of the crystals is able to explain these cases; the result can be expressed by saying that Debye's Θ is not constant, but increases for low temperatures. The real T^3-law should only be expected at the very lowest temperatures. This increase of the specific heat is actually observed. In the classical example of diamond it was possible to deduce the lattice forces from measurements of the elastic constants (Bhaghavantam and Bhimasenachar, 1944), and to account for the behaviour of the specific heat down to the lowest temperatures (H. M. S. Smith, 1946). One might say that at the present time the experimental facts relating to the specific heat of solids have been completely explained.

The quantum theory of the oscillator can be applied in an exactly similar way to *polyatomic gases*. Here also the specific heat, as determined experimentally, increases with temperature in accordance with Planck's formula corresponding to the molecular vibrations. We have seen in Chap. I (§ 5, p. 9) that the classical theory of specific heats has to assume that the electronic motion in the atom does not contribute to the energy, but that the heat content corresponds to the motions of the molecules as rigid structures; the atoms having only translatory motions, while diatomic molecules have two rotational degrees of freedom (round the two axes perpendicular to the central line). The explanation of this fact, which was rather puzzling to pre-quantum physics, is of course the enormous size of the binding energy of the electrons compared with the average energy kT of thermal motion. If we replaced the electrons by virtual oscillators having frequencies ν corresponding to the terms of the line spectra, then the energies $h\nu$ would be large compared with kT. At normal, and even very high temperatures, none of these oscillators would be excited. This explains immediately why they do not contribute to the specific heat.

Similar considerations hold for the rotations of molecules consisting of light atoms. We have seen that the rotational energy is given by $E = j^2 h^2 / 8\pi^2 A, j = 0, 1, 2, \ldots$, corresponding to Bohr's quantum theory (p. 112); wave mechanics replaces j^2 by $j(j+1)$ (p. 143), but in both cases the difference between neighbouring energy levels is $\epsilon = h^2/4\pi^2 A$. Here A means the moment of inertia of the molecule, and ϵ is greatest when A is smallest, i.e. for the molecule H_2. We should expect that for temperatures given by $kT < \epsilon$ the rotational energy of H_2 should vanish; this is really the case, as was first observed by Eucken (1912), who found that at about $40°$ K. the specific heat of H_2 has decreased to the value for a monatomic gas. For other gases the critical temperature is too low to allow an observation of this effect of the complete " freezing in " of the molecular rotations. But the beginning of the process, a decrease of the specific heat, can be observed, and from this the moment of inertia calculated. The results are in good agreement with other observations, as on band spectra (p. 256), and with calculations from theoretical models.

3. Quantisation of Black Body Radiation.

We now return to the law of cavity radiation. We have seen in the preceding section that Planck's hypothesis has been brilliantly successful not only for cavity radiation, but also in the theory of specific heats. The latter success furnishes additional strong support for the quantum theory.

On the other hand, the deduction of the radiation law by Planck's method is somewhat unsatisfactory, in so far as it is based in part on the laws of classical physics and only in part appeals to the quantum theory. The method by which the formula is obtained, connecting the mean energy of an oscillator with the radiation field in a cavity, viz. $u_\nu = (8\pi\nu^2/c^3)\bar{\epsilon}$, is purely classical; in deducing it, the classical laws of absorption and emission by an oscillator are employed.

Can this circuitous route via the absorbing and emitting oscillator be avoided? The idea suggests itself, in view of the methods of Rayleigh-Jeans and Debye, that we should try to deal with the electromagnetic field, within a cubical cavity with reflecting walls, in exactly the same statistical way as with the proper vibrations of crystals in the theory of specific heats. The cavity with reflecting walls possesses precisely the same kind of proper vibrations as a crystal, and we can therefore calculate the number of these vibrations in a definite frequency interval between ν and $\nu + d\nu$ in the same way as before:

$$dz = 2 \frac{4\pi V}{c^3} \nu^2 d\nu.$$

We have here, however, an extra factor 2, since for each possible wave-length and direction of wave normal there are two different waves, corresponding to the two independent directions of polarization.

It is now only consistent to assume that every proper vibration behaves like a Planck oscillator with the mean energy

$$\frac{h\nu}{e^{h\nu/kT} - 1},$$

where we are simply extending the method of calculation and line of thought of the last section, from the proper vibrations of the crystal lattice to the proper vibrations of the cavity. In this way we again obtain Planck's formula

$$u_\nu \, d\nu = \frac{1}{V} \, dz \, \frac{h\nu}{e^{h\nu/kT} - 1}$$

$$= \frac{8\pi h\nu^3}{c^3} \frac{1}{e^{h\nu/kT} - 1} \, d\nu.$$

Although this method of deduction is formally extremely simple, it contains all the same a serious difficulty of principle. The formula used for the mean energy of an oscillator is bound up with the idea that an oscillator of frequency ν can possess not only the energy $h\nu$, but also integral multiples of this energy quantum, the frequency of occurrence of the energy $n h\nu$ in an oscillator being proportional to $e^{-nh\nu/kT}$. Thus we obtained the mean energy of an oscillator as the mean value

$$\frac{\sum\limits_{0}^{\infty} n h\nu e^{-nh\nu/kT}}{\sum\limits_{0}^{\infty} e^{-nh\nu/kT}} = \frac{h\nu}{e^{h\nu/kT} - 1}.$$

If we extend these ideas to the proper vibrations of the cavity, treating them as if they were oscillators, it follows that an electromagnetic proper vibration of frequency ν can have the quantity of energy $n h\nu$. How has n to be interpreted in terms of the light-quantum hypothesis? To assume that one light quantum could have an energy $n h\nu$ would be contrary to experience (photoelectric effect). We have to interpret n as the number of light quanta with energy $h\nu$. But these are supposed to fly about like the free particles in a gas, and it seems impossible to obtain Planck's radiation formula from this picture. This will be elaborated in the next section and it will be shown that the difficulty may be overcome by a refinement of the statistical method.

4. Bose-Einstein Statistics of Light Quanta.

We now try to deduce the radiation formula by assuming that we have to do with a corpuscular system. The obvious suggestion is to apply the methods of the classical Boltzmann statistics, as in the kinetic theory of gases; the quantum hypothesis, introduced by Planck in his treatment of cavity radiation by the wave method, is of course taken care of from the first in the present case, in virtue of the fact that we are dealing with light quanta, that is, with particles (photons) with energy $h\nu$ and momentum $h\nu/c$. It turns out, however, that the attempt to deduce Planck's radiation law on these lines fails, as we proceed to explain.

We can characterize a light quantum by its vector wave number κ. Using the results obtained in Section 2, and remembering that there are two distinct polarizations, we note that the number of states with wave numbers in the range $d^3\kappa$ is

$$dN = 2V d^3\kappa.$$

The number of states in the spherical shell containing wave numbers with magnitudes between κ and $\kappa + \Delta\kappa$ is therefore

$$dN = 2.4\pi V \kappa^2 \, \Delta\kappa$$

which can also be written as

$$dN = 2 \frac{4\pi V}{h^3} p^2 \Delta p = 2 \frac{4\pi V}{c^3} \nu^2 \, \Delta\nu$$

since $p = h\kappa = h\nu/c$. In terms of the notation used in Chap. I, § 6, this number dN is the weight factor g_s of this shell, and we write

$$g_s = dN = 2 \frac{4\pi V}{h^3} p^2 \, \Delta p.$$

We can now see, with this expression before us, that the result could also be obtained by dividing up the whole phase space—i.e. co-ordinate space and momentum space together—into smallest cells of magnitude h^3, and then counting the number of these cells which lie in that region of phase space which corresponds to the spatial volume V and the region between p and $p + \Delta p$ in the momentum space; taking account of the doubling of states due to polarization, we thus find

$$g_s = 2 \frac{4\pi p^2 \Delta p V}{h^3},$$

i.e. the same expression as before. The improvement in the new method of counting, as compared with that of the Boltzmann statistics, is merely this, that the quantum theory assigns a definite size to the smallest cells, into which we have to partition the phase space for the purposes of statistics. In the ordinary kinetic theory of gases—disregarding the case of gas degeneration, which will be dealt with later— the size of the cells played no part, but dropped out in the further course of the investigation. In this case of light quantum statistics, however (as also in the case of gas degeneration, see § 5, p. 230), the size of the cells is of great importance. We may mention further, that the finite size h^3 thus found for a cell corresponds precisely to Heisenberg's uncertainty principle, according to which the position and momentum of a particle cannot be more exactly defined than is consistent with the relation $\Delta p \Delta q \sim h$; in view of this relation, it would in fact be quite meaningless to make a finer division of the phase space, as it is impossible to decide by experiment in which of these cells a particle lies.

We now return to the statistics of light quanta, and begin with a verification of the statement made above, that the idea of light quanta (together with the definition of the size of the cell by quantum considerations) is *not* sufficient to enable us to deduce Planck's formula, if we adhere to the Boltzmann statistics. For, according to Boltzmann's result, the number of quanta in a definite shell—which we shall characterize by a mean ν_s of the frequencies in the shell—is given by the product of the number g_s of the cells in this shell and the Boltzmann factor $Ae^{-\epsilon_s/kT}$, so that

$$n_s = Ag_s e^{-h\nu_s/kT} \qquad (\epsilon_s = h\nu_s).$$

The Boltzmann law of distribution was obtained, let us repeat (§ 6, p. 9), as the most probable distribution of the particles of a gas (in our case the light quantum gas) in the various shells (called cells in our earlier investigation), subject to the two subsidiary conditions $\Sigma n_s = n$ and $\Sigma n_s \epsilon_s = E$ when the number of particles and the total energy are given. For the distribution of energy in our light quantum gas we therefore find

$$hv_s n_s = A g_s h v_s e^{-hv_s/kT},$$

or

$$u_\nu d\nu = \frac{h\nu n}{V} = A \frac{8\pi h \nu^3 d\nu}{c^3} e^{-h\nu/kT},$$

which, apart from the factor A, is Wien's result, instead of Planck's formula, viz.

$$u_\nu d\nu = \frac{8\pi h \nu^3 d\nu}{c^3} \frac{1}{e^{h\nu/kT} - 1}.$$

The question therefore arises of what changes must be made in the classical statistics in order that it may become possible to deduce Planck's radiation law by purely statistical reasoning, without making use of the roundabout road by way of an absorbing and emitting oscillator.

We see in the first place that for T small, i.e. for $e^{h\nu/kT} \gg 1$, our formula would agree with Wien's approximation (valid in this case) to Planck's law, provided we took $A = 1$. What does the condition $A = 1$ signify? If we go back to the proof of Boltzmann's law of distribution, we easily see that the value of the coefficient A follows from the first subsidiary condition (fixed number of particles). If this subsidiary condition were dropped, we would get $A = 1$; the radiation formula which we have just deduced would then agree asymptotically with the correct formula. There is good reason for dropping the first subsidiary condition (fixed number of particles) in the case of light quanta; for of course in every emission process in an atom a new light quantum is formed, and in every absorption process one is absorbed and converted into other forms of energy.

But even this assumption is not sufficient by itself to allow Planck's formula to be established on a statistical foundation. We must in fact completely alter the statistical foundation itself, and set up a new "quantum statistics". The way in which this might be done was shown by Bose (1924) and improved by Einstein. They assume complete *indistinguishability* of the light quanta. The hypothesis is a very plausible one. Suppose the light quanta to be numbered in a certain order. If light quantum 1 is in the cell z_1, and light quantum 2 in the cell z_2, this distribution obviously represents the same state as the distribution in which light quantum 1 is in the cell z_2, and light quantum 2 in the cell z_1, while the distribution of the rest of the light quanta remains as before; for the two light quanta of course only differ from each other in just this, that one of them is in the first cell and the other in the second. Numbering or individualization of the separate light quanta is entirely meaningless, since the state is completely and uniquely described by merely specifying how many light quanta there are in cell z_1, how many in cell z_2, and so on. This means, however, an essential distinction as compared with classical statistics. In the latter, two cases, either of which passes into the other by merely interchanging two light quanta, are counted in the enumeration of states as two different states, whereas now they represent the same state, and in the enumeration of states must be counted as only a single state.

We shall consider the subject from the mathematical standpoint, and speak of particles in general, and not specially of light quanta. We have already seen in connexion with the statement of Pauli's principle (Chap. VI, § 5, p. 177) that there also it does not matter which precise electron is at this place or that; the second half of Pauli's principle was simply this, that the interchange of two electrons does not change the state of the system as a whole, so that two distributions, which only differ in respect of the interchange of two electrons, represent one and the same state.

From the standpoint of wave mechanics every particle, a light quantum included, is described by the specification of its wave function; let the wave function of one particle (the first) be $\psi_k^{(1)}$, that of a second $\psi_l^{(2)}$, of a third $\psi_m^{(3)}$, and so on; where $k, l, m, \ldots$ represent the state of the particle in question (in the case of light quanta, e.g. k stands for the three quantum numbers k_1, k_2, k_3). The state as a whole is then described (to a first approximation at least, neglecting mutual action between individual particles) by the product of these wave functions, i.e. by a wave function

$$\Psi'_{klm\,..} = \psi_k^{(1)}\psi_l^{(2)}\psi_m^{(3)} \ldots .$$

If two particles are now interchanged, e.g. particle 1 and particle 2, another wave function is obtained.

$$\Psi''_{klm\,...} = \psi_k^{(2)}\psi_l^{(1)}\psi_m^{(3)} \ldots ,$$

which obviously corresponds to the same value of the energy of the whole system, viz. to

$$\epsilon_{klm\,...} = \epsilon_k + \epsilon_l + \epsilon_m + \ldots .$$

We obtain other wave functions, for the same value of the energy, by taking an arbitrary linear combination of those wave functions which arise from the one first written down by a permutation of the individual particles among themselves, i.e. by interchange of the arguments 1, 2, 3 of the separate functions:

$$\Psi'_{klm\ldots} = \sum_P a_{123}\ldots \psi_k^{(1)}\psi_l^{(2)}\psi_m^{(3)}\ldots,$$

where the sum is to be taken over all permutations P of these arguments, and the factors $a_{123}\ldots$ represent arbitrary constant coefficients.

In the sense of classical statistics these wave functions give as many different states as there are linearly independent wave functions among them. From the standpoint of the new statistics, however, those cases which arise from one another by mere permutation of particles belong to the same state. Hence the wave function which describes this state does not change when the particles are permuted, or at most it can only change its sign, since only quadratic forms, such as $|\Psi|^2$, are of any account so far as physical interpretation is concerned. Now it is easily seen that the only wave function of the form specified above, which does not change when the particles are permuted, is the one in which all the coefficients are equal to 1, i.e. the *symmetrical wave function*

$$\Psi_s = \sum_P \psi_k^{(1)}\psi_l^{(2)}\psi_m^{(3)}\ldots.$$

Another possibility, in which the sign of the wave function changes, but not the value of its square, is the *skew or antisymmetrical form*

$$\Psi_a = \sum_P \pm\, \psi_k^{(1)}\psi_l^{(2)}\psi_m^{(3)}\ldots,$$

where the $+$ sign is to be taken for an even permutation of the particles, the $-$ sign for an odd permutation. This antisymmetrical form is known from the theory of determinants—it is the expansion of the determinant

$$\Psi_a = \begin{vmatrix} \psi_k^{(1)} & \psi_k^{(2)} & \psi_k^{(3)} & \ldots \\ \psi_l^{(1)} & \psi_l^{(2)} & \psi_l^{(3)} & \ldots \\ \psi_m^{(1)} & \psi_m^{(2)} & \psi_m^{(3)} & \ldots \\ \cdot\ \cdot\ \cdot\ \cdot\ \cdot\ \cdot\ \cdot\ \cdot \end{vmatrix}.$$

No other functions exist which satisfy the requirement of indistinguishability.

We note further a special feature in the case of the antisymmetric function. We know that a determinant vanishes if two rows, or two columns, are the same; hence if two functions ψ_k and ψ_l are equal, the determinant vanishes, and with it the wave function of this state, i.e. this state does not exist. This is the precise expression of Pauli's principle, that two electrons cannot be in the same state (i.e. cannot have their ψ_k the same).

There are therefore only two possible ways of describing a state by a wave function, viz. either by the symmetrical or the antisymmetrical wave function; the second possibility corresponds to Pauli's principle, the first is another and entirely different matter. If we count the possible states on the basis of their wave functions (i.e. of the possible wave functions which are linearly independent), two different statistics present themselves. If we confine ourselves to the symmetrical wave functions (without Pauli's principle), we get the so-called *Bose-Einstein statistics*; if we describe the state by the antisymmetric function (with Pauli's principle), we get the *Fermi-Dirac statistics* (1926). Which of the two statistics we are to use in a particular case must be left to experience to decide. With regard to electrons, we already know that they obey Pauli's principle—we shall therefore deal with them by the Fermi-Dirac statistics (see § 6, p. 234); on the other hand, it turns out that we have to treat light quanta (Bose) and also certain gas molecules (Einstein) according to the Bose-Einstein statistics.

We proceed now to work out the latter type of statistics. In the first place we have to count the number of different states (i.e. of linearly independent wave functions). In carrying out the enumeration, however, we use, not the wave picture, but the corpuscular picture, and have to find the number of distinguishable arrangements of the particles in a shell, for the case of the *Bose-Einstein statistics*. For this purpose, we denote the individual cells of this shell by z_1, $z_2, \ldots, z_{gs}$; the number of them is, by definition, given by the weight factor g_s of this shell. On the other hand, let there be n_s particles in this shell, which for the present we denote individually by $a_1, a_2, \ldots,$ a_{ns}. We have to distribute these particles among the g_s cells of the shell, and determine the number of distinguishable arrangements. To this end, we describe a definite arrangement in the following way. We write down, purely formally, the elements z and a in an arbitrary order, e.g.

$$z_1 a_1 a_2 z_2 a_3 z_3 a_4 a_5 a_6 z_4 z_5 a_7 \ldots ,$$

with the understanding that the particles standing between two z's is in each case supposed to be in the cell which stands to their left in the sequence; the sequence written down above means therefore that the particles a_1 and a_2 are in the cell z_1, the particle a_3 in the cell z_2, the particles a_4, a_5, a_6 in the cell z_3, no particle in z_4, and so on; that being so, the first letter in the symbolic arrangement must obviously be a z. We therefore obtain all possible arrangements, by first setting down a z at the head of the sequence—which can be done in g_s different ways—and then writing down the remaining $g_s - 1 + n_s$ letters in arbitrary order one after the other. The total number of these arrangements is therefore

$$g_s(g_s + n_s - 1)!$$

Distributions which can be derived from one another by mere permutation of the cells among themselves, or of the particles among themselves, do not, however, represent different states, but one and the same state; the number of these permutations is $g_s! \, n_s!$ We thus obtain for the number of distinguishable arrangements in the shell which is characterized by the index s, in the case of the Bose-Einstein statistics,

$$\frac{g_s(g_s + n_s - 1)!}{g_s! \, n_s!} = \frac{(g_s + n_s - 1)!}{(g_s - 1)! \, n_s!}.$$

Altogether, the number of distinguishable arrangements for the case when there are n_1 particles in the first shell, n_2 particles in the second, and so on, is given by the product of expressions of the above type, for the shells:

$$W = \prod_s \frac{(g_s + n_s - 1)!}{(g_s - 1)! \, n_s!}.$$

We call this the " probability " of that distribution of the particles among the various shells which is defined by the numbers $n_1, n_2, \ldots$. It takes the place here of the probability found in the Boltzmann statistics (p. 12), viz.

$$W = \frac{n!}{n_1! \, n_2! \ldots} g_1{}^{n_1} g_2{}^{n_2} \ldots$$

The remaining part of the calculation proceeds as at the place cited. We have to determine the most probable distribution; for this purpose we use Stirling's theorem, and write

$$\log W = \Sigma\{(g_s + n_s) \log(g_s + n_s) - g_s \log g_s - n_s \log n_s\},$$

where we have neglected the 1 in comparison with the large numbers g_s and n_s. We must now make $\log W$ a maximum for variations of n_s, subject to the subsidiary condition

$$\sum_s n_s \epsilon_s = E, \quad (\epsilon_s = h\nu_s).$$

For light quanta, as we have shown above (p. 224), the second subsidiary condition (constancy of number of particles) drops out. Thus we find in the usual way

$$\frac{\partial \log W}{\partial n_s} = \log(g_s + n_s) + 1 - \log n_s - 1 = \log \frac{g_s + n_s}{n_s} = \beta \epsilon_s,$$

or

$$\frac{g_s + n_s}{n_s} = e^{\beta \epsilon_s}.$$

The Bose-Einstein law of distribution for light quanta therefore runs (if we drop the index s)

$$n = \frac{g}{e^{\beta \epsilon} - 1};$$

this gives for the energy density

$$u_\nu \, d\nu = \frac{n h\nu}{V} = \frac{1}{c^3} \frac{8\pi h \nu^3 \, d\nu}{e^{\beta h\nu} - 1}.$$

This is just Planck's radiation formula, if we put $\beta = 1/kT$. The justification for this last step is given by thermodynamics; according to Boltzmann, $S = k \log W$ is to be regarded as the entropy, and it can then be shown (see Appendix XXXV, p. 459) that from the equation $T \, dS = dQ$ we can infer that $\beta = 1/kT$ (dQ is the increment of the heat content, or, at constant volume, of the energy content of the light quantum gas). From the Bose-Einstein statistics, therefore, Planck's radiation law can be deduced in a way to which no objection can be taken.

5. Einstein's Theory of Gas Degeneration.

After the brilliant success of the Bose-Einstein statistics with the light quantum gas, it was a natural suggestion to try it in the kinetic theory of gases also, as a substitute for the Boltzmann statistics. The investigation, which was undertaken by Einstein (1925), is based on the hypothesis that the molecules of a gas are, like light quanta, indistinguishable from each other.

The calculations run exactly as in the light quantum case, except that here a second subsidiary condition appears, on account of the conservation of the number of particles:

$$\Sigma n_s = N.$$

The determination of the probability of a definite distribution $n_1, n_2, \ldots$ follows the same lines as before. The calculation of the most probable distribution leads now, owing to the presence of the second subsidiary condition, to the equation

$$\frac{\partial \log W}{\partial n_s} = \log \frac{g_s + n_s}{n_s} = a + \beta \epsilon_s,$$

or, on dropping the suffix s,

$$n = \frac{g}{e^{a+\beta \epsilon} - 1},$$

where again $\beta = 1/kT$ (see Appendix XXXV, p. 459). Here the number g of cells in a shell can be expressed by the corresponding energy; we have of course

$$\epsilon = \frac{1}{2m} p^2, \quad \text{and} \quad d\epsilon = \frac{1}{m} p \, dp,$$

where p is the momentum of the particles; the expression for g obtained above therefore becomes

$$g = \frac{4\pi V}{h^3} p^2 dp = \frac{4\pi V}{h^3} \sqrt{2m^3 \epsilon} \, d\epsilon.$$

We thus find the *Bose-Einstein law of distribution for atoms*:

$$dN = F(\epsilon) \sqrt{\epsilon} \, d\epsilon$$
$$= \frac{4\pi V}{h^3} \frac{\sqrt{2m^3} \sqrt{\epsilon} \, d\epsilon}{e^{a+\beta \epsilon} - 1}, \quad (\beta = 1/kT),$$

while the law of distribution given by the Boltzmann statistics was (p. 15)

$$dN = V dn = 4\pi V n \left(\frac{m}{2\pi kT}\right)^{\frac{3}{2}} e^{-(m/2)(v^2/kT)} v^2 dv$$

$$= 4\pi N \left(\frac{m}{2\pi kT}\right)^{\frac{3}{2}} \sqrt{\frac{2}{m^3}} e^{-\epsilon/kT} \sqrt{\epsilon} \, d\epsilon.$$

(N is the number of particles in the volume V, n the number per unit volume.)

The quantity a is of course determined from the subsidiary condition

$$\int dN = \int_J^\infty F(\epsilon)\sqrt{\epsilon}d\epsilon = N = nV.$$

The constant a, or more usually $A = e^{-\alpha}$ is called the *degeneracy parameter*, the reason for the name being as follows. If a is very large, so that A is much less than one, we can neglect the 1 in the denominator of the expression for n_s in comparison with $e^{\alpha+\beta\epsilon_s}$, and in this case

$$n_s = g_s e^{-\alpha} e^{-\beta\epsilon}$$

which is the classical distribution law of Maxwell. We also see that in this regime the probability of finding a particle in a cell is much less than one. For a gas of free particles the distribution becomes

$$dN = \frac{4\pi V}{h^3} \sqrt{2m^3} \sqrt{\epsilon}\, d\epsilon\, A\, e^{-\beta\epsilon}.$$

The constant A can be determined at once from the subsidiary condition on the total number of particles. We find (p. 15; Appendix I, p. 359.)

$$A_0 = \frac{nh^3}{(2\pi mkT)^{3/2}}$$

(the subscript 0 indicating that this refers to the limiting case of no degeneracy). Thus if A_0 is very small compared to 1, the Bose–Einstein distribution for the gas reduces to the classical one. We see that this expression for A_0 can be written as

$$A_0 = \left(\frac{3}{2\pi}\right)^{\frac{3}{2}} n\bar{\lambda}^3$$

where $\bar{\lambda} = h/p = h/(3mkT)^{\frac{1}{2}}$ is a typical de Broglie wavelength for the particles in the gas. The degeneracy parameter A is seen to be small when the mean spacing of the particles is much larger than this de Broglie wavelength, which is the size of a typical wave packet. In this case the wave functions Ψ_k, Ψ_l, etc., used in § 4 overlap only slightly and forming a symmetrized wave-function has no physical effect. It is otherwise in the case when the wave packets overlap appreciably and A becomes comparable with 1 (the case of $A > 1$, i.e. $\alpha < 0$ cannot occur, for then the denominator vanishes for the positive particle energy $\epsilon = -\alpha/\beta$ and for smaller values of ϵ becomes negative, so that the whole theory becomes meaningless). If $A \sim 1$, deviations from the classical properties occur; we say that *the gas is degenerate*. In this case the subsidiary condition on the number of particles leads to the equation

$$A_0 = \frac{4A}{\sqrt{\pi}} \int_0^\infty \frac{e^{-x^2} x^2\, dx}{1 - Ae^{-x^2}}.$$

By solving this transcendental equation we obtain A as a function of A_0. The integrand can be expanded in powers of A and integrated term by term with the help of the formulæ in Appendix I. Thus we find

$$A_0 = A\left(1 + \frac{1}{2\sqrt{2}} A + \frac{1}{3\sqrt{3}} A^2 + \cdots\right)$$

and by solving with respect to A,

$$A = A_0\left(1 - \frac{1}{2\sqrt{2}} A_0 + \frac{(3\sqrt{3} - 4)}{12\sqrt{3}} A_0^2 - \cdots\right).$$

By substituting particular values of the constants n, m and T we can decide from this equation whether the gas is degenerate or not. We see in the first place quite generally that A increases, and accordingly the degeneracy becomes greater, as the density increases; on the other hand, it diminishes with increasing temperature and atomic weight. To take a special example: for hydrogen gas under normal conditions (for $T = 300°$K, $n \sim 3 \times 10^9$ cm.$^{-3}$) we have $A \sim 3 \times 10^{-5} \ll 1$; for heavy gases A becomes still smaller. Gases are therefore never degenerate at normal temperatures and pressures, but behave according to the classical laws. Degeneracy would only become noticeable at unattainably low temperatures or at extremely high pressures, that is to say in regions where even according to classical statistics the gases no longer behave as ideal gases (intermolecular forces, condensation of the gas, etc.). Thus, when applied to gases in the regime in which the kinetic theory of gases is valid, the Bose–Einstein statistics leads to no significant differences from the classical Maxwell–Boltzmann statistics. However it was suggested soon after the discovery of superfluidity and the HeI–HeII transition that Einstein's theory might be relevant to the anomalous behaviour of liquid He^4 at very low temperatures (Tisza, 1935; F. London, 1939). A detailed study of the equation of state of a degenerate gas shows that at a very low temperature there is a discontinuity at which a kind of condensation takes place, and at lower temperatures a finite fraction of the atoms are in the lowest quantum state. This model, however, predicts a singular first-order transition at a temperature of $3 \cdot 13°K$, whereas the real liquid exhibits a λ-transition at a temperature $T_c = 2 \cdot 18°K$, with $c_p \sim -\ln |T - T_c|$ close to T_c. Recent microscopic theories (Bogoliubov, 1947; Feynman, 1952) which have led to an understanding of the low-temperature properties of liquid He^4 depend in an essential way on the use of properly symmetrized wavefunctions for the particles in the liquid. The importance of the symmetry of the wave-function, that is to say the statistics obeyed by the particles, is shown very clearly by the differences between the behaviour of liquid He^3 and liquid He^4 at low temperatures. Samples of He^3 do not show any indication of a transition to a superfluid state (at least at temperatures above $10^{-2}°K$), and the specific heat is quite different from that of He^4. These differences in the behaviour of the two liquids, which consist of particles of very similar mass with identical interatomic potentials, are due to the fact that the wave-function of He^4, in which the nuclei have spin $I = 0$, must be symmetric under exchange of particle coordinates, corresponding to Bose–Einstein statistics, while the wave-function of He^3, nuclear spin $I = \frac{1}{2}$, must be antisymmetric, corresponding to another type of statistics which we shall now discuss. At low temperatures they are both quantum fluids, exhibiting the effects of quantization on a macroscopic scale.

6. Fermi-Dirac Statistics.

We have shown in § 4 (p. 228) that the introduction of the principle of indistinguishability into statistics leads to two, and only two, new systems of statistics, one of which, the *Bose-Einstein* statistics, we have discussed in detail in the last two sections (light quanta, gas molecules). We turn now to the second possible statistics, which is based on Pauli's principle, and was introduced by Fermi and Dirac. We have seen in § 4 (p. 228) that this statistics is intimately connected with the employment of Pauli's principle, observing that the proper function of a state in which two electrons have the same partial proper function (with respect to the four quantum numbers, including the spin quantum number) automatically vanishes.

To deduce the nature of this statistics, we shall use the model of a gas consisting of electrons, which, on the experimental evidence of spectra, obey Pauli's principle. In this case again, our first aim is to find the distribution of the electrons over the individual cells, bearing in mind, however, that there are now twice as many cells as in the previous case of gas atoms, in consequence of the two possible settings of the spin; on the other hand, no cell can be occupied by more than one electron, or, in other words, the " occupation numbers " of the cells in this case must be either 0 or 1. (We might of course proceed on the assumption that the number of cells in each sheet is only half as great, but in compensation for this provide two possible places for the electrons in each cell, corresponding to the two directions of spin.)

We begin as before with the *enumeration of the distinguishable distributions*. Let the number of electrons in the sth shell be n_s, these being distributed over the g_s cells of this shell; of these g_s cells, n_s are therefore (singly) occupied (1), and $g_s - n_s$ are empty (0). We characterize such a distribution by assigning to each cell its occupation number:

$$z_1 \quad z_2 \quad z_3 \quad z_4 \quad z_5 \quad z_6 \quad z_7 \quad z_8 \quad z_9 \quad z_{10} \cdots$$
$$1 \quad\; 0 \quad\; 0 \quad\; 1 \quad\; 1 \quad\; 1 \quad\; 0 \quad\; 1 \quad\; 0 \quad\; 0 \;\cdots,$$

or by specifying the cells which are occupied by no particle, and those occupied by 1:

0	1
$z_2 \; z_3 \; z_7 \; z_9 \; z_{10} \cdots$	$z_1 \; z_4 \; z_5 \; z_6 \; z_8 \;\cdots$

Clearly there are $g_s!$ such distributions, corresponding to the permutations of the g_s cells z in this scheme. But the same state (as regards occupation) corresponds to all distributions among these, which only differ from one another by permutation of the n_s occupied cells, or of the $g_s - n_s$ unoccupied cells. Hence the " probability " of a distribution characterized by the occupation numbers $n_1, n_2, n_3 \ldots$ for the individual cells is given by

$$W = \Pi_s \frac{g_s!}{n_s! \, (g_s - n_s)!},$$

or, from Stirling's theorem, by

$$\log W = \Sigma_s \left\{ g_s \log g_s - n_s \log n_s - (g_s - n_s) \log (g_s - n_s) \right\}.$$

As before, we wish to find the most probable distribution subject to the two subsidiary conditions

$$\Sigma_s n_s = N, \quad \Sigma_s n_s \epsilon_s = E.$$

We obtain it in the usual way:

$$\frac{\partial \log W}{\partial n_s} = -\log n_s + \log(g_s - n_s)$$

$$= \log \frac{g_s - n_s}{n_s} = a + \beta \epsilon_s,$$

or

$$n_s = \frac{g_s}{e^{a + \beta \epsilon_s} + 1},$$

i.e. except for the $+$ sign in the denominator, the same formula as in the case of the Bose-Einstein statistics. This difference of algebraic sign, however, carries with it a fundamental distinction between the present case and that of the Bose-Einstein statistics, in that a can now have all values from $-\infty$ to $+\infty$, and the degeneracy parameter $A = e^{-a}$ all values therefore from 0 to $+\infty$; for the denominator of the distribution function is here always greater than 1. If we now substitute the value of g_s (with the factor 2 on account of the two directions of spin), we find, in the same way as above (p. 231), the *Fermi-Dirac law of distribution*:

$$dN = F(\epsilon)\sqrt{\epsilon}\,d\epsilon = \frac{8\pi V}{h^3} \frac{\sqrt{2m^3}\sqrt{\epsilon}\,d\epsilon}{e^{a+\beta\epsilon}+1}, \quad (\beta = 1/kT).$$

The degeneracy parameter is determined as before from the first subsidiary condition:

$$\int dN = \int_0^\infty F(\epsilon)\sqrt{\epsilon}\,d\epsilon = N = nV.$$

This transcendental equation for A can be solved, in the case when $A \ll 1$, by means of a power expansion, as before (p. 233):

$$A = \frac{nh^3}{2} (2\pi mkT)^{-3/2} + \dots \quad (A \ll 1).$$

When A is extremely small, this case, it is easy to see, passes over into the classical statistics. In the case when $A \gg 1$, we find an approximate expansion of the form (which we merely state, without proof):

$$\log A = -a = \frac{h^2}{2mkT} \left(\frac{3n}{8\pi}\right)^{\frac{2}{3}} + \dots \quad (A \gg 1);$$

this is the case of *degeneracy* of the electron gas. A more thorough discussion of these formulæ is given in the next section, where at the same time the most important application of the Fermi-Dirac statistics will be dealt with.

7. Electron Theory of Metals.   Energy Distribution.

In the preceding section we spoke of an " electron gas ", and pictured it to ourselves as a definite number (n per cm.^{3}) of electrons, moving freely, without mutual disturbance. Such a case is of course unrealizable, since in virtue of their electric charge the electrons will always act upon each other; however, to a first approximation we can neglect this disturbing action, owing to the neutralizing effect of the positive ions.

Now such an approximately *free electron gas* exists in the interior of a metal. In proof of this statement, we may cite the high conductivity of metals; and metallic conduction is carried on by electrons only, since, in contrast to electrolytic conduction, it is demonstrably not bound up with transport of matter. In order to secure that the electrons shall be able to react to an external electric field in such a way that the resulting conductivity shall be as high as it is, they must at least approximately be able to move freely in the metal, in contrast to what happens in non-conductors, in which the electrons are firmly bound to the atoms. Founding on these ideas, even the older theories of metallic conduction (Riecke, Drude, Lorentz, since 1900) were able to give a satisfactory explanation of the *Wiedemann-Franz law*, which states that the electrical and thermal conductivities are proportional to one another, and that their ratio is inversely proportional to the absolute temperature. The older theories, however, consistently led to difficulties in regard to the explanation of the specific heat of metals. It is established by experiment that metals obey the Dulong-Petit law, i.e. that their specific heat, referred to 1 mole, is 6 cals/degree. This could be explained at once, if the temperature of the metal were determined solely by the vibrational energy of the atoms in the lattice, since the mean energy per lattice point is $3kT$. But to explain the process of conduction and other related phenomena it is necessary to assume that to every atom (ion) there corresponds approximately one free electron. The free electrons take part in the thermal motion in the metal, and in fact (as shown by the Wiedemann-Franz law) are largely responsible for the high thermal conductivity of metals. According to classical statistics, every free electron in the metal would therefore possess the mean kinetic energy $\frac{3}{2}kT$, and the specific heat of the metal would then be, per atom, not $3k$, but of the order $(3 + 3/2)k$, i.e. referred to 1 mole, 9 cals/degree, which is contrary to the facts.

The solution of this difficulty is due to Pauli and Sommerfeld (1927), who pointed out that the laws of classical statistics ought not to be applied to the electron gas within a metal, since it is bound to behave as a *degenerate gas*. Thus, since the mass of the electron is 1840 times smaller than that of the hydrogen atom, it follows that, at room temperature ($T = 300°$) and an electronic density of $n \sim 3 \times 10^{19}$, corresponding to a gas density at a pressure of 1 atmosphere, the value of the degeneracy parameter A_s for the electron is

$$A_s = A_H \frac{1}{2}(1840)^{3/2} = A_H 4 \times 10^4 \sim 1·2,$$

where A_H is the degeneracy parameter for hydrogen gas under the same conditions; even in this case, therefore, A is of the order of magnitude 1. Values decidedly greater than this are found for the electron gas in metals. In silver the number of atoms per cm.[3] is $n = 5·9 \times 10^{22}$. Since, as we have already remarked, we must assume that roughly speaking there is one free electron for each atom, we find for this value of n by the first approximate formula that A is about 2300; in this case, therefore, the gas is degenerate to a high degree. For a value of A so high as this, it is true, it is not permissible to apply the first approximate formula, and we must use the second; but even this gives the still high value $A \sim 210$. The electron gas in metals is therefore in all cases highly degenerate—its properties are essentially different from those of an ordinary gas.

The most important characteristics of the Fermi-Dirac distribution function are the slightness of the dependence of distribution on temperature, and the occurrence of a zero-point energy. The latter property is closely connected with the Pauli principle. In the classical gas theory the absolute zero is characterized by the fact that the mean kinetic energy of the gas particles vanishes at that temperature, and accordingly the energy of every individual particle also vanishes; classically, therefore, at the absolute zero the gas particles are at rest. It is different in the Fermi-Dirac statistics; here each cell can only have a single occupant; in the state of lowest energy, all the cells of small energy are occupied, and the limit of the " filling up " of the system of cells is given by the number of electrons. We characterize this limit by the momentum p_0 of that cell, up to which the filling up reaches; it is found from the formula deduced above for the number of cells:

$$2 \frac{4\pi V}{3h^3} p_0^3 = N,$$

or
$$p_0 = h \sqrt[3]{\left(\frac{3N}{8\pi V}\right)} = h \sqrt[3]{\frac{3n}{8\pi}}.$$

The limiting energy ϵ_0 is then given by

$$\epsilon_0 = \frac{p_0^2}{2m} = \frac{h^2}{2m}\left(\frac{3n}{8\pi}\right)^{\frac{2}{3}}$$

$$= 5\cdot 77 \times 10^{-27} n^{2/3} \text{ ergs} = 3\cdot 63 \times 10^{-15} n^{2/3} \text{ electron volts.}$$

We therefore obtain the following distribution curve for the electrons at the absolute zero (see fig. 7). Taking the electronic energy ϵ as abscissa, and as ordinate the previously (p. 236) defined distribution function $F(\epsilon)$, whose product by the factor $\sqrt{\epsilon} d\epsilon$ gives the number of electrons with energy values between ϵ and $\epsilon + d\epsilon$, we find for the graph a rectangle; up to the energy value ϵ_0 the cells are com-

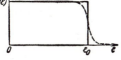

Fig. 7.—Fermi's distribution curve. The continuous, sharp-cornered line corresponds to the absolute zero ($T = 0$), the dotted line to a temperature other than zero

pletely filled, the cells with greater energy values are empty. (This is the same distribution which is used in the Thomas-Fermi model of the electronic cloud of an atom, see Chap. VI, § 9, p. 196.) In this case the degeneracy parameter A, as the approximate formula shows, becomes infinitely great. For finite, but small T, A behaves as $1/T$; comparison with the foregoing formula for the limiting energy shows that we can put approximately

$$a = -\frac{h^2}{2mkT}\left(\frac{3n}{8\pi}\right)^{\frac{2}{3}} = -\frac{\epsilon_0}{kT}.$$

The distribution function, which is approximately valid for large values of A, i.e. for low temperatures, is then

$$F(\epsilon) = \frac{8\pi V}{h^3} \frac{\sqrt{2m^3}}{e^{(\epsilon-\epsilon_0)/kT} + 1},$$

and for the limiting case when $T \to 0$, gives the graph of fig. 7. When $\epsilon < \epsilon_0$, the exponential function in the denominator vanishes when

$T \to 0$, and we have $F(\epsilon) = 8\pi V \sqrt{2m^3}/h^3$; when $\epsilon > \epsilon_0$, however, the exponential function in the denominator becomes infinite, and $F(\epsilon)$ vanishes.

As the temperature increases, the electrons are gradually raised into higher states; but the change in the electronic distribution will at first only take effect at the place where the Fermi function falls away, and that by slowly rounding off the corners of the distribution curve, as indicated in fig. 7. The main body of the electrons, however, is left untouched by the rise of temperature. For not too high temperatures, therefore, only a vanishingly small fraction of the electrons takes part in the thermal motion, so that the specific heat of the electrons is very small. It is only when high temperatures are reached (of the order of 10^4 degrees C.), far above room temperature, that the tight packing of the electrons in the deeper states gradually becomes loosened, and we obtain a noticeable contribution from the electrons to the specific heat of metals.

8. Thermionic and Photoelectric Effect in Metals.

A further proof of the correctness of the idea of free electrons in metals and of the applicability of the Fermi-Dirac statistics is furnished by the phenomenon of thermionic

Fig. 8.—Diagram of the potential relations in a metal; the potential hollow is partly " filled up " with electrons.

emission. It is known (Richardson, 1903) that electrons emerge spontaneously from incandescent metals (e.g. glowing cathodes), and that these electrons in the absence of an applied potential form an electronic cloud or atmosphere round the incandescent body. Their number can be determined by measuring the current set up when an external E.M.F. is applied. Theoretically, the phenomenon of thermionic emission has to be pictured in the following way (fig. 8). Within the metal the electrons can certainly move freely, but in general their escape from the metal is opposed by a potential barrier, the internal potential, ϵ_i. At higher temperatures, however, it may happen that the energy of an electron becomes greater than ϵ_i, so that it can escape from the metal. Using the formulæ of the Fermi-Dirac statistics, we can determine the number of electrons emerging in this way per unit time; the current is found to be

$$i = \frac{4\pi e m}{h^3} (kT)^2 e^{-(\epsilon_i - \epsilon_0)/kT},$$

while the classical statistics gives for it the expression (Richardson, 1902)

$$i = en \sqrt{\frac{kT}{2\pi m}}\, e^{-\epsilon_i/kT},$$

which, regarded as a function of the temperature, differs somewhat from that given by the new theory (see Appendix XXXVI, p. 460). The two formulæ differ both in the power of T before the exponential function, and in the meaning of the constant in the exponent. To test them, it is usual to plot $y = \log i$ against $x = 1/T$, i.e. the function

$$y = A - a\log x - bx \begin{cases} & \text{Classical} & \text{Fermi Statistics} \\ A = \log\left(en\sqrt{k/2\pi m}\right) & \log\left(4\pi emk^2/h^3\right) \\ b = \epsilon_i/k & (\epsilon_i - \epsilon_0)/k \\ a = \tfrac{1}{2} & 2 \end{cases}$$

Now the term $a\log x$ is in general so small compared to the other two terms, that it has been impossible up till now to decide whether the quantum theory formula with $a = 2$ is superior to the classical one with $a = \tfrac{1}{2}$. On the other hand, the constant b is easily found, for, when the term $a\log x$ is omitted, the equation $y = A - bx$ represents a straight line, and b is its gradient. We can therefore determine the difference $kb = \epsilon_i - \epsilon_0$, called the *work function*, experimentally for various metals, and hence calculate ϵ_i by substituting for ϵ_0 the value given by theory, viz. 0 for the classical, and $3 \cdot 63 \times 10^{-15}n^{2/3}$ electron volts for the quantum theory.

By comparing the values thus found for ϵ_i with other measurements, it has been definitely settled that the quantum theory formula is correct, and not the classical one.

We can in fact determine the internal potential ϵ_i in another, quite independent, way, viz. by means of *diffraction experiments with slow electrons*. When a crystal lattice is irradiated with cathode rays, the electrons, as we know, are diffracted by the lattice, the position of the diffraction maxima depending upon the de Broglie wave-length $\lambda = h/p$ of the electrons. When, however, the cathode ray enters the crystal, refraction occurs at the surface, since the kinetic energy of the electrons in the solid is greater, and therefore their wave-length smaller, than in the external space. As in optics, so here we can speak of a refractive index of the crystal with respect to electrons; this in fact, as in the optical case, being given by the ratio of the wave-

lengths inside and outside. If we reckon the energy from the bottom
of the potential trough, we have

$$\lambda_{outside} = \frac{h}{p_0} = \frac{h}{\sqrt{2m(\epsilon - \epsilon_i)}},$$

$$\lambda_{inside} = \frac{h}{p_i} = \frac{h}{\sqrt{2m\epsilon}},$$

so that we find for the refractive index the expression

$$\mu = \frac{\lambda_{outside}}{\lambda_{inside}} = \sqrt{\frac{\epsilon}{\epsilon - \epsilon_i}}.$$

By measurement of the position of diffraction maxima we can now
determine the value of the refractive index and so, knowing ϵ, calculate
the potential ϵ_i. For the sake of greater accuracy, we should choose ϵ
very small, i.e. use slow electrons.

Careful measurements of this kind by Davisson and Germer on
single crystals of nickel give $\epsilon_i \approx 17$ eV (electron volts; see p. 65). On
the other hand, measurements on the thermionic effect give values of
bk in the region of 5 eV, in disagreement with the classical formula,
according to which we should have $bk = \epsilon_i$. The quantum theory,
however, gives $\epsilon_0 = 11\cdot7$ eV, if we assume that in nickel two electrons
per atom are free, in accordance with the fact that nickel has two
valency electrons. This would give $\epsilon_i - \epsilon_0 = 5\cdot3$ eV, in good agreement with the results of measurement of thermionic emission. Re-
liable measurements on other metals are scarce. For Zn it was found
from diffraction experiments that $\epsilon_i = 15\cdot4$ eV, and from the photoelectric effect (see below) $\epsilon_i - \epsilon_0 = 3\cdot8$ eV; hence $\epsilon_0 = 11\cdot6$ eV.
From this the number of effective free electrons can be calculated
and is found to be 2·7. Table V (p. 181) shows that the number of
electrons in the outer shell of Zn 30 is 2. In view of the approximate
character of the theory (no electron is quite free, nor is the potential
exactly constant inside the metal) a better agreement can hardly be
expected. Large discrepancies for gold and silver have been found by
Blackman (1950), which, however, do not seem to be out of reach of
a more refined theory.

The same constants as occur in the theory of thermionic emission
also determine the *photoelectric effect*, which sets in at the frequency
given by $h\nu = \epsilon_i - \epsilon_0$, at which the energy of the incident light
quantum is just sufficient to raise the electron from the limiting point

of the Fermi distribution, ϵ_0, to the height of the internal potential. The same constants also determine the law of the *cold discharge*, in which the following state of affairs occurs. If by means of sharp points very high field strengths (10^6 volts/cm.) are produced at the surface of a metal, electrons issue from the metal even though the temperature has not been raised. The explanation depends on the same principle as is applied in the theory of radioactive disintegration of the nucleus (p. 309, and Appendix XXX, p. 444). The external field implies a potential distribution in which the potential falls linearly from the surface outwards. We have therefore a potential barrier (fig. 9) at the surface of the metal, and we know that according to wave mechanics an electron can make its way through such a barrier. The greater the external field, the narrower becomes the barrier, and therefore the greater becomes the number of electrons issuing per second. This number clearly depends on the height of the original potential above the zero-point level $\epsilon_i - \epsilon_0$, so that the number, if it could be found experimentally, would allow the value of $\epsilon_i - \epsilon_0$ to be deter-mined. The experiment, how-ever, is impracticable, owing to the presence of minute in-equalities and impurities in the surface, which change the magnitude of the field in an uncontrollable way; at every little projection the field is greater than the mean potential gradient. In point of fact, electrons begin to emerge at distinctly smaller field strengths than would be expected if the surface were ideally smooth.

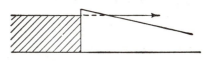

Fig. 9.—Potential barrier at the surface of a metal.

9. Magnetism of the Electron Gas.

Another circumstance which confirms the correctness of our ideas about electrons in metals was pointed out by Pauli (1927). The elec-trons possess on account of their spin a magnetic moment equal to a Bohr magneton. As, according to our present picture, they behave in metals practically like free particles, one might expect, therefore, that they should cause a very strong paramagnetism. Experiment shows, however, that simple metals (e.g. Li, Na) are either not para-magnetic, or only very slightly so. Pauli explains this as follows. We can consider the valency electrons in the metal as free; the ions forming the residue, having an inert gas configuration of electrons, are non-magnetic. Of the free electrons (for $T = 0$), two sit in every cell, and they have opposite spins, so that their magnetic moments exactly balance each other. If an external field H is applied, the electrons will tend to direct their spins parallel to the field, which they cannot do without leaving the doubly occupied cells and jumping into higher states. This increase in kinetic energy goes on until it is compensated by the decrease of potential energy due to the orientation in the field. As only a few electrons jump to higher states, the paramagnetism is much smaller than for systems not satisfying the exclusion principle of Pauli. When the temperature rises, the uppermost sheets of the Fermi distribution, as we have shown in § 7 (p. 240), begin to be " loosened ", individual electrons being lifted out of the doubly occupied cells, so that now there are cells which are only singly occupied. But this gives, as can be shown, only a second order effect. Consequently the paramagnetism of an electronic gas is almost independent of temperature. This is in agreement with experiment. But now the orbital motions of the electrons also produce a magnetic effect—diamagnetic indeed, as may be seen at once. The theory of this phenomenon is not simple, and there has been much discussion about it. From the standpoint of the quantum theory, however, a definite formula results for the magnitude of the diamagnetic moment. If we subtract this from the paramagnetic moment, we get the following table:

TABLE VIII—SUSCEPTIBILITIES OF THE ALKALI METALS

	Na	K	Rb	Cs
$\chi \cdot 10^7$: Theory	4·38	3·40	3·26	3·02
$\chi \cdot 10^7$: Experiment	5·8	5·1	0·6	−0·5

The agreement in order of magnitude is good. The differences, especially in Rb, and in Cs (where the sign does not agree), can be explained as due to the neglect of facts which should be considered. Thus, in the heavy elements the inner electronic shells make an appreciable contribution to the diamagnetism, and in all metals the assumption of entirely free electrons is incorrect, as we proceed to explain more fully.

10. Electrical and Thermal Conductivity. Thermoelectricity.

To explain electrical conductivity, we must suppose the electrons in a metal to have a " free path ". In fact, if we were to adhere in the problem of electrical conductivity to the idea of perfectly free electrons (that this is only a first approximation we have already emphasized above), the result would be an infinitely great conductivity. To explain finite resistance, therefore, we must take into account the fact that the electrons, in the course of their motion through the metal, collide from time to time with the ions of the lattice, and are thus deflected from their path, or are retarded; the average distance which an electron traverses between two collisions with the lattice ions is called, by analogy with the similar case in the kinetic theory of gases, the mean free path.

It has been shown by Sommerfeld (1928) that we can calculate the general behaviour of electrical and thermal conductivities without necessarily making special hypotheses as to the free path. The Wiedemann-Franz law follows from this theory; and we can explain in the same way the Joule heat, the Peltier and Thomson thermoelectric effects, and other phenomena.

The refinements of the theory are treated briefly in Chapter IX on the theory of solids.

CHAPTER VIII

Molecular Structure

1. Molecular Properties as an Expression of the Distribution of Charge in the Electronic Cloud

This chapter will be devoted to the subject of molecular structure. Our object in the first place will be to arrive at clear ideas with regard to two questions, viz. by what properties a molecule, from the physical standpoint, is most conveniently characterized and described, and how these properties can be determined experimentally. Only after these points have been dealt with will we take up the question, how the phenomenon of chemical binding can be understood and explained physically.

A molecule consists of a number of heavy nuclei, the atomic nuclei of the atoms or ions which form the molecule; round these nuclei the electrons revolve. Just as in the case of atoms, so here we can speak of an electron cloud. On account of the great difference between the masses of the electrons and the nuclei, the motion of the electrons is of course much more rapid than that of the nuclei—a circumstance which simplifies the discussion very decidedly. If, for example, we are investigating the motion of the electrons, we can to a first approximation regard the nuclei as at rest, since they move very little during the period of revolution of an electron. On the other hand, if we are examining the motion of the molecule as a whole (rotations), or of the individual atoms in the molecule relative to one another (vibrations), we can obtain a good approximation to the results by replacing all properties of the electronic motion by mean values; for during the time which must elapse before the nuclei have undergone any appreciable displacement from their original positions, the electrons make a great many complete revolutions.

With regard now to special molecular properties, one of the first importance is the *distribution of charge* in the molecule. As regards the total charge, we must distinguish here, as in the case of atoms, between neutral molecules and positive or negative ions. The charge distribution itself is completely characterized by specifying on the one hand the mutual *distances of the nuclei*, and on the other the charge density ρ of the electrons. The latter can be regarded either classically as the mean charge per unit volume obtained by averaging over the motion of revolution of the electrons, or from the standpoint of wave mechanics as the charge density given by probability waves as in the case of the atom. The charge density on the one hand, in the case of equilibrium due to the action of the electrical forces, determines the nuclear distances; on the other hand, by its external boundary it gives the *molecular volume v* (from the standpoint of wave mechanics, in the case of a neutral molecule, exactly as in that of a neutral atom, the charge distribution falls off exponentially outwards, beyond a definite boundary, so that the size of the charge cloud can be assigned with comparative exactness).

We can now consider the question of the electrical centroids of the positive and the negative charges apart, that is to say, the electrical centroid of the nuclei alone, and the electrical centroid of the electron cloud alone. It may happen that the two points coincide, just as they always do in individual atoms, where the positive centroid is identical with the nucleus, and where also the centroid of the negative charge cloud, on account of the central symmetry of its charge distribution, always coincides with the nucleus. In general, however, the two centroids will be distinct from each other; consequently the external action of the molecule is like that of an electric dipole. In this case we speak of a *permanent electric dipole moment*, and denote it by the vector

$$\boldsymbol{p}_0 = \Sigma \overline{e\boldsymbol{r}},$$

where the radius vectors $\boldsymbol{r}$ are the position vectors of the nuclei and the electrons; the bar signifies averaging over the electronic motion, and the sum is taken over all the nuclei and electrons. If $\boldsymbol{p}_0 = 0$, the electrical behaviour of the molecule is determined to a first approximation by the *quadrupole moment*, defined by its components

$$\Theta_{xx} = \Sigma \overline{ex^2}, \ldots, \quad \Theta_{xy} = \Sigma \overline{exy}, \ldots;$$

this is a form exactly analogous to the mechanical moment of inertia; we therefore also speak of the *electrical moment of inertia*. It is a tensor, and as such can be represented by an ellipsoid (ellipsoid of inertia).

If the molecule comes into an electric field, it is deformed, the positive nuclei being attracted in the direction of the lines of force, the negative electrons in the opposite direction. Consequently, even when there is no permanent dipole moment, a dipole moment is induced; to a first approximation its magnitude increases linearly with the field strength, and it can be represented by the relation

$$p = aE,$$

where a is called the deformability or *polarizability*, and, as may easily be seen, has the dimensions of a volume. For spherically symmetrical molecules a is of course a scalar, that is, a constant independent of direction. In general, however, a depends on the direction, and may then be regarded as a tensor and represented by an ellipsoid, the *ellipsoid of polarization*. We may also mention the case where the molecules can rotate freely (gases); the ellipsoids of polarization corresponding to the individual molecules can then assume all possible positions in space, so that when an external field is applied, a mean polarization of the gas arises, which is given by $\overline{p} = \overline{a}E$; the mean value $\overline{a}$ over all directions is all that matters in this case.

As determining elements of molecules, therefore, besides the total charge e, the nuclear distances and the molecular volume, we have also now before us the dipole and quadrupole moments, as well as the polarizability. All these quantities (with the exception of e), however, also depend more or less on the *state of excitation* of the molecule. Just as in the case of atoms, so also in molecules there exist different electronic states, characterized by quantum numbers which in the following sections we shall denote collectively by n. Moreover, the nuclear motion has an effect, consisting on the one hand of a rotation of the whole molecule, determined by the rotational quantum number j, and on the other hand of a vibration of the nuclei relative to one another, described by a vibrational quantum number s. With these different states of motion we shall deal in detail in the pages which follow.

2. Experimental Determination of the Molecular Constants.

We now proceed to explain in detail how the molecular constants enumerated in the preceding section can be determined experimentally. There is first the *molecular volume*, the determination of which, when neutral molecules are in question, can be effected by the methods of the kinetic theory of gases, already referred to in Chapter I (viscosity, free path, diffusion and direct measurement by molecular rays). The following table shows some molecular diameters * so determined, in Ångström units (10^{-8} cm.):

* The concept "molecule" in the kinetic theory of gases includes "monatomic molecule". Compare the concept "mole" (p. 3).

TABLE IX—DIAMETERS OF SOME MOLECULES, IN Å.

He	1·9	H_2	2·3	H_2O	2·6
Ne	2·3	O_2	2·9	CO	3·2
A	2·8	N_2	3·1	CO_2	3·2
Kr	3·2	Cl_2	3·6	C_6H_6	4·1
Xe	3·5			$(C_2H_5)_2O$	4·8

On account of their charge, the volume of ions has to be determined by other means. Two main methods have been used (Wasastjerna (1923), Goldschmidt (1926), Pauling (1927)). One deals with the lattice spacing in ionic lattices, in rock salt, for example. If we assume that the molecules in crystal lattices are packed as tightly as possible, then the lattice spacing gives directly the sum of the radii of the two ions, i.e. in the rock salt cube, $r_{Na+} + r_{Cl-}$; for in the ionic lattices the ions are arranged in such a way that a positive ion is always surrounded by negative ions only, and conversely, so that the lattice spacing is actually equal to the sum of the radii of the two ions. It is always only the sum of the two radii which we obtain in this way, not the radii themselves; if we knew one radius, we could then calculate all the rest. What we do is to measure the lattice spacing in crystals one of whose ions we have reason to believe is very small, for example, Li^+; this has only two electrons, in the K shell, and will therefore be distinctly smaller than, for example, the Cl^- ion with completely filled K and L shells, and a full sub-shell (of 8 electrons) of the M shell. The lattice spacing in the Li^+Cl^- lattice will therefore be approximately equal to the radius of the Cl^- ion.

The second method of determining ionic radii consists in measuring *ionic mobility* in electrolytes; small ions will make their way through the liquid more easily than large ones. The difficulty occurs in this method, however, that water molecules become attached to the ions (hydration), and so produce a deceptive appearance of substantially greater ionic radii. Here we give another table, showing ionic diameters of atomic ions, and for the sake of comparison we repeat the diameters of the inert gases; atoms or ions with similar electronic configurations are placed in the same row or column. We see that the negative ions, which have an inert gas configuration with a smaller nuclear charge than the corresponding inert gas, are larger than the latter, the reason being of course that the electrons in these ions are more loosely bound, so that their orbits have greater radii. A corresponding result, *mutatis mutandis*, holds for the positive ions also.

TABLE X—DIAMETERS OF SOME ATOMIC IONS, IN Å.

		O^{--}	2·6	S^{--}	3·5	Se^{--}	3·8
H$^-$	2·5	F$^-$	2·7	Cl$^-$	3·6	Br$^-$	3·9
He	1·9	Ne	2·3	A	2·8	Kr	3·2
Li$^+$	1·6	Na$^+$	2·0	K$^+$	2·7	Rb$^+$	3·0
Be^{++}	0·7	Mg^{++}	1·6	Ca^{++}	2·1	Sr^{++}	2·5
B^{+++}	—	Al^{+++}	1·2	Sc^{+++}	1·7	Y^{+++}	2·1
C^{++++}	0·4	Si^{++++}	0·8	Ti^{++++}	1·3	Zr^{++++}	1·7
N^{+++++}	0·3	P^{+++++}	0·7				

The second property to be considered is the *mean polarizability* $\bar{a}$. Here we confine ourselves in the first instance to molecules without permanent dipole moment. By definition, $\bar{a}$ represents that mean dipole moment which is induced in a molecule by an electric field of unit strength (at least in the case of molecules which can rotate freely); the total polarization P, per unit volume (containing N molecules), in an external field E, is given by $P = \bar{a}NE$. But, according to electrodynamics, the polarization P is connected with Maxwell's displacement vector D by the relation $D = E + 4\pi P$; on the other hand, by definition, $D = \epsilon E$, where ϵ is the *dielectric constant*. In the case of gases, where we can neglect the mutual action of the molecules, these relations lead to the following equation connecting the dielectric constant and the mean polarizability:

$$\epsilon = 1 + 4\pi N\bar{a}.$$

In liquids, where the induced moments of the molecules influence each other, the relation is somewhat more complicated.

The dielectric constant can easily be measured by well-known methods, for instance by determining the refractive index n of the substance for long waves (infra-red), which by Maxwell's theory is, as we know, connected with ϵ in this limiting case by the relation $n \to \sqrt{\epsilon}$. As has already been remarked, the whole line of argument is valid only for substances free from dipoles. We give a short table of mean polarizabilities for inert gases and atomic ions, forms of similar structure being placed in the same row or column. We can recognize here the same order of succession in the values as in the previous case of the diameters; this was to be expected, since of course to a greater diameter there corresponds a smaller binding force on the outer electrons, and accordingly a greater polarizability. We remark specially that $\bar{a}$ has the dimensions of a volume; it is always in fact of the same order of magnitude as the molecular volume.

TABLE XI—MEAN POLARIZABILITIES OF INERT GASES AND
ATOMIC IONS. The numbers denote $\bar{\alpha} \cdot 10^{24}$ cm.³)

He 0·202 Li⁺ 0·075	F⁻ 0·99 Ne 0·392 Na⁺ 0·21 Mg⁺⁺ 0·12 Al⁺⁺⁺ 0·065 Si⁺⁺⁺⁺ 0·043	Cl⁻ 3·05 A 1·629 K⁺ 0·87	Br⁻ 4·17 Kr 2·46 Rb⁺ 1·81 Sr⁺⁺ 1·42	I⁻ 6·28 Xe 4·00 Cs⁺ 2·79

The polarizability of some neutral atoms (H, Li, K, Cs) has also been determined by a method similar to the Stern-Gerlach experiment (§ 7, p. 185), namely, by measuring the deflexion of a beam of atoms in an inhomogeneous electric field (Stark, 1936). The results do not agree very well with theoretical computations from atomic models.

We go on now to consider molecules with *permanent dipole moment* p_0. Here, in addition to the polarization effect just considered, we have also the influence of the electric field on this permanent moment. In the absence of an external field, the moments of the individual molecules will have all possible directions, so that the gas is unpolarized. If an external field is applied, this tends to turn the individual dipoles round into the field direction (fig. 1); this ten-

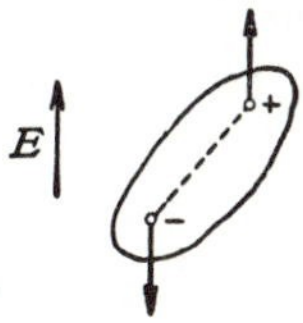

Fig. 1.—Couple exerted by an external electric field on a molecule with permanent dipole moment.

dency is opposed, however, by the thermal motion, which, as we have already frequently remarked, always has a smoothing out effect, and in this case acts in the sense of equal distribution of the dipole directions. Exactly the same relations are present here as in paramagnetism, where, as we have seen, it is a matter of the setting of the magnets in the direction of the magnetic field. For that case (see Appendix XXXVII, p. 462) we find a formula for the mean moment per unit volume, the field strengths being supposed not too great; we can apply that formula directly here, so that the polarization per unit volume due to the permanent dipoles is given by

$$P = \frac{N p_0{}^2}{3kT} E.$$

This is additional to the polarization determined by the polarizability of the molecules, and we therefore obtain a dielectric constant

$$\epsilon = 1 + 4\pi N \left(\bar{a} + \frac{p_0{}^2}{3kT} \right) = \epsilon_0 + \frac{4\pi N p_0{}^2}{3kT},$$

where ϵ_0 stands for the dielectric constant for the case of a vanishing permanent dipole moment. The value of the dielectric constant therefore depends on two concurrent effects, one purely electrostatic and therefore independent of temperature, the other dynamical (orientation of the dipole moments) and a function of the temperature (Debye's law (1912), analogous to Curie's law for paramagnetism). By determining ϵ exactly as a function of the temperature, we can therefore separate these two effects, and so from a series of measurements of the dielectric constant deduce the polarizability and the magnitude of the permanent dipole moment. Here, however, it is taken for granted that ϵ is measured in the electrostatic way. For if we were to proceed as above by determining the refractive index (for infra-red waves), we should always be measuring the first effect only, that is, the one dependent on the polarizability. This is due to the fact that the orientation process in the dipole moments cannot follow the rapid vibrations of the electric vector of the light wave; for this orientation requires a rotation of the whole molecule and accordingly a motion of the atomic nuclei, which on account of the great mass of the nuclei takes place far too slowly to be appreciably affected by the rapidly varying electric forces of the light wave. Hence here also we have

$$n \to \sqrt{\epsilon_0} = \sqrt{1 + 4\pi N \bar{a}},$$

so that a measurement of the refractive index always gives the polarization effect only. This makes it possible to determine p_0 by a particularly simple method; n is measured optically and ϵ_0 found from it, ϵ being then determined by a statical measurement at a known temperature (Debye):

$$\epsilon - \epsilon_0 = \frac{4\pi N p_0^2}{3kT}.$$

We note further that the limitation to long waves is necessary in the determination of the refractive index, in order to keep clear of the region of "anomalous dispersion", which is roughly characterized by the condition that the optical frequencies are of the same order of magnitude as the classical orbital frequencies of the electrons. In the long wave region, however, the electronic motion is much more rapid than the light vibration, so that the action of the light on the electrons depends only on the mean distribution of the electronic charge, or the polarizability.

The results so obtained have been successfully verified for some substances by direct deflection of a molecular beam (§ 7, p. 17) in a non-homogeneous electric field (Estermann, 1928).

The following table contains a small selection from the very large number of published determinations of moments:

TABLE XII—DIPOLE MOMENTS OF MOLECULES (in 10^{-18} e.s.u.)

Carbon monoxide	.. CO	0·12
Carbon dioxide ..	.. CO_2	0·0
Water	.. H_2O	1·8
Methane	.. CH_4	0·0
Methyl chloride ..	.. CH_3Cl	1·9
Methylene chloride	.. CH_2Cl_2	1·6
Chloroform ..	.. $CHCl_3$	1·0
Carbon tetrachloride	.. CCl_4	0·0

With reference to this table we add some remarks. In general, molecules of symmetrical structure, such as CO_2, CH_4, CCl_4, &c., have no dipole moment. Since H_2O possesses a dipole moment, its structure cannot be symmetrical. At the present time it is assumed to have the form of an isosceles triangle. CO, as an asymmetric molecule, has of course a dipole moment. The series of chlorine compounds between methane and carbon tetrachloride is interesting; their

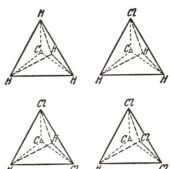

Fig. 2.—Structural formulae of CH_4, CH_3Cl, CH_2Cl_2, and $CHCl_3$. Methane is symmetric, and therefore has no dipole moment; the other three compounds are asymmetric, and more or less electrically polar.

structural formulæ are shown in fig. 2. We see that the highly symmetric forms CH_4 and CCl_4 at the ends of the series have no dipole; in the intermediate cases dipoles are present, the gradations of which can be represented roughly in terms of " vectorial composition " of elementary dipoles.

We have already mentioned above that in the case of a gas free from dipoles ($p_0 = 0$) the electric field arising from a molecule, and consequently the interaction between the molecules, are determined by the *quadrupole moment* (electrical moment of inertia). This therefore makes a contribution in this case to the *cohesive forces* which act between the molecules, and which in gases find their expression in the constants of the equation of state (e.g. van der Waals' equation, p. 20); they can be measured either by means of these constants or from the latent heat of vaporization, &c. We do not consider this further here, however, but merely refer to § 7, p. 275, where we shall return to the matter.

We must add a remark with regard to polarizability. In what precedes we have taken account only of the mean value of a over all directions—a procedure which, in the case of a gas, whose molecules can rotate freely, is certainly permissible as a first approximation. But by suitable experiments we can also determine the anisotropy of the polarizability, and so also form for ourselves a picture of the *anisotropy of the electron cloud*. We have already mentioned (p. 248) that the polarizability is a tensor, and can be represented by the so-called ellipsoid of polarization (see fig. 3). This has the following

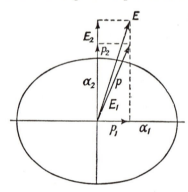

properties. The three principal axes of the ellipsoid, a_1, a_2, a_3, lie in the directions of the least and greatest polarizability, and the direction perpendicular to both of these; if electric field strengths of unit amount act in succession in these three directions, the lengths of the axes give the electrical

Fig. 3.—Diagram of the relative position of field and dipole moment in the case of an anisotropic molecule.

dipole moments corresponding to the respective cases. If we now let the unit electric field strength act obliquely to these three special directions, we can determine the polarization of the molecule by splitting up this field strength into its components in the three special directions, and determining the polarization effects of these three components separately; the total moment is then found by vectorial addition, and it is clear that in general the direction of the acting field strength does not agree with the direction of the induced dipole moment.

This has a marked effect on the polarization relations in the case of the *scattering of light*. Let us consider in the first place (see fig. 4) the case of the isotropic molecule or atom (atoms are in this sense always isotropic). If a light wave falls on this molecule, the electric vector E of the light wave excites in the molecule an electric moment p, which is parallel to the exciting field strength, and in phase with it; a scattered wave is therefore emitted having the same frequency as

the primary light.   If we observe the scattered light in a direction
perpendicular to the incident beam, we find that it is completely
polarized; this is easy to understand, for its electric vector, which is
determined by p, is always parallel to the vector E of the primary
beam. This is not so for the anisotropic molecule, in which, as we
have just seen, the induced dipole moment in general has a different
direction from the exciting electric vector. If in this case again we
observe the scattered light in a direction at right angles to the incident
beam, we find that it is no longer completely polarized, but only
partially; in the scattered light there now occurs a component of the
electric vector at right angles to the incident light vector E. If we
make an experiment with the polarization apparatus, we no longer

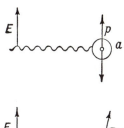

find, as we did before, a position of the
nicol for which the field of view is completely dark. In this case we speak of the
depolarization of the scattered light (Gans,
1912; Born, 1917). By measuring the
degree of depolarization, we can draw conclusions with regard to the anisotropy of
the polarizability; for example, in the case
of axisymmetric molecules (in which two of

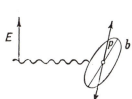

Fig. 4.—Depolarization due to the anisotropy of the polarizability. (*a*) In an isotropic molecule the induced dipole moment
vibrates in the direction of the electric vector of the light wave.
(*b*) In an anisotropic molecule it vibrates obliquely to E, and
causes depolarization in the scattered light.

the axes of the ellipsoid of polarization become equal: $a_1 = a_2$), we
can find, as an exact analysis shows, the value of $a_3 - a_1$. Since we
can determine the mean polarizability by other measurements, we
obtain in this way complete information regarding the lengths of the
axes of the ellipsoid of polarization (Raman, Daure, Cabannes, 1928).
   The same result can be obtained from the *Kerr effect* (1875). In
a static field E, even in the absence of a permanent dipole moment
($p_0 = 0$), anisotropic molecules are subjected to a couple, since the
direction of the induced dipole moment does not fall in the field
direction. This couple is again in the present case opposed by the
thermal motion, which tends to produce a uniform distribution of
direction, so that the molecules become partially oriented in the field
direction, to an extent dependent on the temperature (Langevin, 1905;
Born, 1916). But it can be shown that such a substance behaves
towards light passing through it exactly like a doubly refracting

uniaxial crystal. In this case, again, measurement of the double refraction gives for axisymmetric molecules the value of $a_3 - a_1$ (Gans (1921), Cabannes, Raman, Stuart). We may remark further that the effect in question is much employed in modern technical work; it is the basis of the Kerr cell (Carolus), extensively used in telecommunication technique as a light relay.

We come now to the determination of the distances between the nuclei, the frequencies of the nuclear vibrations, and other molecular properties connected with the nucleus. Here optical methods of a special sort play a part which we shall consider in next section.

3. Band Spectra and the Raman Effect.

We disregard in the first place the relative motions of the nuclei. A diatomic molecule, so far as its mass distribution is concerned, can be pictured as a nearly rigid dumb-bell, since of course the electrons by reason of their vanishingly small mass form an inappreciable factor in the mass distribution. This dumb-bell can turn round an axis fixed in space, and so possesses angular momentum, which according to Bohr must be quantised. If j is the quantum number of this angular momentum, the energy of the rotating dumb-bell is given on Bohr's theory (p. 112) by

$$E_j = \frac{h^2}{8\pi^2 A} j^2 \quad (j = 0, 1, 2, \ldots)$$

or, according to quantum mechanics (p. 145)

$$E_j = \frac{h^2}{8\pi^2 A} j(j + 1).$$

We have called the latter energy term the Deslandres term (as contrasted with the Balmer term). Here A is the moment of inertia of the dumb-bell about an axis through the centroid at right angles to the line joining the nuclei, and is easily found in terms of the nuclear distance and the masses of the two atoms. Thus, if r_1, r_2 are the distances from the centroid of the atoms of masses m_1, m_2, the moment of inertia is by definition $A = m_1 r_1^2 + m_2 r_2^2$; also $m_1 r_1 = m_2 r_2$ and $r_1 + r_2 = r$, where r is the nuclear distance, which we wish to find If then we denote by m the effective mass, i.e. if we take

$$\frac{1}{m} = \frac{1}{m_1} + \frac{1}{m_2}, \quad \text{or} \quad m = \frac{m_1 m_2}{m_1 + m_2},$$

we have for the moment of inertia

$$A = mr^2.$$

The nuclear distance can therefore be found if we know the values of the Deslandres terms, which can be determined from the emitted *rotational band spectra.* We have already pointed out (p. 112) that the spectrum emitted by a rotator consists of a series of equidistant lines. In fact, as we have already repeatedly stated, in simply periodic motions there is a selection rule $\Delta j = \pm 1$, so that the emitted frequency is found as the difference of two consecutive energy terms:

$$\nu = \frac{E_j - E_{j-1}}{h} = \frac{h}{8\pi^2 A} \left\{ j(j+1) - (j-1)j \right\}$$

$$= \frac{h}{4\pi^2 A} j.$$

By measuring the separation of the lines, $\frac{h}{4\pi^2 A}$, we therefore get A, and accordingly r also. It is assumed, however, in this method of deduction that the initial and final states of the electronic system are identical, since a difference in these would involve a change in the nuclear distance also; moreover, we have disregarded any possible oscillations of the nuclei relative to one another. The *purely rotational bands* are not very suitable in practice, however, for the determination of nuclear distances, since they lie in the extreme infra-red. We can easily make a rough estimate of their position. Thus, atomic masses are of the order of magnitude 10^{-21} to 10^{-23} gm., the nuclear distances are about 10^{-8} cm., giving moments of inertia of approximately 10^{-37} to 10^{-39} gm. cm.^{2} We thus find for the frequencies values from 10^9 to 10^{11} sec.$^{-1}$, and therefore fractions of a centimetre for the wave-lengths.

TABLE XIII

NUCLEAR DISTANCES AND MOMENTS OF
INERTIA OF THE HYDROGEN HALIDES

	$r \cdot 10^8$ cm.	$A \cdot 10^{40}$ gm. cm.^{2}
HF	0·93	1·35
HCl	1·28	2·66
HBr	1·42	3·31
HI	1·62	4·31

Above we give a table of a few nuclear distances and moments of inertia found by means of the infra-red bands.

In polyatomic molecules, we have different moments of inertia about different axes, and must determine them separately.

The circumstances become distinctly more favourable, if we also take into account the *nuclear vibrations*. We spoke at the outset of equilibrium of the forces between the nuclei and the mean distribution of the charges of the electrons. Round the position of equilibrium, which of course must be stable, the nuclei can oscillate, and the whole electron cloud pulsates along with it. The whole molecular energy, after deduction of the kinetic energy of the nuclei, is to be regarded as *potential energy*, $V(r)$, *of the nuclear motion*; this therefore includes, besides the pure Coulomb energy of the (positively charged) nuclei, the mean electronic energy, or, more exactly, the averaged energy of the electronic motion, calculated on the supposition that the nuclei are kept fixed. Rotations of the molecule as a whole are in the first instance disregarded. The equilibrium position of the nuclei is defined by the minimum of $V(r)$; hence the equilibrium nuclear distance r_0 is given by $(dV/dr)_{r_0} = 0$. Such a minimum necessarily exists, as otherwise no molecule having a finite nuclear distance could be formed at all. Fig. 5

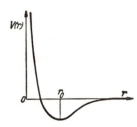

Fig. 5.—Graph of potential as a function of the distance between the two atoms combined in a molecule. The position of equilibrium is at r_0.

shows diagrammatically a potential curve of the type in question. From the minimum at r_0 the potential rises very steeply as r becomes smaller, the Coulomb repulsion between the two nuclei preponderating here; in the direction of greater nuclear distances the potential curve flattens and asymptotically approaches a definite limiting level, which in the diagram has been arbitrarily taken as the zero level; this corresponds to the case of nuclei which are far apart, so that the molecule is practically completely split up into its constituent parts. In this position the potential energy $V(r)$ of the nuclear motion is simply equal to the constant electronic energy of the two separated atoms, and can therefore be normalized to zero; the nuclei (and with them the whole atoms) then move as free particles.

In the potential hollow of $V(r)$ the nuclei can vibrate as *quantum oscillators*. In the neighbourhood of the equilibrium position r_0, the potential curve is approximately parabolic in form; this we see from the Taylor expansion of $V(r)$ at the point $r = r_0$:

$$V(r) = V(r_0) + \frac{(r - r_0)^2}{2} \left(\frac{d^2V}{dr^2}\right)_{r_0} + \cdots$$

For not too great amplitudes, therefore, the nuclei vibrate like harmonic oscillators, since of course in this case the restoring force is proportional to the distance; we can therefore apply the formulæ for the harmonic oscillator, which from the classical standpoint (Bohr) lead to the energy levels

$$E_s = h\nu_0 s \quad (s = 0, 1, 2, \ldots),$$

and from the standpoint of wave mechanics to the formula

$$E_s = h\nu_0(s + \tfrac{1}{2});$$

the first formula was deduced in Chap. V, § 2 (p. 115), the second is obtained by solving the wave equation of the harmonic oscillator (p. 138, or Appendix XVI, p. 396). The proper frequency ν_0 is determined by the restoring force, being given, as we may easily prove, by the equation

$$\nu_0 = \frac{1}{2\pi} \sqrt{\frac{1}{m}\left(\frac{d^2V}{dr^2}\right)_{r_0}}.$$

By a rule of differential geometry $(d^2V/dr^2)_{r_0}$ is proportional to the curvature of the potential curve at the point $r = r_0$, so that the result can also be expressed in the following form: the greater the curvature of the potential curve of the nuclear motion at the position of equilibrium, the greater is the proper frequency, and the higher are the corresponding energy levels.

The foregoing formulæ, as has already been mentioned, only hold for small amplitudes of vibration, or, what comes to the same thing, for the low quantum numbers. In the case of the more highly excited states, the deviation of the potential curve from the parabolic form has the effect of making it no longer allowable to treat the vibration as that of a harmonic oscillator; the formulæ deduced above have to be supplemented by corrections, which alter the values of the terms; when the quantum number s increases, the terms in fact crowd more and more closely together as they approach the so-called *convergence limit* (fig. 6). This limit corresponds to the *dissociation* of the molecule; it requires a quantity of energy equal to the depth of the potential hollow below the asymptotic limiting value of $V(r)$, that is, the quantity $V(\infty) - V(r_0)$; on excitation by this or a greater quantity of energy the molecule splits up into atoms or ions, which then move apart with a definite velocity, given by the excess energy. In the spectrum, this finds its expression in the fact that a band occurs with a convergence limit, which is immediately followed by a continuum (Franck). From the position of the limit we can determine the *dissociation energy*, and that much more exactly than by the thermal measurements used in chemistry. Thanks to this principle, the work of separation is now well known for a great number of molecules, for example,

$$H_2 : 101, \quad N_2 : 210, \quad O_2 : 117\cdot3 \text{ kcal/mole.}$$

To give some idea of the order of magnitude of the frequencies of vibration, we quote here a few funda-

Fig. 6.—Term scheme of a vibrational band. The discrete energy levels converge towards a limit, which corresponds to the dissociation of the molecule; immediately above this a continuum follows, signifying that after dissociation the components of the molecule have kinetic energy, and fly apart.

mental vibrational frequencies $\nu_{osc.}$ (after Czerny), in the form of wave numbers:

$$HF : 4003 \text{ cm.}^{-1},$$
$$HCl : 2907 \text{ cm.}^{-1},$$
$$HBr : 2575 \text{ cm.}^{-1}.$$

For comparison, we give also the corresponding fundamental rotational frequences $\nu_{rot.}$:

$$HF : 41\cdot1 \text{ cm.}^{-1},$$
$$HCl : 20\cdot8 \text{ cm.}^{-1},$$
$$HBr : 16\cdot7 \text{ cm.}^{-1}.$$

It will be observed that the vibrational quanta are very decidedly larger than the rotational quanta.

Up till now, in dealing with vibrations we have disregarded the possible occurrence of molecular rotations. When absorption takes place, rotation and vibration can of course be excited simultaneously; the energy is then given approximately by

$$E_{s,j} = E_s + E_j = h\nu_0(s + \tfrac{1}{2}) + \frac{h^2}{8\pi^2 A_s} j(j + 1);$$

the energy quanta of the rotational motion are for a first approximation simply added to the vibrational energy terms. The frequencies, which arise from differences of such terms (the selection rules $\Delta j = 0, \pm 1$; $\Delta s = \pm1, \pm2, \ldots$ hold here), that is, for instance, for the fundamental vibration

$$\nu = \frac{E_{s,j} - E_{s-1,j-1}}{h} = \nu_0 + \frac{h}{8\pi^2}\left\{\frac{j(j+1)}{A_s} - \frac{j(j-1)}{A_{s-1}}\right\},$$

are the frequencies which give the rotational-vibrational bands; in consequence of the large value of the proper frequency ν_0 they lie in the short wave infra-red. The addition of the vibrational quantum $h\nu_0$ therefore causes the whole spectrum to be displaced in the direction towards shorter waves, i.e. into the region which can be reached fairly easily by experimental methods; still, the law of succession of the rotational lines is more complicated here than in the case of pure rotations, since the moment of inertia, which is determinative for the separation of the lines, depends on the value of the nuclear distance for the time being, and may be different in the initial and final states.

The energy terms $E_{s,j}$ written down above do not give the complete scheme of terms, since we laid down the condition that the electronic state of the molecule does not change in the rotational and vibrational transitions. But the electronic state can be altered by absorption or emission of light; for the molecule, exactly as for the atom, there are excited states, which are distinguished by quantum numbers $n = 1, 2, \ldots$. To every such state there corresponds a particular potential energy of the nuclear motion, $V_n(r)$, which is found by averaging; we therefore obtain different curves for the potential of the nuclei, corresponding to the separate excitation levels of the electronic motion. The horizontal asymptotes $(r \to \infty)$ of these curves give the differences of energy in the end products occurring in dissociation. If, for example, molecular hydrogen H_2 dissociates, it splits up, according to the electronic state of the molecule, either into two hydrogen atoms in the ground state, or into an unexcited and an excited hydrogen atom, or into two excited atoms; to the various energies of the products of dissociation

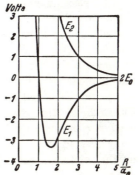

Fig. 7.—Potential curves for homopolar binding (H_2 as example). One curve has a potential minimum (attraction); the other corresponds to a pure repulsion.

there correspond the various horizontal asymptotes of the potential curves for the nuclear vibrations.

In fig. 7 two potential curves are shown, one of which, E_1, possesses a minimum, and therefore makes a stable chemical binding possible, while the other, E_2, steadily falls; the latter of course does not correspond to any chemical binding, but to a repulsion between the atoms, since its lowest state represents the state of atoms infinitely far separated. This double possibility occurs even in the case of the hydrogen molecule (see § 6, p. 271).

In the case of simultaneous excitation of higher electronic states, of vibration and of rotation, the total energy is given approximately by the formula

$$E = E_n + E_s + E_r$$

where E_n represents the pure electronic energy, as determined by the energy difference between the minima of the various potential curves. The presence of E_n has the effect of displacing the bands into the visible or ultra-violet region, since the order of magnitude of the frequency determined by transitions from one electronic state of the molecule into another is the same as that of electronic transitions in the case of atoms; the combination of an electronic jump with a transition in the vibrational and rotational state implies the emission of a rotational-vibrational band in the region of wave-lengths fixed by the electronic transition. The appearance of these bands is the same as that of the pure rotational-vibrational bands, except for the fact that small " perturbations " occur, which are due to the interaction between the electronic and the nuclear motion (for example, alteration of the proper frequency of the nuclear vibration by an electronic jump) (figs. 8 and 9, Plate XI, facing p. 187).

The electronic terms E_n for diatomic molecules admit of similar classification to those of atoms. Here, however, we cannot use the orbital angular momentum for this purpose, as in the case of atoms ($l = 0, 1, 2, \ldots$; $S, P, D \ldots$ terms), since the electron cloud has no longer a fixed total angular momentum, as it has in atoms; for the line joining the nuclei represents a specially distinguished direction in the molecule, and it rotates in space (rotational terms), carrying of course the electron cloud along with it. The *component of the electronic angular momentum* in this special direction is quantised, however, not being affected by the rotation of the molecule as a whole. The quantum number of this angular momentum component is denoted by λ, and, in analogy with the atomic case, the terms are designated by Greek capital letters, corresponding to the value of the quantum number:

$$\lambda = 0 \quad 1 \quad 2 \quad \ldots$$
$$\Sigma \quad \Pi \quad \Delta \quad \ldots \quad \text{terms.}$$

We may also mention that special features appear in the scheme as developed so far, if a molecule consists of two equal nuclei; degeneracies then occur, which express themselves in definite typical alterations in the spectrum (dropping out of certain lines).

Finally, we must refer to a complication which arises from the fact that the nuclei are really not point charges, but have a structure. The most important effect of this is the nuclear spin. The inner angular momenta of the nuclei have to be added vectorially to the angular momentum of the rest of the system, due to the rotational motion of the nuclei and electrons, with their spin. In the case of two equal nuclei, very large degeneracy effects arise from this cause. The simplest case is that of the molecule H_2. Here we have as nuclei two protons, each with the spin $\frac{1}{2}$. Vectorial composition gives $\frac{1}{2} - \frac{1}{2} = 0$, or $\frac{1}{2} + \frac{1}{2} = 1$; in the latter case the resultant nuclear moment 1 can have three settings with respect to the moment of the rest of the motions, determined by the possible components $-1, 0, +1$ of the vector of length 1. These states have all the same probability. Hence, molecules with nuclear moment 1 will be three times more frequent than those with the nuclear moment 0. Further, it is found that practically no transitions take place spontaneously from the one sort of molecule to the other (Heisenberg, Hund); they exist almost independently of each other, and so they have been given names. Molecules with the moment 0 are called *parahydrogen*; those with the moment 1, *orthohydrogen*. Ordinary hydrogen is a mixture of these in the ratio 1 : 3, as is shown by many properties, especially the specific heat (Dennison, 1927); but orthohydrogen, which has the higher energy content, can be converted into parahydrogen by means of catalytic effects, such as adsorption at surfaces (Bonhoeffer and Harteck, Eucken and Hiller, 1929). The difference in respect of energy arises from the fact that the paramolecules with the spin 0 can only have even rotational quantum numbers, and orthomolecules with the spin 1 only odd ones; the lowest state with quantum number 0 belongs therefore to parahydrogen. Since the latter occurs in ordinary hydrogen in the ratio 1 : 3 as compared with the ortho-modification, the intensities of the band lines are in the same ratio. Such a variation in the intensity of band lines occurs in all molecules which consist of two equal atoms with nuclear spin, and makes it possible to determine this spin. The method has already been mentioned (p. 312).

Band research has already developed into a science of considerable dimensions. The investigation and analysis of band spectra gives very far-reaching information regarding the structure of molecules. For the interpretation of the extensive experimental material which has been collected, ingenious mathematical methods, such as group theory, considerations of symmetry, and so on, have been called into service.

To conclude this section, there is one other phenomenon we should like to discuss, viz. the *Raman effect*. Let it be mentioned beforehand, however, that this is not a revolutionary discovery, like, for example, the discovery of the wave nature of the electron, but an effect which was predicted by the quantum theory (Smekal (1923), Kramers-Heisenberg) some years before it was found experimentally, though it can also be explained within the framework of classical physics (Cabannes (1928), Rocard, Placzek); its great importance rests rather on the facility with which it can be applied to the study of molecules, and on the colossal amount of material relating to it which has been accumulated so quickly. The effect was discovered simultaneously (1928) by Raman in India, and by Landsberg and Mandelstam in Russia. They found that scattered light contains, in addition to the frequency of the incident light, a series of other frequencies.

The classical explanation of this effect is as follows. If a light wave $E = E_0 \cos 2\pi\nu t$ falls on a molecule, it produces in this molecule a dipole moment

$$p = aE_0 \cos 2\pi\nu t,$$

where a denotes the polarizability tensor. As we have already remarked, the direction of the induced polarization is not in general the same as that of the exciting field strength. In the present case what we have to consider is the effect produced on the induced polarization by the state of the molecule as regards rotation and vibration. For, when the molecule rotates, the ellipsoid of polarization turns along with it, and the induced dipole moment therefore vibrates in the same rhythm. Similarly, when the nuclei oscillate, the whole electronic system also does so, and this again causes an oscillation of the polarizability in the same rhythm; in fact, since the electronic motion round the nuclei is much more rapid than the oscillations of the nuclei themselves, we can use the averaged electronic distribution when we are considering the effect of the nuclear vibrations on the polarizability. From the Fourier series which represents the influence of the rotations and vibrations on the polarizability we pick out a single term ν_s, and write a in the form

$$a = a_0 + a_1 \cos(2\pi\nu_s t + \delta);$$

here δ denotes an undetermined phase which varies from molecule to molecule. If we substitute this expression in the formula for the dipole moment, we find

$$p = a_0 E_0 \cos 2\pi \nu t + a_1 E_0 \cos 2\pi \nu t \cos (2\pi \nu_s t + \delta)$$

$$= a_0 E_0 \cos 2\pi \nu t$$
$$+ \tfrac{1}{2} a_1 E_0 \{\cos [2\pi(\nu + \nu_s)t + \delta] + \cos [2\pi(\nu - \nu_s)t - \delta]\}.$$

The vibration of the dipole moment may therefore be regarded as due to the superposition of three vibrations with the frequencies ν, $\nu + \nu_s$ and $\nu - \nu_s$. Since the dipole moment is the cause of the scattered light, the latter contains the frequencies $\nu + \nu_s$ and $\nu - \nu_s$ in addition to the frequency ν of the incident light; moreover, since the phase δ is arbitrary, the three vibrations are incoherent. Precisely the same holds good for all the rotational and vibrational frequencies of the molecule.

The spectrum of the scattered light is therefore a sort of band in the neighbourhood of the incident line, from which the rotational and vibrational frequencies can be read off, exactly as in the case of emission and absorption bands. The advantage of the method, however, is that the whole band is situated at a part of the *visible spectrum* which can be chosen at will, its position depending only on the choice of the irradiating frequency. Thus, for example, we can determine the constants of the ground state of the electronic system by observations, in the visible spectral region, of the separations of the lines of the bands, while the corresponding emission bands lie far in the infrared. It should be noted that the displacement of the bands into the visible, by observation of the rotational-vibrational bands in the case when a simultaneous electronic jump is involved, does not give the molecular constants of the unexcited " natural " state, but those of some excited state, which in general does not interest the chemist.

With regard, however, to the intensity relations in the Raman lines, special features are found, which do not admit of explanation on classical lines; for example, the component of the scattered light with the frequency $\nu - \nu_s$ is much stronger than the one with the frequency $\nu + \nu_s$. This phenomenon can be understood at once in the light of the following simple quantum consideration. If an incident light quantum hits the molecule, it may as a first possibility be scattered without loss of energy. It may, on the other hand, excite the molecule, and so give up to it the vibrational energy $h\nu_s$; in the scattered light the light quantum then appears with the energy $h(\nu - \nu_s)$. In rare cases, again, it may happen that a light quantum hits a molecule which is already excited; it may then take energy away from the molecule, the latter falling into the lower state, while the light quantum flies on with energy $h(\nu + \nu_s)$. Hence on the long wave side the Raman lines are strong, but only a few weak lines lie on the short wave side.

The Raman effect has come to be of the greatest importance for molecular research, since it is comparatively simple to observe, and makes it possible to reach exact results regarding molecular structure in many cases, from the mere consideration that certain lines occur or do not occur. The following example serves to illustrate these remarks.

The molecule N_2O, assuming rectilinear arrangement, might have either of the forms NON or NNO. The question is how we can distinguish between these alternatives. Now the first form is obviously symmetrical, the second not. But, among the vibrations taking part in the Raman effect, those belonging to a symmetrical molecular structure differ characteristically (in the number and polarization of the lines) from those belonging to one which is unsymmetrical. Thus we can deduce without ambiguity from experiment that the non-symmetric form NNO is the correct one. Similar considerations can also be adduced with respect to complicated molecules (for instance, the symmetric structures methane, CH_4, and carbon tetrachloride, CCl_4), but we cannot go into this here.

4. Chemical Binding. Classification of Types of Binding.

Hitherto we have considered the electronic system of the molecule as a whole, and confined our attention in the main to properties and effects in relation to which the molecule is regarded as a given structure. We must now proceed to deal with the question which to the chemist is the specially interesting one, viz. how a molecular binding can come about at all, or how our molecular model is produced from the individual separate atoms.

We distinguish several different kinds of chemical binding, but between the principal sorts all possible intermediate stages occur. In our classification we adhere to the distinction drawn by Franck, according to which the decisive criterion is whether a molecule in dissociating tends more readily to split up into ions or into neutral atoms. It is true that the type of dissociation sometimes depends also on the excitation level of the molecule; still Franck's criterion gives in many cases a suitable point of view for the classification of binding. We therefore distinguish the following cases:

I. *The molecule splits up into ions* by thermal dissociation or in consequence of absorption of radiation; this kind of binding is called *ionic binding*.* The extreme case of this type of binding is that in which the atoms are charged even in the molecule, so that they are present as ions; the binding is then explained by their Coulomb attraction. In this case we speak of *polar* (or *heteropolar*) binding and of *electro-valency*. The typical case of this kind of binding is that of NaCl. About a hundred years ago it was conjectured by Berzelius that all chemical forces are really of electrostatic nature. The hypothesis was dropped, however, on account of the difficulty of explaining in this way the binding of atoms of the same kind (e.g. H_2, N_2, . . .), which cannot possibly be of polar nature. It was only after it had become possible, by observation of electrolytic behaviour, of dipole moments and the like, to demarcate polar molecules in some measure from others, that the hypothesis of Berzelius was revived; it holds good for a limited class of bindings only, but within this range, as has been shown by Lewis and Kossel (1916), is capable of yielding important results.

II. *The molecule breaks up into atoms*, as, for instance, in the case of thermal excitation, or in consequence of absorption of radiation; we speak in this case of *atomic binding*. Here there are several sub-cases to be distinguished:

1. Atomic bindings with saturation of valency: *co-valency bonds* (also called *homopolar bindings*); to these belong in the first place the diatomic gases, such as H_2, N_2, O_2, as well as most organic compounds, for instance, CH_4.

2. Loose bindings without saturation of valency, due to the van der Waals forces: *cohesion bindings*.

3. Bindings which are effective, for example, in the case of the lattice formation of metals, and which we comprise under the name *metallic bindings*.

4. A series of bindings which cannot be strictly classified according to the above scheme, as for instance the *benzene binding and other similar binding types*.

Between these various groups there is of course a series of intermediate stages, which will not, however, be considered here, the next sections being confined to a discussion of the most important of the types of binding mentioned above.

* The nomenclature is not quite fixed, but that given here seems to be gradually becoming established. As a matter of history, it may be noted that the distinction between heteropolar and homopolar binding was first drawn by Abegg; the separation into ionic and atomic binding, with the criterion of the dissociation products, was introduced by Franck.

5. Theory of Heteropolar Ionic Binding.

How this binding comes about, we have already indicated above; we shall now consider the question a little more fully (Kossel, Born and Landé, 1918). The atoms which come next the inert gases are always striving to become converted into the inert gas configuration, by taking up or giving up electrons. In the alkali atoms the valency electron moves outside a closed shell and is comparatively loosely bound—the alkalies have a very small *ionization energy I* (see the numerical values of the ionization potential in Table V, pp. 181, 182, 183). Conversely, in the halogens there is one electron too few to make up a closed shell, which, as we know, represents a very stable electronic configuration; they are therefore very ready to pick up an electron in order to complete the shell. We call this the *electronic affinity E*; it is given by the energy which is set free when the electron settles in its place. Strictly connected with these concepts is that of *electrovalency*; positive electrovalency being the number of electrons loosely bound and therefore outside a closed shell in the atom, and negative electrovalency the number of electrons required to complete the inert gas configuration.

The genesis of an ionic binding may be pictured as a process consisting of two steps. In the first, the two reacting atoms become charged in opposite senses; thus, for example, $Na + Cl = Na^+ + Cl^-$. The second step is the attraction between the two ions in accordance with Coulomb's law; the energy of the attraction is $-e^2/r$. By itself, this would lead to the absolute coalition of the two ions; however, at small distances repulsive forces come into play, which can be accounted for by quantum mechanics. To represent them, a law of the form b/r^n has been tried, with good success; quantum mechanics gives approximately an exponential law $be^{-r/\rho}$, which has been found to answer even a little better. The position of equilibrium is then given by the value of r for which the sum of this part of the energy and the Coulomb energy has a minimum value.

In the gas molecule, in consequence of the one-sided action of the electric forces, the electron clouds of the two ions are of course very much deformed, so that the analysis of the ionic binding becomes extremely difficult. The circumstances are simpler in crystals, especially in the highly symmetrical ones of the rock salt (cubic face-centred, fig. 10) and similar types. In these the deformation disappears on account of the symmetry; in a rock salt lattice the same force is exerted on all sides on a chlorine ion by the neighbouring sodium ions, and a similar result holds for the more distant ions. Thus we can calculate the *lattice energy*, that is, the energy U which must be supplied in order to break up the lattice completely. It is given by a sum of the form

$$U = \Sigma\left(\pm \frac{e^2}{r} + \frac{b}{r^n}\right),$$

the summation being taken over all the lattice points; here we adopt the same hypothesis as before with regard to the repulsive forces. The evaluation of this sum presents considerable difficulties, since the first or Coulomb part is only slowly convergent; the acting forces certainly diminish as the distance increases, but on the other hand the number of ions, at the same distance from a given ion on which they act, increases as the square of the distance. Advantageous methods of evaluating such

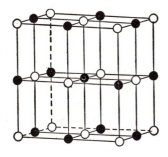

Fig. 10.—Face-centred cubic lattice

lattice sums have been given by Madelung and Ewald (1918); the expression found for U has the form

$$U = -\frac{e^2 a}{\delta} + \frac{\beta}{\delta^n},$$

where δ is the lattice constant, i.e. the distance of a sodium ion from the nearest chlorine ion; a, which is called Madelung's constant, depends on the lattice type, and for NaCl has the value 1·75. To determine the other constants, β and n, which are unknown initially, two equations are available. There is first the equation of equilibrium $dU/d\delta = 0$, expressing that the lattice energy is a minimum; secondly, $d^2U/d\delta^2$ is the force needed to compress the crystal by a certain amount, and so can be determined experimentally. For the alkaline halides values of n between 6 and 10 are obtained in this way. Better results are found by using the exponential law $be^{-r/\rho}$ for the repulsive force, as given by wave mechanics; the constant ρ is found to have approximately the same value for all the alkaline halides, viz. $\rho \sim 0.35$ Å. (Born and Mayer, 1932).

The results predicted by the theory can be tested directly. Thus, an experimental determination of lattice energies by J. Mayer, by means of thermal dissociation, showed good agreement for several salts. It is also possible to test the theory indirectly by calculating the *electronic affinity* of a halogen derived from different salts (say of Cl from the compounds LiCl, NaCl, &c.), with the help of the following cyclic process (Born, 1919; Haber):

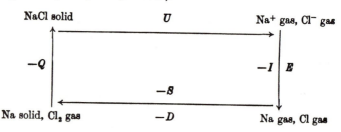

We start with solid rock salt (crystalline). By adding the lattice energy U (energy supplied is to be counted as positive), we break up the crystal into ions. By further addition of the energy E corresponding to the electronic affinity of Cl, we can remove the excess electron from the Cl⁻, whereupon the Na⁺ ion picks this electron up, at the same time giving up the ionization energy I, and so forming neutral Na gas. This gives up the heat of sublimation S of metallic Na, and is converted into solid sodium, while the atomic Cl gas gives up the dissociation energy D of the halogen and is converted into molecular Cl_2 gas. By the action of this gas on the Na metal, crystalline rock salt is formed, the heat of formation Q being given up; and thus the cycle is closed. The energy of course must balance, so that we must have

$$U - I + E - S - D - Q = 0.$$

Here all the quantities are known except the electronic affinity E, viz. I, D, S, Q from thermal and electrical measurements, and U from the lattice theory; and we can therefore calculate E from the

TABLE XIV.—Electronic affinity E of Cl from the alkaline chlorides (in kcal per mole)

LiCl	NaCl	KCl	RbCl	CsCl	Mean
85·7	86·5	87·1	85·7	87.3	86·5

TABLE XV.—Electronic affinities of the halogens (in kcal per mole)

F	Cl	Br	I
95·3	86·5	81·5	74·2

equation. In the result, the same value must be found from all salts of the same halogen, and this, to a good approximation, is found to be the case. From the chlorine salts, for example, the values shown in the accompanying Table XIV were found for the electronic affinity of Cl. The average values found for the different halogens are given in Table XV.

Mayer has succeeded in confirming these results by direct measurement, by means of molecular rays. It may be mentioned further, that for ions in solutions of salts there exists an absorption spectrum, which is due to this process of liberation of the electron from the halogen ion; by taking into consideration the action of the surrounding medium (water), the electronic affinity is verified to a rough approximation (Franck and Scheibe).

The theory indicated here is semi-empirical. From the standpoint of quantum mechanics we have the much more difficult problem to determine the structure and properties of crystals in terms of the nuclear charges alone. This can be done by generalizing the approximation methods developed by Hartree, Fock, &c., for dealing with the electronic clouds of atoms (Chap. VI, § 9, p. 196) to the case of crystals. In this way absolute values of lattice constants, compressibilities, &c., have been calculated in fair agreement with observations (Hylleraas, 1930; Landshoff, 1936; Löwdin, 1948).

6. Theory of Co-valency Binding.

We proceed now to consider the so-called co-valency binding. Here (according to Lewis, 1923) experience shows that such binding is frequently associated with the existence of an *electron pair*, which is such that each of the pair is shared by, or belongs to both of, the two atoms which are combined with each other. The hydrogen molecule is to be regarded as the simplest case of this kind, being built up from two equal nuclei and two electrons.

It is convenient to approach the problem of binding in the H_2 molecule by way of the simpler one of binding in the H_2^+ molecular ion, containing only one electron. When the two protons A and B are far apart there are two equally good approximations to the wave function of the system, namely atomic wave functions centred on proton A or proton B respectively. We now ask for a reasonable approximation when the protons are separated by a distance of only a few Bohr radii. The electron is then in the range of both protons and will no longer be definitely attached to one of them but will have an equal probability of being on either. We are led to the notion of a *molecular orbital*, a wave function for a single electron which runs over both nuclei. A natural approximation to such an orbital in wave mechanics is obtained by superposing the $1s$ atomic orbitals centred on the two protons. In this simple case, by the symmetry of the problem, the two possibilities are

$$\sigma_S^b \sim \phi_A + \phi_B$$
$$\sigma_S^* \sim \phi_A - \phi_B$$

where we have omitted normalization constants. The forms of the functions σ_S^b and σ_S^* and the corresponding charge densities are shown in fig. 1.

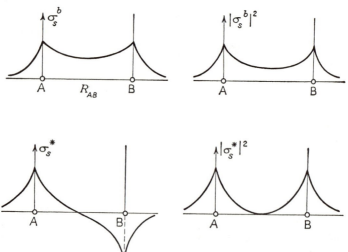

Fig. 1.—Sketches of the wave functions σ_s^b and σ_s^* and the corresponding charge densities.

When the electron is in the state σ_S^b the energy of the system decreases as the distance R_{AB} between the protons decreases from very large values. This is a wave-mechanical effect arising from the interference between the *overlapping* single-atom components of σ_S^b. In this approximation there is a reduction in the kinetic energy (since σ_S^b is somewhat more extended than ϕ_A or ϕ_B) and also a redistribution of charge into the region between the protons so that they are screened from one another. As the protons come still closer, the energy E_b of the system falls to a minimum and then begins to rise again when the protons are so close that they cannot be screened and they repel each other, as shown in fig. 2.

In this approximation the equilibrium distance R_0 is 1·3 Å and the binding energy is 1·76 eV. A more refined calculation using contracted atomic orbitals $\phi' \sim e^{-\lambda r / a a_0}$ yields values appreciably closer to the experimental values $R_0 = 1·06$ Å, $E_b = 2·79$ eV. The wave function σ_S^b is a bonding orbital.

The situation is quite different when the electron is in the state σ_S^*.

We expect from fig. 1 that the protons will never be screened from one another and so repel at all internuclear separations, giving the behaviour of E^* shown in fig. 2. There is no bonding, in fact the reverse, and σ_S^* is an anti-bonding orbital.

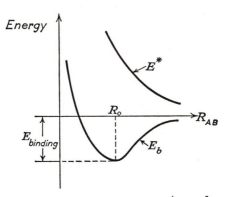

Fig. 2.—The energies of the states σ_s^b and σ_s^* as functions of the distance between the protons.

We can now treat the molecule H_2 in terms of molecular orbitals. In the ground state of H_2 we expect to find both electrons in the bonding orbital σ_S^b, so that the wave function is approximately

$$\Phi_b \, (1\ ;\ 2) = \sigma_S^b \, (1) \, \sigma_S^b \, (2).$$

This wave function is symmetric in the co-ordinates of the two electrons, so the spin part must be anti-symmetric, that is the electrons must be in the singlet spin state with total spin $S = 0$. There is indeed bonding in this state, and it can be used as a rough approximation to the true wave-function. A more exact wave-function would contain a term which introduced correlation between the electrons, to keep them apart and minimize their mutual repulsion. As a function of internuclear distance the energy of this state has the same general behaviour as E_b for H_2^+ in fig. 2.

The lowest state possible when the total spin is $S = 1$ has the anti-symmetric wave function

$$\Phi_a(1, 2) \sim \sigma_S^b(1) \, \sigma_S^*(2) - \sigma_S^b(2) \, \sigma_S^*(1).$$

In this state there is a deficiency of shielding charge between the two protons, which always repel, and the energy depends on R_{AB} in the same general way as E^* in fig. 2.

Before leaving the H_2-molecule we note that an alternative starting point is the wave function suggested by Heitler and London (1927), which is treated in Appendix XXXVIII (p. 466), and a comparison of the two approaches is instructive. When refined both approaches lead to the same result.

The molecular-orbital approach seems more convenient for a qualitative treatment of more complex molecules, and we shall take over the idea that we can form approximations to bonding molecular orbitals by properly superposing atomic orbitals which have appreciable overlap and redistribute some charge into the region between the nuclei. These bonding molecular orbitals can be occupied by up to two electrons. The other possible superpositions (anti-bonding or non-bonding orbitals) describe states of higher energy.

We see in this way that it is not possible to form covalent bonds between two He atoms. Only two of the four electrons can go into the bonding orbitals formed from the $1s$-states of the He atoms. The other two must occupy anti-bonding orbitals of higher energy. There is no net bonding.

Moving on to the case of the molecule N_2 a new feature appears. The atomic configuration is $(1s)^2 (2s)^2 (2p)^3$, and now p-orbitals come into play. There are three of these atomic orbitals, of the form $2p_x = xf(r)$, $2p_y = yf(r)$, $2p_z = zf(r)$, and the bonding combinations are

$$\sigma_z^b \sim 2p_{zA} - 2p_{zB}, \quad \pi_x^b \sim 2p_{xA} + 2p_{xB}, \quad \pi_y^b \sim 2p_{yA} + 2p_{yB}$$

where the z-axis runs along the line joining the two nuclei. The molecular orbitals π_x^b, π_y^b have the same energy by symmetry, and the energy of σ_z^b is slightly higher. The sequence of levels is shown in fig. 3.

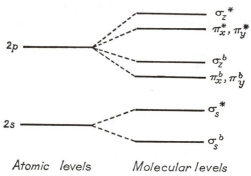

Fig. 3.—Atomic and molecular levels in homonuclear diatomic molecules.

The fourteen electrons in the molecule occupy the lowest orbitals allowed by the Pauli exclusion principle, and the molecular configuration is $(\sigma_{1s}^b)^2\ (\sigma_{1s}^*)^2\ (\sigma_{2s}^b)^2\ (\sigma_{2s}^*)^2\ (\pi_x^b)^2\ (\pi_y^b)^2\ (\sigma_z^b)^2$. There is an extremely strong bonding of about 10eV per molecule from the six electrons in the bonding orbitals.

In the case of O_2, fourteen of the electrons go into the same states as in N_2, but the last two must go into anti-bonding orbitals. In fact they occupy π_x^* and π_y^* with their spins parallel, rather than being paired with their spins anti-parallel. The explanation is the same as for Hund's rule in atoms: in the antisymmetric space state formed from π_x^* and π_y^* the two electrons tend to keep apart, and their repulsion is minimized. It is well known experimentally that the O_2 molecule is paramagnetic and has a magnetic moment corresponding to two unpaired electrons. The simple electron-pairing model of Lewis could not explain this paramagnetism in O_2, and its explanation is one of the notable successes of the theory of molecular structure.

At first sight, the principles used so far seem hardly capable of explaining some of the properties of more complex molecules, of which a good example is CH_4. The atomic configuration of C is $(1s)^2\ (2s)^2\ (2p)^2$ and it might be expected that C would be divalent, forming two bonds with the electrons in p-states. The quadrivalence of C can be understood in terms of *hybrid valence orbitals* on the C atoms which are linear combinations of $2s$ and $2p$ atomic orbitals. The appropriate linear combinations are of the form

$$\psi_h \sim 2s \pm 2p_x \pm 2p_y \pm 2p_z.$$

The charge distribution in the four hybrid orbitals for C is concentrated in single lobes which point away from the carbon atom, each towards a different vertex of a tetrahedron centred on the carbon atom. Putting the last four electrons into these hybrid orbitals does not give a stationary state of an isolated carbon atom, and the mean energy of this state is some 5-7 eV above the ground state of carbon (^{3}P). However, in a molecule like CH_4 this increase is more than offset by the energy gained from covalent bonding with the hydrogen atoms.

The ideas introduced in this Section, refined and elaborated, have enabled chemists to correlate a wide range of data on bond angles, bond strengths, isomerism, stereochemistry, reactions and reaction rates.

7. Theory of van der Waals Forces and Other Types of Binding.

We shall give here only a brief summary of this extensive subject. Forces of the van der Waals type arise from deformations of atoms or molecules. As was pointed out by Debye and Keesom (1921) using classical physics, a molecule with a permanent dipole moment (such as H_2O) can induce a dipole moment on a symmetric molecule and attract it. But from this point of view it is impossible to understand why the inert gases should first condense and then solidify (with the notable exception of He) at low temperatures. The way out of this difficulty was indicated by London (1930), who showed that the interaction of inert gas atoms or symmetrical molecules depended on mutual deformations. At large separations, the leading term in the electrostatic interaction of two neutral systems is

$$H' = \frac{1}{R^3}\left\{ d_1.d_2 - 3\,\frac{d_1.R d_2.R}{R^2} \right\}$$

where d_1 and d_2 are their respective electric dipole moments. The isolated atoms have no mean dipole moment, but they can take advantage of this interaction when the motion of the electrons in the two atoms is correlated so that, in the simplest case of hydrogen, if the electron on one atom is in the region between the protons, the other electron tends to keep away and be on the remote side of its proton. The dependence of this interaction on the distance R between the atoms follows immediately from perturbation theory (see Appendix XXXIX, p. 469). There is no contribution in first order, since neither atom has a mean dipole moment. However there will be a contribution in second order, and the interaction must be attractive and vary as R^{-6}. It is clear that the strength of the interaction is related to the polarizabilities, and that the higher the polarizability the stronger the van der Waals force will be. The detailed treatment of a simple model in Appendix XL, p. 471, bears out these general arguments.

The cohesion of metals cannot be explained by the types of binding dealt with so far. The bonding is not ionic, it involves too many nearest neighbours to be covalent, and it is much too strong to be attributed to van der Waals forces. We must regard the metal as a set of ion cores held together by the almost free conduction electrons. A few cases, in particular the alkali metals, were worked out in the 1930s (Slater, Wigner, Seitz, Fuchs). More complex metals are only now coming within the scope of refined approximation schemes suitable for electronic computers.

Although the tetrahedral co-ordination in crystals such as germanium, silicon and carbon itself, which all have the diamond structure, is a clear indication of directed covalent bonding, it is not easy to describe quantitatively the cohesion in these materials. However, as in the case of metals, reliable approximations are now being developed.

Other systems which require special consideration are the ring structures which occur in organic chemistry, the benzene ring for example. The structural formula is conventionally written as shown in fig. 4 with quadrivalent carbon atoms and alternating double and single bonds. It is well established chemically that the molecule does not have two such distinct localized types of bond. The simplest wave-mechanical model of the situation is that the actual state is the super-position of several wave-functions describing arrangements of the type illustrated.

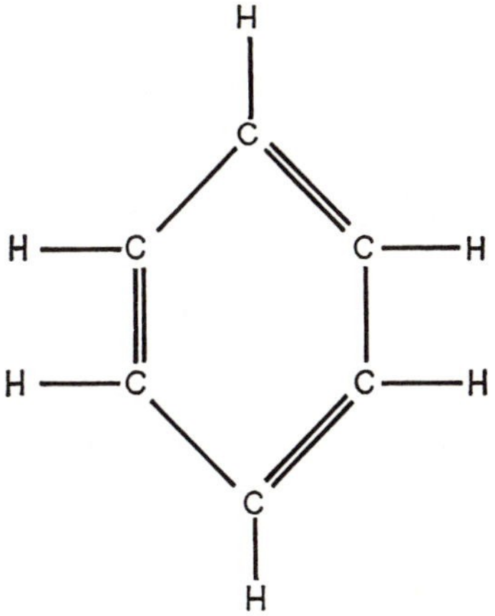

Fig. 4.—A conventional structural formula for benzene.

CHAPTER IX

Quantum Theory of Solids

1. Introduction.

In the chapter on quantum statistics, we have treated some aspects of the properties of metals by supposing that some of the electrons were free to move and carry current, and to contribute to the specific heat and to the paramagnetism. This simple model was able to explain qualitatively a large proportion of the experimental facts, but by no means all of them. The sign of the Hall effect in Be and Zn, for instance, is inexplicable on this theory. Many other questions arise in turn: why are some substances good conductors and others insulators? These general questions, and other more detailed ones, can be answered when the ideas of wave mechanics are brought into play. In the present chapter we shall consider in parallel the three main types of crystal: metals, insulators and semiconductors. To broaden the scope, we mention briefly two of the most striking co-operative phenomena in solids, namely ferromagnetism and superconductivity.

One of the fundamental properties of all crystals is that the atoms or ions of which they consist form a periodic array, so that a crystal can be built up by repeating periodically in space a unit cell, which may itself contain many atoms. This periodic repetition was proposed long ago to explain the occurrence of natural crystals exhibiting character-istic faces which yielded, after repeated cleavage, a " nucleus " whose shape was independent of the initial form of the crystal (Haüy, 1784). However, it was not until the advent of the technique of X-ray dif-fraction (von Laue) that it was possible to elucidate the size, symmetry and constitution of this unit cell. To be more precise, we should say that X-ray measurements tell us the way in which the mean density of the electron cloud is distributed over the unit cell. The unit cell also contains a number of heavy nuclei, the atomic nuclei of the atoms of which the crystal is composed. The electrons in the inner shells will be distributed around their respective nuclei almost exactly as they would in a free atom. The electrons in the outer shells, the less tightly bound electrons, will have a modified distribution, moving as they do in the electric fields due to the surrounding electrons and nuclei. Just as in the case of molecules, we note that the oscillating nuclei (they cannot be actually at rest, being particles confined to a quite small volume) move much more slowly than the electrons around them. This means again that in treating the electrons we can, to a first approximation, treat the nuclei as fixed and, conversely, in treating the motion of the nuclei compute the electrostatic forces on them by using the average value of the electronic density. We put aside the question of nuclear motions for the moment, returning to it in the next Section, and consider them to be in fixed positions. Their equilibrium positions will be those in which, firstly, there is no net force on any nucleus, and secondly the energy of the whole crystal is a minimum. These two conditions are not quite equivalent, for taking the case of a body-centred cubic lattice (fig. 1) as an example, every nucleus in an infinite

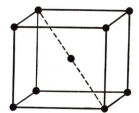

Fig. 1.—The body-centred cubic structure.

crystal of this type is at a centre of symmetry and all the forces balance automatically. However, the energy of the system is still a function of the internuclear distance r and will depend on it in roughly the same way as the energy of a diatomic molecule, shown in fig. 5 (p. 258), depends on the internuclear distance. The energy of the crystal is a minimum when $r = r_0$, and this is the equilibrium spacing. In a few cases (alkali halides, simple metals, diamond) it is possible to argue why a particular crystal symmetry is favoured. The Group IV elements carbon, silicon and germanium, for instance, all crystallize in a structure in which every atom is at the centre of a regular tetrahedron with atoms at its vertices. This is of course appropriate to strong covalent bonding via the directed sp^3 hybrid orbitals which were mentioned earlier in the discussion of molecules containing carbon. Again, metals will tend to crystallize in structures in which an atom has a large number of nearest neighbours. But in general we are content to take the observed structure and lattice constants from experiment and use them simply as data. The difference in minimum energies between two alternative structures may well be extremely small, so that even at temperatures at which kT is still quite small some crystals undergo phase transitions from one structure to another if the respective free energies cross when the temperature rises from absolute zero.

We can now proceed to a discussion of the vibrations of the nuclei.

2. Modes of Lattice Vibration.

As we have mentioned already, the nuclei are oscillating about equilibrium positions. In his theory of lattice specific heat Einstein assumed that all the atoms oscillated independently with the same frequency, whereas Debye treated the crystal as a whole, but as an elastic continuum. The actual behaviour is an amalgam of these aspects, and was first derived by Born and von Kármán (1912) by taking into account the lattice structure of crystals. The amplitude of the nuclear excursions from equilibrium is in any case (because of the large nuclear mass) much smaller than the internuclear spacing, and nuclei also have a very slow speed on the atomic scale. As a nucleus moves relative to neighbours, the shielding changes and its own electron cloud is slightly distorted, as are those of these near neighbours. Consequently the nucleus suffers an electrostatic force and is impelled back towards its equilibrium position, about which it oscillates. If the excursions are small enough, this restoring force will be linear in the displacements, and the crystal will oscillate harmonically. For the moment (even though they have profound effects on the thermal conductivity of the crystal) we shall neglect the effect of restoring forces which depend on the second or higher powers of the nuclear displacements because these anharmonic terms can be treated as perturbations.

Relegating a detailed discussion of the case of a linear chain to Appendix XLI, p. 474, we can use general arguments to discover the main features of the modes of vibration. A crystal will support travelling waves in which the displacement u of the nucleus at the lattice site R in a monatomic crystal is given by

$$u\ (R) = \epsilon\ e^{2\pi i[\kappa.R\ -\nu t]}$$

where ν is the frequency and κ is the wave number. The displacement of a neighbour is given by

$$u\ (R + a) = \epsilon\ e^{2\pi i[\kappa.(R+a)\ -\nu t]} = e^{2\pi i \kappa.a}\ u\ (R).$$

In the case of a mode with a long wave-length, so that κ is correspondingly small, the relative displacement of near neighbours is small. Essentially all the atoms within a given wave-length move in phase.

The mass involved is proportional to this wave-length and consequently the mode has a low frequency; to be precise it is given by $\nu = c \mid \kappa \mid$. We are in the regime of sound waves in an elastic medium. Each half wave-length contains so many unit cells that the lattice structure is irrelevant. It is different when the wave-length is of the order of the lattice constant. For instance, if we take for κ the value $1/2a$ then successive atoms are moving exactly out of phase. Now the frequency, depending as it does on the ratio of restoring force to inertia, attains its highest value, and we also note that the corresponding wave is in fact a standing wave. All possible modes lie between these two limits, for the wave vectors κ and $\kappa + 1/a$ yield exactly the same displacements for *all* the atoms in the crystal. In one dimension we can label all the distinct modes by using wave numbers in the interval $-1/2a < \kappa < 1/2a$. In three dimensions the corresponding range is the interior of a polyhedron in wave-number space: the Brillouin Zone. Also, in three dimensions we have the possibility of both longitudinal and transverse waves according as the nuclei are displaced parallel to the direction of propagation of the wave or perpendicular to it.

The number of modes must always be exactly equal to the number of degrees of freedom they account for. In the case of a linear chain that means N modes, and for a three-dimensional crystal there are N longitudinal modes and $2N$ transverse modes. To check this, we note that applying the Born–von Kármán periodic boundary conditions

$$u(n) = u(n + N), \text{ that is } e^{2\pi i \kappa na} = e^{2\pi i \kappa(N+n)a}$$

to the displacements of the atoms in a linear chain containing N atoms, the allowed values of κ are given by $0, \pm 1/Na, \pm 2/Na \ldots$, just as they were for the elastic string of Chap. VII, § 2. The interval between successive values of κ is $\delta\kappa = 1/Na$ and the number of modes with wave numbers in the Brillouin Zone $-1/2a < \kappa < 1/2a$ is just $1/a\delta\kappa = N$. In exactly the same way, we could have verified that there are exactly $3N$ modes with wave numbers in the polyhedral Brillouin Zone of a three-dimensional crystal. These results generalize simply to the case of crystals containing more than one atom in the unit cell. If there are a atoms in the unit cell, then the Brillouin Zone is still determined by the displacement vectors which carry a unit cell to its neighbours, but now there are a branches in the spectrum. The main features are that in the lowest branch, the so-called *acoustical branch*, the atoms or ions in the unit cell move in phase. In the higher branches they tend to move out of phase. In the case of ionic crystals, like the alkali halides, these high-frequency modes are associated with a large oscillating dipole moment as the positive and negative ions in the unit cell move in opposition. Transverse modes of this type can interact strongly with the electric vectors of light waves, and alkali halide crystals show a strong absorption band at the appropriate frequency. For this reason, these higher-frequency branches have acquired the generic name of *optical* even in crystals which are not ionic. A general picture of the spectrum of lattice vibrations is shown in fig. 2, which

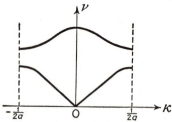

Fig. 2.—The vibrational spectrum of a diatomic linear chain.

represents the spectrum of lattice vibrations in a crystal containing two atoms or ions in the unit cell. The details of scale and shape vary of course from crystal to crystal, and the diagram does not tell us how the relative displacements of the two atoms change in going from the acoustic modes of long wave-length to the modes with wave numbers at the edge of the zone and lastly to the high-frequency optical modes.

3. Quantisation of the Lattice Vibrations.

So far we have made no reference to wave mechanics in discussing the lattice vibrations and their frequencies, but have worked in effect with a collection of classical particles coupled by linear elastic forces. Just as for a simple harmonic oscillator, the effect of wave mechanics is to quantize the energy, so that the possible energies of the crystal mode with wave number κ are $(n_\kappa + \frac{1}{2})h\nu_\kappa$, where ν_κ is the corresponding frequency and n_κ can take the integer values 0, 1, 2. . . .

The total energy of lattice vibration is then

$$E = \sum_\kappa (n_\kappa + \tfrac{1}{2})h\nu_\kappa = E_0 + \sum_\kappa n_\kappa h\nu_\kappa$$

where E_0 is the zero-point energy of the nuclear motions. The statistical mechanics of the lattice vibrations is now very easy, since we can take over the results for Bose–Einstein statistics which were obtained in Chap. VII, § 4. The mean energy of the lattice vibrations is

$$U = E_0 + \sum_\kappa \bar{n}_\kappa h\nu_\kappa = E_0 + \sum_\kappa \frac{h\nu_\kappa}{e^{h\nu_\kappa/kT} - 1}.$$

Just as the electromagnetic radiation in a cavity can be described in terms of photons, that is the number of quanta of electromagnetic energy at each frequency, so here we can describe the lattice vibrations in terms of numbers of *phonons*, that is the number of quanta of lattice energy $h\nu$. In any case we can see from the expression above for U that at low enough temperatures the only modes excited are those of low

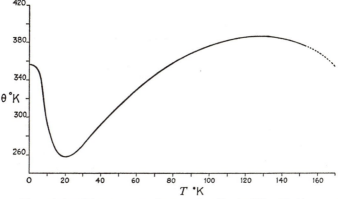

Fig. 3.—The equivalent Debye temperature for germanium (data by Hill and Parkinson, 1952).

frequency, that is long wave-length, and we have an exact correspondence with the Debye theory based on an elastic continuum. In particular the specific heat varies as T^3. At higher temperatures, modes with higher frequencies are excited, and the details of the spectrum become important. One way of plotting experimental data is to define an equivalent Debye temperature $\theta_D(T)$ by fitting the measured values of C_v to a Debye formula. The results for germanium are exhibited in fig. 3. The implication of the dip in $\theta_D(T)$ is that there is a large number of modes with relatively low frequency. This is confirmed by a direct measurement of the phonon spectrum given in fig. 4 (p. 284).

The harmonic approximation gives a good fit to specific-heat data over a wide range of temperatures. One of its predictions is that at high temperatures the specific heat will take the constant classical value of $3R$ per mole. In fact, the specific heat of insulating crystals does take almost this value, but continues to vary slowly. The small temperature-dependent term arises from the thermal energy residing in the part of the internuclear potential energy which corresponds to the non-linear restoring forces which were mentioned earlier (Born and Brody, 1921). This contribution, depending as it does on higher powers of the ratio of the nuclear displacements to the lattice spacing, gives an extremely small contribution at low temperatures and a small but measurable one at temperatures above the Debye temperature. These non-linear forces are also responsible for the expansion of crystals, for if the restoring forces were exactly linear the only effect of increasing the temperature would be to increase the amplitude of the nuclear vibrations but not the mean spacing between atoms: there would be no thermal expansion. Non-linear forces correspond to an interatomic

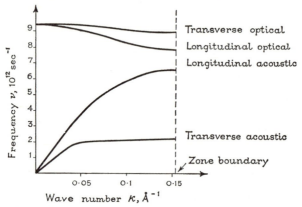

Fig. 4.—The frequencies of lattice vibrations in germanium for wave-number vectors in the [111] direction (after Brockhouse and Iyengar, 1958). (*By permission of the Physical Review.*)

potential which is asymmetric about the minimum, and the mean spacing in general will increase as the amplitude of the nuclear vibrations increases.

The anharmonic forces play an important role in the thermal resistance of crystals. Again, if the forces are linear, the lattice vibrations can be described exactly by travelling-wave modes. If such a harmonic crystal were put into a state in which there is a net flow of energy, that is heat, with more phonons flowing in one direction than the other, that state would persist indefinitely, the phonon wave-packets travelling unhindered through the crystal. We should have then a crystal supporting a flow of heat without any temperature gradient—the thermal conductivity would be infinite. In real crystals, however, if a flow of heat is set up, the phonons carrying this thermal energy are coupled by the anharmonic terms to others, some carrying energy in the opposite direction, and more or less rapidly the flow of heat is brought to zero unless there is a temperature gradient present to provide a driving force. This anharmonic coupling falls to zero very rapidly with diminishing temperature as the nuclear vibrations decrease in amplitude, and if this were the only resistive mechanism the thermal conductivity would become very large below the Debye temperature (Debye; Peierls). In practice, the crystal will contain point defects which deflect the phonons, though even they are inefficient scatterers (Rayleigh's law) of phonons of long wave-length, and at the very lowest temperatures the mean free path of the phonons becomes comparable with the dimensions of the crystal itself, and boundary scattering then takes over. In this range the thermal conductivity is a function of the size of the crystal, the mean free path being a constant of the order of the dimensions of the crystal, and the conductivity then exhibits the simple behaviour to be expected from the kinetic theory of gases, $k \sim c_v \lambda$.

4. Inelastic Scattering of Neutrons.

It is appropriate to mention here one method which has been used extensively to determine the dispersion curves for the frequency of lattice vibrations as a function of wave number. This method employs the scattering of slow neutrons from the crystal. A neutron with wave number κ can be scattered inelastically by the nuclei of the crystal, and in the process can either gain energy from the lattice vibrations or give it up. (It is the neutron analogue of Raman scattering, treated in Chap. VIII on molecules). But the neutron interacts with many nuclei, and we have to do with a diffraction effect. A neutron will only be scattered from the state with wave number κ to wave number κ' with the emission or absorption of a quantum of lattice energy if all the unit cells scatter in phase. What is more, total energy must be conserved in the scattering process, and we have a pair of equations to satisfy simultaneously:

$$\kappa - \kappa' = \pm\, q, \quad \frac{h^2}{2M_n}\,(\kappa^2 - \kappa'^2) = \pm\, h\nu_q.$$

We might express this classically by saying that the two neutron waves then beat in space and time in just the right way to couple to the lattice vibration of wave vector $q = \pm\,(\kappa - \kappa')$ and frequency

$$\nu_q = \pm\, \frac{h}{2M_n}\,(\kappa^2 - \kappa'^2).$$

In practice, a beam of cold neutrons from a nuclear reactor is monochromated so that κ is known. The direction of the outgoing wave is determined by the position at which the detector is placed, and finally the energy of these inelastically scattered neutrons is determined by measuring their speed by a time-of-flight method or by observing Bragg diffraction off an analysing crystal. In this way it is possible to obtain ν_q^i as a function of q for a selection of points in the Brillouin Zone and build up a picture of the dispersion curves, as in fig. 4 (p. 284).

5. The Mössbauer Effect.

In conclusion we mention an effect which has found wide application in recent years. This is the Mössbauer effect (1958), which involves the emission or absorption of a γ-ray by a nucleus at a site in a crystal. We want to contrast this with the emission or absorption of a γ-ray by the same nucleus at rest in free space. When a free nucleus emits a γ-ray the total momentum of the system remains unchanged and the nucleus recoils. The momentum of the γ-quantum is $\dfrac{E_\gamma}{c}$, which though small is important because the recoiling nucleus will have kinetic energy and the energy of the γ-ray will be slightly less than is given by the Bohr equation $E_\gamma = E_o - E_f \equiv W$, where E_o and E_f are the energies of the respective levels of the nucleus at rest. In fact, for a heavy nucleus of mass M emitting a γ-ray with an energy of only a few kilovolts, the energy of the γ-ray is to a very good approximation $E_\gamma^{(e)} = W\left(1 - \dfrac{W}{2Mc^2}\right)$. Now consider the reverse process, the absorption of a γ-ray with excitation of a free nucleus from the ground state to the excited state. We see that in contrast the γ-ray must supply some recoil energy to the nucleus, and the energy condition is only satisfied if

$$E_\gamma^{(a)} = W\left(1 + \frac{W}{2Mc^2}\right).$$

The energy deficit can be made up by using the Doppler shift of a moving source of nuclei in thermal motion or at the tip of a high-speed rotor. In any case, the excited state has a finite lifetime and hence a width (see Breit-Wigner, Chap. X, p. 352) and the reaction will take place over a range of energies near the Bohr energy. It is a question of achieving an appreciable overlap between the two narrow bands of energies. Mössbauer showed that when the nucleus is embedded in a solid, it is possible to observe *recoilless* emission and absorption. When the nucleus emits or absorbs the γ-ray, it still receives an impulse. The question we must ask in wave mechanics is: what final states of the lattice are possible? In fact, as the success of Mössbauer's experiment shows, the state in which the emitting nucleus has received an impulse $\dfrac{W}{c}$ from the γ-ray has a certain overlap with the state of the lattice *unchanged*, so that the energy of the γ-ray is given by $E_\gamma = W$. Conversely, there is the same probability that if this γ-ray interacts with a nucleus in its ground state at an identical lattice site it will be able to excite that nucleus without altering the lattice state, and the energy balance is satisfied. Of course, there is also a certain probability that the lattice state will change, but so far most interest has centred on the " recoilless " processes. To estimate the fraction f of recoilless decays we remember that the emitting nucleus, like all the others in the crystal, is confined to a small part of the unit cell and has some mean square displacement $\overline{(\Delta u)^2}$. From the uncertainty principle that also means that it has a spread of momenta $\overline{(\Delta p)^2}$ given approximately by $\overline{(\Delta p)^2} \sim \dfrac{\hbar^2}{\overline{(\Delta u)^2}}$. We must then expect that the magnitude of f is given approximately by $e^{-W^2/c^2 \overline{(\Delta p)^2}}$, which is a measure of the overlap with the initial wave function after the impulse $\dfrac{W}{c}$ has been applied to the nucleus. The conditions for f to be appreciable are that $\left(\dfrac{W}{c}\right)$ is small compared to $\overline{(\Delta p)^2}$. In the case of the nucleus Fe^{57}, which emits a γ-ray of 15 keV, f may have a value as large as 0·8.

The importance of the Mössbauer effect in solid-state physics lies in the fact that the energies of the emitting and absorbing nuclei will depend slightly on the crystal environment if either of the nuclear states has a magnetic or electric moment (dipole, quadrupole, . . .) which can interact with fields in the crystal. Even with recoil eliminated, the energies of absorption and emission may now differ. This difference can be made up by providing a Doppler shift by moving the source at the appropriate velocity, which, because of the small differences in energy, is of the order of a few mm. sec.$^{-1}$. In this way, much useful information about the internal fields in crystals has been obtained.

The Mössbauer effect also provides us with monochromatic radiation of constant frequency. The fractional width of the line is 10^{-12}, so that we can detect changes in frequency down to a fractional order of 10^{-15}. One of the most striking applications has been to the observation of the gravitational red-shift in the energy of the γ-ray between two points differing in height, and hence in gravitational potential (Pound and Rebka, 1960).

We now leave the subject of lattice vibrations and turn to the electronic properties of solids in § 6.

6. Electrons in a Periodic Lattice: Bands.

In the section on Fermi-Dirac statistics we mentioned some of the problems left unsolved by the Sommerfeld free-electron theory of metals. All these difficulties in the elementary treatment find an answer, at least in principle, when we use wave mechanics to describe the motion of the electrons. Each electron moves in a potential due to the nuclei and the other electrons. To a good approximation we can take together the nuclei and the tightly-bound electrons in the inner shells around them which are hardly affected when a crystal is assembled by bringing together its component atoms or ions. On the other hand the behaviour of the outer, valence, electrons can be profoundly affected. Let us consider the motion of such a valence electron in the periodic potential due to the ion cores and the other valence electrons. The general picture is given already by a one-dimensional model of a crystal containing a large number N of unit cells, with lattice constant a, and a potential $V(x)$ like that in fig. 5.

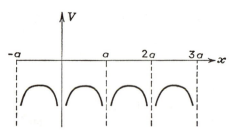

Fig. 5.—Sketch of the potential energy of an electron in a periodic lattice.

The Sommerfeld model assumes that the wave functions of the electrons are simply plane waves, $\dfrac{1}{\sqrt{(Na)}} e^{2\pi i \kappa x}$. Now the probability density associated with a plane wave is $1/Na$, which is a constant throughout the crystal. We must expect that in fact the probability density will not be constant but rather that the states of lower energy will tend to have wave functions which are to some extent peaked in the regions where the potential energy is lowest, that is in the troughs of fig. 5. We can easily construct such states by superposing, for a start, two plane waves.

Consider the wave function

$$\Psi(x) = \frac{1}{\sqrt{(Na)}} \left[A_1 e^{i\phi_1} e^{2\pi i \kappa_1 x} + A_2 e^{i\phi_2} e^{2\pi i \kappa_2 x} \right]$$

which is of just this type. The corresponding probability density is

$$P(x) = |\Psi(x)|^2 = \frac{1}{Na} \left[A_1^2 + A_2^2 + 2 A_1 A_2 \cos \left\{ 2\pi(\kappa_1 - \kappa_2)x + (\phi_1 - \phi_2) \right\} \right].$$

Now no unit cell is distinguished from another, and so this probability density must be periodic with the period of the lattice. Clearly it will be if the two wave numbers satisfy the condition

$$\kappa_1 - \kappa_2 = 0, \pm 1/a, \pm 2/a, \ldots$$

It is sufficient to take the case in which $\kappa_1 - \kappa_2 = 1/a$ and then generalize directly from it. To start with, suppose that κ_1 is small (and of course small and large in this context refer to the scale of the atom) so that consequently $|\kappa_2|$ is much larger than κ_1. In this case what do we expect for the relative size of A_1 and A_2? If the potential were constant, A_2 would have to be zero, but in the presence of a varying potential superposition of the two plane waves modulates the probability density to take advantage of the dips in the potential energy in every unit cell. The price paid is the kinetic energy of the wave with the large wave number κ_2. The mean kinetic energy is

$$\langle T \rangle = A_1^2 \frac{h^2 \kappa_1^2}{2m} + A_2^2 \frac{h^2 \kappa_2^2}{2m}$$

while the mean potential energy is

$$\langle V \rangle = \int_0^{Na} V(x) P(x) \, dx$$

$$= \frac{2 A_1 A_2}{Na} \int_0^{Na} V(x) \cos \left[2\pi(\kappa_2 - \kappa_1)x + (\phi_1 - \phi_2) \right] dx$$

$$+ \frac{(A_1^2 + A_2^2)}{Na} \int_0^{Na} V(x) \, dx.$$

For very small values of A_2 the reduction in the potential energy, being linear in A_2, always outweighs the increase in kinetic energy. The actual value of A_2 occurs at just that point at which any further increase in it would cost more in kinetic energy than can be recouped by a decrease in the potential energy. For small κ_1 the coefficient A_2 is itself small. In this region of κ_1 the wave function is only slightly different from a plane wave and the energy is very nearly $h^2\kappa_1^2/2m$. However the situation changes when κ_1 has increased to a value just less than $1/2a$, for then the wave number $\kappa_2 = \kappa_1 - 1/a$ has a modulus only slightly larger than κ_1 itself, and the two waves being superposed have almost the same kinetic energy. The only consideration now is to take maximum advantage of the potential, which clearly happens when $A_1 \simeq A_2$, and the wave is strongly modulated. In the limiting case of $\kappa_1 = 1/2a$, for a potential like the one in fig. 6, the lowest state has $A_1 = A_2 = 1/\sqrt{2}$, $\phi_1 = \phi_2 = 0$, so that the wave function is $\Psi(x)$ $= \sqrt{\dfrac{2}{Na}}.\,\cos\dfrac{2\pi x}{a}$. It is a standing wave, for which the corresponding probability density $P(x) = |\,\Psi(x)\,|^2$ is largest just where the potential

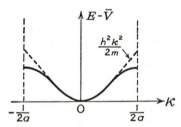

Fig. 6.—The energies of the lowest states
of an electron in a periodic potential.

energy is lowest. The energy of this state is certainly less than that of the free particle. Another important difference is that the wave function is a standing wave. But the group velocity (Appendix XI, p. 382) is $v = \dfrac{\partial v}{\partial \kappa} = \dfrac{1}{h}\dfrac{\partial E}{\partial \kappa}$ and so we expect that the E versus κ curve will have zero slope at the point $\kappa = 1/2a$. We can sketch the curve shown in fig. 6 for the interval $-1/2a < \kappa < 1/2a$, which is of course the Brillouin Zone which was introduced in the section on lattice vibrations.

Having established this much about the behaviour of the energy as a function of wave number, let us now look at the corresponding wave functions. The point to note is that they can be written in the form

$$\Psi(x) = e^{2\pi i\kappa x}\, u_\kappa(x)$$

where the function $u_\kappa(x)$ is periodic in x with period a, that is $u_\kappa(x + a)$ $= u_\kappa(x)$. The wave function is a plane wave modulated by a periodic function (Bloch), and it is characterized by the change in phase from one unit cell to the next. In fact

$$\Psi(x + a) = e^{2\pi i\kappa(x+a)}\, u_\kappa(x + a) = e^{2\pi i\kappa a}\, \Psi(x)$$

so that this phase change is $e^{2\pi i\kappa a}$. All distinct phase changes can be represented by values of κ in the interval $-1/2a < \kappa < 1/2a$, exactly as it was in the case of the change of phase of nuclear displacements from cell to cell in normal modes of lattice vibration. But this wave number is not sufficient to specify the wave function completely, as there are (infinitely) many states with the same κ. For instance we argued that for the potential of fig. 5, superposing the two plane waves $e^{\pm i\pi x/a}$ to form a cosine function with $\kappa = 1/2a$ gave a state of low energy. If we superpose them differently to form the sine function, which also has $\kappa = 1/2a$, we have in contrast an orthogonal wave function which is small where the potential energy is lowest. It therefore has a higher energy than the cosine, and there is an *energy gap* between these two states at $\kappa = 1/2a$. The general behaviour of the energy spectrum can be seen by looking at fig. 7, which is a plot of the

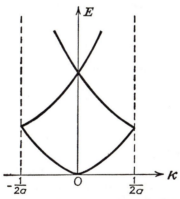

Fig. 7.—The energy spectrum of a free particle with its wave function written in Bloch form.

energies of a free particle with the plane wave always written in the form

$$e^{2\pi i\kappa x}\, u_\kappa(x), \text{ with } -1/2a < \kappa < 1/2a.$$

By superposing plane waves with the same κ in this plot we automatically form states with periodic charge distributions. The most important effects of the periodic crystal potential occur at points where there are two waves with the same kinetic energy; for then, as in the case of $\kappa = 1/2a$ just treated in detail, it costs little kinetic energy to form standing waves with probability densities peaked either where the potential energy is low or where it is high. The effect of the periodic potential is to alter the energy spectrum from the completely continuous one of the plane waves, for which all positive energies are possible, to a spectrum in which the allowed energies form continuous bands, separated by forbidden gaps, and the physical reason for these gaps is that at those points it is possible to form two distinct standing waves which have different probability distributions relative to the crystal potential. This spectrum is shown in fig. 8.

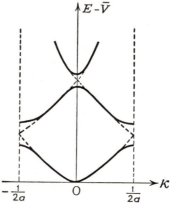

Fig. 8.—The effect of the periodic potential on the plane-wave spectrum shown in fig. 7.

We might point out that the same conclusions can be established from a different and complementary point of view. A crystal can be treated as a large molecule. As an approximation to the " molecular " orbital we take linear combinations of atomic wave functions ϕ_s on the different lattice sites. The appropriate superpositions are

$$\Psi'_{s\kappa}(x) \sim \sum_n e^{2\pi in\kappa a} \phi_s(x - na)$$

for a particular atomic level s. It is clear that the distribution of charge density in the regions between atoms depends on the value of κ, and each atomic level gives rise to a band of levels in the crystal. The width of this band depends on the overlap of atomic wave functions on neighbouring lattice sites, and this point of view is most appropriate when the overlap is small, as it is for tightly bound electrons.

7. Metals and Insulators.

Before leaving this one-dimensional model of a crystal we must determine how the energy levels we have described will be occupied by the electrons. In the ground state the valence electrons, let us say z electrons per unit cell, will occupy the lowest levels they can consistent with the Pauli principle. The question is now: how many states are there in each band? This is a matter of counting the number of distinct values of κ in the interval $-1/2a < \kappa < 1/2a$. We can again apply the Born–von Kármán boundary condition to the wave function and require that $\Psi(x) = \Psi(x + Na)$. Substituting the Bloch wave function $\Psi(x) = e^{2\pi i \kappa x} u(x)$, we see that the condition on κ is $e^{2\pi i N \kappa a} = 1$, i.e. $\kappa = 0, \pm 1/Na, \pm 2/Na, \ldots$ The allowed values of κ are exactly what they were for lattice waves. The spacing between successive κ is $\delta \kappa = 1/Na$ and the number $\mathcal{N}$ of distinct wave functions in a band is given by $\mathcal{N} = 1/(a\delta\kappa) = N$, the number of unit cells. Remembering that electrons have two spin states, we can say, in one dimension, that if there are an odd number $z = 2\nu + 1$ of electrons from each unit cell then in the ground state they will completely fill the lowest ν bands and half-fill the $(\nu + 1)$th band, whereas if there is an even number $z = 2\nu$ they will precisely fill all the states in the lowest ν bands. Thus the application of wave mechanics has led to the conclusion that there are two distinct classes of one-dimensional crystal. How will they differ in their properties, and how do they connect with what we know experimentally? The distinction is precisely that between metals on the one hand and insulators on the other. Crystals in the first class, with the top band half-filled, will have metallic properties. For instance there are states of the whole crystal, with energies only infinitesimally higher than the ground state, which carry electric current. In the ground state there are equal numbers of electrons moving in each direction, and consequently no net current flows. To produce a net current it is simply a matter of redistributing slightly some of the electrons in the highest occupied states, so that more are moving in one direction than the other. A d.c. external electric field (balanced in the steady state by some resistive mechanism) will have this effect and induce a current to flow. It is different in the case of crystals with only filled bands. Again there is no net current in the ground state, with all the individual electronic currents balancing out. The only way to make a state in which there is a net flow of current is to promote some electrons from the last filled band, the valence band, to states in an empty band above it. We note that there will be a contribution to the total current both from the promoted electrons and from the holes, the latter occurring because the currents of the electrons left behind no longer cancel. Since a full band carries no net current, the current carried by the valence band with holes is just the negative of the current that would have been carried by the missing electrons. This begins to suggest that holes behave as carriers of positive charge. However such a state as we have described cannot be achieved by applying a d.c. field. (We neglect the possibility of wave-mechanical tunnelling processes (Zener) which only become important at extremely high fields.) To promote one electron requires an energy of at least E_{gap}, and to do this, as in the photoelectric effect, the applied field must have a frequency of at least $\nu = E_{gap}/h$. We see then that crystals in this second class are insulators at $0°K$.

In one dimension we have a sharp distinction: a crystal with an odd number of electrons per unit cell is a metal, and a crystal with an even number is an insulator. Now we see how far these conclusions can be carried over to crystals in three dimensions.

The basic ideas remain valid. In the presence of a periodic crystal potential the wave functions of the electrons have the Bloch form $\Psi(\mathbf{r}) = e^{2\pi i\kappa \cdot \mathbf{r}} u_\kappa(\mathbf{r})$, where now the (vector) wave number takes values inside the polyhedral Brillouin Zone appropriate to the crystal structure. The corresponding energies $E(\kappa)$ form bands, and there are gaps between bands at some places in the Brillouin Zone where there would have been none in the case of free particles. Again the physical reason is that it is possible to form superpositions which differ in the way the probability density is distributed over the unit cell. It is easy to show that each band contains one state per unit cell, or two including spin.

A new possibility is that bands may overlap. A typical circumstance is that if the energy gaps at the faces of the Brillouin Zone are small, then states with wave numbers near the extreme vertices of the zone in one band may have a higher energy than states of the next band with wave numbers somewhat nearer the origin of the zone. All we can say now is that a crystal with an odd number of electrons in the unit cell (Na, Cu, Al) will be a metal. If there is an even number the crystal will be an insulator if there is no overlap (diamond, sodium chloride, etc.) and a metal if there is overlap (Mg, Zn, Sn, etc.).

We now consider the two classes separately.

8. Metals

In this case we can carry over many of the ideas of the Sommerfeld theory. Electrons are still labelled by a wave number κ, but now the corresponding energy is some, often complicated, function $E(\kappa)$ rather than $h^2\kappa^2/2m$ as it was for free particles. In particular, this energy will be a function of the direction of κ as well as its magnitude. The equation $E(\kappa) = $ constant describes a surface distorted to a larger or smaller extent relative to the spherical shape for free particles. At the absolute zero of temperature the electrons occupy the lowest states, filling them up to a boundary surface of constant energy in the Brillouin Zone. This surface is called the *Fermi surface*. In recent years it has become possible to build up a detailed picture of the Fermi surface in many metals. The actual shape of the surface for copper is shown in fig. 9. All the states

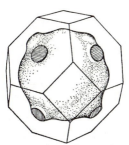

Fig. 9.—The Fermi surface and Brillouin Zone for copper, which has the face-centred cubic structure (after E. Fawcett).

within this surface are occupied, and all those outside are empty. It has a volume which is just half of the volume of the Brillouin Zone itself. The occurrence of necks in the $\langle 111 \rangle$ directions shows that the energies of states in these regions are somewhat lower than for free particles.

At finite temperatures, the electrons are distributed statistically over the states according to the Fermi–Dirac distribution, so that the probability of a state being occupied is

$$f = 1/[e^{(E(\kappa)-\mu)/kT} + 1].$$

As in the Sommerfeld model (Chapter VII) there is a smooth transition from states being almost completely occupied if they are more than kT in energy below the Fermi surface to states essentially unoccupied if they are more than kT in energy above it. The electronic specific heat is still linear in T, but the coefficient depends on the details of the energy contours.

Experimental determinations of the Fermi surface depend on the way in which the electrons respond to external electric and magnetic fields. The result for electric fields is so simple that we derive it here. The effect of an electric field $\mathscr{E}$ is to accelerate an electron, changing its wave number. In a short time δt the wave number will change from κ to $\kappa' = \kappa + \dot{\kappa}\,\delta t$, and the energy by the amount

$$E(\kappa') - E(\kappa) = \delta t\,\dot{\kappa}\cdot\frac{\partial E}{\partial \kappa}.$$

But this change must be equal to the work $\delta W = q\mathscr{E}.\delta x$ done by the external field in this time. The displacement is just $\delta x = v\,\delta t$ so that $\delta W = q\delta t\,v.\mathscr{E}$. But the group velocity (see Appendix XI) is given by $v = \dfrac{1}{h}\dfrac{\partial E}{\partial \kappa}$. Thus, equating the two terms, we deduce that $h\dot{\kappa} = q\mathscr{E}$. The semi-classical result when there is also a magnetic field present is

$$h\dot{\kappa} = q\left[\mathscr{E} + \frac{1}{c}v \wedge B\right].$$

This equation is particularly interesting for a magnetic field alone. The behaviour of the wave number κ as a function of time can be seen directly. The rate of change of κ is always perpendicular to the field B, and also is directed along an energy surface, the magnetic field doing no work on the electron. As time evolves the vector wave number periodically traces out an orbit along a constant-energy contour in a plane perpendicular to the external magnetic field. These orbits however cannot be of arbitrary dimensions. Being periodic, they can be quantized using the Bohr–Sommerfeld rule (Chap. V, § 2, p. 117). The effect of quantizing the orbits is to produce oscillations in the magnetic susceptibility as a function of the strength of the applied magnetic field. The period of these oscillations depends on the extremal cross-sectional area of the Fermi surface. By measuring this area for a series of orientations of the external field with respect to the crystal axes it is possible to construct an accurate picture of the Fermi surface. This has been done for many metals, and the method is now being applied to ordered alloys.

So far as the effect of an electric field goes, a perfect metal would have zero resistivity—a current once established would persist indefinitely. In real metals the current would be dissipated as the net flow of electrons is brought to zero by the scattering from impurities in the metal (which gives a resistance independent of temperature), and also by interaction with phonons. As the ion cores oscillate they produce fluctuations in potential which can scatter electrons from one Bloch state to another. This component of the resistance, depending as it does on the amplitude of the vibrations and also on the net deflection of the electrons at each collision, is strongly temperature-dependent, varying as T^5 at low temperatures and as T at high temperatures.

9. Superconductivity.

Out of the variety of phenomena exhibited by metals we mention two which lie outside this simple picture of electrons moving independently in the metal. The first, and one which has only recently found a satisfactory theoretical explanation, is that of superconductivity, which occurs in many metals at temperatures below the so-called critical temperature T_c, which is of the order of a few degrees Kelvin. On cooling the metal through this temperature there is a reversible transition from the normal state to one in which the resistance vanishes completely (Kamerlingh Onnes, 1911). Furthermore, if a metal in a sufficiently weak magnetic field is cooled down, the flux inside the metal is expelled when the temperature reaches T_c (Meissner, 1933). It is also observed experimentally that at temperatures below T_c the specific heat is no longer linear in T but instead varies as $e^{-\alpha/T}$. This exponential has the appearance of a Boltzmann factor, and occurs because there is a gap in energy between the ground state and the excited states, in contrast to a normal metal in which there are excited states available immediately above the Fermi surface. The basic idea of the theory (Bardeen, Cooper and Schrieffer, 1957) is that in the superconducting state the electrons are not all moving independently of one another but are correlated in a very specific way, namely that there is a dynamical pairing of electrons, $\{\kappa \uparrow ; -\kappa \downarrow\}$, so that if the state with wave number κ and spin up is occupied by an electron, then so is the state with wave number $-\kappa$ and spin down. Furthermore, these pairs are superposed in phase. At least for the majority of superconductors found so far, and perhaps for all, the interaction which favours this pairing is—paradoxically—the interaction with lattice vibrations which is responsible for most of the resistance of pure normal metals. In simple terms we can say that a superposition of two Bloch waves can beat and drive lattice oscillations which then react on other superpositions of Bloch waves. The net effect is a weak interaction acting between all pairs with wave numbers in a shell near the Fermi surface. In the ground state a certain number of the electrons in this shell, say $2\mathcal{N}$, are paired and the energy of this state is lower in energy than the normal state with no pairing by an amount $W = \dfrac{2\mathcal{N}^2}{N(0)}$ where $N(0)$ is the density of states at the Fermi surface. This form for W shows clearly the co-operative nature of the state. The simplest excited state is one in which $2(\mathcal{N} - 1)$ electrons are paired, i.e. a pair has been broken. Now there are only $(\mathcal{N} - 1)$ pairs and the energy of this new state is a small but finite amount $2\Delta = \dfrac{4\mathcal{N}}{N(0)}$, above that of the ground state— there is a gap. Bardeen, Cooper and Schrieffer were able to show that their pairing wave function could explain quantitatively a wide range of phenomena exhibited by superconductors, including the Meissner effect. The recent discovery of alloys which remain superconducting up to exceptionally high critical fields has made it feasible to build solenoids with coils of superconducting wire which eliminate the Joule heating. Such coils can provide stable and adjustable fields up to a maximum strength of 100kG with none of the problems of heat dissipation which arise in an ordinary electromagnet.

10. Ferromagnetism.

Many metals in the transition and rare-earth groups exhibit ferromagnetism, antiferromagnetism and still more complex types of magnetic ordering. In ferromagnetism, the only type we consider here, there is an alignment of spins on the atoms so that a crystal has a net magnetic moment. The occurrence of ordinary unmagnetized iron was explained by Weiss (1907) who postulated that such a sample contained many magnetized domains which, however, were distributed over such a range of orientations that the sample had no net moment. The magnetization of iron is then a process of aligning domains. The existence and size of these domains is now well established by various methods, powder patterns, etc.

A complete and universally valid microscopic theory of ferromagnetism is still lacking. We can mention only a few of the points which seem to be fundamental. The first remark to be made is that the dipole-dipole interaction between the magnetic moments of the electrons themselves is not strong enough to be responsible for ferromagnetism in these metals, which have transition temperatures T_c of the order of hundreds of degrees Kelvin, implying that the ordering energy must be at least kT_c, that is $\sim 0{\cdot}1\text{eV}$ per spin. Such an energy is of a magnitude that we must associate with Coulomb forces, and this was first suggested by Heisenberg (1928). Because the wave function of the electrons must be totally antisymmetric with respect to both the spin and the space parts, there can be a difference between the Coulomb energies of states of different spin. This is shown already in atoms, in Hund's rule that the term of maximum spin should lie lowest. In that case, the space part of the wave function is antisymmetric and either vanishes or is very small where the co-ordinates of two electrons overlap. The electrons tend to keep apart and the energy of repulsion between them is minimized. A similar situation occurs in the hydrogen molecule, although in this case the state of lowest spin ($S = 0$) lies lowest because then the electrons both benefit by spending a large fraction of the time in the region between the ions, as has been mentioned in the chapter on molecules. Even granted this spin dependence of the Coulomb energy, there is still the question of why iron is a ferromagnet, but sodium is not. This is undoubtedly related to the fact that magnetic ordering occurs only in elements in particular parts of the periodic table, namely in elements whose atoms have partially occupied $3d$ and $4f$ shells. The corresponding wave functions are nodeless and relatively compact. Such levels in the interior of atoms are affected relatively little when the atoms come together to form a crystal, and they give rise to a narrow band in which the energy of the Bloch states depends only slightly on the value of the wave number. It is possible to occupy many of these states without paying too high a price in band energy. The contributions to the spin-dependent Coulomb energy can arise from interactions within a given cell, from direct interactions between electrons in d or f states in different cells, or be mediated by interaction of these electrons with conduction electrons in the wide bands which arise from the atomic s and p states. It is quite different in sodium. To form a magnetic state it would be necessary to promote electrons to states in the upper half of the band, and the energy required is very large, and outweighed by what would be a small gain in ordering energy.

Even though the mechanisms responsible for the occurrence of ferromagnetism are not yet all clarified in detail, it is possible to understand some of the properties of ferromagnets, such as the way the magnetization and (magnetic) specific heat depend on temperature. In the ground state, the state of the crystal at 0°K, the spins are aligned and there is a net moment. One possible excited state involves reversing an electron spin. In fact, the true excited states are spin waves, labelled once again by a wave number which gives the change in phase from cell to cell. In ferromagnets, the energy of these spin waves is of the form $h\nu_\kappa = \beta\kappa^2$. As the temperature of the crystal increases, the number of spin waves increases. Thus there is a spin-wave contribution to the specific heat, with a temperature dependence $c_v \sim T^{\frac{3}{2}}$.

Another effect is a reduction in the magnetization of the crystal. The moment is $M = M_0 - \mu_B \sum_\kappa \bar{n}_\kappa$, and for temperatures much below the Curie point this behaves as $M = M_0(1 - aT^{\frac{3}{2}})$ (Bloch). At temperatures just below the Curie point, the magnetization decreases more rapidly. The molecular field which is responsible for alignment itself depends on the aligned moment, and we have another example of a co-operative phenomenon. Above the Curie point there is no net moment, but some short-range order persists as is shown by the curve in fig. 10 showing the variation of specific heat with temperature for a typical ferromagnet.

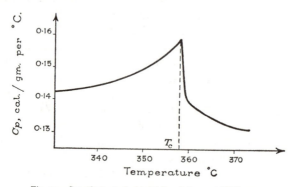

Fig. 10.—Specific heat of nickel (after Sykes and Wilkinson).

There is a large number of non-metallic substances with magnetic ordering (e.g. Fe_3O_4 is a ferrite, having a net moment from two interpenetrating sub-lattices with oppositely directed magnetizations of different magnitude) but this is not an aspect we shall have occasion to pursue in the section which follows.

11. Insulators and Semiconductors.

In the case where the highest occupied band is full, and does not overlap with any others, a crystal at $0°K$ is an insulator. The application of an electric field will polarize the crystal (it is a dielectric) but there is no steady flow of current (again we are neglecting wave-mechanical tunnelling in the Zener process). However, at finite temperatures there will be some thermal excitation of electrons to states in the lowest empty band, the conduction band, and of course an equal number of unoccupied states, that is holes, will be left in the previously filled valence band. The number of electrons is easily computed using

Fermi–Dirac statistics (Chap. VII, § 6), for which the probability f of a state being occupied can be written

$$f = \frac{1}{e^{(E(\kappa) - \mu)/kT} + 1}.$$

The states in question are those within an energy $\sim kT$ of the top of the valence band, and within the same range of the bottom of the conduction band. For the sake of simplicity we will assume that the energies of states near the maximum in the valence band have energies given by

$$E_v(\kappa) = -\frac{h^2\kappa^2}{2m_r}$$

and that similarly the energies of states near the minimum of the conduction band have energies given by

$$E_c(\kappa) = E_g + \frac{h^2\kappa^2}{2m_c}.$$

The parameters m_v and m_c are determined by the curvature of the bands and are known as effective masses. Using the results of Chap. VII, § 5, it follows that the densities of states in these portions of the two bands are given by

$$g_c(E) = 2(4\pi V/h^3)\sqrt{2m_c^3}\sqrt{E - E_g}$$

and

$$g_v(E) = 2(4\pi V/h^3)\sqrt{2m_v^3}\sqrt{- E}$$

respectively.

The number of electrons in the conduction band is

$$n = \int_{E_g}^{\infty} dE\, g_c(E) f(E) = \int_{E_g}^{\infty} dE\, g_c(E)/[e^{(E - \mu)/kT} + 1]$$

and the number of holes in the valence band is

$$p = \int_{-\infty}^{0} dE\, g_v(E)\,[1 - f(E)] = \int_{-\infty}^{0} dE\, g_v(E)/[e^{(\mu - E)/kT} + 1].$$

Considering fig. 11, we note that if kT is small compared to both μ and $E_g - \mu$, as it usually is, the exponential terms will always outweigh the unit in the denominators and

$$n = e^{\mu/kT} \int_{E_v}^{\infty} dE\, g_c(E)\, e^{-E/kT},$$

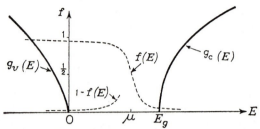

Fig. 11.—The Fermi-Dirac distribution function f and the densities of states in the valence and conduction bands.

which is nothing more than the Maxwell–Boltzmann formula: the electrons in the conduction band are distributed classically. This integral was also needed in Chap. I and is evaluated in Appendix I (p. 359). The result is

$$n = 2(2\pi m_c \, kT/h^2)^{\frac{3}{2}} \, e^{(\mu - E_g)/kT}.$$

Similarly, the number of holes is

$$p = 2(2\pi m_v \, kT/h^2)^{\frac{3}{2}} \, e^{-\mu/kT}.$$

Combining these two formulæ it follows that

$$np = 4(2\pi kT/h^2)^3 \, (m_v m_c)^{\frac{3}{2}} \, e^{-E_g/kT} \equiv n_i^2$$

whatever the value of μ, provided the approximations remain valid.

In the case of a pure (" intrinsic ") crystal, the position of the Fermi level μ (and it can easily be shown that μ is also the chemical potential, that is, the amount by which the free energy of the system increases when an electron is added to it) is determined by the condition that the number of electrons shall be equal to the number of holes. Then

$$n = p = n_i = 2(2\pi kT/h^2)^{\frac{3}{2}} \, (m_v m_c)^{\frac{3}{4}} \, e^{-E_g/2kT}$$

and

$$\mu_i = \tfrac{1}{2}E_g + \tfrac{3}{4}kT\ln\!\left(\frac{m_v}{m_c}\right).$$

For germanium, which has an energy gap $E_g \simeq 0.7$ eV, the value of n_i at $T = 300°$K is 2.5×10^{13} cm.$^{-3}$. In comparison, a material with the same effective masses but an energy gap of 2.1 eV would have an intrinsic carrier concentration of 10^3 cm^{-3} at this temperature. The magnitude of the energy gap is seen to be of profound importance, for the current a pure crystal can transport is proportional to this carrier density. The current density can be written

$$j = |q| \{n\mu_n + p\mu_p\} \mathscr{E}$$

where q is the charge on an electron and the parameters μ_n and μ_p (not to be confused with the chemical potential) are the *mobilities* of the electrons and holes respectively. The latter are the drift velocities acquired by the carriers under the action of a field of unit strength. The value of the mobility is a sensitive function of the temperature and purity of a crystal; the more effective the scattering of the carriers by lattice vibrations and impurities the lower the mobility. Typical values at room temperature are $\mu \sim 3000$ cm²/volt-sec for germanium and $\mu \sim 50$ cm²/volt-sec for silver chloride. Crystals with large energy gaps and/or low mobilities can have resistivities up to 10^{14} ohm-cm and are effectively insulators. Materials such as germanium, silicon, indium antimonide and lead telluride, with fairly small gaps have resistivities in the range 10^{-2}–10^9 ohm-cm and are known as *semiconductors*. The difference is only one of degree and not of a fundamental nature.

We shall now consider some further aspects of semiconductors, properties which have had enormous technical significance in recent years.

First we note some of the properties of impurities in a typical semiconductor like germanium (which has an energy gap of about 0·7 eV). The structure of germanium is the same as that of diamond. In this structure each atom is surrounded by four nearest neighbours at the vertices of a regular tetrahedron, which is the arrangement to be expected for covalent bonding by a quadrivalent atom in Group IV of the Periodic Table. In chemical terms, all the bonding orbitals are occupied, and in physical terms the valence band is filled. Thermal excitation can be thought of either as excitation to the conduction band, leaving holes, or as promotion of some electrons to higher orbital states in which they can jump from atom to atom, and neighbouring electrons in a bonding orbital can jump to fill a vacancy. If an element such as antimony or arsenic from Group V of the Periodic Table is substituted for a germanium atom, one of its electrons must go into these higher orbitals. However, this fifth electron is not entirely free. The arsenic atom is neutral, and the fifth electron moves in the attractive Coulomb field of an As^+ core. This binding force is considerably reduced by the dielectric constant of the germanium itself. The states of the fifth electron are a series of hydrogen-like levels, up to a continuum starting at the bottom of the conduction band. The energy of the lowest state is about 0·01 eV below this edge, so that at 300°K a large fraction of arsenic atoms will be ionized and contributing an electron to the conduction band. The chemical potential must now lie somewhere close to this energy, just below the conduction band. The number of holes is much reduced because, as we showed earlier, $np = n_i^2$. For germanium at room temperature $n_i = 2 \cdot 5 \times 10^{13}$ cm^{-3}, and a typical value of n for a lightly doped crystal is $n = 2 \cdot 5 \times 10^{15}$ cm^{-3} and in that case $p = 2 \cdot 5 \times 10^{11}$ cm^{-3}. Such a crystal, in which most of the conductivity is due to electrons donated by impurities is known as an n-type extrinsic semiconductor. It is clear that if a pure germanium crystal is doped with an element such as gallium or indium from Group III of the Periodic Table, there will be one electron too few to form the four covalent directed bonds—there is a hole which is weakly bound to a Ga$^-$ core. In such a crystal the chemical potential is just above the maximum in the valence band. There are very few electrons in the conduction band, and the current is carried predominantly by holes left behind when the Group III element accepts an electron from the valence band. Such a crystal is known as a p-type extrinsic semiconductor. The technique of growing crystals with a definite concentration of donors and acceptors has now been brought to a fine art.

However, there is still another property of semiconductors which makes them so useful. This is the fact that the conductivity of semiconductors can be modulated by carrier injection. In this process the local conductivity is altered by producing locally a non-equilibrium density of carriers. Of course an equal number of electrons and holes must be introduced in order to maintain local charge neutrality. (Any deviation from neutrality would be cancelled immediately by a flow of charge from other parts of the crystal under the influence of the powerful Coulomb forces.) This change in local conductivity can be very large. Since it involves a departure from thermal equilibrium, there will be a tendency to return to the equilibrium distribution. In the absence of any· disturbing forces the excess electrons will fall back into unoccupied states in the valence band, either directly or by way of impurity states: the excess electrons and holes recombine. This process can be characterized by the recombination time τ, which in a typical sample of germanium of the purity that might be used in a transistor is of the order of 10^{-6}–10^{-4} sec. In this time excess carriers can diffuse a mean distance $L = (D\tau)^{\frac{1}{2}}$ in random Brownian motion before they recombine. The constant D is the diffusion constant, which is related to the mobility μ by the Einstein relation $D = \mu kT/|\,q\,|$. At room temperature, taking $\mu = 3 \times 10^3$ cm.^{2}/volt-sec., that is $D \simeq 75$ cm.^{2} sec.$^{-1}$, and $\tau = 10^{-4}$ sec., one finds $L \simeq 1$ mm.

All these ideas are illustrated by considering the operation of a p-n junction as a rectifier (Bardeen, Brattain, Shockley, 1950). The corresponding band diagram is as shown in fig. 12.

On the left the material is n-type, with the chemical potential close to the conduction band. On the right the crystal is p-type, with the chemical potential close to the valence band. But in thermal equilibrium the chemical potential must be the same throughout the crystal, for if it were not the free energy could be reduced by transferring electrons from one side of the junction to the other. At the junction itself there are very few carriers (the chemical potential is close to neither band) and the donors and acceptors are ionized, producing a double layer of charge, and hence a step in the potential which has the

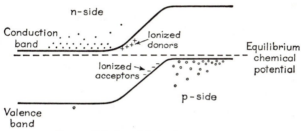

Fig. 12.—Behaviour of the bands at a p-n junction.

effect of raising the energies of states on one side by the requisite amount relative to the same type of band state on the other side of the junction, which has a thickness $d \sim 10^{-4}$ cm.

In thermal equilibrium there is no net flow of carriers through the sample. There are a few electrons on the n-side which approach the barrier with enough energy to carry over it and reach the p-side where they recombine. This current j_{nr} is balanced by a continual generation of pairs by thermal fluctuations near the junction on the p-side, and some of the electrons produced diffuse back down the gradient into the n-type region giving a current $j_{ng} = j_{nr}$. Similarly for the holes, in the opposite sense. Now suppose that the energy of the electron states on the left is lowered by an amount qV relative to those on the right by applying a voltage V across the specimen. It is sufficient to consider the electrons alone. The height of the barrier has increased by an amount qV. Remembering the Boltzmann distribution factor, it is clear that the current of electrons over the barrier will now be $j'_{nr} = j_{nr}e^{-qV/kT}$. The rate of spontaneous generation will not have changed, because it depends on local properties of the crystal near the junction. There is thus a net current density across the junction, given by

$$j_n = j''_{nr} - j_{ng} = j_{ng} (e^{-qV/kT} - 1).$$

The holes behave in an exactly similar way. The two currents add, and the total current density is

$$j = j_n + j_p = (j_{ng} + j_{pg}) (e^{-qV/kT} - 1).$$

The junction is a rectifier for applied voltages larger than a few times kT/q. In the forward high-current direction, carriers are being injected and there is no limit to the current. In the back low-current direction, the current is limited to that carried by the few carriers which are generated within a diffusion length L of the junction and swept through it by the junction field.

The operation of a transistor involves the use of two junctions, for example in the sequence n-p-n. If one junction is biased in the back direction, its conductance can be changed considerably by collecting carriers which have diffused to it after injection at the other junction. The carriers are injected through the low forward impedance and collected through the high reverse impedance, so that there can be a large gain. The use of transistors which firstly need no filament heating, and secondly are small and reliable, has revolutionized the techniques of electronic circuitry in recent years.

CHAPTER X

Nuclear Physics

1. The Size of the Nucleus and α-Decay.

The basic facts about nuclei have been explained in Chap. III, and are repeatedly used in the following chapters of this book. We have now to investigate how far modern quantum mechanics can account for these elementary properties of nuclei and for many other observations of a more complicated kind.

The simplest property is the size of the nucleus which, as we have seen, is of the order 10^{-13} cm. This alone is sufficient to exclude the assumption that electrons are constituent parts of the nuclear structure; for, to confine an electron within nuclear dimension of the order 10^{-13} cm. would, because of the uncertainty relation lead to an uncertainty in its momentum of the order $h/10^{-13}$. Corresponding to this would be a kinetic energy of the order 30 MeV and so an extremely powerful potential would be needed in order to keep the electron within the nucleus. Such a potential would not be in agreement with experimental observation on nuclear potentials. This conclusion is confirmed by other evidence (Chap. III, § 4, p. 66, and p. 311). The question now arises whether we cannot obtain more detailed information about the " radii " of different nuclei.

If we could assume that ordinary mechanics holds for nuclei, and that these behave like rigid spheres, the cross-section which could be observed by scattering experiments (as described in Chap. III, § 3, p. 60) would be identical with the geometrical cross-section πR^2, where R is the radius. Both assumptions are, in general, not fulfilled. We can, however, satisfy them approximately by using fast neutrons as projectiles; for, if there is no charge, only short-range forces come into play, and for high velocities the de Broglie wave-length λ of the projectiles is short compared with the nuclear radius R, hence ordinary mechanics will hold with good approximation.

In this way it was found that the radius of heavy elements like Pb or U is 10^{-12} cm., whereas elements in the middle of the periodic table have a considerably smaller radius, 6×10^{-13} cm.

Another method of estimating the radius consists in considering pairs of nuclei which differ only through the exchange of a proton and a neutron, so-called " mirror nuclei ", like

$$(_1H^3,\ _2He^3,)\ (_3Li^7,\ _4Be^7),\ (_7N^{15},\ _8O^{15}),$$

and many others. It is found that the two components of a pair show a difference in binding energy such that the energy of the nucleus with higher charge is the greater. Now it can be assumed that the short-range nuclear forces (§ 8, p. 49) are the same for any pair of nucleons (neutrons or protons); then the difference in binding energy between the two mirror nuclei is just due to the change of electrostatic energy produced by the addition of a proton. The corresponding energy is $W = \frac{6}{5}Ze^2/R$ (see Appendix XXIX, p. 444).

Using this expression and the observed values of W (deduced from mass defects), we obtain numerical values for R which can be represented by the empirical formula

$$R = 1 \cdot 5 \times 10^{-13}\ A^{1/3}\ \text{cm.,}$$

where A is the mass number (nearest integer to atomic weight; see Chap. III, § 4, p. 63), i.e. the total number of nucleons. More refined calculations on this problem taking into account exchange effects and deviations of the density distribution from that for a uniform sphere indicate, however, that

$$R = (1 \cdot 3 \pm 0 \cdot 1) \times 10^{-13}\ A^{1/3}\ \text{cm.}$$

This smaller value has been supported during the last ten years or so by experiments on the Coulomb scattering of high-energy electrons by nuclei. For an electron with energy of, say, 200 MeV, the de Broglie wave-length is of the order 10^{-13} cm., and such a particle is therefore ideal for exploring the detailed structure and shape of a nucleus. A great deal of this work has been carried out by Hofstadter and his collaborators using the Stanford linear accelerator, and their results for a wide range of nuclei show that the density distribution of the nucleus is well represented by the formula

$$\rho(r) = \rho_0 \Big/ \Big(1 + \exp \frac{r - c}{z}\Big)$$

where c and z are parameters. Such a distribution differs from that of, for example, the density of a billiard ball and implies the existence of a sizeable surface region in which the density is decreasing. It is found that a value for z of about $2 \cdot 25 \times 10^{-13}$ cm. fits the scattering data rather well. This corresponds to a surface region having a thickness of the order $\frac{1}{3}R$. The formula also implies that the radius R of an equivalent uniform distribution can be expressed in the form $R = r_0 \times A^{1/3}$ cm., where $r_0 = 1 \cdot 32 \times 10^{-13}$ cm. for A less than 50, and $r_0 = 1 \cdot 21 \times 10^{-13}$ cm. for A greater than 50. A similar value for R is also obtained from the properties of μ-mesic atoms (i.e. atoms in which μ-mesons occupy bound states similar to those occupied by electrons). If this result is written in the form $4\pi R^3/3A = \text{const.}$, it shows that each nucleon occupies roughly the same space, and this very reasonable result confirms the assumption made about the forces. Furthermore, the formula is in fair agreement with the results of neutron scattering (mentioned above) and, if extrapolated to high atomic numbers, with another method applicable to radioactive elements.

This method leads us at once to the most fundamental application of quantum mechanics to nuclei—the explanation of radioactive decay. As we have seen, there are two different types of natural emission, α- and β-decay. Here we have to do only with the emission of α-particles, postponing the discussion of the β-decay for a later section.

An important quantitative law of α-decay was experimentally discovered by Geiger and Nuttall (1911); it gives a relation between the velocity v of the α-particle and the life period T of the emitting nucleus, namely,

$$\log T = A + B \log v,$$

where A and B are constants.

The explanation, which was given independently by Gamow, and by Condon and Gurney (1928), depends on a deep-seated distinction between quantum mechanics and ordinary mechanics, which is of importance in other cases also (as in cold electron emission, p. 243). In order to get a mechanical picture of the binding of an α-particle to the rest of the nucleus, we must imagine a field of force which holds it fast, and whose potential therefore has the form of a crater (fig. 1). We know the form of the outside walls of the crater, for the scattering experiments (§ 3, p. 63) tell us that Coulomb's law holds down to very short distances—for uranium, certainly to less than $r_0 = 3 \times 10^{-12}$ cm. The crater edge is therefore certainly higher than $2e^2(Z - 2)/r_0$ (remember the double charge of the α-particle), and, for $Z \sim 90$, this is 14×10^{-6} erg. But the energy of the escaping α-particles is less than half this value. If therefore an α-particle is originally inside the crater, according to the laws of classical mechanics it could never get out.

It is different in quantum mechanics. Here the motion of the particle within the crater is represented by a wave—a standing wave in fact. Consider an optical analogy—the total reflection of a ray of light which passes through glass and meets at a sharp angle the plane surface separating glass and air. According to the wave theory, however, there exists in the air also a sort of wave disturbance, which, it is true, carries off no energy, and only penetrates a few wavelengths into the air space. But if we now set a second piece of glass with its face parallel to the face of the other, and at such a short distance that the disturbance in the air-gap reaches it with intensity

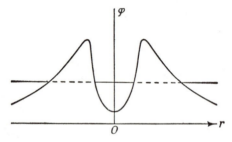

Fig. 1.—Crater form of the potential of the field of force which binds an α-particle to the rest of the nucleus

not too much impaired, then a small amount of energy at once passes into the second body, and the incident wave is propagated through it, though of course very much weakened. The weakening obviously diminishes rapidly, as the air-gap is made narrower, until in the limit when the two pieces of glass come into contact, the ray goes through with intensity unchanged.

In wave mechanics, we have quite a similar state of affairs at the walls of our crater. If these are of finite height and thickness, the standing waves inside the crater are by no means completely obliterated at the wall, but are propagated in it in weakened form, and emerge at the outside as progressive waves. Since, however, the square of the amplitude $|\psi|^2$ gives the probability of occurrence of the particles, we must conclude that the particles can penetrate through the wall with a frequency which is given by $|\psi|^2$. This will clearly diminish extremely rapidly with increasing thickness of the rampart; here the important matter is the thickness at the height which corresponds to the energy of the α-particles (see fig. 1 above). Fast α-particles will therefore have a much greater probability of getting out than slow ones; this corresponds qualitatively to Geiger and Nuttall's law. The exact analysis (Appendix XXX, p. 444) leads to a somewhat complicated formula of similar character; to obtain this, detailed assumptions about the interior of the energy crater are not required—its radius at the upper edge is really all that matters. For this radius a value in the neighbourhood of 8×10^{-13} cm. is found in all heavy radioactive nuclei, and it is just this constancy of the effective radius for this group of nuclei which forms the foundation of Geiger and Nuttall's formula, with fixed constants A and B. The penetrability of the potential wall explains the fact (§ 6, p. 73) that the disintegration of lithium nuclei by protons can be observed for so small velocities that the distance of nearest approach is much larger than the radius of the Li nucleus. The proton wave penetrates into the interior of the Li nucleus, and determines a small probability for the presence of a proton inside the nucleus, which becomes unstable and breaks into two α-particles.

We see from this example how Gamow's model of the nucleus can be used for explaining certain effects of collisions of other particles with the nucleus. But in general the model is much too simplified to account for all phenomena of this kind; the effect of bombardment, elastic scattering, or penetration into the interior with subsequent emission of another particle, depends in a very complicated and characteristic way on the energy of the particle and the nature of the nucleus.

We shall take up this question later (p. 350). Here it suffices to state again the result that this method leads to $R \sim 8 \times 10^{-13}$ cm., whereas the formula derived from electron scattering gives, say for uranium, $A = 238$, the value $R \approx 7 \cdot 5 \times 10^{-13}$ cm., in fair agreement.

We can compare with this quantity the de Broglie wave-length of a nucleon bound in the nucleus. We had (Chap. IV, § 6, p. 93) the formula $\lambda = h/\sqrt{(2meV)}$, where m is the mass of the particles considered and eV the energy which they have acquired. Taking for the latter the average binding energy of a particle in the nucleus, 8 MeV, and for m the mass of a nucleon, we obtain $\lambda = 9 \times 10^{-13}$ cm., a reasonable result, as it is of the same order as the diameter of the nucleus.

2. Angular Momentum and Magnetic Moment.

The next simple property of the nuclei is their angular momenta. We expect, of course, that these obey the laws of quantum mechanics (Chap. VI, § 5, p. 177), and this is actually the case.

The first way of demonstrating this was by investigating the hyperfine structure of spectral lines. It had long been known that the lines of some elements, for instance mercury, consist of aggregates of very fine lines which could not be explained with the help of the electronic structure. Systematic investigation has revealed that there are two kinds of hyperfine structure. One kind arises from the presence of several isotopic nuclei for a given chemical element; the slight mass differences cause displacements of the spectral lines which appear in the isotopic mixture as a fine structure. The second kind is much more important; it is connected with the orientation of the nucleus. It must be assumed that the nucleus acts on the outer electrons not merely as a point charge, but with forces which depend on the quantum state of its angular momentum, so that a transition between these states affects the total energy emitted or absorbed.

The quantum-mechanical treatment is obvious. The nuclear angular momentum I (in units $\hbar$) and the electronic angular momentum J combine vectorially into the total angular momentum F, and since each of these quantum numbers is an integer or half-integer, definite combinations of lines arise in the same way as in the multiplets of ordinary fine structure. We can also call in the aid of the Zeeman effect, and find out how the hyperfine structure changes in a magnetic field. A simple counting of the numbers of fine structure lines suffices to find I.

In this way many I-values have been determined; others have been found by different methods to be discussed presently. A few of the results are shown in Table VI, p. 316. The elements with even Z and even A show no splitting ($I = 0$); the moments of the nucleons seem to neutralize each other in pairs. Elements of odd Z have mostly half-integral angular momenta, though nitrogen is a notable exception, with $I = 1$.

From this material we can draw a new and most convincing argument against the presence of electrons in the nucleus. Consider nitrogen, for example, with $Z = 7$; if it were composed of 14 protons and 7 electrons, each of which has the spin $\frac{1}{2}$, the total angular momentum of these 21 particles would be necessarily half-integral, whereas actually it is 1. Conversely, the isotopes of Cd, Hg, and Pb with odd A would have an odd number of electrons (since Z is even), and therefore an even number of particles altogether; but the total angular momentum is half-integral. On the other land, the facts are in complete agreement with Heisenberg's hypothesis of a structure consisting of neutrons and protons, each with spin $\frac{1}{2}$. If the nitrogen nucleus consists of 7 neutrons and 7 protons, i.e. of 14 particles, its total momentum must be integral, as it actually is; the same holds for Cd, Hg, and Pb, and for all other cases.

There is another spectroscopic method for the determination of the nuclear angular momentum, based upon intensity relations in the band spectra of molecules (Chap. VIII, § 3, p. 263); but we cannot deal with them here.

There will be a magnetic moment μ produced by the rotation of the nucleus. The assumption suggests itself that it is the magnetic field of the nuclear moment μ which causes the change in the electronic cloud observed as hyperfine structure. Hence it must be possible to calculate the term displacements from μ, or vice versa. In this way, by a relatively simple mathematical theory, the μ-values for many nuclei could be determined, though not with very great precision.

The most important case is the proton. We should expect a magnetic moment 1840 times smaller than the Bohr magneton $e\hbar/(2mc)$ obtained by replacing in this formula the electronic mass m by the proton mass M, the *nuclear magneton*

$$\frac{e\hbar}{2Mc}.$$

Stern (1933) applied his method of molecular beams to this problem by refining the Stern-Gerlach apparatus (p. 186) in such a way that the sensitivity was increased by the enormous factor needed. The astonishing result was that the moment of the proton is almost three times larger than expected.

The method of molecular beams was then very much improved by Rabi and collaborators (from 1936), by combining it with the principle of magnetic resonance, as described already (Chap. VI, § 2, p. 163; see also the resonance method applied to the fine structure of the H-atom, Chap. VI, § 3, p. 168).

The idea is to influence the Larmor precession of the nucleus in a magnetic field by an electromagnetic wave in resonance. There not even ultra-short radar is needed, as the nuclear precession frequencies in fields of reasonable intensity are, on the optical scale, extremely small and actually in the range of easily produced electromagnetic vibrations. The instrument (fig. 2) has two inhomogeneous magnetic fields which deflect the molecular beam successively in opposite directions; in this way a kind of focusing effect is produced. Near the middle of the path a strong constant magnetic field is applied which produces a definite Larmor precession of each nucleus according to its momentum, and in the same region a weak alternating field is superposed. If this is in resonance with the precession of one of the nuclei, a strong reaction occurs which throws the particle out of its normal path. By changing either the constant field for a given frequency or vice versa, we can obtain all possible Larmor precessions; each precession actually present appears as a sudden drop in the intensity of the atomic beam.

In this way the number of terms and their distances, i.e. the angular momentum I and the magnetic moment μ, can be determined with high accuracy.

The result for the proton confirmed Stern's determination; the correct value is 2·79275 nuclear magnetons. Hence a satisfactory

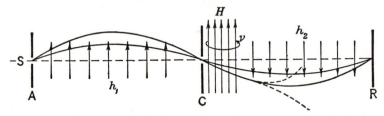

Fig. 2.—Schematic representation of Rabi's apparatus. The atoms emitted by the source S pass through the entrance slit A and are deflected by the inhomogeneous field h_1. They reach the collimator slit C on different paths according to their velocity; they are then deflected in the opposite direction by the inhomogeneous field h_2, and are thus focused on the receptor R. A strong magnetic field H behind C is superposed on a vibrating field; if the frequency ν of this is in resonance with the Larmor frequency produced by H, the atoms are thrown out of focus (dotted lines).

theory of the proton cannot be obtained by just inserting another mass into Dirac's equation.

Alvarez and Bloch (1940) then succeeded in measuring the magnetic moment of the neutron by using the powerful neutron emission from a cyclotron. As the great penetrating power of neutrons does not allow a method based on deflection of narrow beams with the help of slits, they had to be content with a wider beam produced by a cadmium tube as collimator and protected by large blocks of paraffin against stray neutrons. The deflection was replaced by " polarization ". Bloch (1936) had found that the intensity of a neutron beam sent through two parallel ferromagnetic plates in succession depends on the angle between the directions of magnetization of the plates; hence these can be used as polarizer and analyser, just like the Nicol prism in optics. If now a strong constant magnetic field, together with a weak vibrating field, are interposed between the polarizing and analysing plate, the number of emerging neutrons will be sharply reduced if their Larmor precession is in resonance with the frequency of the vibration. By reflecting neutrons from magnetized sheets of cobalt it has been found possible to produce even more highly polarized beams of neutrons.

More recent accurate measurements (Cohen, Corngold and Ramsay, 1956) have given for the magnetic moment of the neutron the value $\mu = -1.913148$ nuclear magnetons, where the negative sign means that the direction of the magnetic moment is opposite to that of the angular momentum—a most interesting result which shows the hopelessness of trying to account for the properties of elementary particles by simple mechanical models.

We should expect to obtain the magnetic moment of the deuteron by adding those of the proton and the neutron; in fact the sum $2.793 + (-1.913) = 0.880$ is in fair agreement with the experimental value of 0.857 found for the deuteron. Yet this slight difference is important. The additivity of the moments can only be assumed if there is no orbital contribution; the small deviation shows that there must be a relative orbital motion of the two nucleons, or in other words, the ground state cannot be a pure S-state $(l = 0)$. We shall return to this question presently (p. 319).

Nuclear moments can also be measured for matter in bulk with the help of resonance methods, and indeed the most accurate values for nuclear magnetic moments have been determined in this way. Purcell, Torrey and Pound (1946) developed a method in which the absorption of energy from an oscillating magnetic field by nuclei subject to a fixed magnetic field of a few thousand gauss enables the Larmor precessional frequency and hence the nuclear g-factor to be determined. The sample under investigation is surrounded by a few turns of wire producing an oscillating magnetic field much smaller than 1 gauss and perpendicular to the fixed field. Either the frequency of the oscillating field or the strength H of the fixed field can be varied, although the latter is more convenient. The resonance frequency is then obtained by noting the value of H for which maximum power is absorbed from the oscillating field.

A related method due to Bloch, Hansen and Packard (1946) detects the resonance by the maximum in the strength of signals induced in a detector coil placed perpendicular to both the fixed field and the oscillating field.

Spin and magnetic moment are the simplest indication of anisotropy of a structure. But there are other effects of higher order connected with anisotropy. This can be seen in a simple way by regarding

TABLE XVI

NUCLEAR ANGULAR MOMENTUM I, MAGNETIC MOMENT μ, AND ELECTRIC QUADRUPOLE MOMENT q OF SELECTED NUCLEI

Z	Element	A	I	μ	$q(10^{-24}\text{cm.}^2)$
0	n	1	$\frac{1}{2}$	$-1\cdot91315(7)$	
1	H	1	$\frac{1}{2}$	$+2\cdot7926(1)$	
	D	2	1	$+0\cdot85735(1)$	$+0\cdot00274(2)$
2	He	3	$\frac{1}{2}$	$-2\cdot127414(3)$	
		4	0		(0)
3	Li	6	1	$+0\cdot82189(4)$	
		7	$\frac{3}{2}$	$+3\cdot2559(1)$	$(+0\cdot02)$
4	Be	9	$\frac{3}{2}$	$-1\cdot77(1)$	$(0\cdot02)$
5	B	11	$\frac{3}{2}$	$+2\cdot6886(3)$	$+0\cdot036(2)$
6	C	12	0		
		13	$\frac{1}{2}$	$+0\cdot7023(2)$	
7	N	14	1	$+0\cdot40365(3)$	$+0\cdot02$
		15	$\frac{1}{2}$	$-0\cdot28299(3)$	
8	O	16	0		
9	F	19	$\frac{1}{2}$	$+2\cdot6273$	
17	Cl	35	$\frac{3}{2}$	$+0\cdot8219(2)$	$-0\cdot07894(2)$
29	Cu	63	$\frac{3}{2}$	$+2\cdot2262(4)$	$-0\cdot157$
		65	$\frac{3}{2}$	$+2\cdot3845(4)$	$-0\cdot145$
48	Cd	111	$\frac{1}{2}$	$-0\cdot5949(1)$	
		113	$\frac{1}{2}$	$-0\cdot6224(1)$	
		110	(0)		
		112	(0)		
		114	(0)		
		116	(0)		
63	Eu	151	$\frac{5}{2}$	$+3\cdot6$	$+1\cdot2$
		153	$\frac{5}{2}$	$+1\cdot6$	$+2\cdot5$
70	Yb	173	$\frac{5}{2}$	$-0\cdot65$	$+3\cdot9(4)$
80	Hg	199	$\frac{1}{2}$	$+0\cdot50413(3)$	
		201	$\frac{3}{2}$	$-0\cdot5990(1)$	$+0\cdot6$
		198	(0)		
		200	(0)		
		202	(0)		
		204	(0)		
82	Pb	207	$\frac{1}{2}$	$+0\cdot58750(7)$	
		204	(0)		
		206	(0)		
		208	(0)		
83	Bi	209	$\frac{9}{2}$	$+4\cdot082$	$-0\cdot4$

the electrostatic potential Φ as being produced by an arbitrary distribution of charge at a large distance r. In first approximation, Φ will be the Coulomb potential, proportional to $1/r$; but higher terms with $1/r^2$, $1/r^3$, ... will follow, so that we have, for a nucleus of atomic number Z, and axial symmetry,

$$\Phi(r, \theta) = \frac{e}{r}\left(Z + \frac{p}{r} P_1(\cos \theta) + \tfrac{1}{2}\frac{q}{r^2} P_2(\cos \theta) + \ldots\right),$$

where θ is the angle between the radius vector and the spin axis, and P_1, P_2, ... the so-called harmonic functions (Legendre polynomials). The coefficients p, q, ... depend on the charge distribution; p is the dipole moment and q the quadrupole moment. For illustration, assume the system to be a homogeneously charged ellipsoid of rotation, the semi-axis of symmetry being a, the other axis b, we then have

$$q = \tfrac{2}{5}Z(a^2 - b^2).$$

The effect of nuclear quadrupole moments shows up in many spectroscopic phenomena, for example, in deviations from equidistance of hyperfine structure lines and many quadrupole moments have been deduced from effects of this kind. Similarly nuclear quadrupole moments have been deduced from their perturbing effects in beam experiments. A direct measurement of q for the deuteron yields $q = 2\cdot74 \times 10^{-27}$ cm^2. This means a very small anisotropy; for from the last formula we get for small $(a - b)$ and $Z = 1$,

$$\frac{a - b}{a} = \tfrac{5}{2} \cdot \frac{q}{2a^2} = \frac{3\cdot41 \times 10^{-27}}{3^2 \times 10^{-26}} \simeq 0\cdot04,$$

where we have taken for a approximately 3×10^{-23} cm. In spite of the smallness of this quantity its existence is decisive for the theory of nuclear forces, as will be seen presently.

3. The Deuteron and Nuclear Forces.

We have now to consider those properties of the nucleus which depend on its structure: how it is built up from nucleons, and by what forces.

A general remark may be made in advance. The fact that we can venture to study the construction of the nucleus at all rests on the recognition that its bricks are heavy particles, not electrons or positrons. These light particles would move so fast under the action of nuclear forces that relativistic effects would play a dominating part

A thoroughly satisfactory relativistic quantum mechanics of strongly interacting particles is, however, not yet available—Dirac's theory of the electron (§ 8, p. 191), for example, being successful only for weak interactions. With heavy particles, on the other hand, the velocities remain so small that non-relativistic quantum mechanics is a sufficient approximation. We shall, in the course of the investigation, find new evidence that the nucleons are in fact the constituent parts of the nucleus.

The simplest nucleus is the deuteron, consisting of a proton and a neutron. Let us first make the simplest possible assumption: that the two particles are acting on one another with a central force, having the potential energy $V(r)$. Then we have to do with a 2-body problem of quantum mechanics, analogous to the hydrogen atom, with two essential differences—the masses are almost equal, and the range of the force is very small. Although the wave equation for the relative motion is the same as that given in Chap. V, § 7, p. 145, the nature of the solution is totally different. The mass has to be replaced by the reduced mass $M_n M_p/(M_n + M_p)$, which is very near to $\frac{1}{2}M$, where M is the mean mass of a nucleon. We have already mentioned that from the scattering of neutrons by protons, it can be inferred that the range of force is small compared with the de Broglie wave-length corresponding to the mean energy of a nucleon. Hence the exact dependence of the potential V on the distance will not matter much. Let us first try a law for $V(r)$ which leads to a very simple solution, namely, a " box-shaped " potential, where

$$V(r) = \begin{cases} -V_0 \text{ for } r < a, \\ 0 \text{ for } r > a. \end{cases}$$

Assuming that in the ground state the orbital angular momentum is zero $(l = 0)$, the wave equation for $u(r) = r\psi(r)$ is (Appendix XXXI, p. 447)

$$\frac{d^2u}{dr^2} + \frac{2M}{\hbar^2}[E - V(r)]u = 0.$$

Here the factor $[E - V(r)]$ in the second term is constant in both domains of r, being equal to $E + V_0$ for $r < a$, and to E for $r > a$. Hence the solution is easily found (Appendix XXXI). The significant result is that a relation between the range and the depth of box potential is obtained, namely,

$$\cot\left[\frac{a}{\hbar}\sqrt{M(V_0 - W)}\right] = -\sqrt{\frac{W}{V_0 - W}},$$

where $W = -E$ is the binding energy. By assigning a reasonable value of a, and using the experimental value of W, the depth V_0 is fixed; its value is $V_0 = 21$ MeV. The wave function thus determined differs only slightly from those obtained by choosing other types of potentials like $e^{-(\eta r)^2}$, $e^{-\eta r}$, or $e^{-\eta r}/r$, which is Yukawa's assumption (Chap. II, § 8, p. 50).

So far no account has been taken of the spin of the particles. Now all nucleons have spin $\frac{1}{2}$, whereas the total angular momentum of the deuteron is 1 (see Table VI, p. 316); hence the ground state of the deuteron is characterized by the spins of the two nucleons being parallel (in spectroscopic language: it is a triplet state). Further, we have seen on p. 315 that this state cannot be a pure S-state, but must have an orbital angular momentum (since the total magnetic moment is not exactly the sum of that of a neutron and a proton, and a quadrupole moment exists). From the latter fact it can be inferred that the forces cannot be central; for in this case the state of lowest energy would be necessarily an S-state. An account of the qualitative features of the two-body system under non-central forces is given later on (p. 322). Here it suffices to remark that by assuming a proper mixture of central and non-central forces of box shape we can quantitatively explain all properties of the ground state.

Two nucleons in interaction can be in many different states (characterized by definite quantum numbers); so far we have only learned something about the ground state of the deuteron. There are other possible states. Each nucleon has a spin quantum number $m_s = -\frac{1}{2}, +\frac{1}{2}$ (in units of $\hbar$) where m_s refers to the component of spin in a given direction. At this point it is convenient to introduce another quantum number for describing the charge state of a nucleon. This is done by ascribing a fictitious spin—*isobaric spin*—to the nucleon, which is completely analogous to, but quite unrelated to, ordinary spin. It is supposed that the isobaric spin of a nucleon is $\frac{1}{2}$ and that associated with this spin is an isobaric spin operator t (cf. s in the case of ordinary spin). Then in the same way that s can be related to the Pauli spin operators $\boldsymbol{\sigma}$ by $s = \frac{1}{2}\hbar\boldsymbol{\sigma}$, so t can be related to an operator $\boldsymbol{\tau}$ ($t = \frac{1}{2}\boldsymbol{\tau}$), having exactly the same properties as $\boldsymbol{\sigma}$. Thus the eigenvalues of t_z are $\pm \frac{1}{2}$ and of τ_z are ± 1. It is conventional to describe a proton as a nucleon in the state with $t_z = -\frac{1}{2}$, and a neutron as corresponding to $t_z = +\frac{1}{2}$.

There is further the orbital angular momentum quantum number $l = 0, 1, 2 \ldots$ There are many combinations of these quantum numbers, but they are restricted by the Pauli exclusion principle, according to which the wave function must be anti-symmetric in *all* the co-ordinates of the two nucleons. For instance, if $\tau_{z1} = -1$, $\tau_{z2} = +1$, and $m_{s1} = \frac{1}{2}$, $m_{s2} = \frac{1}{2}$, which means a combination of neutron and proton with parallel spins (as in the ground state of the deuteron), the wave function is anti-symmetric in the charges, and symmetric in the spin. Hence it must be symmetric in the orbital motion, from which it can be inferred that l must be even ($l = 0, 2, \ldots$). Carrying out this procedure for other combinations of the quantum numbers, the following table of allowed fully anti-symmetric states can be constructed.

Spin	Charge	l	Typical term
Triplet	Triplet	Odd	^3P
	Singlet	Even	^3S
Singlet	Triplet	Even	^1S
	Singlet	Odd	^1P

There is no reason to assume that the nuclear forces as functions of the distance are the same in any of these different states; even if the potential is box-shaped in every case, the depth may differ. That this is really the case has been shown by scattering experiments. The observed value for the cross-section of a slow neutron scattered by a proton extrapolated to zero energy is $20 \cdot 36 \times 10^{-24}$ cm.^{2}. Now if it is assumed that this scattering takes place in a ^3S-state (corresponding to the ground state of the deuteron) then from the properties of the deuteron, it is easy to deduce that the cross-section should be about $2 \cdot 2 \times 10^{-24}$ cm.^{2}. There is clearly a large discrepancy between these two figures. In 1935 Wigner suggested how this discrepancy could be resolved. He pointed out that in proton-neutron scattering there can be an interaction both in the triplet and in the singlet S-states and that the total cross-section σ is given by

$$\sigma = \tfrac{1}{4}\sigma \text{ (singlet)} + \tfrac{3}{4}\sigma \text{ (triplet)}$$

where the factors $\frac{1}{4}$ and $\frac{3}{4}$ allow for the multiplicities of the $S = 0$ and $S = 1$ spin states. If it is assumed then that the singlet S-state interaction is quite different from the triplet S-state interaction, the discrepancy can be resolved.

The implication of all this is that nuclear forces are spin-dependent. Thus if neutron-proton scattering below about 10 MeV is analysed in terms of square-well potentials, for example, it turns out that the triplet potential has a range of about 2×10^{-13} cm. and a depth of about 35 MeV, whilst the range of the singlet potential is something like $2 \cdot 8 \times 10^{-13}$ cm. and its depth is about 12 MeV. In the case of proton-proton scattering, the charge is necessarily in a triplet state and therefore at low energies the protons can only interact in the singlet S-state. Analysis of this scattering is of course complicated by the Coulomb force between the two protons. However, if this effect is subtracted, then it turns out that the strength of the ^1S proton-proton scattering force is essentially the same as the ^1S neutron-proton scattering force. This is an indication that nuclear forces are charge-independent, and other investigations do indeed indicate that charge independence holds to within about 1%.

During the last 10–20 years, a great deal of effort has been put into investigating neutron-proton and proton-proton scattering at very much higher energies than 10 MeV. At such energies, higher orbital angular momenta can contribute and the analysis becomes much more complicated. However, it is becoming clear that it is possible to formulate a force between two nucleons which is able to account for the main features of the scattering. One feature which emerges is that the nuclear force has an exchange character. By this is meant that in addition to the potential depending just on the distance r between the particles, say $V(r)$, there are other terms of the form $PV(r)$, where P is an exchange permutation operator. Such an operator has the property that when it operates on a wave function $\psi(1, 2)$ of two sets of variables, it exchanges some properties between them, $P\psi(1, 2) = \psi(2, 1)$. There are two independent types of these operators, namely, P_σ and P_τ, which exchange the spin and the charge of the two particles respectively. The third possibility, exchange of position, can be expressed in terms of P_σ and P_τ; for all three operators together correspond to a complete exchange of two particles, hence, according to Pauli's principle, $P_x P_\sigma P_\tau = -1$, and as $P_\sigma^2 = P_\tau^2 = 1$, $P_x = -P_\sigma P_\tau$. We can take for P either P_τ (suggested by Heisenberg, 1932) or P_σ (Bartlett, 1936), or finally $P_\sigma P_\tau$ (Majorana, 1933). The most general form of a central potential is a linear combination of all of them, including the ordinary force, $V_0(r) + P_\tau V_1(r) + P_\sigma V_2(r) + P_\sigma P_\tau V_3(r)$.

That the exchange of charge actually happens can be seen directly from neutron-proton scattering experiments. If the scattering processes are examined in, say, a Wilson cloud chamber or in nuclear emulsion, the following phenomenon is frequently seen to take place. A beam of fast neutrons (which remains invisible) in hydrogen produces tracks of protons by head-on collisions. We could expect that a fast neutron would be strongly deflected at nearly 90°, whereas the proton goes off parallel to the neutron beam. But just the opposite is the case: the visible tracks, produced by protons, are deflected through large angles to the direction of the neutron beam. This shows that crudely speaking charge has jumped from one particle to the other.

Another feature which emerges is that there is a non-central component to the nuclear force. This has already been referred to in connection with the ground-state structure of the deuteron. A non-central potential is one in which the magnitude of the potential experienced by two nucleons depends not only on the separation of the nucleons, but also on the relative orientation of the spins of the nucleons to the line joining them. Specifically, this part of the potential can be written in the form

$$\left[\tfrac{1}{3}\boldsymbol{\sigma}_1 \cdot \boldsymbol{\sigma}_2 - \frac{(\boldsymbol{\sigma}_1 \cdot \boldsymbol{r})(\boldsymbol{\sigma}_2 \cdot \boldsymbol{r})}{r^2}\right] f(r).$$

In the above expression $\boldsymbol{\sigma}_1$, $\boldsymbol{\sigma}_2$ are the Pauli spin operators for the two nucleons and r is the internucleon distance. Since $\boldsymbol{\sigma}_1$ and $\boldsymbol{\sigma}_2$ are operators, represented by the 2-dimensional Pauli matrices (Chap. VI, § 8, p. 189), the wave function must have several (four) components, and the wave equation really consists of a set of coupled equations as explained for a single electron in Chap. VI, § 8. It can be seen without actual calculation that this (so-called) tensor potential will give qualitatively correct results for the ground state of the deuteron. We know that in this ground state the spins of the two nucleons are parallel to one another. Thus for a given r the potential has a minimum equal to $-\tfrac{2}{3}f(r)$ (attraction) when $\boldsymbol{r}$ is parallel to the spin direction and has a maximum equal to $\tfrac{1}{3}f(r)$ when $\boldsymbol{r}$ is perpendicular to the spin direction (repulsion). This non-central interaction, together with the central interaction, will tend to make the deuteron cigar-shaped, with axis parallel to the total spin—which is just what is required by the experiments, as under these circumstances the orbital motion will contribute an additional magnetic moment and a quadrupole moment with the correct sign.

With the introduction of a non-central force, the ground state of the deuteron is no longer an S-state, but a mixture of states of different orbital momentum. As the total angular momentum is quantised and is found experimentally to be 1 (in units of $h/2\pi = \hbar$), and as the total spin is 1, the orbital angular momentum can only be either 0 or 2, as follows from the rule of addition of angular momentum $J = S + L$ We shall not go here into further details of solution of the wave equation, but be content to remark that with a suitable proportion of central and non-central interaction, all physical properties mentioned regarding the ground state of the deuteron can be well explained.

Measurements on the way in which nucleon spins are affected in scattering processes make it clear that there is also a spin orbit component to the inter-nucleon potential. Such a potential has the property that its strength varies according to the way in which the nucleon spins are oriented with respect to the orbital angular momentum. It has the simple form

$$(\sigma_1 + \sigma_2) \cdot L \; V(r)$$

where L is the orbital angular momentum operator.

The final feature of the nucleon-nucleon force which should be mentioned is that it contains a repulsive core. The force is of course predominately attractive at large distances, otherwise nuclei would not bind. However, when two nucleons become very close to one another (within about 0.5×10^{-13} cm.) then the force becomes strongly repulsive.

4. Nuclear Structure and Nuclear Saturation.

The same forces which are responsible for nucleon-nucleon scattering, and which keep the two particles of the deuteron together, must also be responsible for the cohesion of the other nuclei. Here a new feature appears, namely, the so-called *saturation* of the forces, very similar to the well-known property of chemical bonds—two hydrogen atoms, for example, bound together have a negligible affinity to a third one.

The total binding energy (B) of the nucleus turns out to be roughly proportional to A, the number of nucleons. (This is apparent from fig. 5 (Chap. III) where B/A is seen to be roughly constant at about 8 MeV.) There are important deviations, of which we shall speak later, but with a fair approximation we can define a mean binding energy per particle. Now, in interpreting this binding energy in terms of the nucleon force between pairs of particles, we would at first sight expect to obtain a binding energy roughly proportional to the total number of pairs of particles in the nucleus, i.e. proportional to $\frac{1}{2} A(A-1)$. Thus we have a binding energy roughly proportional to A^2 rather than to A as is observed experimentally. This feature of the inter-nucleon force, which leads to a binding energy proportional to A, is referred to as *saturation* and what it means, roughly, is that each particle in the nucleus is only interacting strongly with a limited number of other nuclear particles.

A *possible* explanation of the saturation of nuclear forces was first put forward by Heisenberg, who proposed that the saturation property followed from an exchange character of nuclear forces. It can be seen from fig. 5 (Chap. III) that the binding energy increases as 2, 3, 4 nucleons are put together, and then saturates as more nucleons are added. This means that the exchange force must have the characteristic that it allows strong binding up to and including helium 4, but that beyond that the binding energy per particle must remain roughly steady. Now the ground state of helium 4 is symmetric in the space co-ordinates of all four particles (it is an S-state). The Pauli exclusion principle allows such a situation since we have two types of particles, neutrons and protons, and within each pair nucleon spins will be opposite to one another. However, if another nucleon is added, then the exclusion principle will require it to go into a different orbital state. So an exchange force is needed which leads to attraction when the nucleons are in the same orbital state, but for which the attraction is significantly decreased when the nucleons are in different orbital states. In this way, any given nucleon will only be able to interact strongly with those few other nucleons which are in similar orbital states.

In the last section, discussion was given of the different types of exchange force, and it is straightforward to show that the Majorana force is one which will have significantly different values when nucleons are in similar spatial states and when they are in different spatial states. A detailed investigation of the saturation problem shows in fact that if the strength of the Majorana force is greater than about four times the strength of the ordinary force, then saturation results. However, high-energy nucleon-nucleon scattering experiments show that these two forces have roughly the same strength and so, although exchange forces in principle can account for saturation, the actual exchange character of the nuclear force is not of the correct form to lead to saturation.

Another factor which can contribute to the saturation effect is the existence of a repulsive core to the nuclear force, and present-day thinking about saturation is that both the exchange character of the force and perhaps more importantly, the repulsive core, lead to the phenomenon of saturation.

Turning now to other properties of a nucleus, the problem we are faced with is to solve the Schrödinger equation for a system of Z protons and N neutrons interacting with each other through a highly complicated nuclear force. Unfortunately to obtain a full wave-mechanical description of such a system is impossible with present-day mathematical techniques. Therefore the approach adopted in order to give some theoretical account of the more detailed nuclear properties has been to devise models of the nucleus which may be regarded as crude approximations to the real physical nucleus. Such models are suggested by the known properties of the nucleus and other semi-classical considerations.

One particular nuclear feature which has had considerable influence in this way is the existence of the so-called " magic number " nuclei. It is found that nuclei for which Z or $A - Z$ or both are equal to 2, 8, 20, 28, 50, 82, 126 have very distinctive properties. Among them may be mentioned the following:

  (i) Such nuclei have much higher cosmic abundances.
  (ii) Nuclear binding energies are greatest at magic number nuclei.
  (iii) Nuclear quadrupole moments change sign at magic numbers.

The existence of particularly stable nuclei characterized by certain numbers of neutrons or protons is at once reminiscent of the situation which obtains in the case of atomic structure where certain atoms (the inert elements) are particularly stable. In the atomic case this stability can be traced to the filling of shells of electrons and at once suggests that in the nuclear case the magic numbers may again be indicative of a shell structure. It is from this viewpoint that the nuclear-shell model emerged.

5. The Nuclear Shell Model.

The basic idea of the nuclear shell model is that the interaction of any one nucleon within the nucleus with the remaining nucleons can be mainly represented by a static potential well whose shape and spatial extension (fig. 3) is expected to be similar to that of the nuclear density distribution. It can be naïvely argued that because of the short-range nature of nuclear forces, a nucleon experiences a potential energy roughly proportional to the number of nucleons in its immediate vicinity, that is, pro-portional to the nuclear density.

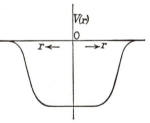

Fig. 3.—Nucleon potential energy $V(r)$

In such a potential well there will be a series of single particle energy levels $E(n, l)$, where n is the principal quantum number and l the orbital angular-momentum quantum number (cf. App. XVIII, p. 399), whose relative spacing depends on the depth, size and shape of the well. The arrangement of these levels for such a well with a depth of about

40 MeV and radius 8×10^{-13} cm. is shown on the left of fig. 4. (This size of well corresponds to a heavy nucleus near lead; for a lighter nucleus with smaller spatial extension the level separation is greater and there are fewer bound states. Correspondingly, there are fewer nucleons to accommodate in the well.) It is then supposed that these

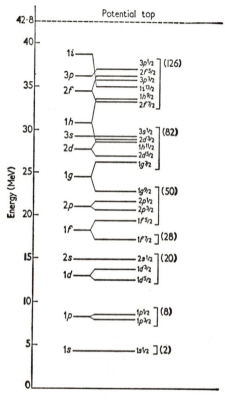

Fig. 4.—Energy levels in a potential well of the type shown in fig. 3. The usual spectroscopic notation is adhered to and the convention used that the value of the quantum number n is the number of nodes minus one occurring in the radial function (not counting nodes at the origin). On the extreme right of the diagram are the magic numbers representing the number of nucleons that completely fill all the preceding levels.

levels are filled by neutrons and protons, the number of particles going into each level being determined by the exclusion principle. It can be seen that certain large gaps occur in the energy level scheme and that the total number of neutrons or protons that can be included below the lower gaps (remember that $2(2l + 1)$ particles can go into each level) corresponds exactly with the magic numbers 2, 8, 20. However, the higher magic numbers do not appear.

It was in 1948 that Mayer and Jensen showed how the higher numbers could be produced in a natural way. They pointed out that, because of the spin dependence of nuclear forces, it is to be expected that a nucleon in the nucleus experiences a strong spin-orbit potential of the form $C\boldsymbol{l} \cdot \boldsymbol{s}$ where $\boldsymbol{l}$ is the orbital, $\boldsymbol{s}$ the spin angular momentum vector of a nucleon and C an approximate constant. Now if $\boldsymbol{j}$ is the total angular momentum vector ($\boldsymbol{j} = \boldsymbol{l} + \boldsymbol{s}$), then

$$j^2 = l^2 + s^2 + 2\boldsymbol{l} \cdot \boldsymbol{s}$$

so that, writing $j(j+1)$, $l(l+1)$, $s(s+1)$ for j^2, l^2 and s^2 (p. 162) we have

$$C\boldsymbol{l} \cdot \boldsymbol{s} = \tfrac{1}{2}C[j(j+1) - l(l+1) - s(s+1)].$$

But for a nucleon $s = \tfrac{1}{2}$ and $j = l \pm \tfrac{1}{2}$, so that for the two possible values of j the spin-orbit energy has the magnitudes

$$j = l + \tfrac{1}{2}; \quad \tfrac{1}{2}Cl; \quad j = l - \tfrac{1}{2}; \quad -\tfrac{1}{2}C(l+1).$$

This means that a given level (n, l) will be split into two levels (n, l, j) with $j = l \pm \tfrac{1}{2}$ such that the separation between the levels is $\tfrac{1}{2}C(2l+1)$. Thus, if C is negative and sufficiently large, a level scheme such as that shown on the right of fig. 4 is obtained.

Each level of angular momentum j can contain $2j + 1$ particles, and filling the levels up according to the exclusion principle it can be seen that all the magic numbers now occur in a natural way immediately below relatively large gaps. What has happened as a result of introducing the spin orbit coupling is that for the higher groups of levels the splitting of the state with the highest orbital angular momentum is so great that the low component ($j = l + \tfrac{1}{2}$) of the resulting doublet joins the next lowest group of levels.

We now have to consider to what extent such a naïve nuclear model can account for the actual properties of nuclei. Its success, in fact, is very great. Consider first of all the question of nuclear spins. To proceed further we have to take account of the fact that the nucleons do not move completely independently of one another. Although the forces between the nucleons are partially accounted for by the average potential well that has been introduced, there are still residual nuclear interactions representing the difference between the well potential energy and that actually experienced by a nucleon. From the known nature of nuclear forces it is to be expected that these interactions are such that like nucleons in the different levels tend to pair off with one another so that their total angular momentum is zero. Under these circumstances there is no contribution to the angular momentum of a nucleus having an even number of neutrons and an even number of protons, since their angular momenta are all paired off. This is in agreement with the observation that the so-called even-even nuclei have spin zero. However, for an odd A nucleus, in the last neutron or proton energy level to be filled there is either an odd neutron or an odd proton which is unpaired. This is the only particle contributing to the nuclear spin, which should therefore be equal to the angular momentum j of this nucleon.

Thus O^{17} has 8 protons and 9 neutrons; the last odd neutron has $j = \frac{5}{2}$ and this is the observed spin of O^{17}. Similarly Co^{59} has 27 protons and 32 neutrons; the last odd proton has $j = \frac{7}{2}$ which is again in agreement with the experimental spin value. It is, however, significant that in the cases where high spins are to be expected (e.g. when the $j = \frac{11}{2}$ state is being filled) they are not observed. The reason for this is that the residual pairing forces are much more effective in states of high angular momentum so that particles like to pair off in such states, even at the expense (energetically) of lifting a nucleon out of a lower angular momentum state in order to do this. In this case the nucleus has its spin equal to the lower angular momentum. Taking this effect into account, there are only very few odd A nuclei whose spins are in disagreement with the simple shell model predictions. As far as odd-odd nuclei are concerned (i.e. those with odd Z and odd N) the shell model in the form so far described makes no prediction other than that the spin should be an integer.

The shell model also predicts the parities of nuclei. The concept of parity was introduced earlier (p. 144) and is said to be " even " or " odd " according as the nuclear wave function remains the same or changes sign under reflection of axes. In particular, we saw that a particle with angular momentum l has even or odd parity according as l is even or odd. Since in the shell model the nuclear properties are essentially vested in the last odd particle, it follows that the parity of an odd A nucleus should be that of the odd particle; that is, it should be determined by the value of l for this particle. Again, theory and experiment agree very well.

Perhaps one of the most striking successes of the shell model is its ability to describe the general behaviour of the magnetic moments of odd A nuclei. For these nuclei the magnetic moment is predicted to be that of the last odd neutron or proton as the case may be. This is because the contributions to the magnetic moment from the other nucleons exactly cancel due to the pairing effect. Because of the strong spin orbit coupling, the situation obtaining in the case of the nuclear shell model is identical in form with that describing the behaviour of a one-electron atom in a weak magnetic field (p. 161). The magnetic moment of the nucleus is thus given by

$$\mu = gj\,\frac{e\hbar}{2Mc}$$

where M is the proton mass and $e\hbar/2Mc$ is called the nuclear magneton (n.m.), j is the angular momentum of the last odd nucleon, and g the Landé g-factor for the odd nucleon.

$$g = \tfrac{1}{2}\left[(g_l + g_s) + (g_l - g_s)\,\frac{l(l+1) - \tfrac{3}{4}}{j(j+1)}\right]$$

and for an odd proton $g_l = 1$, $g_s = 5\cdot586$, whilst for an odd neutron $g_l = 0$, $g_s = -3\cdot826$. The theoretical and experimental results are presented in the so-called Schmidt diagrams (figs. 5 and 6) for odd proton and odd neutron nuclei. On the horizontal axis is plotted the nuclear spin j in units of $\hbar$ and on the vertical axis the magnetic moment in nuclear magnetons. The full lines correspond to the theoretical predictions of the above formulæ for the two possibilities $j = l \pm \tfrac{1}{2}$ and of course only have meaning at the allowed half-integer spin values at which are plotted the experimental points. The qualitative agreement is surprisingly good and can be considerably improved by refining the shell model in various ways.

When we come to consider nuclear electric quadrupole moments, Q, however, there is violent disagreement (Table VI, p. 316). This arises mainly in the approximate regions $150 < A < 185$, $225 < A$, where the experimentally observed values of Q are sometimes as much as 30 times larger than the shell model prediction. These predictions are calculated in the same way as the magnetic moments by attributing Q to the last odd particle. In particular, since a neutron has zero charge, an odd neutron nucleus is expected to have $Q = 0$, but this is never observed. The fact that quadrupole moments having many times the single particle value occur is significant since it indicates that many particles are contributing and that we have here a collective effect of some kind. This will be discussed in detail a little later (p. 333).

Of the other nuclear properties, the nature of excited states must be considered. In a nucleus such as O^{17}, for example, there are 8 protons and 8 neutrons each in a " magic " and therefore very stable closed shell, together with an odd neutron in a state with $j = \tfrac{5}{2}$. It is then to be expected that excited states could most simply arise (as in the analogous case of the sodium atom, for example) by leaving the closed shells undisturbed and by exciting the odd neutron into one of the higher states; thus among the low excited states we should expect one with $j = \frac{3}{2}$ ($1d_{\frac{3}{2}}$) and one with $j = \frac{1}{2}$ ($2s_{\frac{1}{2}}$). This is in agreement with experiment and there are many cases such as this where the low-lying excited states of nuclei can be interpreted in a very simple fashion.

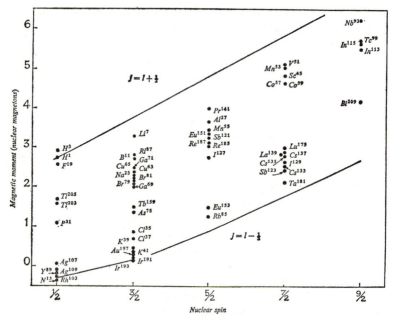

Fig. 5.—Schmidt diagram for odd proton nuclei

However, the case quoted was ideal and unambiguous. In more complicated nuclei away from " magic " closed shells the energy level scheme cannot be deduced so simply, and recourse must be had to detailed calculations taking into account the residual interactions between nucleons, the energy levels then being interpreted as re-arrangements of the nucleons within incomplete shells. By choosing this interaction to be a mixture of ordinary, Majorana, Heisenberg, and Bartlett forces (p. 321) it has been found possible to explain both ground and excited-state properties of many nuclei in considerable detail. Right away from closed shells, however, the necessary calculations are unfeasible, and it is just in these regions that the collective model suggested and developed by Aage Bohr (1952) has had considerable success. During the last few years another approach to the problem of more complex nuclei has been introduced. This is to approximate the nuclear force by one which carries the main features of the force, but which at the same time is such that the resulting many-body

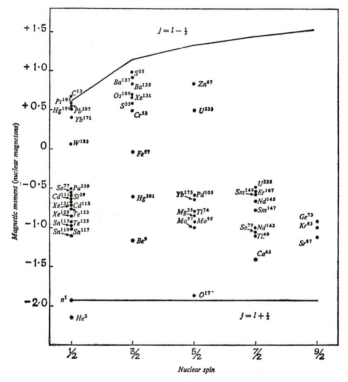

Fig. 6.—Schmidt diagram for odd neutron nuclei

Schrödinger equation is much more amenable to solution. Two features of the force which it is important to preserve are: (1) its pairing characteristic; this property has already been referred to on page 328 where it is pointed out that like nucleons tend to pair off with one another so that their total angular momentum is zero; (2) the force must be such that it can in suitable cases favour the considerable quadrupole distortion of the nucleus which is required to explain the very large quadrupole moments which occur. Using such a force, it has been found possible to account for various nuclear properties in complex nuclei, such as ground-state deformation, energies of low-lying states, magnetic moments and so on. The pairing character, in particular, also allows a very simple explanation to be given of the fact that in even-even nuclei there is a sizeable energy gap between the ground and first excited states, whilst in odd-A nuclei, a gap of this kind is not observed. Here the situation is very similar to that occurring in the superconductive state of solids and the basic ideas of Bardeen, Cooper and Schrieffer referred to on p. 297 have been taken over into nuclear physics. The main difference is that in superconductivity the pairing leads to pairs having zero linear momentum, whilst in the nucleus the pairs have zero angular momentum.

One final point has to be discussed in connection with the shell-model and that is why it works so well. The basis of the shell model is the assumption that the nucleons move approximately independently of one another in a smooth potential well. Plausibility arguments were given for this assumption; but these arguments completely ignored the fact that, since nuclear forces are so strong and have such a short range, a nucleon moving through the nucleus would in fact experience violent fluctuations in potential energy as it passed near other nucleons, in complete contradiction with our assumption of roughly independent particle motion.

The presence of such effects is confirmed in a great many nuclear reactions in which a neutron, for example, strikes a nucleus and is immediately captured to form a highly excited state of the nucleus. Such capture would not take place if the neutron moved through the nucleus in the way suggested by the shell model, since motion independent of the other particles implies that the nucleus is approximately " transparent ". The reason why, in spite of this apparently contrary evidence, the shell model might still be expected to work, was first given by Weisskopf (1952). He pointed out that for nucleons deep down in the potential well the violent fluctuations are quite ineffective. This is because the nucleus is a highly degenerate system where all the lower states are occupied and most transitions resulting from a collision between two nucleons are forbidden in virtue of the Pauli exclusion principle. Thus for the bulk of the nucleons, the violent changes in potential have little effect. On the other hand, in the case of a nuclear reaction, the incoming neutron has an energy at least 8 MeV greater than that of any nucleon in the nucleus, since nucleons only fill the potential well to within about 8 MeV of the top. This means that it has sufficient energy to excite particles, even fairly deep down in the potential well, up into unoccupied states. The incoming particle therefore interacts strongly with the nucleus, although the bulk of the constituent particles of the nucleus move approximately independently of one another. Recently Brueckner (1954) and Bethe (1956) have shown that these ideas can be formulated in a quantitative way and enable the basic ideas of the shell model of the nucleus to be fully justified.

6. The Nuclear Collective Model.

It was pointed out in the previous section that the quadrupole moments of a number of nuclei are much larger than the single particle values predicted by the shell model. These nuclei are all those which are well away from closed shells and in which there is a large number of " loose " particles, some of which have high angular momenta. Such particles will exert a strong disturbing force on the surface of the nucleus tending to distort its shape from approximately spherical to much more ellipsoidal, thus giving it a large quadrupole moment. This effect is collective in nature in that a great many nucleons participate in the overall nuclear distortion.

In the case of nuclei with such large distortions it might be expected that excited states of a rotational character should be observed (Bohr, A., 1952). These states would be similar to those which arise in the case of diatomic molecules (see p. 256) and are states of motion in which the nucleus rotates as a whole about an axis perpendicular to its axis of symmetry. The energies of such states for even-even nuclei would be

$$E_j = \frac{\hbar^2}{2I} j(j + 1)$$

where $j(=0, 2, 4 \ldots)$ is the angular momentum quantum number and I the moment of inertia. It is a remarkable success of such a simple theory that energy levels agreeing very well with this formula are observed just where they are to be expected (the regions $150 < A < 185$, $A > 225$).

An interesting feature of rotational energy levels of this kind is that the moment of inertia I deduced from the level spacing is roughly equal to one-half the moment of inertia that would be obtained if the nucleus rotated rigidly. This means that the nucleus behaves in some sense like a drop of liquid (see p. 346) and that the rotation can be pictured as one in which the central region of the nucleus remains stationary and the external region moves round like a wave.

Collective motion of nuclear matter can also be obtained in regions where Q is not so large in the form of harmonic vibrations about the equilibrium shape. This sort of motion is characterized by a series of equally spaced energy levels (p. 138).

$$E_n = (n + \tfrac{1}{2})h\nu$$

where ν is the natural period of oscillation of the nucleus. Again it is satisfying that energy levels of this type have been observed for a number of nuclei. In fact, there appears to be an abrupt transition at $N = 88$ in that for nuclei with $N > 88$ the characteristic level spacing for rotational levels is observed, whilst for $N < 88$ the excited states are vibrational in nature.

In all, the shell model and the collective model give an excellent semi-quantitative account of most nuclear properties and at the present time are the most useful ways of describing the nucleus. However, there have recently been a number of developments in techniques for the direct mathematical treatment of the formidable problem presented by the many-body nature of the nucleus (Brueckner, 1954; Bethe, 1956). Such calculations have as one of their main objectives the prediction of the various parameters that we have already introduced empirically for the models such as the nuclear size, density distribution, binding energy, and potential well depth. The situation at present (1969) is that it now seems possible to obtain a value for the binding energy per particle of a nucleon in nuclear matter in fairly good agreement with that measured experimentally.

7. β-Decay and K-Capture.

We have already outlined (Chap. III, § 4, p. 67) the general features of β-decay and their theoretical explanation.

The main facts of observation are a well-defined (and by nuclear standards long) lifetime combined with a continuous range of the β-spectrum (Chadwick, 1914), which has a rather sharp upper energy limit as shown in fig. 7 for the nucleus H^3.

The theory suggested by Pauli and developed by Fermi (1934) rests on the assumption of an interaction between nucleons and electrons, as a consequence of which nucleons may make transitions from a proton state to a neutron state and vice versa, and thereby emit an electron, provided that this process is energetically possible. As there is plenty of evidence that the residual nucleus, as well as the initial nucleus, is in a definite state with a definite energy and a definite angular momentum, the simultaneous emission of a neutrino of spin $\tfrac{1}{2}\hbar$ must be assumed. The upper limit of the β-spectrum then corresponds to the case where all the energy of the nuclear transition is given to the electron and none to the neutrino. It should therefore (divided by c^2, and provided that the rest mass of the neutrino is negligible) be equal to the mass difference of the parent and daughter nuclei. This has been confirmed in many cases where from other considerations the mass differences were known with sufficient accuracy.

In order to calculate the transition probability per unit time we

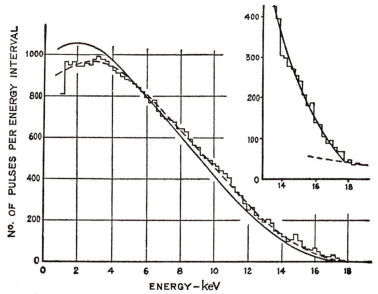

Fig. 7.—The β-spectrum of tritium observed by Curran, Angus and Cockroft (1949). The full line is the theoretical spectrum, corrected for the finite resolution of the proportional counter used for this measurement.

(By permission of the Philosophical Magazine.)

have to use the general formula in Chap. V, § 7, p. 151 (see also Appendix XXVII, p. 435).

$$P_{if} = \frac{2\pi}{\hbar} \mid H'_{if} \mid^2 \rho_f(W).$$

Here i denotes the initial, f the final state of a system composed of two parts : part (1) has a discrete spectrum, part (2) a continuous one of density ρ ; W means the energy difference of (1) in the states i and f. In the present case the system (1) is the nucleus ; i is the state (A, Z), f the state $(A, Z \pm 1) + \beta^\mp + \nu$ (ν stands for the neutrino), and W/c^2 is the mass difference between parent and daughter nucleus.

Now the conservation of energy demands

$$W = E_e + E_\nu.$$

Hence there is still an infinity of transitions for a given W, and we have to replace $\rho_f(W)$ by a differential $\rho_f(W, E_e)\, dE_e$. We have therefore to use the slightly modified formula

$$dP_{if} = \frac{2\pi}{h} \mid H'_{if} \mid^2 \rho_f(W, E_e)\, dE_e.$$

Some aspects of the theory can be greatly simplified if we remember that the de Broglie wave-length of the electron, as well as that of the neutrino (it is assumed that the neutrino is at least not heavier than the electron), is very large in comparison with the nuclear radius. This implies that H'_{if} should not depend strongly on the distribution of energy between the light particles, since in nuclear dimensions the wave functions of the electron and neutrino will be nearly constant and independent of the energy. We therefore obtain an adequate picture by putting

$$H'_{if} = gM_{if}/V,$$

where g is the Fermi constant describing the strength of the coupling between the nucleons and the light particles, M_{if} is a quantity characteristic for the transition $i \to f$ of the nuclear system, and V is a volume in which the functions of electron and neutrino are conveniently normalized. The factor $1/V$ then represents the product of the values of the normalized wave functions of neutrino and electron inside the nucleus.

After these simplifications have been made we expect that the form of the β-spectrum should derive from the level density $\rho_f(W, E_e)$ alone. As we have already seen when discussing the Thomas-Fermi model of the atom, there is one quantum state for each cell of size $dp\, dq = h$ in the pq-plane for each pair of conjugate variables. Hence, for a particle confined in a volume V, the number of states in $dp_x dp_y dp_z$ is $V\, dp_x dp_y dp_z/h^3$, or, if we introduce polar co-ordinates in momentum space, $Vp^2\, dp\, d\Omega/h^3$, where $d\Omega$ is the element of the solid angle (Chap. VII, § 4, p. 223). If we assume that there is no correlation between the directions of emission of the two particles, we have, for the number of states in which the electron has a momentum in dp_e and the neutrino has a momentum in dp_ν, the expression

$$V^2 \frac{(4\pi)^2}{h^6} p_e{}^2\, dp_e p_\nu{}^2\, dp_\nu.$$

If here p_e and p_ν are expressed in terms of E and W, we obtain $\rho_f(W, E_e)\, dW\, dE_e$. Now

$$c^2 p_e{}^2 = E_e{}^2 - m_e{}^2 c^4,\quad c^2 p_\nu{}^2 = E_\nu{}^2 - m_\nu{}^2 c^4,$$

hence

$$p_e{}^2 dp_e = \frac{1}{c^3} E_e \sqrt{E_e{}^2 - m_e{}^2 c^4}\, dE_e,\quad p_\nu{}^2\, dp_\nu = \frac{1}{c^3} E_\nu \sqrt{E_\nu{}^2 - m_\nu{}^2 c^4}\, dE_\nu.$$

Replacing E_ν by $W - E_e$, hence dE_ν by dW, we obtain

$$\rho_f(W, E_e) = \left(\frac{4\pi V}{c^3 h^3}\right)^2 E_e(W - E_e) \sqrt{E_e{}^2 - m_e{}^2 c^4}\,\sqrt{(W - E_e)^2 - m_\nu{}^2 c^4}.$$

It is convenient to introduce the dimensionless quantities

$$\epsilon = E_e/m_e c^2,\quad w = W/m_e c^2,\quad \mu = m_\nu/m_e,$$

and to define a function $P(w, \epsilon)$ by

$$dP_{if} = \frac{2\pi}{\hbar} \mid H'_{if} \mid^2 \rho_f\, (W, E_e) dE_e = P(w,\, \epsilon)\, d\epsilon.$$

Then we obtain

$$P(w,\, \epsilon) = C^2 \epsilon(w - \epsilon) \sqrt{\epsilon^2 - 1}\,\sqrt{(w - \epsilon)^2 - \mu^2},$$

where

$$C^2 = \frac{m_e c^2}{h}\left(\frac{8\pi^2 g m_e{}^2 c}{h^3}\right)^2 \mid M_{if} \mid^2.$$

$P(w,\, \epsilon)$ represents the number density of the β-spectrum in the energy scale where the self-energy of the electron is taken as unity. P vanishes for $\epsilon = 0, \pm 1$ and again for $\epsilon = w$ and $\epsilon = w \pm \mu$; only the interval $1 < \epsilon < \epsilon_m$ has a physical significance, $\epsilon_m = w - \mu$ representing the end of the spectrum. The term $-\mu$ corresponds obviously to the energy necessary to create a neutrino. Information about this energy $m_\nu c^2$ can be obtained from the experimental curve (see fig. 7, p. 335) near its upper end. If we introduce the distance from the end, $x = \epsilon_m - \epsilon$, we have

$$P = C^2(\epsilon_m - x)(\mu + x) \sqrt{(\epsilon_m - x)^2 - 1}\,\sqrt{x(2\mu + x)}.$$

This shows that, for finite μ, P behaves near $x = 0$ like $\sqrt{x}$, i.e. the density curve should have a vertical tangent. Of course experiments at the high-energy end of the density curve are difficult since the intensity is so low, however it is possible from investigations in that region to set an upper limit for the mass of the neutrino. Measurements by Curran (1948) and Langer and Moffat (1952) give an upper limit of

1/2,000th of the electron mass. Now the end of the spectrum is $\epsilon_m = w$ and P has the form

$$P(w, \epsilon) = C^2 \epsilon (w - \epsilon)^2 \sqrt{\epsilon^2 - 1},$$

which agrees with the experimental results.

If instead of $P(\epsilon, w)$ the quantity

$$F(\epsilon, w) = \sqrt{\left(\frac{P(\epsilon, w)}{\epsilon \sqrt{\epsilon^2 - 1}}\right)}$$

is used, we have

$$F(\epsilon, w) \propto (w - \epsilon).$$

Hence F plotted against ϵ (so-called Kurie plot) should give a straight line which cuts the ϵ-axis at w. Roughly speaking experimental measurements are essentially in agreement with the prediction of theory. However, exact agreement is not to be expected, since no account has been taken of the Coloumb interaction between the outgoing electron or positron and the nucleus. A wave-mechanical treatment of this interaction bears out simple classical considerations and shows that it tends to favour the emission of slow negative electrons and to suppress the emission of slow positrons. When it is taken into account very good agreement between theory and experiment is obtained.

The total decay probability—the reciprocal of the mean lifetime t —is obtained by integrating $P(\epsilon, w)$ over ϵ from 1 to w. It has the form

$$\frac{1}{t} = \Sigma \, P_{if} = \int_1^w P(\epsilon, w) \, d\epsilon = C^2 f(w),$$

where

$$f(w) = \int_1^w \epsilon (w - \epsilon)^2 \sqrt{\epsilon^2 - 1} \, d\epsilon.$$

This function varies rapidly with w—for $w \gg 1$ as $w^5/30$. In the present rough approximation $tf(w)$ should be a constant (equal to C^{-2}) provided that the matrix element M_{if} is the same for all nuclei. M_{if} depends on the characteristics of the nuclear transitions. In particular there are nuclei for which we can assume that the orbit of the β-emitting nucleon is almost unchanged in the process of β-decay (mirror nuclei, see § 1, p. 308), and for these M should be essentially the same and of the order unity. Some such cases are listed in the following table.

TABLE VII

Process	t(sec.)	w(MeV)	ft
$n \to p + e^- + \nu^0$	702	1·294	1180
$H^3 \to He^3 + e^- + \nu^0$	$3·92 \times 10^9$	0·019	1137
$O^{15} \to N^{15} + e^+ + \nu^0$	124	2·855	4400
$F^{17} \to O^{17} + e^+ + \nu^0$	66	2·770	2330
$Ca^{39} \to K^{39} + e^+ + \nu^0$	0·95	6·6	4150

Although t is seen to vary over an enormous range, ft is remarkably constant. The values given in the table are corrected for the electrostatic forces between nucleus and electron. It is to be noted that the neutron lifetime fits well with the lifetimes of the other β-emitters.

A more detailed discussion of the matrix element M_{if} based on the relativistic transformation properties of the lepton wave functions leads to selection rules similar to those holding for the emission of light by atomic electrons. M_{if} will depend on the change of angular momentum of the nuclear system in such a way that, in general, large changes of angular momentum are less likely. They are, however, not entirely forbidden, if we remember that the wave functions of the leptons are only approximately constant inside the nucleus. Expanding these functions into a power series in kr where $k = (p_e + p_\nu)/\hbar$ we obtain a series of new matrix elements $M_{if} = M_{if}^{(0)}, M_{if}^{(1)}, M_{if}^{(2)}, \ldots,$ where $M_{if}^{(n)}/M_{if}^{(n-1)}$ is of the order kR. R is the nuclear radius. Since for all known β-emitters kR is small (of the order $\frac{1}{10}$ to $\frac{1}{100}$), the " forbidden " transition (i.e. the transition for which $M_{if}^{(0)} = 0$) will be less likely than the allowed transition by a factor $k^2 R^2$. The forbidden β-spectra will in general differ from the allowed ones, since the factor $k^2 R^2$ depends on the distribution of energy between electron and neutrino.

The selection rules which arise in β-decay can be understood in the following simple physical terms. In the first place transitions can be divided into two classes: those in which the electron and neutrino are emitted in a singlet state with their spins anti-parallel (Fermi transitions) and those in which they are emitted in a triplet state with their spins parallel (Gamow–Teller transitions). An " allowed " transition is one which is governed by the matrix element $M_{if}^{(0)}$ and corresponds to the emission of the electron-neutrino system with zero orbital angular momentum. In the case of a Fermi transition then, zero angular momentum is carried away, since the electron and neutrino are emitted in a singlet state, whilst in a Gamow–Teller allowed transition, one unit of angular momentum is carried away, since the electron and neutrino are in a triplet state. So the selection rule for an allowed transition in β-decay is that the nuclear spin can change by 1 unit at most. In addition there is no change in the parity of the nucleus. For a first forbidden transition, we deal with the matrix element $M_{ij}^{(1)}$ and in this case one unit of orbit angular momentum is carried away by the electron–neutrino system. Thus for a first forbidden transition, adding together the orbital angular momentum and the spin carried away by the electron neutrino system, we can see that the maximum change in the nuclear spin is two units. In this case there is a change of parity. Similarly for higher forbidden transitions.

The detailed form of the β-decay interaction is still subject to much discussion, but is gradually being understood. At the present time (1969) it seems fairly clear that some parts of the interaction are completely analogous to the interactions responsible for the electromagnetic properties of nucleons. In other words there is a correspondence between the emission of electrons and neutrinos by a nucleus and the emission of γ-rays. This correspondence has been referred to as the " conserved vector current theory." This in fact is the part of the interaction responsible for Fermi transitions. The Gamow–Teller part of the interaction is less well understood, but even there a rather elegant interaction theory seems to be emerging.

From $C^{-2} = ft$ the constant g can be determined. It turns out to be extremely small, $g \approx 10^{-49}$ erg cm.^{3} A remarkable numerical coincidence is to be noticed at this point, namely, that if the decay of the μ-meson (Table III)

$$\mu^{\pm} \rightarrow e^{\pm} + \nu^0 + \nu^0$$

is described by an interaction of the same form as that responsible for β-decay, then the relevant coupling constant also has the value $g \approx 10^{-49}$ erg cm.^{3} Further, in the process of μ-capture, when a μ-meson is captured by a proton

$$\mu^- + p \rightarrow n + \nu^0$$

the probability of the process is again consistent with a coupling constant $g \approx 10^{-49}$ erg cm.^{3} This equality of coupling constants suggests that the interaction responsible for β-decay, μ-decay, and μ-capture may be a universal interaction between the various combinations of fermions involved. One difficulty which comes in the way of the formulation of a universal interaction theory for leptons and nucleons, is that it might be expected that the effect of strong interactions would destroy the equality between the different coupling constants. Now in the case of electromagnetism we have a similar situation where the interaction of photons with all charged particles is essentially the same. Putting it in another way, the charge of the proton and the positron are identical with one another, even though the proton experiences strong nuclear interactions while positrons do not. Now it is well known that this equality of charge follows from the fact that the electromagnetic current is conserved, and in a similar way it is possible to formulate β-decay so that the effects of strong interactions do not show up, at least in allowed Fermi transitions. Hence the " conserved vector current theory " just referred to. It might also be expected that the universality could be extended to include hyperons. Decay of hyperons into a nucleon and electron and a neutrino has indeed been observed in many cases, but the coupling constant here seems to be smaller than 10^{-49} erg cm.[3] by a factor of about 3. This means that if there is a universality, it is not of a very simple kind. In fact recent work dealing with the classification of elementary particles has led to a more extended version of universality which does indeed lead to a difference in the coupling constants for hyperons as compared with nucleons.

One other outstanding feature of these weak four-fermion processes is that parity (p. 144) is not conserved. That parity non-conserving processes might occur in nature was first suggested by Lee and Yang (1957) who were seeking an explanation for the different decay modes of the K-mesons (Table III, p. 53). They suggested that all weak decay processes may be processes in which parity is not conserved. This would manifest itself in β-decay, for example, in electrons being emitted with their spins pointing predominantly along their direction of motion and in a forward-backward asymmetry when the electrons were emitted from polarized nuclei (i.e. nuclei whose spins are preferentially in one direction). Experiment soon showed that this was indeed the case both for β-decay and μ-decay and in various other weak decay processes. So far, however, there has been no indication of parity non-conservation in strong interactions such as those responsible for nuclear forces or in the weaker electromagnetic interactions, apart from effects which can be attributed to virtual contributions of weak interactions (Boehm, Abov, 1964).

The theory outlined so far also gives an explanation for the phenomenon of K-capture discussed in Chap. III, § 4, p. 68. We have to remember that with Dirac's theory of the positron the emission of a positron should be equivalent to the absorption of an electron. Thus a nucleus which is energetically capable of emitting a positron should also be capable of absorbing an electron. Now every nucleus is surrounded by atomic electrons, and the wave functions of the K-electrons penetrate most deeply into the nucleus. It is therefore most likely that the absorbed electron should be a K-electron. To determine the lifetime of this process of K-capture,

$$(A, Z) + \beta_K \to (A, Z - 1) + \nu^0,$$

in which the atom $(A, Z - 1)$ is left with a hole in the K-shell, we have to apply the original formula for P_{ij} (see p. 335), since the continuous spectrum in the final state corresponds only to one particle emitted. We remember that the factor $V^{-1/2}$ in the expression for H'_{ij} was due to the value of the electronic wave function inside the nucleus. For a K-electron this value is $Z^{3/2}a_1^{-3/2}\pi^{-1/2}$ (Appendix XVIII, p. 403), where $a_1 = h^2/4\pi^2me^2$ is the radius of the first Bohr orbit. Hence in the case of K-capture,

$$H'_{ij} = gZ^{3/2}a_1^{-3/2}\pi^{-1/2}V^{-1/2}M_{ij}.$$

$\rho_f(W)$ now represents the density of states for the one free particle of the final state—the neutrino. Hence

$$\rho_f(W) = \frac{4\pi V}{h^3}p_\nu^2\frac{dp_\nu}{dE_\nu} = V\frac{4\pi m^2c}{h^3}\epsilon^2,$$

where $\epsilon = E_\nu/mc^2$. The energy of the neutrino follows from the conservation of energy. If w (in units of mc^2) is the change of energy of the nuclear system, and K (in the same units ranging from 0 to about 0·2) the ionization energy of the K-shell, then $\epsilon = w + 1 - K$. Inserting this into the expression for the transition probability, we obtain

$$P_{ij}^{(K)} = \left(\frac{C}{2\pi}\right)^2 \left(\frac{2\pi Z}{137}\right)^3 (w - 1 + K)^2$$

by remembering that $hc/2\pi e^2 = 137$. K-capture should therefore compete with β-decay, and the ratio of the probabilities should be

$$P_{ij}^{(\beta+)}/P_{ij}^{(K)} = 4\pi^2\left(\frac{137}{2\pi Z}\right)^3 \frac{f(w)}{(w - 1 + K)^2}.$$

Since, for large values of w, $f \to w^5/30$, β^+-decay should be the dominating process for sufficiently high energies. There are now very many experimental examples of the competition between β^+-decay and K-capture and they are all essentially in agreement with theory.

In many cases β^+-decay may be energetically impossible, because the change in nuclear energy does not suffice to make up for the rest energy of the positron: $w < 1$. In this case K-capture will still be possible provided that $K - 1 < w$. It follows that in the energy range

$$K - 1 < w < 1$$

K-capture is the only form of β-activity which is energetically possible. The check of the theory in this range of energy is difficult, since the energy of the neutrino is unknown in the absence of competing processes. However the general trends predictable from theory, as, for example, the rapid decrease of lifetime with increasing atomic number, are well established.

8. Nuclear Electromagnetic Interactions.

In §§ 5 and 6 of this chapter some discussion has been given of the magnetic dipole and electric quadrupole moments of nuclei. For these two cases the concern is with the interaction of a nucleus with a static magnetic or electric field. In this section we wish to make some remarks about the emission and absorption of electromagnetic radiation by nuclei. Now in § 8 of Chap. V discussion has been given of the emission and absorption of radiation by an *atomic* system. Here we are dealing with radiation which is usually in the visible or ultra-violet regions, and in which the wavelength λ is 100 or even a 1,000 times larger than the size of an atom. For this reason it is generally possible to restrict consideration to electric dipole radiation (see Appendix XXVIII). In the case of nuclear radiation and the emission of γ-rays, the energies we have to deal with are in general of the order of several MeV, and here the wavelengths are much shorter than in the atomic case. However, comparing them with nuclear dimensions, they are again at least an order of magnitude larger. Even so it turns out in the field of nuclear physics that only in special cases is the electric dipole approximation sufficient to account for the situation, and in general it is necessary to consider higher multipoles. The way in which these different multipoles appear has already been illustrated in § 2 of this chapter, where an expansion is made of the electrostatic potential in terms of the electric dipole moment, the electrical quadrupole moment and so forth, of the nucleus. In the case of electromagnetic radiation where we have to deal with oscillating, rather than static, fields, a similar sort of expansion can be made, both for the electric fields and for the magnetic fields present. In the case of the magnetic field, we have to do with an expansion in terms of a magnetic dipole moment, magnetic quadrupole moment, magnetic octupole moment, etc. There are corresponding selection rules associated with the emission or absorption of radiation by these various electric and magnetic multipoles. For electric dipole radiation we have already seen in Chap. V, § 8, that the selection rules for a transition of a single electron are

$$\Delta l = \pm 1 \quad \text{and} \quad \Delta m = 0, \pm 1.$$

The implication of these selection rules was that unless they were satisfied then an electric dipole transition cannot take place. In the nuclear case, exactly analogous rules apply, but they now have to be expressed in terms of changes in nuclear spin I and can best be written in the following way:

$$\Delta I = 0, \pm 1, \text{parity changes.}$$

This is to say that in order that an electric dipole transition should take place in a nucleus, the nuclear spin can change by at most one unit and also there must be a change in parity (that is, the parities of the two states involved must be opposite to one another). In the case of an electric quadrupole transition, the selection rules are:

$$\Delta I = 0, \pm 1, \pm 2; \text{ no change in parity.}$$

For magnetic transitions, the selection rules are essentially the same, except that the parity rule is opposite; so for a magnetic dipole transition the selection rules are:

$$\Delta I = 0, \pm 1; \text{ no change in parity.}$$

It is conventional to denote electric multiple transitions by E1, E2, E3, . . . and magnetic multiple transitions by M1, M2, M3, . . . In Table VIII the selection rules for all these different multipole transitions are collected together.

One of the reasons why electric dipole transitions do not play such a crucial part in nuclear physics as they do in atomic physics, is that very frequently low-lying nuclear states all have the same parity, so that transitions between these states cannot have the nature of an electric dipole transition, since the parity selection rule would not be satisfied. Furthermore, the spins of nuclear states frequently differ quite considerably, and so again low multipole transitions are not relevant. Roughly speaking, successive multipole transitions of a given class (i.e. electric or magnetic) are at least an order of magnitude less probable than the preceding one. Also, a magnetic transition is in general at least an order of magnitude less probable than the corresponding electric transition of the same multipolarity. This means that if the most important multipole contribution to a transition is electric, then any competition from a magnetic multipole (which in general will be one order higher) will be negligible. On the other hand if the most important multipole is magnetic, then there may be substantial competition from a competing electric multipole one order higher.

TABLE VIII

SELECTION RULES FOR ELECTROMAGNETIC MULTIPOLE TRANSITIONS

Multipole	ΔI	Parity Change
E1	$0, \pm1$	Yes
E2	$0, \pm1, \pm2$	No
E3	$0, \pm1, \pm2, \pm3$	Yes
M1	$0, \pm1$	No
M2	$0, \pm1, \pm2$	Yes
M3	$0, \pm1, \pm2, \pm3$	No

If the wave functions of the states between which a transition takes place are known, then it is possible to calculate the transition probability and therefore the half-life of the excited state. Of course we never know a nuclear wave function exactly, but by using the various models which have already been discussed, fairly good agreement between experimental measurements of lifetimes and theoretical predictions is obtained. In fig. 8 we see a diagram in which the half-life for the different multipole transitions is plotted as a function of the energy of the emitted γ-ray. These results were obtained by Blatt and Weisskopf (1952) assuming that the transitions could be described in terms of just the change of state of a single particle. They give a good guide to the order of magnitude of transition probabilities but do not give detailed agreement with experiment. It can be seen from fig. 8 that for high-multipolarity transitions there can be extremely long lifetimes. Nuclear states having such long lifetimes are said to be *isomeric* states.

Associated with the emission of electromagnetic radiation is another phenomenon known (for historical reasons) as *internal conversion*. The essential point is that if a nucleus is in an excited state, then it can decay electromagnetically from this state to another, either by emitting radiation in the form of γ-rays or by transferring its energy through the electromagnetic field to one of the surrounding atomic electrons which is then emitted. It is important that this process should *not* be regarded as one in which the nucleus emits a γ-ray photon which then knocks out an electron. Thus the process of internal conversion does not compete with γ-ray emission, it is something additional. This viewpoint was demonstrated conclusively by Goldhaber and Wilson (1957), who showed that the rate of decay of Tc[99] could be altered if its surrounding electronic configuration was altered by changing the

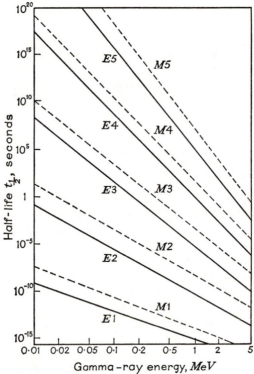

Fig. 8.—Lifetime-energy relations for gamma-radiation according to the single-particle formula of Blatt and Weisskopf (1952).

chemical composition of the Tc compound used. The internal conversion process is responsible for the discontinuous spectrum of β-particles which is frequently superimposed on the continuous spectrum discussed in the foregoing section. It is a process which has been found extremely useful in analysing γ-ray transitions. Calculations of the internal conversion coefficient, defined to be the ratio of the probability of electron emission to the probability of γ-emission in a given nuclear decay process, show that this coefficient is sensitive to the multipolarity of the transition but not to the details of the nuclear wave function. Hence its measurement can lead to clear information about this multipolarity.

In one particular case, namely that of an $I = 0$ to $I = 0$ transition, γ-ray emission is completely forbidden since a γ-ray photon must carry away at least one unit of angular momentum. In this case there are two ways in which the nucleus can de-excite. One is for it to give up its energy to an atomic K-electron which can penetrate into the nucleus, and the other is that an electron-positron pair can be created (this process was discussed in Chap. VI, p. 194). However these processes can only occur if the two $I = 0$ states have the same parity. If they have different parity, then highly complicated processes have to take place involving for example the emission of 2 γ-ray quanta.

9. The Drop Model, Nuclear Reactions and Fission.

In our preliminary survey of nuclear phenomena (Chap. III, § 6, p. 70) we have seen that ordinarily stable nuclei can be transformed by bombarding them with different kinds of particles (protons, neutrons, deuterons ; neutrons being particularly effective). The process can be best understood by assuming that the bombarding particle is first absorbed by the nucleus, which in this way becomes unstable and breaks up, emitting one or two of its constituent nucleons, or an α-particle. The remaining nucleus may still be not quite stable, and may after some time emit another particle (artificial radioactivity). How can these processes be understood ?

They may be compared with the ionization of an atom produced by bombarding it with electrons ; here also a particle (electron) is knocked out from the electronic cloud, and a new system, the ion, created. Yet this comparison is rather superficial ; the mechanical, or better quantum-mechanical features are very different in the two cases, for the electronic ionization is an instantaneous elementary process (described by one quantum-mechanical transition probability) —the incoming electron interacts with one of the bound electrons, and both fly away simultaneously. In the case of nuclear transformations, however, the situation is in general quite different, as Niels Bohr (1936) was the first to emphasize. By virtue of the enormously close packing of fast-moving nucleons in the nucleus, an incoming particle can hardly be assumed to react with a definite nucleon, and even if it did so, this nucleon has no chance to preserve the momentum and energy acquired from the projectile, but would share them with its neighbours (if we can speak in these terms), so that they would spread through the whole nucleus. The energy thus accumulated will then later lead to the emission of another nuclear particle.

The question arises whether this qualitative reasoning can be transformed into a quantitative theory and used for the detailed description of nuclear reactions.

We have already seen that the density of nuclear material is almost constant (p. 309) and that the binding energy is roughly proportional to the number of nucleons (p. 323); we have connected this fact with the very short range of nuclear forces. The state of the nucleus is therefore comparable to that of solid or liquid matter, to which the same rules apply. But, as Bohr has pointed out, there is more likeness to a liquid than to a solid (or a very large molecule), as we cannot assume that the particles in a nucleus are arranged in a regular structure, like a crystal lattice. The reason is that their zero-point energy (see p. 138, also pp. 395–6, 471) is too strong in comparison with the potential energy of the forces, and keeps them in a state of disordered motion. We can therefore compare the nucleus with a drop of a liquid, provided we bear in mind that in the liquid the heat motion is responsible for the disorder, whereas the nucleus is in its normal (lowest-energy) state and must be considered to be " cold ". There is in fact one ordinary substance which behaves in the same way, namely helium, which remains liquid down to zero-temperature. It is probable that the reasons are also the same: the weakness of the binding forces acting on particles of very low weight produces a strong zero-point energy which prohibits solidification. Bohr's picture of a nuclear reaction then is that a bombarding particle is captured by the nucleus to form what he called a compound nucleus.

This idea of the liquid-drop structure of the nucleus leads at once to an important consequence if we remember that the drop has a surface and is electrically charged. Both these facts influence the total energy, i.e. the mass. Let us apply well-known elementary laws to estimate the mean binding energy ϵ, i.e. the work necessary to remove one particle from the nucleus.

If a is the mean binding energy per particle in the interior due to non-electric forces (i.e. nuclear forces), the total binding energy would be aA, where A is the mass number (number of nucleons), provided the effect of the surface could be neglected. In fact there is a surface energy which has the opposite sign, since the surface particles have fewer neighbours than those in the interior, and need therefore less work to be removed. Since each nucleon occupies about the same space (constant density), the surface is roughly proportional to $A^{2/3}$

and the surface energy can therefore be written as $-bA^{2/3}$. Finally, the Coulomb energy, which, being a repulsion, also diminishes the interaction, can be written in the form $-cZ^2A^{-1/3}$, where Z is the number of protons (atomic number), and the mean distance is taken proportional to $A^{1/3}$.

By adding the three contributions we obtain, for the binding energy per particle, the sum divided by A,

$$\epsilon = a - bA^{-1/3} - cZ^2A^{-4/3};$$

or, if we use the rough approximation $Z = \tfrac{1}{2}A$,

$$\epsilon = a - bA^{-1/3} - \tfrac{1}{4}cA^{2/3}.$$

Thus we should expect a small variation of the mean binding energy through the periodic system, and, as the two variable terms are negative, one decreasing with A, the other increasing, there should be a maximum of ϵ somewhere in the middle. Now this is the case. To compare the theoretical formula with experiment, we introduce instead of the mass M (in atomic units, oxygen mass number $= 16$), the so-called packing fraction

$$f = \frac{M - A}{A},$$

which is easily seen to be a measure of the mean binding energy. For if m_n and m_p are the masses of neutron and proton, we have for the total mass

$$M = m_n(A - Z) + m_pZ - \epsilon A,$$

and, with the approximation $Z = \tfrac{1}{2}A$,

$$M = A[\tfrac{1}{2}(m_n + m_p) - \epsilon];$$

hence

$$f = \frac{M}{A} - 1 = [\tfrac{1}{2}(m_n + m_p) - \epsilon] - 1,$$

where the values $m_n = 1\cdot00898$, $m_p = 1\cdot00813$ should be used. f should therefore show the same dependence on A as $-\epsilon$.

The plot of fig. 9 represents f as a function of A (from measurements by Aston, Dempster, Mattauch, a.o.). In fact, f is not a function of A alone, but also of Z; there are nuclei with the same A and different Z (isobars), and vice versa (isotopes). The curve in the figure is drawn from the lowest f-value in a group of isobars; it has a minimum at about $A = 50$.

The theoretical curve can, by choosing the constants a, b, c, be fairly well fitted to the experimental points, which is all we can expect. A more refined theory (v. Weizsäcker, 1935) can be developed which gives f as a function of A and Z ; it is able to account, to some degree, for the binding energies of all stable nuclei, except the very lightest, which have to be treated by more individual considerations (as the theory of the deuteron, § 3, p. 317).

The existence of the minimum in the $-\epsilon(A)$ curve signifies that in principle all nuclei are more or less unstable, except those at the minimum (which, by the way, is near the extremely abundant iron isotope $_{26}Fe^{56}$), the nuclei below the minimum tending to fuse, those

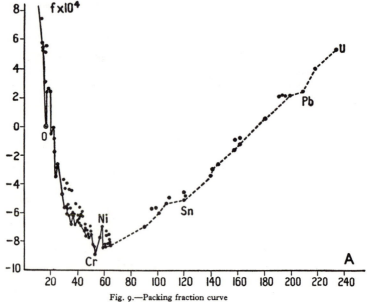

Fig. 9.—Packing fraction curve

From " Theory of the Atomic Nucleus and Nuclear Energy Sources " by Gamow and Critchfield,
by permission of the publishers, The Clarendon Press, Oxford

above to disintegrate. The existence of matter in its variety of forms seems therefore to be a rather precarious affair. In fact the reaction velocities for these transformations are so small that they can be completely neglected for all practical purposes. There are only two exceptions known. The first is the fusion of nucleons to form a helium nucleus, a process which is assumed to go on in the interior of stars. The second case is the fission of some very heavy nuclei (uranium, plutonium) induced by neutrons. Both of these processes will be discussed presently.

We now return to the problem of nuclear transformations by the impact of other particles, especially of neutrons which are not repelled by a barrier of Coulomb forces (as are protons and deuterons), and can therefore easily penetrate into the nucleus.

If this happens, the energy of the arriving neutron will in general be quickly dispersed through the whole nucleus. We can say that the nucleus is " heated up ". It will take a considerable time (large compared with that of a simple collision, which is given by the ratio: diameter of nucleus divided by the velocity of the particle, which is of the order 10^{-21} or 10^{-22} sec.) before a sufficient fraction of the excess energy is concentrated again by a random fluctuation on to one of the particles to enable it to escape. The position is similar to that of a small liquid drop evaporating at low temperature, which is a very slow process, even if the total heat content of the drop is much larger than the energy necessary to liberate a single molecule. In this way we can understand how a nucleus formed by the impact and absorption of a neutron or proton by a smaller nucleus may live for a considerable time before emitting again, and behaves in many respects like a stable nucleus, having almost sharp energy levels. These states of a nucleus are called *compound states* and it is possible to set up a theory of nuclear reactions in terms of them. This was first done by Breit and Wigner (1936) who considered the energy dependence of a nuclear reaction in a region where only one of these compound states was important.

In discussing nuclear reactions it is convenient to introduce the concept of nuclear cross-section σ, the idea being roughly speaking that the larger the effective nuclear cross-sectional area, the greater the probability of the bombarding particle interacting with the nucleus. In other words, the nuclear cross-section is directly proportional to the probability of a nuclear reaction taking place. Consider as an example the following possible nuclear reactions:

$$\mathrm{Na^{23} + a \rightarrow Al^{27*} \rightarrow} \quad \begin{array}{l} \mathrm{Mg^{26} + p} \\ \mathrm{Al^{26} + n} \\ \mathrm{Al^{27} + \gamma} \\ \mathrm{Na^{23} + a} \end{array}$$

In the above processes the asterisk on the Al^{27} implies that this is a compound state. On the extreme right are various possible decay modes of this compound state. Here it should be mentioned that the probability of the different decay modes is taken to be quite independent of the way in which the state is formed. Roughly speaking, we could say that the lifetime of the compound state is so long that it "forgets" how it was formed. Reverting now to the theoretical description of such a process, let us denote the configuration on the left-hand side, namely $Na^{23} + a$, by a and one of the configurations on the right hand side by i. Then what Breit and Wigner showed was that the cross-section $\sigma(a, i)$ for the reaction starting with the configuration a and ending with the configuration i is given by the following formula:

$$\sigma(a, i) = \frac{\pi}{k^2} \frac{\Gamma_a \Gamma_i}{(E - E_c)^2 + \frac{1}{4}\Gamma^2}$$

Here k is the wave number of the bombarding particle and E the corresponding energy; E_c is the energy of the compound state; Γ is the level width of the compound state (level width is related to the lifetime of the state by the uncertainty relation $\Gamma\tau = \hbar$); Γ_a and Γ_i are "partial level widths" and represent the probability that if the compound state is formed, it will decay into the configuration a or i respectively. This formula is the famous Breit-Wigner formula and has been found to give a very good account of nuclear reactions when they are dominated by a single compound level.

In many nuclear reactions, however, the single compound nucleus description does not form the most natural way of describing the reaction and frequently we have to deal with what have become known as "direct" reactions. Such reactions are much more akin to the situation discussed a little earlier in connection with atomic ionization by bombardment with electrons. The situation then is that the bombarding particle interacts strongly with only one or two nucleons in the target nucleus, and the reaction does not proceed at all through a single compound nuclear state. During the last fifteen years or so, reactions of this kind have been investigated very thoroughly indeed and have been found particularly useful in giving information about individual nucleon states in the nucleus.

We now turn to another phenomenon which was first observed when attempts were being made to develop elements of atomic number higher than 92 by bombarding thorium and uranium with neutrons. The search for these trans-uranic elements led to the discovery of a strange phenomenon, the breaking up, or *fission*, of heavy nuclei into two parts of comparable size (Hahn and Strassmann, 1938), for instance, $_{92}U \rightarrow _{56}Ba + _{36}Kr$, with liberation of enormous energy. Meitner and Frisch (1939) suggested a simple explanation on the lines of Bohr's droplet model, which we have already discussed and which shows that all nuclei are in principle unstable, those at the ends of the periodic table like Th and U particularly strongly. If a neutron is captured, the life is reduced to such a degree that the nucleus disintegrates into fragments of about equal size which fly apart with a surplus energy of the order of 200 MeV (as compared with up to 27 MeV in ordinary nuclear reactions). Bohr and Wheeler (1939) followed this up with predictions about the effective cross-section of the nucleus for varying neutron velocities. One of these predictions opens possibilities of inestimable importance for the future not only of physics, but of mankind as a whole.

The process of fission is in fact accompanied by the emission of neutrons (either simultaneously, or later as disintegration products of the unstable fragments). Now these neutrons may be absorbed and produce fission in other neighbouring nuclei, neutrons being again emitted, so that an avalanche develops. If it were possible to produce such a *chain reaction*, the enormously condensed nuclear energy could be utilized for practical purposes, as driving power for engines or as explosive for super-bombs.

The main facts which led to the realization of these ideas in the so-called " atomic bomb " rest on the following considerations. The normal thorium $_{90}$Th and uranium $_{92}$U, and also protoactinium $_{91}$Pa nuclei, behave for fast neutrons in the same way; but for slow neutrons the uranium nucleus has a very much greater effective cross-section (i.e. a greater probability) for fission than thorium. Ordinary U^{238} cannot be responsible for it, and Bohr suggested that it was due to the presence of a small amount (0·7 per cent) of the isotope $_{92}U^{235}$. This prediction was completely confirmed by Niers, Heydenburg, and Lawrence in 1941.

But it is a different thing to separate enough of the rare isotope for an experiment on nuclear cross-sections and to separate enough for producing an explosive, as all known processes for separating chemically identical atoms of nearly equal mass are extremely inefficient and slow (see § 5, p. 41). However, the methods of diffusion and of magnetic separation have been used for producing U^{235} in large quantities.

Another possibility was to build up *trans-uranic elements*. The nucleus $_{92}U^{239}$ formed when a neutron is captured by the ordinary uranium nucleus $_{92}U^{238}$ is likely to be unstable and to disintegrate by successive emissions of electrons into new elements according to the formulæ:

$$_{92}U^{238} + {}_0n^1 \rightarrow {}_{92}U^{239} + \gamma; \quad {}_{92}U^{239} \rightarrow {}_{93}Np^{239} + \beta^-;$$
$$_{93}Np^{239} \rightarrow {}_{94}Pu^{239} + \beta^-,$$

where Np is the chemical symbol for the element No. 93, *neptunium*; Pu that of the following element No. 94, *plutonium*. It seemed probable that plutonium 239 would react on the capture of a neutron by fission. All these predictions were confirmed by experiment. The neutrons of a certain range of velocity produce plutonium from U^{238}, whereas those slowed down to thermal velocity keep up the chain by fission of U^{235}.

The method used in these experiments as well as in the final process of production of plutonium consists in the building of a " reactor " or " pile ", a regular arrangement (lattice) of lumps of uranium (metal or oxide) separated by a light material (graphite, heavy water, etc.) which serves as " moderator " for reducing the velocity of the neutrons emitted.

To avoid the loss of a high proportion of the neutrons through the surface, dimensions of the pile must be large. In order to keep the chain reaction under control, strips of cadmium are used which can be more or less deeply inserted in the pile; this substance has a high absorbing power for slow neutrons, and allows a delicate adjustment of the power developed. In view of the explosive character of the reaction (as used in the atomic bomb) it may appear strange that such an adjustment of the reaction velocity is possible. It is due to the fact that a small fraction of the neutrons liberated by the fission process do not escape instantaneously but are delayed (about 1 per cent are retarded by periods up to one minute). This is just sufficient to keep control of the chain reaction near the critical stage of the pile, with the help of the absorbing cadmium strips.

The first pile, designed under the direction of Fermi, came into operation in Chicago on 2nd December, 1942. The plutonium thus produced turned out to have the properties expected. It is extracted by chemical methods which are simpler, faster, and more efficient than the methods for isotope separation (see Chap. II, § 5, p. 41).

Both the extraction of the rare uranium isotope 235 and the synthesis of plutonium were industrially developed in the United States during the last period of the war of 1939–45 with the purpose of producing a super-explosive. The characteristic feature of the nuclear explosive is that there must be a sufficient accumulation of fissionable material, above a " critical mass "; for a subcritical mass, the number of neutrons escaping through the surface is too large to allow the start of a chain reaction. The bomb therefore consists of at least two sub-critical masses which have to be brought together in an extremely short time. The actual detonation in this moment of contact is due to stray neutrons which are always present. How this sudden assembly of the parts of the bomb is performed is one of the main technical secrets which envelop the whole field of nuclear explosives. The importance of this new weapon for power politics has had the most fatal influence on the freedom of scientific research and communication of ideas, and the use of the bomb for mass destruction of whole cities and their civil population is a still more regrettable consequence of discoveries intended originally solely for the widening and deepening of knowledge. It will need great efforts to rebuild the international structure of a free science, unfettered by political and military restrictions, as a means of progress and not of destruction.

The pile itself shows many new ways for scientific and technical progress. The most obvious is the production of energy. But this is rather restricted, apart from technical difficulties, by the rareness of the fuel. However, a transformation of ordinary uranium into plutonium 239 (which can be used as fuel) can be performed in the pile itself, a process referred to as " breeding ". This means a multiplication of the raw material available by a factor bigger than 100, and changes the technical and economic position in favour of nuclear-energy production. In this case also the rather more abundant element thorium can be used as fuel.

Other peaceful applications of the pile are the production of isotopes of all the elements in considerable amounts. It has been already mentioned (p. 39) that trans-uranium elements up to Nobelium 102 have been discovered. The isotopes H^3 and He^3 have been produced in sufficient accumulation to study their physical properties. Radioactive isotopes of many elements are extracted and used as " tracers " in chemical and biological investigations. The whole aspect of these sciences has been changed through the new methods. But these problems are outside the scope of this book.

There is another field in nuclear reaction physics which has attracted a lot of attention during the last few years, namely those nuclear reactions which are believed to be responsible for energy production in stars. It has become quite clear that the source of energy is neither chemical nor gravitational, but rather stems from interactions between atomic nuclei. Inside a typical star (e.g. the sun) the internal temperature may be as high as 10^7 °K. At such temperatures, the kinetic energies of thermal motion are of the order 1 keV and are such that nuclear reactions can take place. One of the basic sets of reactions has become known as " hydrogen burning " and corresponds to the combination of four protons to form helium. Now the atomic mass of four hydrogen atoms is 4·13258, whilst the atomic mass of one helium atom is 4·00387. This means that in the " burning " process, a mass of 0·02871 is converted into energy. Taking account of the fact that one gram of stellar matter contains about 10^{23} protons, this would lead to an energy supply of about 10^{18} ergs per gramme. This is extremely high and is consistent with the energy needed. The basic nuclear reactions which are believed to take place in hydrogen burning are as follows:

$$H^1 + H^1 \rightarrow H^2 + \beta^+ + \nu,$$
$$H^1 + H^2 \rightarrow He^3 + \gamma,$$
$$He^3 + He^3 \rightarrow He^4 + 2H^1.$$

This combination of four protons to form helium is known as *fusion*, and the sequence of reactions is known as the *hydrogen cycle*. Another set of reactions known as the *carbon-nitrogen cycle* is also known to contribute to the energy production of many stars. Recent interest in cycles of reactions of this kind has stemmed from the fact that if they could be controlled (controlled thermonuclear fusion) in some form of man-made machine, then we would have a very prolific source of terrestial energy. Unfortunately these reactions will not take place, except at extremely high temperatures, and so far the containment of a plasma in which these reactions are occurring, has proved difficult if not impossible to achieve. The only example of a man-made thermonuclear fusion process is in the form of the hydrogen bomb, where the high temperatures are produced by means of an initiating atomic (i.e. fission) explosion.

10. Conclusion by M. Born.

We have now reached the final stage of our task. In the preceding chapters we have in the main confined our account to the *positive results of research*, and have not dealt in detail with problems which are not yet completely cleared up. The impression might thus be given that physics had attained to a certain degree of finality. That, however, is by no means the case. There still remain many perplexing questions, all strictly connected with one another, and all in the last resort hanging upon the question of the nature and interaction of the ultimate particles. There are different points of view in classifying these particles. One is their permanence: nucleons and electrons form atoms and thus ordinary matter, whereas other particles, as photons and mesons, are short-lived, and seem to serve as transmitters of force between the particles of the first type. But this classification is not rigorous, as electrons can be generated or destroyed, in pairs with positrons, or in other ways. A different classification would be that according to the value of the rest mass. Here photons, neutrinos, and electrons form one group, having vanishing or very small mass; the mesons form another group, the nucleons a third. But why just these masses and no others? And is there a connexion of the value of the mass to the stability of the particle? The latter is determined by the mutual interaction, hence there will be a relation of the mass of one particle to all possible interactions with other particles. This shows clearly that for a deeper understanding it is impossible to deal with different kinds of particles separately. A future theory must be unitary in a very wide sense; it must connect all the present theories of particles and their interactions in a single rational system. It has indeed been found possible to make a start in assigning the observed particles and resonances to " multiplets " associated with symmetry groups (in particular those of the group SU3) but the underlying dynamics is still not understood. Much effort is now being directed to the construction of self-consistent " bootstrap " theories which could yield the masses and characteristic symmetries of these particles and resonances. There seem, however, to be no reasons to doubt that the future theory will satisfy the principle of relativity and appear as a generalization of the present form of quantum mechanics, with its indeterministic and statistical features.

We look to the physics of the future for the solution of the enigma of inorganic matter. But the principal results of modern physics reach far beyond the domain in which they are won. As Niels Bohr first pointed out, the new views with regard to causality and determinism, which have arisen as a result of the quantum theory, are also of great significance for the biological sciences and for psychology. If even in inanimate nature the physicist comes up against absolute limits, at which strict causal connexion ceases and must be replaced by statistics, we should be prepared, in the realm of living things, and emphatically so in the processes connected with consciousness and will, to meet insurmountable barriers, where mechanistic explanation, the goal of the older natural philosophy, becomes entirely meaningless. But this has the effect of completely changing the philosophical import of research. Physicists of to-day have learned that not every question about the motion of an electron or a light quantum can be answered, but only those questions which are compatible with Heisenberg's principle of uncertainty. There is a hint here for the biologist and the psychologist, that they should search for natural limits to causal explanation in their domains also, and mark out those limits with the same precision as the quantum theory is capable of doing by means of Planck's constant h.

This is a programme of modesty, but at the same time one of confident hope. For what lies within the limits is knowable, and will become known; it is the world of experience, wide, rich enough in changing hues and patterns to allure us to explore it in all directions. What lies beyond, the dry tracts of metaphysics, we willingly leave to speculative philosophy.

APPENDICES

I. Evaluation of Some Integrals Connected with the Kinetic Theory of Gases (p. 15).

Integrals of the type

$$I_v = \int_0^\infty v^r e^{-\lambda v^2} dv,$$

where $\lambda = m\beta/2 = m/2kT$, frequently occur in the kinetic theory of gases. The general form of the integral may at once be obtained from the particular cases I_0 and I_1 by differentiation with respect to λ. Thus, for example,

$$I_2 = -\frac{dI_0}{d\lambda}, \quad I_3 = -\frac{dI_1}{d\lambda},$$

$$I_4 = +\frac{d^2I_0}{d\lambda^2}, \quad I_5 = +\frac{d^2I_1}{d\lambda^2}.$$

I_1, the second of the two fundamental integrals, can be evaluated by elementary methods:

$$I_1 = \int_0^\infty v e^{-\lambda v^2} dv = \frac{1}{2\lambda}.$$

I_0 is Gauss's well-known probability integral

$$I_0 = \int_0^\infty e^{-\lambda v^2} dv = \frac{1}{2}\sqrt{\frac{\pi}{\lambda}}.$$

The values of the succeeding integrals obtained, as stated above, by differentiation are as follows:

$$I_2 = \int_0^\infty v^2 e^{-\lambda v^2} dv = \frac{1}{4}\sqrt{\frac{\pi}{\lambda^3}},$$

$$I_3 = \int_0^\infty v^3 e^{-\lambda v^2} dv = \frac{1}{2\lambda^2},$$

$$I_4 = \int_0^\infty v^4 e^{-\lambda v^2} dv = \tfrac{3}{8}\sqrt{\frac{\pi}{\lambda^5}},$$

and so on.

Application to the integrals occurring in the kinetic theory of gases.—
For n, the total number of molecules, we have

$$n = 4\pi A \int_0^\infty v^2 e^{-\frac{1}{2}\beta mv^2} dv = 4\pi A I_2 = A \sqrt{\frac{\pi^3}{\lambda^3}}.$$

Similarly, the total energy E is equal to

$$4\pi A \int_0^\infty \tfrac{1}{2} mv^4 e^{-\frac{1}{2}\beta mv^2} dv = 2\pi A m I_4 = \tfrac{3}{4} m A \sqrt{\frac{\pi^3}{\lambda^5}}.$$

The expressions for A and β given in the text (p. 15) are obtained by
combining these two formulæ.

Integrals of this type also occur in the calculation of mean values.
As according to Maxwell the number of molecules having a velocity
between v and $v + dv$ is

$$n_v \, dv = 4\pi n \left(\frac{m}{2\pi kT}\right)^{\frac{3}{2}} v^2 e^{-\lambda v^2} dv,$$

the "mean velocity" $\bar{v}$ is given by

$$\bar{v} = \frac{\displaystyle\int_0^\infty n_v v \, dv}{\displaystyle\int_0^\infty n_v \, dv} = \frac{I_3}{I_2} = \frac{2}{\sqrt{(\pi\lambda)}}.$$

Again, the "mean square velocity" $\sqrt{(\overline{v^2})}$, which is frequently used
in the kinetic theory of gases, is given by

$$\overline{v^2} = \frac{\displaystyle\int_0^\infty n_v v^2 \, dv}{\displaystyle\int_0^\infty n_v \, dv} = \frac{I_4}{I_2} = \frac{3}{2\lambda}.$$

$\sqrt{(\overline{v^2})}$ is a little larger than $\bar{v}$, their ratio being $\sqrt{(3\pi/8)} = 1{\cdot}085$.
With these quantities we may compare the "most probable velocity"
v_p, which corresponds to the maximum of the Maxwell distribution
curve. It is given by the equation $\dfrac{dn_v}{dv} = 0$, or

$$\frac{d}{dv}(e^{-\lambda v^2} v^2) \equiv 2v e^{-\lambda v^2}(1 - \lambda v^2) = 0;$$

hence

$$v_p = \frac{1}{\sqrt{\lambda}}.$$

In the Maxwell distribution curve given in the text (fig. 5, p. 15), the three values v_p, $\bar{v}$, and $\sqrt{v^2}$ are shown in their order and approximately proper positions.

Multiplying the expression for v_p above and below by the square root of Avogadro's number N_0 and remembering that $N_0 k = R$ and $N_0 m = \mu$, we have

$$v_p = \sqrt{\frac{2kT}{m}} = \sqrt{\frac{2RT}{\mu}};$$

$$\bar{v} = \frac{2v_p}{\sqrt{\pi}} = \sqrt{\frac{8RT}{\pi\mu}};$$

$$\sqrt{(v^2)} = v_p\sqrt{\frac{3}{2}} = \sqrt{\frac{3RT}{\mu}}.$$

II. Heat Conduction, Viscosity, and Diffusion (p. 19).

In this appendix we shall give a combined account of three phenomena by means of which the mean free path in a gas can be determined experimentally. These phenomena are *heat conduction, viscosity,* and *diffusion.* In all these phenomena there is a variation in some physical property of the molecules of the gas from point to point, which, however, tends to disappear as a result of the movements of the molecules.

Heat conduction occurs when external conditions of any kind give rise to a temperature gradient in a gas, i.e. when the molecules of gas at different points of space have different mean kinetic energies. Heat transference takes place owing to molecules from the warmer regions moving into cooler regions and giving up their surplus energy there, while slower-moving molecules move into the warmer regions and diminish the kinetic energy of the faster molecules there.

The circumstances are similar in the case of *viscosity.* This manifests itself in the form of a resistance acting on the faster-moving parts of the gas under consideration. According to the kinetic theory of gases, this resistance is due to molecules from slower-moving regions moving into faster-moving regions; as they then have a smaller mean velocity of flow than their surroundings, they will on the average be accelerated as a result of collision with the surrounding molecules, while the latter will be retarded, i.e. will be subject to a resistance. We shall not go into the experimental methods for determining viscosity, but merely mention the result (due to Maxwell), which perhaps at first sight is surprising: the viscosity, like the thermal conductivity, is within wide limits independent of the pressure of the gas.

The third phenomenon which we shall consider here is *diffusion*. If in a mixture of two gases the concentration of one gas varies from point to point, i.e. if the ratio of the concentrations of the two gases differs at different points of the region occupied by the gaseous mixture (it is assumed that the pressure is everywhere the same, i.e. that the total number of gas molecules is the same), then it is clear that the molecules of one gas will gradually move from the regions where the concentration of this gas is greatest into regions where it is less; and similarly for the molecules of the other gas.

These three phenomena can be discussed mathematically in a very simple way provided we confine ourselves to a qualitative survey. In order to treat them together, we assume that the property A varies from point to point of space, e.g. has a non-vanishing derivative dA/dz in the direction of the z-axis. Then for heat conduction A is to be taken as the mean kinetic energy of a single gas molecule, for viscosity as the mean velocity of translation of the molecule in the direction of flow, and for diffusion as the number of molecules of a particular gas in a cubic centimetre of gaseous mixture. Now this variation of the property A from point to point gives rise to a transference of the property; a definite number of molecules cross unit area normal to the z-axis per second in either direction, and this number is given, at least approximately, by the product $n\bar{v}$, where n is the number of molecules per cubic centimetre and $\bar{v}$ their mean velocity. But the molecules crossing the surface from one side possess the property A to a greater or less extent than those crossing the surface from the other side, so that there is a transference of the property across the surface. The quantity $M(A)$ crossing per second can easily be estimated if we note that any molecule has a mean free path l between two successive collisions with other molecules, i.e. in the interval of time between successive collisions it describes on the average a path whose length is of the order l; for the purposes of this approximation we are not concerned with the exact numerical factor. For $M(A)$ we readily obtain the expression

$$M(A) \sim n\bar{v}\{A(z_0 - l) - A(z_0 + l)\},$$

where z_0 is the co-ordinate of the element of area considered. Expanding, we have (apart from numerical factors)

$$M(A) \sim -n\bar{v}l \left(\frac{dA}{dz}\right)_{z_0}.$$

The amount of the property A transferred is therefore proportional to the " gradient of A ", and also to the number of molecules per cubic centimetre, their mean velocity, and their mean free path. This equation is known as the *transport equation*.

We note that $M(A)$ is independent of the pressure, provided that A itself denotes a property of the gas molecules which is independent of the pressure (Maxwell). For the pressure of a gas is given by $p = nkT$, i.e. at constant temperature depends only on the number of molecules per cubic centimetre. It is true that n appears as a factor in the transport equation; but this factor is compensated by the occurrence in the formula of the mean free path, which is inversely proportional to n and to the cross-section of the molecule. This independence of the pressure accordingly results from the fact that though more molecules take part in the transference of A at the higher pressures, they do not on the average travel so far.

We shall now make particular application of the transport equation to the three phenomena mentioned above. We begin with *heat conduction*. Here A stands for the kinetic energy of a molecule, i.e. $E_{\text{kin}} = \text{const.} + c_v mT$, where $c_v m$ is the specific heat of the gas at constant volume, for a single molecule (c_v is the specific heat per mole). The quantity of heat Q which crosses unit area per second is then given by

$$Q \sim -n\bar{v}lc_v m \frac{dT}{dz}.$$

We see that it is proportional to the temperature gradient; the factor of proportionality $\kappa = n\bar{v}lc_v m$ is called the *thermal conductivity*.

In the case of the *viscosity*, as we saw above, A stands for mu, the mean linear momentum of a molecule resulting from the flow of the gas. Then the momentum transferred per second (per unit area of the surface of contact between faster-moving portions of gas and slower-moving), i.e. the resistance R, is given by

$$R \sim -n\bar{v}lm \frac{du}{dz};$$

the quantity $\eta = n\bar{v}lm$ is called the *coefficient of viscosity*. We see that the quotient $\kappa/\eta c_v$ is a constant of the order of unity. Theoretically this constant must be the same for all molecules of the same structure, i.e. the quotient must have a constant value for all mona-tomic gases, another constant value for all diatomic gases, and so on. Our qualitative discussion of course does not enable us to obtain the exact value of the constant.

As we have already emphasized, the discussion given here is of course only a rough sketch. Improvements and refinements in the theory have been made by Boltzmann, Maxwell, and others, by considering the mechanism of collision and the distribution of velocities in greater detail; these improvements, however, yield no new principle, but merely lead to greater accuracy in the numerical factors. Here, however, we shall not go into the matter further. It remains for us to point out that the above theory is not valid unless the mean free path is small compared with the dimensions of the vessel containing the gas. If this is not the case (at atmospheric pressure l is of the order of 10^{-6} cm., but is equal to about 10 cm. at the pressure in an X-ray tube (10^{-4} mm. of mercury)), the laws which hold are quite different. The molecules then fly practically straight from one wall of the containing vessel to the other without colliding with other molecules. Thus, for example, if there is a difference of temperature between two opposite parts of the wall of the container, they carry heat energy directly from one wall to the other. The quantity of heat transferred is then proportional to the number of molecules; n no longer disappears from the transport equation, as the mean free path no longer enters into it. The laws of heat conduction, viscosity, &c., at low pressures have been especially studied by Knudsen. They are of great practical importance, for instance, in connexion with the working of air pumps (such as Gaede's rotary molecular pump and the diffusion pump).

We shall now briefly consider the problem of *diffusion*. We imagine a mixture of two gases in dynamical equilibrium, i.e. the temperature and the pressure, hence n, the total number of molecules per cubic centimetre, are to be the same throughout. Here the property A is n_1/n, the concentration of one kind of molecule, or n_2/n, the concentration of the other kind of molecule. Then the transport equation gives the number (Z_1) of molecules of the first kind, or the number (Z_2) of molecules of the second kind, that diffuse through unit area in unit time:

$$Z_1 \sim -n\bar{v}l \, \frac{d(n_1/n)}{dz} = -\bar{v}l \, \frac{dn_1}{dz}, \quad Z_2 \sim -\bar{v}l \, \frac{dn_2}{dz}.$$

If the phenomenon is a steady one, $n_1 + n_2$ must be equal to a constant n, i.e. $dn_1/dz = -dn_2/dz$; then the total flow $Z_1 + Z_2$ is zero. The two kinds of molecule have the same diffusion constant $\delta = \bar{v}l$, which in virtue of the factor l is inversely proportional to n, the total number of molecules.

A different kind of diffusion, called thermal diffusion, was theoretically predicted by Enskog (1912) and Chapman (1916), and experimentally found for a mixture of CO_2 and H_2 by Chapman and Dootson (1917). It arises if the initial concentration is uniform and the pressure constant, but the temperature varies from point to point. In this case the average velocity $\bar{v}$ of the molecules which is proportional to $\sqrt{T}$ depends on the position, and the transport equation has to be modified accordingly. A complete and rigorous theory of the phenomenon has been given by Chapman (1939), an elementary treatment by Fürth (1942). The result is the appearance of a term in the diffusion formula proportional to the gradient of temperature. The coefficient of thermal diffusion depends in a complicated way on the masses and the diameters of the molecules involved.

III. Van der Waals' Equation of State (p. 20)

As compared with the equation of state for perfect gases, van der Waals' equation of state for actual gases, given in the text (p. 20), contains two correction terms, a volume correction and a pressure correction. Here we shall seek to show, at least qualitatively, how these terms arise.

(1) The fact that in the equation of state, as is stated in the text, exactly four times the total volume of the molecules themselves must be subtracted from the total volume of the gas, may be explained as follows. In § 6 (p. 9) we investigated the probability that n molecules should be distributed in a given way among the cells ω_1, ω_2, ... We found it to be the product of the number of ways in which a definite distribution, prescribed by the numbers n_1, n_2 ... of the molecules occupying the individual cells, can be realized and the a priori probability of the occurrence of this distribution. It is with this a priori probability that we are concerned here. If we inquire into the probability that n molecules will be found in a definite portion of volume v, we begin by assuming, as was done in § 6 (p. 9), that the probability is proportional to v^n. This is assuredly the case so long as we can neglect the finite magnitude of the gas molecules, as, e.g., in rarefied gases. It is not so, however, for high pressures, where the gas molecules are so tightly packed together that their own volume is actually comparable with that of the space available for them. Here we obtain

the desired result in the following way. Let v_m be the volume of a molecule (i.e. for a spherical molecule $v_m = \frac{4}{3}\pi(\frac{1}{2}\sigma)^3$, where σ denotes the diameter of the molecule). Now the centres of two molecules cannot approach within a distance equal to the diameter of a molecule; hence each molecule has an effective volume of magnitude $\frac{4}{3}\pi\sigma^3 = 8v_m$, independent of the particular shape which the molecule happens to have.

The probability that a molecule will be found in a definite portion of volume v is of course proportional to v, as above. If we introduce a second molecule into this region, the space available for it is only $v - 8v_m$; the space left for a third molecule is $v - 2 \times 8v_m$, and so on. The probability P_n of finding n molecules in v, then, is proportional, not to v^n, but to the product

$$v(v - 8v_m)(v - 2 \times 8v_m) \ldots (v - (n - 1)8v_m);$$

v has accordingly to be replaced by $\sqrt[n]{P_n}$. An approximate expression for P_n can be obtained by forming

$$\log P_n = \log v^n(1 - \gamma)(1 - 2\gamma) \ldots (1 - (n - 1)\gamma)$$
$$= n \log v + \sum_{k=1}^{n-1} \log (1 - k\gamma),$$

where $\gamma = 8v_m/v$. Now the total volume of all molecules nv_m is assumed to be small compared with the volume v of the vessel, hence $n\gamma \ll 1$, and in all terms of the sum $k\gamma \ll 1$. Therefore $\log (1 - k\gamma) \sim - k\gamma$ and

$$\log P_n = n \log v - \gamma \sum_{k=1}^{n-1} k = n \log v - \gamma \frac{(n - 1)n}{2}.$$

Here $n - 1$ can be replaced by n, so that

$$\log P_n = n \log v - \tfrac{1}{2}\gamma n^2$$

and

$$P_n = v^n e^{-\gamma n^2/2} = (ve^{-n\gamma/2})^n.$$

Taking the n-th root of this we see that v must be replaced by $ve^{-\frac{1}{2}n\gamma}$, or as $n\gamma \ll 1$, by

$$v(1 - \tfrac{1}{2}n\gamma) = v\left(1 - n\frac{4v_m}{v}\right) = v - b,$$

where b stands for $4nv_m$, four times the actual volume of the molecules contained in v.

(2) The term a/v^2 added to the pressure may be explained as

follows. If there are forces of cohesion acting between the molecules, one element of volume acts on another of equal size with a force which is proportional to n^2, where n is the number of molecules per cubic centimetre. The pressure exerted by the gas on an external body is correspondingly less than it would be if there were no forces of cohesion. Hence in the equation of state p must be replaced by $p + An^2$. Now let $nV = N$ be the total number of molecules in the volume of gas V; this number of course does not vary if the volume is changed, so that the additional term An^2 may be written in the form $A\left(\dfrac{N}{V}\right)^2 = \dfrac{a}{V^2}$. The constant a accordingly is closely related to the energy of cohesion of the gas, and we may mention that the magnitude of the latent heat of evaporation of liquids also depends on it. Van der Waals' equation is derived only for small densities; if it is used nevertheless for higher densities it accounts formally for the phenomenon of condensation and represents even the properties of the liquid in a rough way. The problem of a rigorous treatment of liquefaction and the equation of state of liquids is very difficult, as the interactions of more than two molecules have to be considered. Ursell (1927) showed a way of doing this which has been worked out by Mayer (1937) and mathematically improved by others (Uhlenbeck and Kahn, 1938; Born and Fuchs, 1938).

IV. The Mean Square Deviation (p. 21).

All phenomena involving deviation from a mean value depend on the formula

$$\overline{\Delta n^2} = \bar{n}.$$

Here n may denote, e.g., the number of particles in a definite fixed portion of volume of a gas. If we stick to this example, we know that this number is not always the same but varies with the time. There is, however, a time-average ($\bar{n}$) of n, about which the number of particles varies. If we could observe the actual number n at any instant (as if in a snapshot), we should obtain varying values $n_1, n_2, \ldots$, deviating from the mean value $\bar{n}$ by the amounts $\Delta n_1 = n_1 - \bar{n}$, $\Delta n_2 = n_2 - \bar{n}, \ldots$. The sum of these deviations for a large number of observations, divided by their number, must of course vanish; a result differing from zero, however, is obtained if we average the squares of these deviations over a large number of observations. We thus obtain the mean square deviation $\overline{\Delta n^2}$, which according to the assertion made above is equal to the mean value $\bar{n}$.

To prove this, we start from the fundamental formulæ of the kinetic theory of gases, according to which the probability that of N molecules in a total volume V the fraction n will be found in the portion of volume v is given by the formula

$$W_n = \frac{N!}{n!(N-n)!} \left(\frac{v}{V}\right)^n \left(\frac{V-v}{V}\right)^{N-n}$$

(cf. p. 12: instead of the distribution among the cells ω_1, ω_2, ... , we here have the distribution between the two regions v and $V-v$ with "occupation numbers" $n_1 = n$ and $n_2 = N-n$). The sum of all the probabilities of the various distributions is $\sum\limits_{n=0}^{N} W_n = 1$ (by the binomial theorem). The mean number $(\bar{n})$ of molecules in v is obtained by working out the sum

$$\bar{n} = \sum_{n=0}^{N} n W_n.$$

If we put $v/V = x$, this sum becomes

$$\bar{n} = \sum_{n=0}^{N} \frac{nN!}{n!(N-n)!} x^n (1-x)^{N-n}$$

$$= Nx \sum_{n=1}^{N} \frac{(N-1)!}{(n-1)![(N-1)-(n-1)]!} x^{n-1} (1-x)^{(N-1)-(n-1)}.$$

According to the binomial theorem the sum here is again equal to 1, so that for $\bar{n}$, the mean number of particles, we obtain the expression

$$\bar{n} = Nx = \frac{Nv}{V},$$

as might be expected. In order to calculate $\overline{n^2}$, we note that the mean value $\overline{n(n-1)}$ may be found in exactly the same way as $\bar{n}$. In fact,

$$\overline{n(n-1)} = \sum_{n=0}^{N} n(n-1) W_n$$

$$= N(N-1)x^2 \sum_{n=2}^{N} \frac{(N-2)!}{(n-2)![(N-2)-(n-2)]!} x^{n-2} (1-x)^{(N-2)-(n-2)},$$

whence

$$\overline{n(n-1)} = N(N-1)x^2.$$

Hence we at once have

so that
$$\overline{n^2} = \overline{n(n-1)} + \bar{n} = N(N-1)x^2 + Nx,$$

$$\overline{\Delta n^2} = \overline{(n - \bar{n})^2} = \overline{n^2} - \bar{n}^2 = Nx - Nx^2 = N\frac{v}{V}\left(1 - \frac{v}{V}\right).$$

If we confine ourselves to small values of x, i.e. to relatively small volumes v, we at once obtain the deviation formula

$$\overline{\Delta n^2} = \bar{n}$$

give above. Many other mean values can be calculated in exactly the same way as $\bar{n}$ and $\overline{n^2}$.

These formulæ are applied as follows. Spontaneous deviations of the density of the molecules from the mean value are associated with a change in almost all the physical properties of the gas. For example, variations of density in a gas are associated with variations in the refractive index. By observing the variations of measurable properties we can determine the mean square deviation. For example, the variation of the refractive index gives rise to a scattering of the transmitted light, which is proportional to the mean square deviation (to this, according to Lord Rayleigh, is due the blue colour of the sky). In the Brownian movement, too, fine particles in suspension execute certain motions as a result of variations in the density of the surrounding medium, and these motions are a measure of the mean square deviation of the density of the surrounding medium. If we know the way in which the phenomenon in question is related to the density, the mean value of the number of particles ($\bar{n}$) can be determined by measuring the deviations and calculating the mean square deviation.

An important field of application of the theory of deviations is the subject of radioactivity. Here it is a question of counting the number of particles emitted by a radioactive preparation per second (e.g. by means of a Geiger counting apparatus (p. 33)). If we are dealing with a long-lived substance, for which the average number ($\bar{n}$) of particles emitted per second may be regarded as constant, the actual number (n) of particles emitted in individual periods of one second will differ from $\bar{n}$. The fact that the deviations *are* in accordance with the law stated above, $\overline{\Delta n^2} = \bar{n}$, forms a convincing proof of the statistical nature of the disintegration process. Corresponding formulæ can also be deduced for short-lived radioactive substances.

The preceding deviations rest on the assumption that the particles considered are independent. This is not the case for a degenerate gas (Chap. VII, §§ 4, 5, 6) when we have to do either with the Bose-Einstein or the Fermi-Dirac statistics. In this case the deviation formula has to be modified (Fürth, 1928).

V. Theory of Relativity (p. 27).

In classical mechanics it is proved that an observer who experiments only within a closed system cannot determine whether this system is at rest or is in uniform motion. In fact, the Newtonian equations of motion $m\,d^2x/dt^2 = F$ (where m is the mass, F the force, x the co-ordinate of a particle, and t the time) remain unchanged if we pass to a moving co-ordinate system by the transformation $x' = x - vt$, provided the force depends only on the position of the particle relative to the co-ordinate system (since $x_1' - x_2' = x_1 - x_2$). This principle of relativity in mechanics has to be modified when electromagnetic pro-

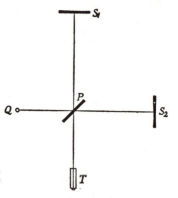

Fig. 1.—Diagram of Michelson's interferometer. Q, source of light; P, semi-transparent, silvered glass plate; S_1, S_2, mirrors; T, telescope.

cesses, light waves for example, are taken into account. Since these waves move in vacuo, it was the custom to assume a carrier, the æther. The earth moves through it, so that an observer on the earth must perceive an "æther wind", which retards or accelerates the light waves according to their direction. The experiment was carried out by means of Michelson's interferometer (fig. 1). A light ray from the source Q was divided into two parts by partial reflection at the thinly silvered glass plate P; the two partial rays were reflected at the mirrors S_1, S_2 and united again at the plate P. Interference fringes are seen in the telescope T. If the apparatus is then turned round so that first PS_1, then PS_2, fall in the direction of the æther wind, the interference fringes ought to be displaced. The result of the experiment was negative: the æther wind is not really there.

To explain this fact Einstein developed his theory of relativity The leading idea is that the customary combination of space and time —kinematics—must be abandoned. There is no absolute time, but just as every moving system has its "proper" co-ordinates x, y, z, so it has also a proper time t, which has to be transformed as well as the co-ordinates when we pass to a new system. The equations defining this so-called Lorentz transformation for two systems moving in the x-direction with the relative velocity v are

$$x' = \frac{x - vt}{\sqrt{(1 - v^2/c^2)}}, \quad y' = y, \quad z' = z, \quad t' = \frac{t - vx/c^2}{\sqrt{(1 - v^2/c^2)}},$$

where c is the velocity of light. It is easily verified that these equations give

$$x'^2 - c^2 t'^2 = x^2 - c^2 t^2.$$

This formula suggests that x, y, z, ict $(i = \sqrt{-1})$ may be interpreted (Minkowski) as co-ordinates in a four-dimensional space, in which $x^2 + y^2 + z^2 + (ict)^2$ represents the square of the distance from the origin; a Lorentz transformation then represents a rotation round the origin in this space. Minkowski's idea has developed into a geometrical view of the fundamental laws of physics, culminating in the inclusion of gravitation in Einstein's so-called general theory of relativity.

Physically, the equation last written expresses the fact that $x = ct$ implies $x' = ct'$; or that the velocity of light is independent of the motion of the observer. The negative result of Michelson's experiment is thus explained.

Further, we see that if the distance of two points is $x_1 - x_2$ at the same time t in one system (x, t), then in a second system (x', t') their distance is

$$x_1' - x_2' = (x_1 - x_2)/\sqrt{(1 - v^2/c^2)}.$$

Thus

$$x_1 - x_2 = (x_1' - x_2')\sqrt{(1 - v^2/c^2)}.$$

Hence lengths in the second system appear from the first system to be shortened (the Fitzgerald-Lorentz contraction). On the other hand, if t_1, t_2 are the times of two events at the same place x in the first system, then

$$t_1' - t_2' = (t_1 - t_2)/\sqrt{(1 - v^2/c^2)},$$

so that the time between the two events seems longer from the second system (cf. p. 372).

It must be assumed that there is no velocity greater than the velocity of light—otherwise the theory would become meaningless. It follows that the basic ideas of mechanics must be so altered that the motion of a body can never be accelerated up to the velocity of light.

We can arrive at this result from the consideration that the velocity defined by the components dx/dt, dy/dt, dz/dt (or the momentum obtained from this velocity on multiplication by the mass) cannot, in view of Lorentz's transformation, be regarded as a vector, since the differential dt in the denominator is also transformed. We obtain a serviceable "covariant" definition if we replace dt by dt_0, where dt_0 is the element of the "proper" time of the particle, i.e. the time measured in the system of reference in which the particle is at rest. The relation between dt and dt_0 is found by taking the derivative of t',

$$\frac{dt'}{dt} = \frac{1 - \dfrac{v}{c^2}\dfrac{dx}{dt}}{\sqrt{(1 - v^2/c^2)}},$$

and putting $dx/dt = v$; dt' then becomes dt_0, and we have

$$dt_0 = \sqrt{(1 - v^2/c^2)}\,dt.$$

The component momentum is now defined to be

$$m_0\frac{dx}{dt_0} = \frac{m_0}{\sqrt{(1 - v^2/c^2)}}\frac{dx}{dt} = m\frac{dx}{dt},$$

where m_0 is a constant, the rest mass. The mass m is therefore given by the formula of the text, $m = m_0/\sqrt{(1 - v^2/c^2)}$. The formula has been confirmed, not only by the experiments with cathode rays already mentioned (p. 27), but also by certain details in the behaviour of spectral lines, particularly those of hydrogen. In fact, since such lines are emitted by rapidly moving electrons, they reflect the mechanical properties of these electrons (p. 119). The energy is

$$E = mc^2 = \frac{m_0c^2}{\sqrt{(1 - v^2/c^2)}},$$

and the momentum is

$$p = mv = \frac{m_0v}{\sqrt{(1 - v^2/c^2)}}.$$

From these two expressions we find

$$m_0 = \frac{1}{c^2}\sqrt{(E^2 - p^2c^2)},$$

a relation frequently used in the text, e.g. in the theory of the β-decay of nuclei (p. 337). The values of E and p therefore determine the rest mass. If the latter vanishes, as it does for a light quantum (photon, § 2, p. 81), E and p are connected by the relation

$$cp = E.$$

VI. Electron Theory (p. 55).

Maxwell's equations, for an isotropic substance, are:

$$\frac{1}{c}\frac{\partial D}{\partial t} - \operatorname{curl} H = -\frac{4\pi}{c}\,i, \quad \operatorname{div} D = 4\pi\rho, \quad D = \epsilon E;$$

$$\frac{1}{c}\frac{\partial B}{\partial t} + \operatorname{curl} E = 0, \qquad \operatorname{div} B = 0, \qquad B = \mu H.$$

We assume the reader to be acquainted with the vector notation, and the meaning of the vectors $E, H, D, B, i,$ as also with the energy theorem

$$-\frac{dW}{dt} = \int S_n\, d\sigma + A,$$

where $W = (1/8\pi)\iiint(\epsilon E^2 + \mu H^2)\,dx\,dy\,dz$ is the electromagnetic energy; $A = \iiint i.E\,dx\,dy\,dz$ is the rate of doing work by the electric force on the currents; and $S = (c/4\pi)E \wedge H$ is the Poynting vector, of which use is made in Appendix VIII, p. 376. ($\wedge$ means vector product.)

Lorentz's theory of the electron considers fields in vacuo only, and therefore puts $\epsilon = \mu = 1$; the current is assumed to be a convection current, $i = \rho v$, the charge being rigidly bound to the electron. To connect the theory with mechanics, it is further necessary to assume that the field exerts the mechanical force

$$F = \iiint \rho\left\{E + \frac{1}{c} v \wedge H\right\} dx\,dy\,dz$$

on the electron.

Mie (1912) has proposed a generalization of Lorentz's theory where the field equations of Maxwell remain unchanged, while the simple relations $D = \epsilon E$, $B = \mu H$ are replaced by very general non-linear ones (involving also current and charge density, and the vector and solar potential). Using this formalism the author of this book developed a " unified " theory which allowed the description of the electron as a point charge in the field with finite total energy. We have to assume ϵ and μ to be the following simple functions * of B and E (in the text H is written in place of B):

$$\mu = \frac{1}{\epsilon} = \sqrt{\left\{1 + \frac{1}{b^2}(B^2 - E^2)\right\}}.$$

* We consider here the theory in its original form; it has been modified by adding a term $\frac{1}{b^4}(EB)^2$ under the square root, which has no influence on the static solution.

The energy is

$$W = \frac{1}{4\pi} \int\!\!\int\!\!\int \{\epsilon E^2 + b^2(\mu - 1)\}\, dx\, dy\, dz,$$

the Poynting vector

$$S = \frac{c}{4\pi}\, D \wedge B = \frac{c}{4\pi}\, E \wedge H.$$

In the static case we have $H = B = 0$, so that

$$D = \epsilon E = \frac{E}{\sqrt{\left(1 - \dfrac{1}{b^2}\, E^2\right)}}$$

For a point singularity, it follows from $\operatorname{div} D = 0$ that

$$D = \frac{e}{r^2},$$

and hence that

$$E = \frac{e}{\sqrt{(r^4 + a^4)}}, \quad a^2 = \frac{e}{b}.$$

The field E therefore remains finite even at the centre of the electron, where it has the value

$$E_0 = \frac{e}{a^2} = b.$$

Since the field equations reduce to $\operatorname{curl} E = 0$, a potential ϕ exists, which is found from $E = -\partial\phi/\partial r$ to be

$$\phi = \frac{e}{a} \int_{r/a}^{\infty} \frac{dx}{\sqrt{(1 + x^4)}}.$$

For $r \gg a$, ϕ is easily seen to reduce to the Coulomb potential e/r; but for $r = 0$ it is finite,

$$\phi(0) = 1{\cdot}8541\, \frac{e}{a}.$$

If we substitute the expression for E in the energy formula

$$W = \frac{1}{4\pi} \int\!\!\int\!\!\int \left\{ \frac{E^2}{\sqrt{(1 - E^2/b^2)}} + b^2(\sqrt{1 - E^2/b^2} - 1) \right\} dx\, dy\, dz$$

we find

$$W = \tfrac{2}{3} e\phi(0) = 1{\cdot}2361\, \frac{e^2}{a}.$$

Einstein's theorem $W = mc^2$ is here rigorously true, but the proof would take us too far. Equating mc^2 to the value of W given above, we get $a = 1 \cdot 2361 \, e^2/mc^2$, which, on introducing the empirical values for the charge and the mass of the electron, comes out as $a = 3 \cdot 50 \times 10^{-13}$ cm. The absolute field is $b = e/a^2 = 9 \cdot 18 \times 10^{15}$ e.s.u. It can be shown that the equations of motion given by Lorentz in the classical theory,

$$\frac{d}{dt}(mv) = e(E + \frac{1}{c} \, v \wedge H),$$

hold here rigorously for a constant external electromagnetic field, and approximately for a field which does not vary appreciably over the "diameter of the electron" $2a$. For quickly vibrating fields there is, as Schrödinger (1942) has shown, a radiation force which corresponds exactly to the classical formula derived from Hertz' theory (see Appendix VIII, p. 376).

The adaptation of these ideas to the principles of quantum theory and the introduction of the spin has however met with no success.

VII. The Theorem of the Inertia of Energy (p. 58).

Here we give the mathematical proof corresponding to the ideal experiment given in the text. Classical electrodynamics, in agreement with experiment, yields the result that the momentum transferred by light of energy E to a surface absorbing it is E/c (cf. Appendix XXXIII, p. 451); hence the momentum transferred by recoil to the box during the emission is of the same magnitude. If the box (fig. 1, p. 57) has the total mass M, it acquires a recoil velocity v to the left, given according to the theorem of the conservation of momentum by $Mv = E/c$. The box continues to move during the time required by the light to traverse the distance l between the transmitter (I) and the receiver (II). If we neglect terms of higher order, this time t is equal to l/c. During this time the box moves a distance $x = vt = El/Mc^2$ to the left. In order not to contradict the fundamental principle of the centre of gravity we must assume, as was explained in the text, that the transference of energy from I to II is accompanied by a simultaneous transference of mass in the same direction. If we denote this meanwhile unknown mass by m, then the total moment due to displacement of mass after the phenomenon is given by $Mx - ml$; by the fundamental principle of the centre of gravity this must be zero. Substituting for x the value found above, we obtain Einstein's formula for m:

$$m = M\frac{x}{l} = \frac{E}{c^2}.$$

VIII. **Calculation of the Coefficient of Scattering for Radiation by a Free Particle** (pp. 59, 149, 196).

In order to calculate the coefficient of scattering we start from the familiar formulæ for the radiation from a dipole antenna. According to Hertz, the field of a dipole of moment p is given by

$$|E| = |H| = \frac{|\ddot{p}|}{c^2 r} \sin \chi.$$

The electric vector is at right angles to the magnetic vector and both are at right angles to the direction in which the radiation is propagated; χ is the angle between the direction of propagation of the radiation and the direction in which the dipole is oscillating, and r the distance from the dipole. A necessary assumption for the validity of this formula is that r must be very much greater than the wavelength of the radiation emitted, or, in other words, the above formula is valid only for the so-called wave zone.

To apply the formula to our case of a vibrating particle, we note that this represents an oscillating electric dipole moment $p = es$, where s is the displacement of the electron at any instant from its position of equilibrium. Then the energy radiated per second in a specified direction is given by the Poynting vector, orthogonal to E and H and of magnitude

$$I_r(\chi) = |S| = \frac{c}{4\pi} |E| |H| = \frac{e^2 \ddot{s}^2}{4\pi c^3 r^2} \sin^2\chi.$$

This formula still contains the displacement of the oscillating electron, or, more accurately, its second derivative with respect to the time. This is determined by the equation of motion of the electron under the influence of the incident radiation; if no other force is acting we have

$$m\ddot{s} = eE_0.$$

Hence s is perpendicular to the incident ray. If this is taken as polar axis, the plane through the point of observation as zero plane of the azimuth, we have $\cos\chi = \cos a \sin\theta$, where a is the azimuth of s, and θ the polar angle of the direction of observation. Hence we have $\sin^2\chi = 1 - \cos^2 a \sin^2\theta$. For unpolarized incident light this must be averaged over a (giving $\cos^2 a \rightarrow \frac{1}{2}$); hence $\sin^2\chi$ is to be replaced by $1 - \frac{1}{2}\sin^2\theta$.

Further, E_0, the electric vector of the primary wave, is connected with the intensity of the incident radiation by the equation

$$I_0 = \frac{c}{4\pi} |E_0| |H_0| = \frac{c}{4\pi} E_0^2.$$

If we substitute these expressions in the equation above, the formula for $I_f(\chi)$ assumes the alternative form (J. J. Thomson)

$$I_f(\theta) = \left(\frac{e^2}{mc^2}\right)^2 \frac{1 - \frac{1}{2}\sin^2\theta}{r^2} I_0.$$

By integrating over the surface of a sphere of radius r, and using the integral

$$\int \sin^2\theta \, d\sigma = \int_0^\pi \int_0^{2\pi} \sin^2\theta \, r^2 \sin\theta \, d\theta \, d\phi$$

$$= 2\pi r^2 \int_0^\pi \sin^3\theta \, d\theta = \tfrac{8}{3}\pi r^2,$$

we obtain the total energy scattered by a free particle,

$$I_f = \frac{2}{3}\frac{e^2}{c^3} \mid \ddot{s} \mid^2 = \frac{8\pi}{3}\left(\frac{e^2}{mc^2}\right)^2 I_0.$$

As the mass of the scattering particle occurs squared in the denominator, we see that a proton or an atomic nucleus scatters some millions of times less than an electron.

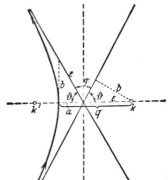

IX. Rutherford's Scattering Formula for α-rays (p. 62).

According to Rutherford, the nucleus of an atom (with charge Ze) and an α-particle (with charge E and

Fig. 2.—Hyperbolic path in the scattering of an α-ray by a nucleus; a and b are the semi-axes, ϵ the distance between the centre and the focus. The angle of deflection is $\phi = \pi - 2\theta$.

mass M) repel one another with the Coulomb force ZeE/r^2. If we consider the heavy nucleus as at rest, the α-particle describes an orbit (fig. 2) which is one branch of a hyperbola, one of whose foci coincides with the nucleus K. Let b be the distance of the nucleus from the asymptote of the hyperbola which would be described by the α-particle if there were no repulsive force. Further, we denote the distance of the nucleus K from the vertex of the hyperbola by q; then

$$q = \epsilon(1 + \cos\theta),$$

where ϵ denotes the linear eccentricity OK (i.e. the distance between the centre O and the focus K), and θ the angle between the asymptote and the axis. From fig. 2 we readily see that

$$\epsilon = \frac{b}{\sin\theta}$$

and hence that

$$q = \frac{b(1 + \cos\theta)}{\sin\theta} = b\cot\frac{\theta}{2};$$

b is obviously equal to the length of the minor semi-axis of the hyperbola.

In the first place we shall seek to find a relation between the "collision parameter" b and the angle of deviation ϕ, which, as we see from the figure, is equal to $\pi - 2\theta$. We accordingly apply the laws of motion to the a-particle. First we have the theorem of the conservation of energy: the sum of the kinetic energy and the potential energy is constant. At a very great distance from the nucleus the a-particle has kinetic energy only; let its velocity there be v. If we equate this energy to the total energy at the instant when the electron is passing the vertex of the hyperbola, we have

$$\tfrac{1}{2}Mv^2 = \tfrac{1}{2}Mv_0{}^2 + \frac{ZeE}{q},$$

or, if we divide by $\tfrac{1}{2}Mv^2$ and for short put $k = ZeE/Mv^2$,

$$\frac{v_0{}^2}{v^2} = 1 - \frac{2k}{b}\frac{\sin\theta}{1 + \cos\theta}.$$

Further, we have the theorem of the conservation of angular momentum, in virtue of which

$$Mvb = Mv_0q,$$

or

$$\frac{v_0}{v} = \frac{b}{q} = \frac{\sin\theta}{1 + \cos\theta},$$

$$\left(\frac{v_0}{v}\right)^2 = \frac{\sin^2\theta}{(1 + \cos\theta)^2} = \frac{1 - \cos\theta}{1 + \cos\theta}.$$

Substituting this value in the above equation and carrying out a few transformations, we have

$$\frac{b}{k} = \frac{\sin\theta}{\cos\theta} = \tan\theta,$$

or, since $\phi = \pi - 2\theta$,

$$b = k \cot \frac{\phi}{2}.$$

This gives the angle of deviation as a function of b, the distance of the rectilinear prolongation of the path of the α-particle (the asymptote) from the nucleus.

It is now easy to calculate how many α-particles in an incident parallel beam are deviated by a specified amount. We imagine a

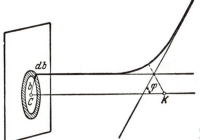

plane E at right angles to the incident beam and at a great distance from K; C is the foot of the perpendicular from K to E (fig. 3). It is obvious that all the α-particles which pass through a ring of E formed by

Fig. 3.—Relative frequency, in a definite angular region, of scattering of α-particles by a nucleus.

two circles with radii b and $b + db$ will be subject to a deviation between ϕ and $\phi + d\phi$. If one particle passes through one square centimetre of E in one second, the number passing through the ring in question will be

$$dn = 2\pi b\, db,$$

where

$$db = k d(\cot \phi/2) = -\frac{k\, d\phi}{2 \sin^2 \phi/2}.$$

Hence

$$|\, dn\, | = \pi k^2 \frac{\cos \phi/2}{\sin^3 \phi/2}\, |\, d\phi\, |.$$

This is the number of particles deviated through an angle between ϕ and $\phi + d\phi$; they are uniformly distributed over a zone of a unit sphere, the area of the zone being $2\pi \sin \phi\, d\phi$. Hence $W(\phi)$, the number of α-particles deviated into a unit of area of the unit sphere, is $dn/2\pi \sin \phi\, d\phi$, so that the probability of deviation per unit solid angle is

$$W(\phi) = \tfrac{1}{4} k^2 \frac{1}{\sin^4 \phi/2} = \left(\frac{ZeE}{2Mv^2} \right)^2 \frac{1}{\sin^4 \phi/2}.$$

This is Rutherford's scattering formula. Each relationship (between

Z, M, v, ϕ) contained in it can be tested experimentally by counting the scattered a-particles; the v-relation, it is true, can only be investigated experimentally over a small region, as the range of velocity available with naturally-occurring a-particles is small. In general the agreement between the formula and experimental results is extremely accurate; with light atoms, of course, we must take into account the recoil of the nucleus K after collision with the a-particle, which is easily done. Marked discrepancies have been found only in the case of almost central collisions (deviations of almost 180°) of light atoms (which have low nuclear charges, so that the incident a-particle approaches the nucleus very closely); here, however, we shall not go into further details.

As the charge and mass of the a-particles are known (they are He^{++} ions, for which $M = 4M_H$, $E = 2e$) and their velocities can be determined from deflection experiments, the formula can be used to find the nuclear charge Z. For this we need only know the number of scattering atoms per unit volume and count the number of a-particles in the presence and absence of the scattering layer. For example, accurate experiments by Chadwick gave the following values for Z:

Platinum	Silver	Copper
77·4	46·3	29·3,

while according to the periodic table the numbers should be

78	47	29.

The excellent agreement between the two sets of figures confirms the basic assumption that the nuclear charge and the atomic number are identical.

X. The Compton Effect (p 87).

Here we shall investigate the collision between a light quantum and an electron, on the assumptions of the special theory of relativity. This procedure is appropriate for our purpose, as it does not make the calculations any more complicated, while on the other hand the result obtained is then valid for the scattering even of very hard radiation.

The calculations are based on the theorems of the conservation of energy and of momentum. The energy of the light quantum before the collision is $h\nu$, its momentum $h\nu/c$; the corresponding quantities after the collision we shall call $h\nu'$ and $h\nu'/c$. For the sake of simplicity we shall regard the electron as at rest previous to the collision; in this case its energy will be equal to the rest-energy m_0c^2 corresponding by Einstein's formula to the rest-mass m_0, while its momentum will be zero. Let v be the velocity of the electron after the collision; then its mass m will be $\dfrac{m_0}{\sqrt{(1-v^2/c^2)}}$, its energy mc^2 will be $\dfrac{m_0c^2}{\sqrt{(1-v^2/c^2)}}$, and its momentum mv will be $\dfrac{m_0v}{\sqrt{(1-v^2/c^2)}}$. We may also say that after collision the electron has the "kinetic energy"

$$(m - m_0)c^2 = m_0c^2\left(\frac{1}{\sqrt{(1-v^2/c^2)}} - 1\right)$$

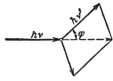

(which, as we easily see by expanding in powers of v/c, agrees with the formula $\frac{1}{2}m_0v^2$ of non-relativistic mechanics when the velocity is small), while before collision this "kinetic energy" is zero.

Fig. 4.—Momentum diagram for the Compton effect

Then if ϕ is the angle of deviation of the light quantum and ψ the angle of deviation of the electron, the theorems of the conservation of energy and of momentum take the following forms (see fig. 4):

Conservation of energy: $h\nu + m_0c^2 = h\nu' + mc^2$;

Conservation of momentum: $\begin{cases} \dfrac{h\nu}{c} = \dfrac{h\nu'}{c}\cos\phi + mv\cos\psi, \\[2mm] 0 = \dfrac{h\nu'}{c}\sin\phi - mv\sin\psi. \end{cases}$

Eliminating ψ from the last two equations, we obtain

$$m^2v^2c^2 = h^2\{(\nu - \nu'\cos\phi)^2 + (\nu'\sin\phi)^2\} = h^2(\nu^2 + \nu'^2 - 2\nu\nu'\cos\phi).$$

Again, from the energy equation we have

$$m^2c^4 = \{h(\nu-\nu') + m_0c^2\}^2 = h^2(\nu^2+\nu'^2 - 2\nu\nu') + 2m_0c^2h(\nu - \nu') + m_0^2c^4.$$

Since $\qquad m^2 = \dfrac{m_0^2}{1 - v^2/c^2}$, i.e. $m^2\left(1 - \dfrac{v^2}{c^2}\right) = m_0^2$,

by definition, subtracting one of these equations from the other gives

$$m_0^2c^4 = -2h^2\nu\nu'(1 - \cos\phi) + 2m_0c^2h(\nu - \nu') + m_0^2c^4$$

or
$$(1 - \cos\phi) = \frac{m_0 c^2}{h} \frac{\nu - \nu'}{\nu\nu'} = \frac{m_0 c^2}{h}\left(\frac{1}{\nu'} - \frac{1}{\nu}\right).$$

If for convenience we denote the quantity

$$\frac{h}{m_0 c} = 0\cdot0242 \text{ Å.}$$

(which is in general referred to as the Compton wave-length) by λ_0, the above equation can also be written in the form

$$\Delta\lambda = \lambda' - \lambda = c\left(\frac{1}{\nu'} - \frac{1}{\nu}\right) = (1 - \cos\phi)\,\frac{h}{m_0 c} = 2\lambda_0 \sin^2\frac{\phi}{2},$$

which was given in the text.

XI. Phase Velocity and Group Velocity (p. 90.)

In order to give a strict proof of the relationship $U = \partial\nu/\partial\kappa$ given in the text, we in the first instance consider the most general form of a group of waves; it must have the form of a Fourier integral

$$u(x,\ t) = \int a(\kappa)e^{2\pi i(\nu t - \kappa x)}d\kappa,$$

where $\nu = \nu(\kappa)$ is to be regarded as a function of the wave-number κ.

We now assume that the group is very narrow, so that in the integral there occur only those waves of finite amplitude whose wave-numbers differ from the mean wave-number κ_0 by a very small amount. If we put $\kappa = \kappa_0 + \kappa_1$, $\nu(\kappa) = \nu_0 + \nu_1(\kappa_1)$, and $a(\kappa_0 + \kappa_1) = b(\kappa_1)$, the wave-group may be written in the form

$$u(x,\ t) = A(x,\ t)e^{2\pi i(\nu_0 t - \kappa_0 x)},$$
where

$$A(x,\ t) = \int b(\kappa_1)e^{2\pi i(\nu_1 t - \kappa_1 x)}d\kappa_1.$$

Hence the wave-group may be regarded as a single wave of frequency ν_0, wave-number κ_0, and amplitude $A(x,\ t)$ varying from point to point and moment to moment. This assumption is justified, as according to our assumption $A(x,\ t)$ is a function which varies only slowly compared with the exponential function $e^{2\pi i(\nu_0 t - \kappa_0 x)}$; to a first approximation it varies in the rhythm of a mean of the beat frequencies ν_1, which are very small compared with ν_0.

The velocity with which a definite value of the amplitude $A(x,\ t,)$

e.g. its maximum, advances with the wave is called the group velocity. This is accordingly found from the relation

$$\frac{\partial A}{\partial x}\frac{dx}{dt} + \frac{\partial A}{\partial t} = 0$$

obtained by differentiating the equation $A(x, t) = $ const. with respect to the time. If we call the group velocity U to distinguish it from the phase velocity, we have

$$U = \left(\frac{dx}{dt}\right)_{A=\text{const.}} = -\frac{\partial A/\partial t}{\partial A/\partial x}.$$

Now obviously

$$\frac{\partial A}{\partial t} = 2\pi i \int b(\kappa_1)\nu_1 e^{2\pi i(\nu_1 t - \kappa_1 x)} d\kappa_1,$$

$$\frac{\partial A}{\partial x} = - 2\pi i \int b(\kappa_1)\kappa_1 e^{2\pi i(\nu_1 t - \kappa_1 x)} d\kappa_1.$$

As we have assumed that the group is confined to a very narrow range of wave-length, we can expand $\nu_1(\kappa_1)$ in powers of κ_1:

$$\nu_1(\kappa_1) = \nu(\kappa) - \nu_0 = \left(\frac{d\nu}{d\kappa}\right)_0 \kappa_1 + \dots.$$

Hence

$$\frac{\partial A}{\partial t} = - \left(\frac{d\nu}{d\kappa}\right)_0 \frac{\partial A}{\partial x},$$

and for the group velocity we accordingly have

$$U = \frac{d\nu}{d\kappa},$$

while the phase velocity is given by

$$u = \frac{\nu}{\kappa}.$$

XII. Elementary Derivation of Heisenberg's Uncertainty Relation (p. 98).

We consider a wave packet of finite width. For the sake of simplicity we represent its amplitude at any moment by a Gauss error function (such as actually occurs in the ground state of the harmonic oscillator, Appendices XVI, p. 396, and XXXIX, p. 471):

$$f(x) = Ae^{-x^2/a^2};$$

then Δx the half-width is given by

$$\Delta x = \sqrt{(x^2)} = \sqrt{[\int x^2 f^2(x)dx / \int f^2(x)dx]} = \tfrac{1}{2}a.$$

The Fourier representation of $f(x)$ is

$$f(x) = \int_{-\infty}^{\infty} \phi(\kappa) e^{2\pi i \kappa x} d\kappa,$$

where

$$\phi(\kappa) = \int_{-\infty}^{\infty} f(x) e^{-2\pi i \kappa x} dx$$

is the amplitude of a partial harmonic wave with the wave number κ. Introducing the expression of $f(x)$ in $\phi(\kappa)$ we have

$$\phi(\kappa) = A \int_{-\infty}^{\infty} e^{-(x^2/a^2 + 2\pi i \kappa x)} dx$$

$$= A e^{-\pi \kappa a)^2} \int_{-\infty}^{\infty} e^{-(x/a + \pi i \kappa a)^2} dx.$$

This integral is transformed by the substitution $x/a + \pi i \kappa a = y$ into the Gauss integral

$$a \int_{-\infty}^{\infty} e^{-y^2} dy = a \sqrt{\pi}.$$

Therefore

$$\phi(\kappa) = A a \sqrt{\pi} e^{-(\pi \kappa a)^2} = A a \sqrt{\pi} e^{-\kappa^2/b^2},$$

where

$$b = \frac{1}{\pi a}.$$

The distribution of the elementary waves composing the wave packet $f(x)$ is again a Gauss function with the half-width $\Delta \kappa = \frac{1}{2} b$. Hence we have $\Delta x \cdot \Delta \kappa = \frac{1}{4} a b = \dfrac{1}{4\pi}$, and introducing the momentum $p = h\kappa$ (p. 89):

$$\Delta x \cdot \Delta p = \frac{h}{4\pi} = \frac{1}{2} \hbar$$

which is the exact expression of Heisenberg's uncertainty law for the special wave packet (see Appendix XL, p. 471). It is evident that with respect to order of magnitude the relation $\Delta x \cdot \Delta p \sim h$ holds for any form of the wave packet. We shall prove the inequality with an exact numerical coefficient later (see Appendix XXVI, p. 433).

XIII. Hamiltonian Theory and Action Variables (p. 118).

Here we shall briefly consider the mechanics of multiply periodic motions and the corresponding quantum conditions. According to Hamilton, the motion of a system is described completely by stating the energy as a function of the co-ordinates q_k and the momenta p_k, the so-called Hamiltonian function $H(q_1, q_2, \ldots, p_1, p_2, \ldots)$. (For ordinary rectangular co-ordinates the momentum p_k becomes $m_k \dot{q}_k$;

here, as in what follows, the dot denotes differentiation with respect to the time.) The equations of motion are then

$$\dot{q}_k = \frac{\partial H}{\partial p_k}, \quad \dot{p}_k = -\frac{\partial H}{\partial q_k}.$$

From these the theorem of the conservation of energy follows immediately; for if we form the total differential coefficient of H with respect to the time, and make use of the equations of motion, we have

$$\frac{dH}{dt} = \sum_k \left\{ \frac{\partial H}{\partial q_k} \dot{q}_k + \frac{\partial H}{\partial p_k} \dot{p}_k \right\} = 0,$$

that is, the energy of the system, $H(q_k, p_k) = E$, is constant.

In general we can replace the pair of variables (p_k, q_k) by an arbitrary pair of canonically conjugate variables, where we define canonical variables by the fact that they satisfy equations of the type of the equations of motion given above. The investigation of such canonical transformations leads to a mathematical problem (the so-called Hamilton-Jacobi differential equation) which in many cases can be solved. Here we shall assume that a solution can be found. Then the problem of integrating the equations of motion may be stated in the following form: to find new canonical variables (J_k, w_k), such that the energy depends on the quantities J_k only, not on the quantities w_k. Then by the equations of motion $\dot{J}_k = -\frac{\partial H}{\partial w_k} = 0$, so that J_k is constant throughout the motion. On the other hand, $\dot{w}_k = \nu_k = \frac{\partial H}{\partial J_k}$ is also constant as time goes on, owing to the constancy of the quantities J_k, and w_k increases linearly with the time: $w_k = \nu_k t$. Thus the integration problem is solved in the new co-ordinates, and it only remains to make the reverse transformation.

A system is said to be singly or multiply periodic if the variables just defined can be found such that each rectangular co-ordinate is periodic in the quantities w_k, i.e. can be represented as a Fourier series in terms of these variables $w_1, \ldots, w_k$; so that, $\tau_1, \tau_2, \ldots$ being positive or negative integers,

$$x = \sum_{\tau_1, \tau_2 \ldots} a_{\tau_1, \tau_2 \ldots} e^{2\pi i (w_1 \tau_1 + w_2 \tau_2 + \ldots)} = \sum_{\tau_1, \tau_2 \ldots} a_{\tau_1, \tau_2 \ldots} e^{2\pi i (\tau_1 \nu_1 + \tau_2 \nu_2 + \ldots)t}.$$

The coefficients $a_{\tau_1, \tau_2 \ldots}$ are then functions of the quantities J_k. If the motion is periodic, the quantities J_k are said to be *action variables* and the quantities w_k *angle variables*. We have already had an example of this kind in the rotator, where the azimuth ϕ and the angular momentum p were taken as canonical co-ordinates. Then the Hamiltonian function is $H = \dfrac{p^2}{2A}$, whence there follow the equations of motion $\dot{\phi} = \dfrac{\partial H}{\partial p} = \dfrac{p}{A} = \omega,\ \dot{p} = -\dfrac{\partial H}{\partial \phi} = 0$. These have the solutions $p =$ const. and $\phi = \omega t$. The rectangular co-ordinate is then given by

$$x = ae^{i\phi} = ae^{i\omega t}.$$

Ehrenfest has proved that the action variables J_k are adiabatic invariants, i.e. that the quantities J_k can be quantised. We now postulate the *quantum conditions*

$$J_k = hn_k \qquad (n_k \text{ an integer}).$$

We can, however, only set up as many conditions of this kind as there are incommensurable frequencies in the motion. This may be seen as follows. By way of example we consider the case of two degrees of freedom; the index in the Fourier series then contains the sum $\nu_1\tau_1 + \nu_2\tau_2$, where $\tau_1,\ \tau_2$ are integers. If, e.g., $\nu_1 = k\nu_2$, where k is an integer, $\nu_1\tau_1 + \nu_2\tau_2 = \nu_2(k\tau_1 + \tau_2)$. Now $k\tau_1 + \tau_2$ can take any integral value, since τ_1 and τ_2 are integers. In reality, therefore, we have only a single periodic motion and a single periodic Fourier series, and it is clear that in these circumstances only one quantum condition can exist. This case is referred to as *degeneracy*.

The canonical variables J_k and w_k can be conveniently determined in the case where the system is separable, i.e. once the equations of motion have been solved each p_k depends only on the corresponding q_k:

$$p_k = p_k(q_k).$$

Here it may be shown that the action variable J_k corresponding to the kth period is given by

$$J_k = \oint p_k dq_k,$$

where the integral is to be taken over the whole period. Hence in the case of separable co-ordinates the quantum condition may be stated at once:

$$\oint p_k dq_k = hn_k.$$

Without further examination these conditions do not reveal whether degeneracy is present or not. Hence before applying them it is necessary to investigate carefully what the actual number of incommensurable periods in the system is. Frequently, however, it is convenient to take no notice of this (Sommerfeld) and simply to write down super-fluous quantum conditions. Then the physically observable quantities like energy, momentum, &c., depend only on certain combinations of the quantum numbers n_1, n_2, ... , so that they may be expressed by a smaller number of integers; thus contradictions cannot arise. This method has the following additional advantage: if a degenerate system is perturbed, e.g. by an external electric or magnetic field, the degeneracy is in general removed and there appears a new fundamental period, incommensurable with the former periods. If the unperturbed system has been discussed by means of suitable variables involving a super-fluous period, and the perturbations are small, the quantum conditions can be taken over direct from the unperturbed case. We shall meet with an example of this in connexion with the Kepler problem.

XIV. Quantisation of the Elliptic Orbits in Bohr's Theory (pp. 104, 119).

The quantisation of the elliptic orbits in the Bohr atom is carried out here by means of the method, described in the previous section, of over-quantising the system by using more periods to describe the motion than it actually has.

We have first to write down the classical laws of motion for two particles of masses m and M and charges $-e$ and Ze which attract one another according to Coulomb's law. (These correspond exactly to the laws of motion in astronomy, except that there the attracting force is the force of gravitation.) If (x_1, y_1, z_1) and (x_2, y_2, z_2) are the co-ordinates of the electron and the nucleus respectively, W, the energy of the motion, is given by

$$W = \tfrac{1}{2}m(\dot{x}_1{}^2 + \dot{y}_1{}^2 + \dot{z}_1{}^2) + \tfrac{1}{2}M(\dot{x}_2{}^2 + \dot{y}_2{}^2 + \dot{z}_2{}^2) - \frac{Ze^2}{r}.$$

If we regard the centre of gravity of the system as at rest, then $mx_1 + Mx_2$ is zero; introducing the relative co-ordinates (x, y, z), where $x = x_2 - x_1$, &c., as new variables, we have

$$x_1 = \frac{-M}{M+m}x, \quad x_2 = \frac{m}{M+m}x.$$

Similarly for the corresponding components of the velocity. The corresponding term of the kinetic energy is then

$$\tfrac{1}{2}m\dot{x}_1{}^2 + \tfrac{1}{2}M\dot{x}_2{}^2 = \tfrac{1}{2}\mu\dot{x}^2,$$

where μ is the so-called effective mass $mM/(m+M)$; cf. the discussion of the motion of the nucleus in § 1, p. 108). Hence the problem may be treated as a Kepler problem for the relative motion about the centre of gravity with co-ordinates (x, y, z) and effective mass μ.

For such motions about a fixed centre the conservation of angular momentum (theorem of areas) holds as well as the conservation of energy. The former implies in the first place that the motion is all in one plane. We take this plane as the xy-plane and transform to polar co-ordinates by means of the equations $x = r \cos \phi$, $y = r \sin \phi$. The momentum theorem then gives

$$\mu r^2 \dot{\phi} = p_\phi = \text{const.}$$

In polar co-ordinates the energy is

$$W = \tfrac{1}{2}\mu(\dot{r}^2 + r^2\dot{\phi}^2) - \frac{Ze^2}{r} = \text{const.}$$

Eliminating ϕ between these two equations and noting that

$$\dot{r} = \dot{\phi} \frac{dr}{d\phi} = \frac{p_\phi}{\mu r^2} \frac{dr}{d\phi},$$

we have

$$W = \tfrac{1}{2} \frac{p_\phi^2}{\mu} \left\{ \frac{1}{r^4}\left(\frac{dr}{d\phi}\right)^2 + \frac{1}{r^2} \right\} - \frac{Ze^2}{r}.$$

Here it is advantageous to introduce the new variable $\rho = 1/r$; for ρ we then have the differential equation

$$W = \tfrac{1}{2} \frac{p_\phi^2}{\mu} \left\{ \left(\frac{d\rho}{d\phi}\right)^2 + \rho^2 \right\} - Ze^2\rho.$$

This may readily be solved by differentiating once more with respect to ϕ; rejecting the factor $d\rho/d\phi$, we have the differential equation of the second order

$$\frac{d^2\rho}{d\phi^2} + \rho - \frac{Ze^2\mu}{p_\phi^2} = 0.$$

As is well known, its general solution is

$$\rho = \frac{Ze^2\mu}{p_\phi^2} + C\cos(\phi - \phi_0),$$

containing the two constants C and ϕ_0; by suitable choice of ϕ the latter can always be made to vanish. If for brevity we also put

$$q = \frac{p_\phi^2}{e^2 Z \mu}, \quad \epsilon = Cq = \frac{Cp_\phi^2}{e^2 Z \mu},$$

we have
$$r = \frac{1}{\rho} = \frac{q}{1 + \epsilon \cos\phi}.$$

As we know, this equation represents a conic. In order that the motion may be periodic, i.e. in order that the conic may be entirely at a finite distance, the denominator in the above polar equation must never vanish; that is, $|\epsilon|$ must be less than 1. Perihelion, i.e. the minimum value of r, corresponds to the azimuth $\phi = 0$; then $r_1 = \frac{q}{1+\epsilon}$. Aphelion corresponds to $\phi = \pi$: $r_2 = \frac{q}{1-\epsilon}$. From these we obtain the expression

$$a = \tfrac{1}{2}(r_1 + r_2) = \frac{q}{1 - \epsilon^2}$$

for the major semi-axis a. For $\phi = \pm\pi/2$, r is equal to the semi-latus-rectum q of the ellipse (see fig. 8, p. 119). Further, we readily see that ϵ is the numerical eccentricity and that the minor semi-axis is given by $b = \frac{q}{\sqrt{(1 - \epsilon^2)}}$. The energy of the motion is obtained by substituting from the equation of the ellipse in the previous expression for W; we have

$$W = \tfrac{1}{2}\frac{p_\phi^2}{\mu}\left\{\left(\frac{d\rho}{d\phi}\right)^2 + \rho^2\right\} - Ze^2\rho$$
$$= \tfrac{1}{2}\frac{p_\phi^2}{\mu}\left\{C^2 + \left(\frac{Ze^2\mu}{p_\phi^2}\right)^2\right\} - \frac{Z^2e^4\mu}{p_\phi^2} = -\frac{e^2Z}{2a},$$

which is the same as the result for a circular orbit (p. 104) if we replace the radius by the major semi-axis.

Now, although there is only one period, we prescribe two quantum conditions
$$\oint p_r\, dr = n'h, \quad \oint p_\phi\, d\phi = kh.$$

Owing to the constancy of p_ϕ, the second condition at once gives
$$p_\phi = \frac{h}{2\pi}k = \hbar k.$$

The first integral, however, is not so easy to calculate; here we have
$$n'h = \oint p_r\, dr = \oint \mu\dot{r}\, dr = \oint \mu\frac{dr}{d\phi}\,\dot\phi\,\frac{dr}{d\phi}\, d\phi$$
$$= p_\phi \oint \frac{1}{r^2}\left(\frac{dr}{d\phi}\right)^2 d\phi.$$

Accordingly, we have to evaluate the integral

$$I = \int_0^{2\pi} \frac{1}{r^2} \left(\frac{dr}{d\phi}\right)^2 d\phi = \int_0^{2\pi} \frac{1}{\rho^2} \left(\frac{d\rho}{d\phi}\right)^2 d\phi = \int_0^{2\pi} \frac{\epsilon^2 \sin^2 \phi}{(1 + \epsilon \cos \phi)^2} \, d\phi$$

for a complete revolution, i.e. ϕ runs from 0 to 2π. Integrating by parts, we have

$$I = \left[\frac{\epsilon \sin \phi}{1 + \epsilon \cos \phi}\right]_0^{2\pi} - \int_0^{2\pi} \frac{\epsilon \cos \phi}{1 + \epsilon \cos \phi} \, d\phi = -\int_0^{2\pi} \frac{\epsilon \cos \phi}{1 + \epsilon \cos \phi} \, d\phi.$$

Multiplying the second expression by $+2$ and the first by -1 and adding, we obtain

$$I = -\int_0^{2\pi} \frac{\epsilon^2 \sin^2 \phi + 2\epsilon \cos \phi (1 + \epsilon \cos \phi)}{(1 + \epsilon \cos \phi)^2} \, d\phi$$

$$= -\int_0^{2\pi} \frac{1 + 2\epsilon \cos \phi + \epsilon^2 \cos^2 \phi}{(1 + \epsilon \cos \phi)^2} \, d\phi + \int_0^{2\pi} \frac{1 - \epsilon^2}{(1 + \epsilon \cos \phi)^2} \, d\phi.$$

The first of these integrals can be evaluated at once and gives -2π; the second, in virtue of the equation of the ellipse, may be written in the form

$$\int_0^{2\pi} \frac{1 - \epsilon^2}{(1 + \epsilon \cos \phi)^2} \, d\phi = \frac{1 - \epsilon^2}{q^2} \int_0^{2\pi} r^2 d\phi.$$

This form of expression shows that apart from the factor $\dfrac{(1 - \epsilon^2)}{q^2}$ the integral represents twice the area of the elliptical orbit, which, as we know, is πab, where a and b are the semi-axes of the ellipse. Substituting the values previously found for the latter and collecting terms, we find that the first quantum condition gives

$$n'h = -2\pi p_\phi + p_\phi \frac{1 - \epsilon^2}{q^2} 2\pi \frac{q^2}{(1 - \epsilon^2)^{3/2}} = -hk + \frac{2\pi p_\phi}{\sqrt{(1 - \epsilon^2)}},$$

or

$$\frac{4\pi^2 p_\phi{}^2}{1 - \epsilon^2} = 4\pi^2 e^2 Z \mu a = h^2(n' + k)^2.$$

If we put

$$n' + k = n,$$

we have

$$a = \frac{h^2 n^2}{4\pi^2 e^2 \mu Z} = \frac{\hbar^2 n^2}{e^2 \mu Z},$$

so that the energy is

$$E = -\frac{e^2 Z}{2a} = -\frac{2\pi^2 e^4 \mu Z^2}{h^2 n^2} = -\frac{Rh}{n^2}.$$

Thus by an argument based on Bohr's theory we have obtained the Balmer term with the correct coefficient.

In deducing this we began by introducing two quantum conditions and hence two quantum numbers, the radial quantum number n' and the azimuthal quantum number k. As the orbit has only one period, however, it is found unnecessary to use both quantum numbers in finding the energy levels, as the latter involve only the sum of the two quantum numbers. This we call the *principal quantum number*, as in the unperturbed motion it alone determines the positions of the terms.

The significance of the two other quantum numbers only becomes evident if the degeneracy is removed by some perturbation (due to deviations from the Coulomb field, introduction of the relativistic variation of mass, presence of an external field, or some other cause). We can, however, gain an idea of the meaning of the quantum numbers from the purely geometrical point of view by considering the elliptic orbit. If, as in § 1 (p. 111), we denote the radius of the first circular Bohr orbit for $Z = 1$ by

$$a_1 = \frac{h^2}{4\pi^2 \mu e^2} = \frac{\hbar^2}{\mu e^2},$$

the major semi-axis of the ellipse is

$$a = \frac{a_1}{Z} n^2.$$

Using the formulæ above, we similarly obtain the values

$$b = \frac{a_1}{Z} nk, \quad q = \frac{a_1}{Z} k^2$$

for the minor semi-axis and the semi-latus-rectum. The ratio of the axes, b/a, is accordingly equal to k/n. If $n = k$, we obtain the circular orbits of the Bohr atom, for $k = 0$ the so-called pendulum orbits (straight lines through the nucleus); but, as was emphasized in the text (p. 120), the latter are to be excluded.

XV. The Oscillator according to Matrix Mechanics (p. 130).

We proceed to illustrate the fundamental ideas of matrix mechanics by means of an example, namely, the linear harmonic oscillator. We start from the classical expression for the energy,

$$W = \tfrac{1}{2}\frac{p^2}{m} + \tfrac{1}{2}fq^2,$$

which leads to the equations of motion

$$\dot{p} = -fq, \; \dot{q} = \frac{p}{m}, \quad \text{or} \quad \ddot{q} = -\frac{f}{m}q = -\omega_0^2 q, \quad \left(\omega_0 = \sqrt{\frac{f}{m}}\right).$$

The difference between classical mechanics and quantum mechanics, as was explained in the text, lies in the fact that the quantities p and q are no longer regarded as ordinary functions of the time, but stand for matrices, the element q_{nm} of which denotes the quantum amplitude associated with the transition from one energy-level E_n to another, E_m. Its square, just like q^2, the square of the amplitude in classical mechanics, is a measure of the intensity of the line of the spectrum emitted in this transition. When we introduce the matrices into the classical equations of motion, we must also bring in the commutation law

$$pq - qp = \frac{h}{2\pi i} = \frac{\hbar}{i},$$

as quantum condition.

These equations are easily solved. If the matrix (q_{nm}) is to satisfy the equation of motion $\ddot{q} + \omega_0^2 q = 0$, this equation must be satisfied by each element of the matrix independently:

$$\ddot{q}_{nm} + \omega_0^2 q_{nm} = 0.$$

Hence if we make the natural assumption

$$q_{nm} = q^{(0)}{}_{nm}\, e^{i\omega_{nm}t},$$

we must have

$$(\omega_0^2 - \omega_{nm}^2)q_{nm} = 0;$$

that is, either $q_{nm} = 0$ or $\omega_{nm} = \pm\omega_0$. Thus all the quantities q_{nm} vanish except those for which $\omega_{nm} = +\omega_0$ or $\omega_{nm} = -\omega_0$. As we are still free to arrange the numbering of the matrix elements as we please, we make the following statement: the frequency $\omega_{nm} = +\omega_0$ corresponds to the passage from the n-th state to the $(n-1)$-th (emission),

$\omega_{nm} = -\omega_0$ to the passage from the n-th state to the $(n + 1)$-th (absorption), so that

$$q_{nm} = 0 \text{ if } m \neq n \pm 1,$$
$$q_{nm} \neq 0 \text{ if } m = n \pm 1.$$

This choice of order will be found convenient later, as individual rows (or columns) of the matrices with higher suffixes correspond to states of higher energy. We would, however, expressly emphasize the fact that this fixing of the order does not destroy the generality of the solution in any way.

The co-ordinate matrix accordingly has the form

$$(q_{nm}) = \begin{bmatrix} 0 & q_{01} & 0 & 0 & \cdots \\ q_{10} & 0 & q_{12} & 0 & \cdots \\ 0 & q_{21} & 0 & q_{23} & \cdots \\ \cdot & \cdot & \cdot & \cdot & \cdot & \cdot \end{bmatrix}.$$

The momentum matrix has an analogous form; from $p = m\dot{q}$ it follows that

$$p_{nm} = im\omega_{nm}q_{nm},$$

and hence that

$$(p_{nm}) = im \begin{bmatrix} 0 & \omega_{01}q_{01} & 0 & 0 & \cdots \\ \omega_{10}q_{10} & 0 & \omega_{12}q_{12} & 0 & \cdots \\ 0 & \omega_{21}q_{21} & 0 & \omega_{23}q_{23} & \cdots \\ \cdot & \cdot & \cdot & \cdot & \cdot & \cdot & \cdot & \cdot \end{bmatrix}$$

$$= im\omega_0 \begin{bmatrix} 0 & -q_{01} & 0 & 0 & \cdots \\ q_{10} & 0 & -q_{12} & 0 & \cdots \\ 0 & q_{21} & 0 & -q_{23} & \cdots \\ \cdot & \cdot & \cdot & \cdot & \cdot & \cdot & \cdot \end{bmatrix}$$

in virtue of the above statement about the quantities ω_{nm}.

What we are chiefly interested in is the question of the energy levels. We accordingly calculate the energy matrix, using the co-ordinate matrix and the momentum matrix just given, from the classical energy function

$$W = \tfrac{1}{2}\frac{p^2}{m} + \tfrac{1}{2}fq^2 = \frac{1}{2m}(p^2 + m^2\omega_0^2q^2).$$

To do this we have to form the squares q^2 and p^2. This is carried out according to the rule given in the text (p. 129) for the multiplication of matrices. If a and b are two matrices, the elements of their product $c = ab$ are given by

$$c_{nm} = \sum_k a_{nk}b_{km}.$$

Using this rule, we readily obtain the two expressions

$$q^2 = \begin{bmatrix} q_{01}q_{10} & 0 & q_{01}q_{12} & \cdots \\ 0 & q_{10}q_{01} + q_{12}q_{21} & 0 & \cdots \\ q_{21}q_{10} & 0 & q_{21}q_{12} + q_{23}q_{32} & \cdots \\ \cdot & \cdot \ \cdot \ \cdot \ \cdot \ \cdot \ \cdot \ \cdot \ \cdot & \cdot & \end{bmatrix},$$

$$p^2 = m^2\omega_0^2 \begin{bmatrix} q_{01}q_{10} & 0 & -q_{01}q_{12} & \cdots \\ 0 & q_{10}q_{01} + q_{12}q_{21} & 0 & \cdots \\ -q_{21}q_{10} & 0 & q_{21}q_{12} + q_{23}q_{32} & \cdots \\ \cdot & \cdot \ \cdot \ \cdot \ \cdot \ \cdot \ \cdot \ \cdot \ \cdot & \cdot & \end{bmatrix}.$$

Substituting these matrices in the expression for the energy, we obtain the energy matrix

$$W = m\omega_0^2 \begin{bmatrix} q_{01}q_{10} & 0 & 0 & \cdots \\ 0 & q_{10}q_{01} + q_{12}q_{21} & 0 & \cdots \\ 0 & 0 & q_{21}q_{12} + q_{23}q_{32} & \cdots \\ \cdot & \cdot \ \cdot \ \cdot \ \cdot \ \cdot \ \cdot \ \cdot \ \cdot & \cdot & \end{bmatrix}.$$

The first thing which strikes us about this matrix is that all the elements vanish except those in the principal diagonal. This, however, is equivalent in meaning to the energy theorem, namely, the energy of a given state is constant as time goes on. Indeed, the time factors disappear from all the diagonal terms; for example, the first term would have the time factor $e^{i\omega_{01}t}e^{i\omega_{10}t}$, but this is equal to unity, as $-\omega_{01} = \omega_{10}$.

The individual terms of the principal diagonal represent the energies of the individual states, and in fact the element W_{nn} gives the energy of the n-th state. The energy matrix involves in the first place the elements of the not yet completely determined co-ordinate matrix These may, however, be obtained by means of the commutation law $pq - qp = h/2\pi i$, and the energy levels may then be found.

If, following the above rules, we substitute the co-ordinate matrix and the momentum matrix in the commutation law, we readily obtain the matrix

$$(pq - qp) = -2im\omega_0 \begin{bmatrix} q_{01}q_{10} & 0 & 0 & \cdots \\ 0 & q_{12}q_{21} - q_{10}q_{01} & 0 & \cdots \\ 0 & 0 & q_{23}q_{32} - q_{21}q_{12} & \cdots \\ \cdot & \cdot \ \cdot \ \cdot \ \cdot \ \cdot \ \cdot \ \cdot \ \cdot & \cdot & \end{bmatrix}.$$

This matrix must be equal to $h/2\pi i$, or, more accurately, equal to $h/2\pi i$ times the unit matrix

$$\begin{bmatrix} 1 & 0 & 0 & \cdots \\ 0 & 1 & 0 & \cdots \\ 0 & 0 & 1 & \cdots \\ \cdots & \cdots & \cdots & \cdots \end{bmatrix}.$$

The two matrices must accordingly have identical elements; this leads to the system of equations

$$q_{01}q_{10} = \frac{h}{4\pi m\omega_0},$$

$$q_{12}q_{21} - q_{01}q_{10} = \frac{h}{4\pi m\omega_0},$$

$$q_{23}q_{32} - q_{21}q_{12} = \frac{h}{4\pi m\omega_0},$$

$$\cdots \cdots \cdots \cdots$$

These equations can be solved in succession, giving

$$q_{n,\,n+1}\,q_{n+1,\,n} = |\,q_{n,\,n+1}\,|^2 = (n+1)\,\frac{h}{4\pi m\omega_0}.$$

If we substitute these expressions in the energy matrix, the general diagonal term becomes

$$W_{nn} = E_n = m\omega_0^2(q_{n,\,n+1}\,q_{n+1,\,n} + q_{n,\,n-1}\,q_{n-1,\,n})$$

$$= \frac{h\omega_0}{4\pi}\,(2n+1) = h\nu_0(n+\tfrac{1}{2}) \qquad (n = 0, 1, 2, \ldots).$$

Here again, therefore, we obtain for our term scheme an equidistant succession of energy levels, as in Bohr's theory. The sole difference lies in the fact that the whole term diagram of quantum mechanics is displaced relative to that of Bohr's theory by half a quantum of energy. Although this difference does not manifest itself in the spectrum, it plays a part in statistical problems. In any case it is important to note that the linear harmonic oscillator possesses energy $\tfrac{1}{2}h\nu_0$ in the lowest state, the so-called *zero-point energy*.

XVI. The Oscillator according to Wave Mechanics (p. 138).

In this section we shall obtain the solution of the wave equation for the linear harmonic oscillator. The equation is

$$\left\{ \frac{d^2}{dq^2} + \frac{8\pi^2 m}{h^2}(E - \tfrac{1}{2}fq^2) \right\} \psi = 0.$$

If we put

$$\lambda = \frac{8\pi^2 mE}{h^2}, \quad a = \frac{2\pi m\omega_0}{h} = \frac{4\pi^2 m\nu_0}{h}, \quad (\omega_0 = 2\pi\nu_0 = \sqrt{(f/m)})$$

the equation becomes

$$\left\{ \frac{d^2}{dq^2} + \lambda - a^2 q^2 \right\} \psi = 0.$$

Here λ is the proper-value parameter and we have to find the values of λ for which the equation has a finite and unique solution throughout all space.

We can write down one solution of the equation at once, namely, the so-called Gaussian error function

$$\psi = a_0 e^{-\frac{1}{2}aq^2} = a_0 e^{-2\pi^2 m\nu_0 q^2/h},$$

where the proper-value parameter is $\lambda_0 = a$, as we readily see by substitution in the wave equation. The corresponding energy is then given by

$$E_0 = \frac{h^2 a}{8\pi^2 m} = \tfrac{1}{2}h\nu_0.$$

This, we may state in anticipation, is the term of lowest energy, corresponding to the ground state. Here, therefore, we obtain the result given by the matrix method, namely, the ground state possesses a zero-point energy amounting to half Planck's quantum of energy.

In order to find the remaining solutions of the wave equation, it is convenient to assume that ψ is of the form

$$\psi = e^{-\frac{1}{2}aq^2}v(q).$$

If we substitute the above expression in the differential equation, a brief calculation gives the following differential equation for v:

$$\frac{d^2 v}{dq^2} - 2aq\frac{dv}{dq} + (\lambda - a)v = 0.$$

The factor $e^{-iaq'}$ disappears, on account of the homogeneity of the equation.

This differential equation has the following important advantage over the wave equation.

We can assume that the function v can be expressed as a power series in q of the type

$$v = \sum_{\nu=0}^{\infty} a_\nu q^\nu.$$

If then we substitute this series in the differential equation, and equate the coefficients of the various powers of q to zero, we obtain the recurrence relation

$$(\nu + 2)(\nu + 1)a_{\nu+2} + \{\lambda - a(2\nu + 1)\}a_\nu = 0,$$

involving two coefficients only; whereas if we attempt to satisfy the wave equation by a power series for ψ, we are led to a recurrence relation involving three coefficients. The recurrence relation just given enables us to calculate the coefficients of the power series.

On detailed examination, however, we find that in general the power series obtained in this way diverges more rapidly than $e^{iaq'}$ when $q \to \infty$, so that in this case the wave functions would increase beyond all bounds at infinity. The wave functions do not converge, i.e. the power series does not increase more slowly than $e^{iaq'}$ when $q \to \infty$, unless the power series terminates. This happens for certain special values of the parameter λ; in fact, as we readily see from the recurrence relation, the series breaks off at the term $\nu = n$ if the factor multiplying a_n vanishes, i.e. if

$$\lambda = \lambda_n = a(2n + 1).$$

Recalling the meaning of λ, in this case we obtain the expression

$$E_n = h\nu_0(n + \tfrac{1}{2})$$

for the energy. It is possible to give a rigorous proof that these values for E are the only values for which the differential equation has a solution of the required type. The actual solutions take the form of a product of the exponential factor stated above and a finite polynomial in q of order n; in mathematical literature these polynomials are known as Hermite's polynomials. For our purposes it is important to note that wave mechanics and matrix mechanics both give the same energy levels for the harmonic oscillator, namely, a succession of equidistant terms, the characteristic deviation from Bohr's theory being the displacement of the terms by $\frac{1}{2}h\nu_0$, half the interval between them.

XVII. The Vibrations of a Circular Membrane (p. 139).

The solution of the differential equation for the vibrations of a circular membrane,

$$\nabla^2\psi + \lambda\psi = 0 \qquad \left(\nabla^2 = \frac{\partial^2}{\partial x^2} + \frac{\partial^2}{\partial x^2}\right),$$

is easily obtained if we use polar co-ordinates. In this case, as we know, the differential equation takes the form

$$\frac{\partial^2\psi}{\partial r^2} + \frac{1}{r}\frac{\partial\psi}{\partial r} + \frac{1}{r^2}\frac{\partial^2\psi}{\partial\phi^2} + \lambda\psi = 0.$$

We see that the solution may be taken as the product of a function R depending on r only and a function Φ depending on ϕ only; i.e. in polar co-ordinates the variables are separable. The differential equation can then be split up into two equations with the single independent variables r, ϕ respectively, by means of a separation parameter, which we shall call m^2:

$$\frac{d^2R}{dr^2} + \frac{1}{r}\frac{dR}{dr} + \lambda R - \frac{m^2}{r^2}R = 0,$$

$$\frac{d^2\Phi}{d\phi^2} + m^2\Phi = 0.$$

The second equation gives

$$\Phi = \begin{Bmatrix} \cos \\ \sin \end{Bmatrix} (m\phi).$$

It is essential that the wave function shall be one-valued; but this is not the case unless m is an integer, for otherwise an increase of 2π in the value of ϕ would give a different value for the wave function; hence $m = 0, \pm 1, \pm 2, \ldots$.

The differential equation in r is the well-known equation defining the Bessel function $J_m(\sqrt{\lambda}r)$ of the m-th order with argument $\sqrt{\lambda}r$. Here, however, we must take the boundary conditions into account. As a particular case, we assume that the membrane is fixed at the circumference, so that for all points on the boundary $R(\rho) = 0$, where ρ is the radius of the membrane. Now the Bessel function of any order has an infinite number of zeros (in fact if the value of the argument is not too small its graph is similar to a sine wave); the zeros of the Bessel function $J_m(z)$ we shall call z_0, z_1, z_2, ... In order that the boundary condition may be satisfied, the argument of the Bessel functions must coincide with one of these zeros when $r = \rho$. Hence not all values of the parameter are possible, but only those for which

$$\sqrt{\lambda}\rho = z_0, \text{ or } z_1, \text{ or } z_2, \ldots.$$

In the first case the wave function has no zeros other than those for $r = \rho$ and (if $m \neq 0$) for $r = 0$; in the second case it has one other zero, in the third case two other zeros; hence in general for the proper value

$$\sqrt{\lambda_n}\,\rho = z_n$$

there are n other zeros (circular nodal lines; see fig. 20, p. 139).

XVIII. Solution of Schrödinger's Equation for the Kepler (Central Force) Problem (p. 139).

Schrodinger's equation for the Kepler problem is

$$\nabla^2\psi + \frac{8\pi^2 m'}{h^2}\left(E + \frac{e^2 Z}{r}\right)\psi = 0.$$

Changing to three-dimensional polar co-ordinates r, θ, ϕ, we obtain the equation

$$\left\{\frac{\partial^2}{\partial r^2} + \frac{2}{r}\frac{\partial}{\partial r} + \frac{1}{r^2}\frac{\partial^2}{\partial \theta^2} + \frac{\cot\theta}{r^2}\frac{\partial}{\partial \theta} + \frac{1}{r^2\sin^2\theta}\frac{\partial^2}{\partial \phi^2}\right\}\psi$$
$$+ \frac{8\pi^2 m'}{h^2}\left(E + \frac{e^2 Z}{r}\right)\psi = 0.$$

This equation is likewise separable; if we put

$$\psi = R(r)\Theta(\theta)\Phi(\phi),$$

the equation may be split up into the three equations

$$\left\{\frac{d^2}{dr^2} + \frac{2}{r}\frac{d}{dr} + \frac{8\pi^2 m'}{h^2}\left(E + \frac{e^2 Z}{r}\right) - \frac{\lambda}{r^2}\right\}R = 0,$$

$$\left\{\frac{d^2}{d\theta^2} + \cot\theta\frac{d}{d\theta} + \lambda - \frac{m^2}{\sin^2\theta}\right\}\Theta = 0,$$

$$\left\{\frac{d^2}{d\phi^2} + m^2\right\}\Phi = 0,$$

where m (as before) and λ are the separation parameters. The solution of the third equation we know already:

$$\Phi = \begin{Bmatrix} \cos \\ \sin \end{Bmatrix} (m\phi), \quad \text{or} \quad \Phi = e^{im\phi},$$

where m must be an integer, otherwise Φ is not one-valued.

The second equation is the equation defining the spherical harmonic $P_l{}^m(\cos\theta)$, when λ has the value $l(l+1)$ and $|m| \leq l$; for other values the equation has no finite one-valued solution. We shall prove this generally by combining the θ- and ϕ-relationships in the general surface harmonic $Y_l(\theta, \phi)$. If for brevity we introduce the notation

$$\Lambda = \frac{1}{\sin\theta} \frac{\partial}{\partial\theta} \sin\theta \frac{\partial}{\partial\theta} + \frac{1}{\sin^2\theta} \frac{\partial^2}{\partial\phi^2},$$

so that

$$\nabla^2 = \frac{\partial^2}{\partial r^2} + \frac{2}{r} \frac{\partial}{\partial r} + \frac{\Lambda}{r^2},$$

$Y_l(\theta, \phi)$ must satisfy the differential equation

$$\Lambda Y_l + \lambda Y_l = 0.$$

A general solution of this equation is obtained in the following way. We consider a homogeneous polynomial U_l of the l-th degree in x, y, z, which satisfies Laplace's equation

$$\nabla^2 U_l = 0.$$

If we now define a function Y_l of the ratios x/r, y/r, z/r, i.e. of the angles θ and ϕ only, by means of the equation

$$U_l = r^l Y_l,$$

substitute this in Laplace's equation $\nabla^2 U_l = 0$, and carry out the differentiation with respect to r, we obtain the equation

$$\left(\frac{\partial^2}{\partial r^2} + \frac{2}{r} \frac{\partial}{\partial r} \right) r^l Y_l + \frac{\Lambda}{r^2} r^l Y_l \equiv r^{l-2} \{ \Lambda Y_l + l(l+1) Y_l \} = 0;$$

that is, the functions defined above are solutions of the differential equation $\Lambda Y_l + \lambda Y_l = 0$ provided that

$$\lambda = l(l+1).$$

It is possible to prove that no other values of λ give finite and continuous one-valued solutions of the differential equation. The proper values of the equation

$$\Lambda Y_l + \lambda Y_l = 0$$

are accordingly equal to $l(l+1)$. Moreover, it is easy to determine the number of arbitrary parameters in any function of the l-th degree. The most general homogeneous polynomial of the l-th degree in x, y, z contains $\frac{1}{2}(l+1)(l+2)$ arbitrary constants (it contains one term in x^l, two in x^{l-1}, three in x^{l-2}, and so on, and finally $(l+1)$ terms not containing x at all). These constants, however, are connected by certain relations depending on the condition $\nabla^2 U_l = 0$; this equation is equivalent to $\frac{1}{2}l(l-1)$ equations for determining the coefficients, for $\nabla^2 U_l$ is a homogeneous function of the $(l-2)$-th degree, which must vanish identically. Hence U_l contains

$$\tfrac{1}{2}\{(l+1)(l+2) - l(l-1)\} = 2l+1$$

independent coefficients. Accordingly there are $(2l+1)$ linearly independent spherical harmonics of the l-th degree. If we write them in the usual form $Y_l^{(m)} = P_l^m e^{im\phi}$, they correspond to the $(2l+1)$ possible values of the third (magnetic) quantum number m.

We now pass on to the differential equation for the radial function R:

$$\left\{\frac{d^2}{dr^2} + \frac{2}{r}\frac{d}{dr} + \frac{8\pi^2 m}{h^2}\left(E + \frac{e^2 Z}{r}\right) - \frac{l(l+1)}{r^2}\right\} R = 0.$$

Its solution must be finite and continuous for all values of r from zero to infinity. Here we are chiefly interested in the magnitude of the proper values E for which this equation has a solution satisfying the prescribed conditions. In particular, we shall only discuss the case where $E < 0$. This corresponds to the elliptical orbits in Bohr's theory; energy must be supplied in order to remove the electron to an infinite distance from the nucleus. The case where $E > 0$ would correspond to Bohr's hyperbolic orbits.

For the sake of simplicity we shall introduce rational units. We shall measure the radius in multiples of the Bohr radius $\dfrac{h^2}{4\pi^2 me^2 Z}$ and the energy in multiples of the ground state of the Bohr atom, $\dfrac{-2\pi^2 me^4 Z^2}{h^2}$, i.e. we put (p. 111)

$$r = \rho\,\frac{h^2}{4\pi^2 me^2 Z}, \quad E = \epsilon\left(\frac{-2\pi^2 me^4 Z^2}{h^2}\right).$$

The wave equation then takes the simpler form

$$\left\{\frac{d^2}{d\rho^2} + \frac{2}{\rho}\frac{d}{d\rho} - \epsilon + \frac{2}{\rho} - \frac{l(l+1)}{\rho^2}\right\} R = 0.$$

We begin by investigating the behaviour of the function R for very large values of ρ. We accordingly omit the terms in $1/\rho$ and $1/\rho^2$ in the differential equation, so that the behaviour at infinity is given by the equation $\left\{\dfrac{d^2}{d\rho^2} - \epsilon\right\} R_\infty = 0$. Its solution is $R_\infty = e^{\pm\rho\sqrt{\epsilon}}$; the upper sign, however, is impossible, as then the wave function would increase exponentially beyond all bounds as r and ρ increased and therefore could not be regarded as a proper function.

The second special region is that of the origin. We obtain an approximation by omitting terms of the differential equation which tend to infinity more slowly than $1/\rho^2$ when $\rho \to 0$. The equation is then

$$\left\{\frac{d^2}{d\rho^2} + \frac{2}{\rho}\frac{d}{d\rho} - \frac{l(l+1)}{\rho^2}\right\} R_0 = 0.$$

Here the solutions are $R_0 = \rho^l$ and $R_0 = \rho^{-l-1}$; the second of these is impossible, for it becomes infinite at the origin.

We now know how the desired function behaves at the two singularities $\rho = 0$ and $\rho = \infty$. It is natural to assume that the complete function R is of the form

$$R = e^{-\rho\sqrt{\epsilon}}\rho^l f(\rho),$$

where f is a function of ρ which must of course behave regularly at these two points (i.e. must not increase more rapidly than $e^{+\rho\sqrt{\epsilon}}$ at infinity) and which determines the nature of the complete function in the region intermediate between the regions of validity of the power, ρ^l, and of the exponential function. Substituting this expression in the differential equation, we readily obtain the following differential equation for f:

$$\frac{d^2f}{d\rho^2} + \frac{2(l+1)}{\rho}\frac{df}{d\rho} - 2\sqrt{\epsilon}\frac{df}{d\rho} + \frac{2}{\rho}(1 - \sqrt{\epsilon}(l+1))f = 0.$$

We attempt to solve it by means of a power series in ρ (or, better, in $2\rho\sqrt{\epsilon}$) and accordingly write

$$f = \sum_0^\infty a_\nu(2\rho\sqrt{\epsilon})^\nu.$$

Substituting this in the differential equation and rearranging the terms somewhat, we obtain

$$\sum_{\nu=0}^\infty a_\nu(2\rho\sqrt{\epsilon})^{\nu-2}\nu(\nu+2l+1) - \sum_{\nu=0}^\infty a_\nu(2\rho\sqrt{\epsilon})^{\nu-1}\left(\nu+l+1-\frac{1}{\sqrt{\epsilon}}\right) = 0.$$

This series must vanish identically, so that we obtain the recurrence relation

$$a_{\nu+1}(\nu+1)(\nu+2l+2) = a_\nu\left(\nu+l+1-\frac{1}{\sqrt{\epsilon}}\right)$$

for the coefficients. At the origin the function f is of course finite and equal to the initial term a_0. At infinity, on the other hand, f becomes infinite, and in fact, as more detailed analysis shows, to a higher order than $e^{+\rho\sqrt{\epsilon}}$, provided the series for f does not terminate. If, however, the series does terminate, R vanishes at infinity, despite the fact that f becomes infinite, in virtue of the exponential factor $e^{-\rho\sqrt{\epsilon}}$. The condition that the series shall terminate is obtained from the recurrence relation; the series breaks off after the n_r-th term if

$$n_r+l+1 = \frac{1}{\sqrt{\epsilon}}$$

that is, $1/\sqrt{\epsilon}$ must be a positive integer, or

$$\epsilon = \frac{1}{n^2},$$

where $n = n_r+l+1$; n is known as the principal quantum number, n_r as the radial quantum number.

We see, therefore, that solutions of the differential equation which satisfy the conditions of finiteness, continuity, and one-valuedness can be found only for certain discrete values of the parameter ϵ, namely the values $\epsilon = 1/n^2$. Hence certain definite energy levels are alone possible, namely

$$E = -\frac{2\pi^2me^4Z^2}{h^2n^2} = -\frac{hR_0Z^2}{n^2},$$

which are those introduced in Bohr's theory.

For $n=1$ we have $l=0$, $n_r=0$ and f reduces to a constant. The wave function ψ is independent of θ and ϕ, and becomes simply

$$\psi = \frac{1}{\sqrt{\pi}}\left(\frac{Z}{a_1}\right)^{3/2}e^{-Zr/a_1}, \quad a_1 = \frac{h^2}{4\pi^2me^2},$$

if a normalization factor is added which makes $\int\psi^2 d\tau = 1$. It is seen that ψ drops off rapidly if $r > a_1/Z$. a_1/Z is the radius of the first Bohr-orbit in an atom with charge Ze. It is equal to the mean value (see Appendix XXV) $\bar{r} = \int\psi^2 r\,d\tau$ of the electronic cloud in the state $n=1$. These expectation values can be evaluated for all the higher states with $l=0$, and are found to coincide with the major semi-axes of Bohr's theory.

We may add the polynomials f are known as Laguerre's poly-nomials; these, however, we shall not discuss in more detail here, except to mention that their zeros determine the position of the nodal surfaces $r = $ const.; in fact, R has n_r nodes, not counting the zeros for $r = 0$ (in the case where $l > 0$) and $r = \infty$.

XIX. The Orbital Angular Momentum (p. 143).

In the text we gave the operators

$$\mathbf{L}_x = \frac{h}{2\pi i}\left(y\frac{\partial}{\partial z} - z\frac{\partial}{\partial y}\right)$$

. . . .

corresponding to the components of the angular momentum. The square of the resultant angular momentum is

$$\mathbf{L}^2 = \mathbf{L}_x{}^2 + \mathbf{L}_y{}^2 + \mathbf{L}_z{}^2,$$

where the expression $\mathbf{L}_x{}^2$ implies the repetition of the operator $\mathbf{L}_x$; that is, if we have operated with $\mathbf{L}_x$ on a function ψ, the application of $\mathbf{L}_x{}^2$ to the function ψ means the formation of $\mathbf{L}_x(\mathbf{L}_x\psi)$.

For the first component, for example, we accordingly have

$$\mathbf{L}_x{}^2\psi = -\left(\frac{h}{2\pi}\right)^2\left(y\frac{\partial}{\partial z} - z\frac{\partial}{\partial y}\right)\left(y\frac{\partial}{\partial z} - z\frac{\partial}{\partial y}\right)\psi,$$

or, carrying out the differentiations,

$$\mathbf{L}_x{}^2\psi = \left(\frac{h}{2\pi}\right)^2\left(-y^2\frac{\partial^2\psi}{\partial z^2} + 2yz\frac{\partial}{\partial y}\frac{\partial\psi}{\partial z} - z^2\frac{\partial^2\psi}{\partial y^2} + y\frac{\partial\psi}{\partial y} + z\frac{\partial\psi}{\partial z}\right);$$

hence the operator $\mathbf{L}_x{}^2$ is identical with the operator

$$\left(\frac{h}{2\pi}\right)^2\left(-y^2\frac{\partial^2}{\partial z^2} - z^2\frac{\partial^2}{\partial y^2} + 2yz\frac{\partial}{\partial y}\frac{\partial}{\partial z} + y\frac{\partial}{\partial y} + z\frac{\partial}{\partial z}\right).$$

Adding the three components, we have

$$\mathbf{L}^2 = \left(\frac{h}{2\pi}\right)^2\left\{-\left[(y^2+z^2)\frac{\partial^2}{\partial x^2} + (z^2+x^2)\frac{\partial^2}{\partial y^2} + (x^2+y^2)\frac{\partial^2}{\partial z^2}\right]\right.$$
$$+ 2\left[xy\frac{\partial}{\partial x}\frac{\partial}{\partial y} + yz\frac{\partial}{\partial y}\frac{\partial}{\partial z} + zx\frac{\partial}{\partial z}\frac{\partial}{\partial x}\right]$$
$$\left.+ 2\left[x\frac{\partial}{\partial x} + y\frac{\partial}{\partial y} + z\frac{\partial}{\partial z}\right]\right\},$$

or, rearranging the terms and noting that

$$\left(x\frac{\partial}{\partial x}\right)^2 = x\frac{\partial}{\partial x}\,x\frac{\partial}{\partial x} = x^2\frac{\partial^2}{\partial x^2} + x\frac{\partial}{\partial x},$$

$$\mathbf{L}^2 = \left(\frac{h}{2\pi}\right)^2\left\{-r^2\left(\frac{\partial^2}{\partial x^2} + \frac{\partial^2}{\partial y^2} + \frac{\partial^2}{\partial z^2}\right) + \left(x\frac{\partial}{\partial x} + y\frac{\partial}{\partial y} + z\frac{\partial}{\partial z}\right)^2\right.$$

$$\left. + \left(x\frac{\partial}{\partial x} + y\frac{\partial}{\partial y} + z\frac{\partial}{\partial z}\right)\right\}.$$

As we know, $\dfrac{x}{r}\dfrac{\partial}{\partial x} + \dfrac{y}{r}\dfrac{\partial}{\partial y} + \dfrac{z}{r}\dfrac{\partial}{\partial z} = \dfrac{\partial}{\partial r}$, so that

$$\mathbf{L}^2 = \left(\frac{h}{2\pi}\right)^2\left\{-r^2\Delta + r\frac{\partial}{\partial r}\,r\frac{\partial}{\partial r} + r\frac{\partial}{\partial r}\right\}.$$

If we now change to polar co-ordinates, we see (cf. Appendix XVIII, p. 400) that

$$\mathbf{L}^2 = \left(\frac{h}{2\pi}\right)^2\left\{-r^2\left(\frac{\partial^2}{\partial r^2} + \frac{2}{r}\frac{\partial}{\partial r}\right) - \Lambda + r\frac{\partial}{\partial r}\,r\frac{\partial}{\partial r} + r\frac{\partial}{\partial r}\right\}$$

$$= -\frac{h^2}{4\pi^2}\Lambda.$$

This simply means that for every state with azimuthal quantum number l, M^2 has the proper value $\dfrac{h^2}{4\pi^2}\,l(l+1)$. The magnitude of the resultant angular momentum is accordingly quantised and has the value $\dfrac{h}{2\pi}\,\sqrt{l(l+1)}$.

Further, in polar co-ordinates with polar axis z we have

$$\mathbf{L}_z = \frac{h}{2\pi i}\left(x\frac{\partial}{\partial y} - y\frac{\partial}{\partial x}\right) = \frac{h}{2\pi i}\frac{\partial}{\partial\phi}.$$

Applying this operator to the one-valued function $e^{im\phi}$ (where $m = 0$, ± 1, ± 2, &c.), we have

$$\mathbf{L}_z e^{im\phi} = \frac{h}{2\pi}\,m e^{im\phi};$$

that is, the component M_z of the angular momentum is also quantised, its proper values being integral multiples of $h/2\pi$, Bohr's unit of angular momentum. We cannot go into the behaviour of the other two components ($\mathbf{L}_x$, $\mathbf{L}_y$) here, but merely mention that the matrices corresponding to them can be evaluated by the method explained in Appendix XXI (p. 409).

XX. Deduction of Rutherford's Scattering Formula by Wave Mechanics (pp. 140, 147).

We consider Schrödinger's differential equation

$$\left\{ -\frac{h^2}{8\pi^2 m}\Delta + V + \frac{h}{2\pi i}\frac{\partial}{\partial t} \right\}\psi = 0,$$

and with it the conjugate imaginary equation

$$\left\{ -\frac{h^2}{8\pi^2 m}\Delta + V - \frac{h}{2\pi i}\frac{\partial}{\partial t} \right\}\psi^* = 0.$$

If we multiply the first equation by ψ^*, the second by ψ, and subtract, we obtain

$$-\frac{h^2}{8\pi^2 m}(\psi^*\Delta\psi - \psi\Delta\psi^*) + \frac{h}{2\pi i}\left(\psi^*\frac{\partial\psi}{\partial t} + \psi\frac{\partial\psi^*}{\partial t} \right) = 0,$$

or

$$\frac{\partial}{\partial t}(\psi\psi^*) = \frac{hi}{4\pi m}\left\{ \frac{\partial}{\partial x}\left(\psi^*\frac{\partial\psi}{\partial x} - \psi\frac{\partial\psi^*}{\partial x} \right) + \ldots \right\}.$$

Thus, defining the current vector j (number of particles per square centimetre per second) by the components

$$j_x = \frac{h}{2\pi i}\frac{1}{2m}\left(\psi^*\frac{\partial\psi}{\partial x} - \psi\frac{\partial\psi^*}{\partial x} \right), \quad j_y = \ldots, \quad j_z = \ldots,$$

we have

$$\frac{\partial |\psi|^2}{\partial t} = -\operatorname{div}j;$$

hence

$$\frac{d}{dt}\int |\psi|^2 dv = -\int \operatorname{div}j\, dv = -\int j_n d\sigma.$$

Since the last integral represents the total number of particles leaving the bounding surface per second, this relation is consistent with the assumption that

$$\int |\psi|^2 dv$$

represents the number of particles within the volume v. If ψ, and therefore s, vanishes at the boundary (for example, at infinity), then

$$\int |\psi|^2 dv = \text{const.},$$

i.e. the total number of particles remains constant.

For a stationary process ($\psi \sim e^{-(2\pi i/h)Et}$), in which particles with large energy come from infinity, we have approximately, as in the text (p. 146),

$$\left(\frac{h^2}{8\pi^2 m}\Delta + E\right)\psi = V e^{(2\pi i/h)p z}$$

or

$$\Delta\psi + k^2\psi = F(x, y, z)e^{ikz},$$

where

$$k = \frac{2\pi}{h}\,p = \frac{2\pi}{h}\,\surd(2mE), \quad F = \frac{8\pi^2 m}{h^2}\,V.$$

The general solution, which corresponds to superposition of the incident wave e^{ikz} with an outgoing spherical wave, is, for central forces,

$$\psi = e^{ikz} - \frac{1}{4\pi}\int\int\int \frac{e^{ikR}}{R}\,F(r')e^{ikz'}\,dx'\,dy'\,dz',$$

where R is the distance between $P(x, y, z)$ and $P'(x', y', z')$.

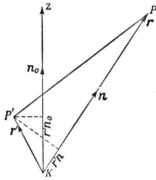

Fig. 5.—Diagram for Rutherford's scattering formula

It is sufficient to have the solution for very great distances $r = \surd(x^2 + y^2 + z^2)$ from the nucleus K (fig. 5); we have then approximately

$$R = r - r' \cdot n,$$

where r' is the vector KP', and n is the unit vector in the direction KP. We can also put

$$z' = r' \cdot n_0,$$

where n_0 is the unit vector in the z-direction. Hence

$$\psi = e^{ikz} - \frac{e^{ikr}}{4\pi r}\int\int\int e^{ikr' \cdot (n_0 - n)}F(r')\,dx'\,dy'\,dz'.$$

On introducing polar co-ordinates a, β round the vector $n_0 - n$ as axis, we find

$$\psi = e^{ikz} - \frac{e^{ikr}}{r} f(\theta);$$

here

$$f(\theta) = \frac{1}{4\pi} \int_0^{2\pi} d\beta \int_0^{\pi} \sin a \, da \int_0^{\infty} r'^2 dr' e^{iKr' \cos a} F(r'),$$

with

$$K = k \, | \, n_0 - n \, | = k\sqrt{n_x^2 + n_y^2 + (1 - n_z)^2}$$

$$= k\sqrt{2(1 - n_z)} = k\sqrt{2(1 - \cos\theta)} = 2k \sin\frac{\theta}{2},$$

where θ is the angle between n, i.e. KP, and the z-axis. The integrations with respect to a and β can be carried out:

$$f(\theta) = \frac{8\pi^2 m}{h^2} \int_0^{\infty} r'^2 V(r') \, dr' \cdot \frac{\sin Kr'}{Kr'}.$$

The intensity of the incident wave is

$$j_z^0 = \frac{h}{2\pi i} \left(\frac{ik}{m} \right);$$

that of the scattered wave (for r great) is found from the secondary wave:

$$j_r = \frac{h}{2\pi i} \frac{1}{2m} \left(\psi^* \frac{\partial\psi}{\partial r} - \psi \frac{\partial\psi^*}{\partial r} \right)$$

$$= \frac{h}{2\pi i} \frac{\{f(\theta)\}^2}{r^2} \left(\frac{ik}{m} \right).$$

Hence the scattering probability per unit solid angle, or "*differential cross-section*", is

$$w = r^2 \frac{j_r}{j_z^0} = \{f(\theta)\}^2.$$

The potential can be represented approximately by

$$V = ZeE \frac{e^{-r/a}}{r},$$

where the exponential function represents the screening effect of the bound electrons; a is of the order of magnitude of atomic radii, 10^{-8} cm.

The r-integration can then also be effected, and we get

$$f(\theta) = \frac{8\pi^2 m}{h^2} \frac{ZeE}{K^2 + 1/a^2}.$$

Now $K = 2k \sin \dfrac{\theta}{2} = \dfrac{4\pi}{\lambda} \sin \dfrac{\theta}{2}$, where λ is the de Broglie wave-length; hence $1/a^2$ is always small compared with K for particles with velocities at all rapid, except for the immediate neighbourhood of the direction of the primary beam $(\theta \sim 0)$. If $1/a^2$ is neglected, the effect of the screening disappears entirely, and we find

$$f(\theta) = \frac{8\pi^2 m}{h^2} \frac{ZeE}{K^2} = \frac{2\pi^2 m ZeE}{h^2 k^2 \sin^2 \theta/2},$$

or, with

$$k = \frac{2\pi}{h} p = \frac{2\pi}{h} mv \qquad (v = \text{velocity}):$$

$$f(\theta) = \frac{ZeE}{2mv^2} \frac{1}{\sin^2 \theta/2}.$$

On squaring, this gives exactly Rutherford's scattering formula, p. 62 (apart from the notation: m instead of M, θ instead of ϕ).

Whatever the law for $V(r)$ may be, a closed formula can be obtained (Bethe) just as easily as in the case before us, this law being expressed by an integral taken over the appropriate screening charge density, the so-called atomic form factor (see Appendix XXIV, p. 419).

XXI. Deduction of the Selection Rules for Electric Dipole Radiation (p. 150).

As was shown in § 7 (p. 149), the radiation associated with a quantum jump is given essentially by the matrix element of the co-ordinate, which is related to the wave-mechanical mean value of the electrical dipole moment in the way indicated in the text:

$$p_x = \int x \psi_{n'l'm'}^* \psi_{nlm} \, dx \, dy \, dz,$$

and so on. Here we shall prove that a large number of these integrals vanish, and that it is only for certain combinations of the quantum numbers (l, m) and (l', m') satisfying the " selection rules " that non-zero values are obtained for these integrals.

To do this we must investigate the proper functions for the various states. We deduced these in Appendix XVIII (p. 399), and showed that they may be written in the form

$$\psi_{n,\,l} = R_{n,\,l}(r)\,Y_l(\theta,\,\phi),$$

where $R_{n,\,l}(r)$ is a function of the radius only and $Y_l(\theta,\,\phi)$ denotes a general surface harmonic satisfying the differential equation

$$\Lambda Y_l + l(l+1)Y_l = 0 \quad \left(\Lambda = \frac{1}{\sin\theta}\frac{\partial}{\partial\theta}\sin\theta\frac{\partial}{\partial\theta} + \frac{1}{\sin^2\theta}\frac{\partial^2}{\partial\phi^2}\right).$$

We recall that the functions Y_l are obtained by removing the factor r^l from a homogeneous polynomial of the l-th degree, $U_l(x, y, z)$, which satisfies Laplace's equation $\nabla^2 U_l = 0$. If we introduce this form of the wave functions into the matrix element above, the integral is split up into two factors, an integral over the elementary solid angle $d\omega = \sin\theta\,d\theta\,d\phi$, which is of the form

$$J_{ll'}^{(x)} = \int \frac{x}{r} Y_{l'}^* Y_l d\omega$$

and therefore does not depend on r, and the radial integral

$$\int R_{n'l'}^* R_{nl} r^3 dr$$

(for $dx\,dy\,dz = r^2 dr\,d\omega$). Here we shall consider the first integral only, as it gives the selection rules by itself, the second integral merely determining the gradations in the intensity of the radiation for the various transitions.

We now assert that $J_{ll'}^{(x)}$ is always zero unless the selection rule $l' = l \pm 1$ is satisfied. To prove this some preliminaries are required. In the first place, it is easy to see that the integral

$$N_{ll'} = \int Y_{l'}^* Y_l d\omega$$

is zero except when $l = l'$. For the differential equation

$$Y_{l'}^* \Lambda Y_l - Y_l \Lambda Y_{l'}^* = [l'(l'+1) - l(l+1)]Y_{l'}^* Y_l$$

when Λ is written out in full becomes

$$\frac{1}{\sin\theta}\frac{\partial}{\partial\theta}\left[\sin\theta\left(Y_{l'}^* \frac{\partial Y_l}{\partial\theta} - Y_l \frac{\partial Y_{l'}^*}{\partial\theta}\right)\right] + \frac{1}{\sin^2\theta}\frac{\partial}{\partial\phi}\left(Y_{l'}^* \frac{\partial Y_l}{\partial\phi} - Y_l \frac{\partial Y_{l'}^*}{\partial\phi}\right)$$
$$= [l'(l'+1) - l(l+1)]Y_{l'}^* Y_l;$$

if we now multiply both sides by $d\omega = \sin\theta\, d\theta\, d\phi$ and integrate over the whole range of θ and ϕ, the left-hand side vanishes, the first term on account of the vanishing of $\sin\theta$ at the limits, the second term on account of its periodicity in ϕ. Hence the integral of the right-hand side must vanish also, and the fact that

$$[l'(l'+1) - l(l+1)] \int\int Y_{l'}^* Y_l \, d\omega = 0$$

leads to the result stated above, since the factor $l'(l'+1) - l(l+1)$ is not zero unless $l' = l$ (l and l' are not less than zero); if $l' \neq l$, therefore, the integral must vanish.

The proof of the selection rules is now completed as follows: as we shall show immediately, an expression of the form $\dfrac{x}{r} Y_l$ may always be represented as the sum of two general surface harmonics of orders $(l+1)$ and $(l-1)$, that is,

$$\frac{x}{r} Y_l = Y_{l+1} + Y_{l-1}.$$

If we substitute this in the integral $J_{ll'}^{(x)}$, it follows from the relation just proved that the integral vanishes identically except when $l' = l \pm 1$.

The proof of the relation just used follows from the theorem that every homogeneous polynomial $F(x, y, z)$ of the n-th degree can be reduced in one way only to an expression of the form

$$F = U_n + r^2 U_{n-2} + r^4 U_{n-4} + \ldots + r^{2h} U_{n-2h} + \ldots,$$

where the functions U_n are the potential functions introduced above. If $n = 0$ or 1 the theorem is trivial, as every polynomial of zero or first degree is already a potential function. We shall prove by induction that it is true in general. If we assume that the theorem is true for all polynomials of degree less than n, it is certain that ΔF, which is a polynomial of the $(n-2)$-th degree, can be written in the form

$$\nabla^2 F = U_{n-2}^* + r^2 U_{n-4}^* + \ldots + r^{2h-2} U_{n-2h}^* + \ldots,$$

where the functions U_n^* are potential functions different from the U_n's. We assert that the general solution of this equation, subject to the condition that F is of degree n, is

$$F = G + \frac{r^2 U_{n-2}^*}{2(2n-1)} + \frac{r^4 U_{n-4}^*}{4(2n-3)} + \ldots + \frac{r^{2h} U_{n-2h}^*}{2h(2n-2h+1)} + \ldots,$$

where G is an arbitrary function of the nth degree, which satisfies Laplace's equation $\nabla^2 G = 0$; for if we pick out an arbitrary term of this series and make use of Euler's theorem for homogeneous polynomials, we have

$$
\begin{aligned}
\nabla^2(r^{2h}U^*_{n-2h}) &= r^{2h}\nabla^2 U^*_{n-2h} + 2\,\mathrm{grad}\,r^{2h}\,\mathrm{grad}\,U^*_{n-2h} + (\nabla^2 r^{2h})U^*_{n-2h} \\
&= 0 + 2\,.\,2hr^{2h-2}(n-2h)U^*_{n-2h} + 2h(2h+1)r^{2h-2}U^*_{n-2h} \\
&= 2h(2n-2h+1)r^{2h-2}U^*_{n-2h},
\end{aligned}
$$

which gives the result stated. The polynomial F is therefore reduced to the required form if we put $U^*_{n-2h} = 2h(2n - 2h + 1)U_{n-2h}$ and $U_n = G$.

The proof of the relation stated follows at once: xU_l is a polynomial of the $(l + 1)$th degree and may therefore be expressed in the form given above: as $\nabla^2(xU_l) = 2\dfrac{\partial U_l}{\partial x}$ is itself a potential function $\left(\text{for } \nabla^2\dfrac{\partial U_l}{\partial x} = \dfrac{\partial}{\partial x}\nabla^2 U_l\right)$ of degree $l - 1$, only the first term of the series, namely U_{l-1}, appears in it, so that xU_l, involves only the first two terms of the series: hence

$$
xU_l = U_{l+1} + r^2 U_{l-1},
$$

or, using the functions Y_l and dividing the equation by r^{l+1},

$$
\frac{x}{r} Y_l = Y_{l+1} + Y_{l-1},
$$

as we asserted above.

Now that we have given a general proof of the validity of the selection rule $l' = l \pm 1$ for the orbital quantum number, we shall go on to consider the case where the levels with differing magnetic quantum numbers are split up, e.g. by an external magnetic field, and seek to find the selection rules which apply to transitions between these various levels.

For this purpose we change to the usual notation for the wave functions, in which the magnitude of the angular momentum about a particular axis may be recognized directly, viz. the general surface harmonic $Y_l(\theta, \phi)$ is replaced by the special function $P_l^m e^{im\phi}$; then the angular momentum about the polar axis is equal to $mh/2\pi$. The integral $J_{ll'mm'}$ then splits up into two integrals involving θ alone and ϕ alone; here we are interested only in the integral in ϕ, the integral in θ merely giving the selection rule just obtained for l.

We accordingly have to consider the three integrals

$$I_x = \int_0^{2\pi} x e^{i(m-m')\phi} \, d\phi,$$

$$I_y = \int_0^{2\pi} y e^{i(m-m')\phi} \, d\phi,$$

$$I_z = \int_0^{2\pi} z e^{i(m-m')\phi} \, d\phi,$$

which are readily evaluated if we introduce polar co-ordinates instead of x, y, and z. Since $z = r\cos\theta$, the third integral is zero except when $m' = m$. We combine the two others in the following way:

$$I_x \pm iI_y = \int_0^{2\pi} (x \pm iy) e^{i(m-m')\phi} \, d\phi = r\sin\theta \int_0^{2\pi} (\cos\phi \pm i\sin\phi) e^{i(m-m')\phi} \, d\phi,$$

$$= r\sin\theta \int_0^{2\pi} e^{i(m-m'\pm 1)\phi} \, d\phi.$$

This gives the selection rule $m' = m \pm 1$.

What, then, is the meaning of these two different selection rules in the case of the matrix elements $\bar{x}, \bar{y}, \bar{z}$ for the emitted radiation? We saw above that apart from the factor e (the elementary charge) this matrix element means the wave-mechanical average (probability) of the electrical dipole moment for a transition from the state nlm to the state $n'l'm'$. According to the correspondence principle, however, the existence of a dipole moment variable in time implies the emission of electromagnetic radiation; also, the radiation field of an oscillating dipole exhibits the peculiarity that there is no emission in the direction of its vibration. If, therefore, the selection rule $m' = m$ holds only for the z-component of the dipole moment, this means that the radiation corresponding to the transition $m \to m$ only appears when the electrical charge-cloud vibrates in the special direction (the z-direction). Hence this radiation can only be observed obliquely (at right angles) to the special direction. It is otherwise in the case of the two transitions $m \to m \pm 1$, which can be observed in all directions. By making observations in the x-direction, for example, we see the radiation due to the vibration of the dipole in the y-direction, which in fact is the direction of vibration of the electric vector of the radiation. If we make observations in the z-direction, we still observe the x- and y-components of the radiation, but they are circularly polarized; as we see at once from the discussion above, the transition

$m \to m + 1$ corresponds to a vibration of the charge-cloud with dipole moment $x + iy$, i.e. a circular vibration in the positive direction about the z-axis; similarly for the transition $m \to m - 1$.

These theoretical statements can be tested directly in the case of the normal Zeeman effect. As is well known, on transverse observation (at right angles to the magnetic field) we see the normal Lorentz triplet, i.e. a splitting-up into three components. Of these the central one, which corresponds to the transition $m \to m$ and is therefore not displaced, oscillates in the direction of the magnetic field, while the two other components, corresponding to the transitions $m \to m \pm 1$, vibrate transversally. In longitudinal observations the undisplaced component disappears and we see only the two displaced components, which, as theory requires, are circularly polarized.

XXII. Anomalous Zeeman Effect in the D Lines of Sodium (p. 157).

We shall now deduce the splitting pattern of the D lines of sodium in the anomalous Zeeman effect. As we stated on p. 156, the D_1 line corresponds to the transition from a p-term with inner quantum number $\frac{1}{2}$, i.e. from $(l = 1, j = \frac{1}{2})$, to an s-term, i.e. to $(l = 0, j = \frac{1}{2})$; the D_2 line represents a transition from $(l = 1, j = \frac{3}{2})$ to $(l = 0, j = \frac{1}{2})$.

We begin by determining Landé's splitting factors for the three terms in question. We have $s = \frac{1}{2}$; the formula

$$g = \tfrac{3}{2} + \frac{s(s + 1) - l(l + 1)}{2j(j + 1)}$$

then gives the following results:

$$l = 0, \ j = \tfrac{1}{2}: \quad g = \tfrac{3}{2} + \frac{\tfrac{1}{2} \times \tfrac{3}{2}}{2 \times \tfrac{1}{2} \times \tfrac{3}{2}} = 2;$$

$$l = 1, \ j = \tfrac{1}{2}: \quad g = \tfrac{3}{2} + \frac{\tfrac{1}{2} \times \tfrac{3}{2} - 1 \times 2}{2 \times \tfrac{1}{2} \times \tfrac{3}{2}} = \tfrac{2}{3};$$

$$l = 1, \ j = \tfrac{3}{2}: \quad g = \tfrac{3}{2} + \frac{\tfrac{1}{2} \times \tfrac{3}{2} - 1 \times 2}{2 \times \tfrac{3}{2} \times \tfrac{5}{2}} = \tfrac{4}{3}.$$

In the two following diagrams we collect the values of the separations of the terms, taking the separation for the normal Zeeman effect as unit, that is, we write down the values mg for the upper and lower terms of the two lines. The values of m, like j, must be halves of odd numbers, as they are equal to $-j, -j + 1, \ldots, j$.

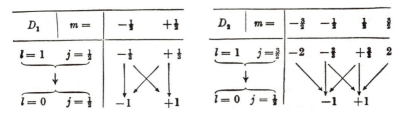

The arrows indicate the possible transitions, i.e give the positions of the lines in the Zeeman effect. Here the selection rules for the magnetic quantum number m must be taken into account. These can be deduced from the correspondence principle in exactly the same way as at p. 122 [see also Appendix XXI (p. 409)]. As m denotes the precessional motion about the direction of the field, the transition $\Delta m = \pm 1$ corresponds to the classical vibrations at right angles to H; these components of the radiation are called σ-components. When observed in the longitudinal direction (in the direction of the field), they appear circularly polarized (as predicted by the classical theory); when observed in the transverse direction, they appear linearly polarized, oscillating at right angles to the field. The transitions $\Delta m = 0$ are also permissible; they correspond to the classical vibrations in the direction of the field. These, known as π-components, are visible as vibrations parallel to the field, on transverse observation only (in the direction of vibration of a dipole, i.e. in the direction of the magnetic field in the present case, the radiation is zero).

These selection rules at once determine the positions of the components in the Zeeman splitting of the D lines. We measure their displacements on either side from a central zero position, taking as before the separation in the normal Zeeman effect, i.e. ν_L in the frequency scale, as unit. The π-components are shown above the horizontal axis and the σ-components below. We thus obtain the splitting diagram of fig. 6,

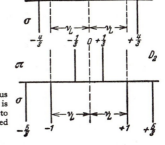

Fig. 6.—Splitting of the sodium D lines in the anomalous Zeeman effect. The splitting (ν_L) in the normal effect is shown as unity. The π-components polarized parallel to the field are shown above; the σ-components polarized perpendicular to the field, below.

which is found to be in complete agreement with that obtained experimentally (see fig. 2b, Plate X, p. 148).

XXIII. Enumeration of the Terms in the case of two p-Electrons (p. 179).

Here we consider an example of the enumeration of the terms of an atom in the case of two valency electrons (Hund); a knowledge of the number of terms is of great importance in connexion with the analysis of the corresponding spectrum and is also required in problems of a statistical nature.

For example, we suppose that the values of the terms are determined by two p-electrons and inquire what the number of terms is. Here we must consider whether these two electrons are equivalent or not, i.e. whether their principal quantum numbers are the same or not; we have of course assumed that their azimuthal quantum numbers are both equal to 1.

We shall take the second case first as it is the simpler; here we do not need to trouble about the exclusion principle, seeing that the principal quantum numbers are different. According to the rules for combining angular momenta in quantum theory, the resultant orbital momentum l may have the three values 0, 1, 2 (l_1 and l_2 may be parallel ($l = 2$) or anti-parallel ($l = 0$) or may be inclined to one another at an angle such that their vector sum is 1). Combination of the two spin moments gives resultant spin 1 for the parallel position, resultant spin 0 for the anti-parallel position. Hence both a triplet system and a singlet system occur. Moreover, owing to the three possibilities for l there are S, P, and D terms. The following terms accordingly appear:

$$^1S, \ ^1P, \ ^1D; \quad ^3S, \ ^3P, \ ^3D.$$

As the ^3S term is single [see the discussion in the text (pp. 156 and 176)], in this case there are ten different terms in all.

If an external field is applied, however, this number is considerably increased owing to the various settings of the resultant angular momentum relative to the special direction; a term with resultant angular momentum j has $2j + 1$ possible settings in the field. We thus obtain the scheme shown at head of next page. In an external field there are accordingly 36 different energy levels in all, for the case of two non-equivalent p-electrons.

The enumeration becomes more complicated in the case where the electrons are equivalent, as here the exclusion principle must be taken into account. Hence it is necessary to write down the complete system of eight quantum numbers and strike out all the cases where all the

Term	l	s	j	$2j + 1$
^1S	0	0	0	1
$^1P_1$	1	0	1	3
$^1D_2$	2	0	2	5
^3S	0	1	1	3
$^3P_0$	1	1	0	1
$^3P_1$	1	1	1	3
$^3P_2$	1	1	2	5
$^3D_1$	2	1	1	3
$^3D_2$	2	1	2	5
$^3D_3$	2	1	3	7
			Total ..	36

quantum numbers of the two electrons are the same. In the first instance, therefore, we imagine the directional degeneracy removed by an external field, so that there is a meaning in stating the components of the angular momenta in the special direction. As quantum numbers we shall use m_{l1}, m_{l2}, the projections of the orbital momenta, and m_{s1}, m_{s2}, the projections of the spin momenta, along the special direction, in addition to $n_1 = n_2$ and $l_1 = l_2 = 1$.

In the following table we have collected all the possible combinations for these four quantum numbers, omitting all those combinations which would contradict the exclusion principle. Further, we have not written down those combinations which arise from a given set merely by interchanging the quantum numbers of the two electrons; for since we cannot distinguish one electron from the other these of course are identical and are associated with the same energy term. Hence we have the following table:

m_{l1}	m_{l2}	m_{s1}	m_{s2}	m_l	m_s
1	1	$\frac{1}{2}$	$-\frac{1}{2}$	2	0
1	0	$\pm\frac{1}{2}$	$\pm\frac{1}{2}$	1	1, 0, 0, -1
1	-1	$\pm\frac{1}{2}$	$\pm\frac{1}{2}$	0	1, 0, 0, -1
0	0	$\frac{1}{2}$	$-\frac{1}{2}$	0	0
0	-1	$\pm\frac{1}{2}$	$\pm\frac{1}{2}$	-1	1, 0, 0, -1
-1	-1	$\frac{1}{2}$	$-\frac{1}{2}$	-2	0

In the fifth and sixth columns we have inserted the sums $m_l = m_{l1} + m_{l2}$ and $m_s = m_{s1} + m_{s2}$. We see, therefore, that in the magnetic field there are only 15 terms in all (magnetic splitting), while, as we saw above, in the case of non-equivalent electrons there are 36 terms in all in the magnetic field.

We are, however, not so much interested in the splitting of the terms in the magnetic field as in values for the undisturbed atom. We must therefore combine into one term of the undisturbed atom those terms in the magnetic field which have the same inner quantum number j and the same orbital angular momentum l; for we know from the above that one term with inner quantum number j is split up in a magnetic field into $2j + 1$ terms. Now from the above table we see that there must be at least one term with orbital angular momentum $l = 2$ (a D term), as there are components of this momentum with values 2 and -2 in the specified direction. We see at the same time that the corresponding spin quantum number s must vanish, as in these terms the spin component $\sigma = 0$ alone appears; we accordingly have a ^1D term with inner quantum number $j = \overrightarrow{l} + \overrightarrow{s} = 2$, which must split up into five terms in the magnetic field, terms for which $m_s = 0$ and $m_l = -2, -1, \ldots, +2$. Of the remaining ten terms, we may likewise readily convince ourselves that nine correspond to P terms ($l = 1$) with spin quantum numbers $m_s = -1, 0, 1$, while the last term is an ^1S-term. We accordingly have the following arrangement:

$$
\begin{array}{lccl}
\text{Multiplet} & & & \text{Term Notation} \\
\begin{cases} m_l = -2, -1, 0, 1, 2 \\ m_s = 0 \end{cases} & \text{corresponds to} & \begin{array}{c} \text{,,} \end{array} & \left.\begin{array}{l} l = 2 \\ s = 0 \end{array}\right\} {}^1D \\[2ex]
\begin{cases} m_l = -1, 0, 1 \\ m_s = -1, 0, 1 \end{cases} & \begin{array}{c} \text{,,} \\ \text{,,} \end{array} & & \left.\begin{array}{l} l = 1 \\ s = 1 \end{array}\right\} {}^3P \\[2ex]
\begin{cases} m_l = 0 \\ m_s = 0 \end{cases} & \begin{array}{c} \text{,,} \\ \text{,,} \end{array} & & \left.\begin{array}{l} l = 0 \\ s = 0 \end{array}\right\} {}^1S
\end{array}
$$

Here (in the case of equivalent p-electrons), therefore, there are only five terms in the absence of a magnetic field, as compared with ten terms in the case of non-equivalent electrons. The terms ^3S, ^1P, and ^3D found above fall out here (owing to the exclusion principle).

An analogous enumeration for the case of a greater number of equivalent p-electrons is given in the following table. As regards

Number of p Electrons	Terms
6	^1S
1 or 5	^2P
2 or 4	^1S, ^3P, ^1D
3	^4S, ^2P, ^2D

the energies of the terms, a quantum-mechanical estimate agrees with experiment in giving the result that the term with the greatest multiplicity, i.e. with the greatest value of s, is always the lowest; that is, if there are two or four p-electrons the ground state is a ^3P term, while if there are three p-electrons it is a ^4S term. If there are several terms of the same multiplicity, that with the greatest value of l is the lowest.

XXIV. Atomic Form Factor (p. 197).

The scattering of light by a free electron has been treated classically in Appendix VIII, p. 376. Here the same problem shall be considered from the standpoint of quantum theory for an arbitrary motion (wave function) of the electron.

For this purpose we have to generalize the wave equation of the electron for the case of an external magnetic field, since in light waves electric and magnetic fields are of equal strength. This is done by following the recipe of Chap. V, § 4, p. 134, to establish first the classical Hamiltonian H and then to translate into the language of quantum mechanics by interpreting the momenta as differential operators.

It is useful to represent the electromagnetic field by a scalar potential Φ and a vector potential A, thus

$$E = -\text{grad } \Phi - \frac{1}{c}\frac{\partial A}{\partial t}, \quad H = \text{curl } A.$$

It has been found that the action of the electromagnetic field can simply be taken into account by replacing the momentum vector p by $p - \frac{e}{c}A$, so that the ordinary Hamiltonian of the electron (p. 137) becomes

$$H = \frac{1}{2m}\left(p - \frac{e}{c}A\right) \cdot \left(p - \frac{e}{c}A\right) + e\Phi,$$

where Φ is the total scalar potential (including that of the nucleus $- e^2 Z/r$ and that of the field of the light wave).

Relativistic effects are here neglected (though it is quite easy to apply the same method to the relativistic generalization). Then to be consistent we have to neglect terms in which A and its derivatives appear in higher order than the first, and write

$$H = \frac{1}{2m}p^2 - \frac{e}{mc}p \cdot A + e\Phi.$$

In order to prove that this expression of H is correct in the frame of classical mechanics, we have to show that it leads to the well-known equations of motion, i.e. the canonical equations

$$\frac{dx}{dt} = \frac{\partial H}{\partial p_x}, \ldots$$

$$\frac{dp_x}{dt} = -\frac{\partial H}{\partial x}, \ldots$$

must be identical with

$$m\frac{dv}{dt} = e\left[E + \frac{1}{c}(v \wedge H)\right],$$

where v is the velocity vector. The expression on the right-hand side is the so-called Lorentz force, which acts on an electron in an electromagnetic field (E, H).

That this is so can be seen directly by performing the prescribed differentiations on the given expression H. Differentiating with respect to p_x gives

$$\frac{dx}{dt} = \frac{1}{m}\left(p_x - \frac{e}{c}A_x\right);$$

when differentiating with respect to x, y, z, we have to take account of the fact that Φ and the components of A depend on them. Thus we get

$$\frac{dp_x}{dt} = \frac{e}{mc}\left(p_x\frac{\partial A_x}{\partial x} + p_y\frac{\partial A_y}{\partial x} + p_z\frac{\partial A_z}{\partial x}\right) - e\frac{\partial \Phi}{\partial x}.$$

Here we have to introduce from the previous equation

$$p_x = m\frac{dx}{dt} + \frac{e}{c}A_x, \ldots$$

and

$$\frac{dp_x}{dt} = m\frac{d^2x}{dt^2} + \frac{e}{c}\left(\frac{\partial A_x}{\partial t} + \frac{\partial A_x}{\partial x}\frac{dx}{dt} + \frac{\partial A_x}{\partial y}\frac{dy}{dt} + \frac{\partial A_x}{\partial z}\frac{dz}{dt}\right), \ldots$$

If now the two expressions for dp_x/dt are equated, and square terms in A neglected, we obtain

$$m\frac{d^2x}{dt^2} = -e\left(\frac{\partial \Phi}{\partial x} + \frac{1}{c}\frac{\partial A_x}{\partial t}\right) + \frac{e}{c}\left[\frac{dy}{dt}\left(\frac{\partial A_y}{\partial x} - \frac{\partial A_x}{\partial y}\right) - \frac{dz}{dt}\left(\frac{\partial A_x}{\partial z} - \frac{\partial A_z}{\partial x}\right)\right],$$

which, by remembering the expressions of the field vectors in terms of the potentials, is seen to be the first component of the Lorentz equation,

$$m \frac{dv_x}{dt} = e\left[E_x + \frac{1}{c}\left(\frac{dy}{dt}H_z - \frac{dz}{dt}H_y\right)\right].$$

Hence our expression for the Hamiltonian H is correct. It amounts to adding the *magnetic Hamiltonian*

$$H_m = -\frac{e}{mc}\,\boldsymbol{p}\cdot\boldsymbol{A}$$

to the ordinary one given on p. 137. Now the canonical equations imply that the total Hamiltonian is constant; its value represents the total energy. Hence H_m should represent the magnetic contribution to the energy. But in the theory of the Zeeman effect (Chap. VI, § 2, p. 158), we have used a different expression for E_{magn} which can be written, in vector symbols, as the scalar product $E_m = -\boldsymbol{H}\cdot\boldsymbol{M}$, where $\boldsymbol{M}$ is the vector magnetic moment. It can, however, easily be seen that our expression for H_m is identical with that for a homogeneous magnetic field; for the vector potential of such a field, with constant H_x, H_y, H_z, can be written in the form

$$\boldsymbol{A} = \tfrac{1}{2}(\boldsymbol{H}\wedge\boldsymbol{r}) = \tfrac{1}{2}(H_y z - H_z y, \ldots),$$

hence

$$\boldsymbol{p}\cdot\boldsymbol{A} = \tfrac{1}{2}\boldsymbol{p}\cdot(\boldsymbol{H}\wedge\boldsymbol{r}) = \tfrac{1}{2}\boldsymbol{H}\cdot(\boldsymbol{r}\wedge\boldsymbol{p}) = \tfrac{1}{2}\boldsymbol{H}\cdot\boldsymbol{m},$$

where the definition of the angular momentum (Chap. V, § 5, p. 141) is used. If this is introduced into H_m, we obtain

$$H_m = -\frac{e}{2mc}\boldsymbol{H}\cdot\boldsymbol{m} = -\boldsymbol{H}\cdot\boldsymbol{M},$$

since, according to p. 158, the factor $e/(2mc)$ just transforms the mechanical momentum $\boldsymbol{m}$ into the magnetic moment $\boldsymbol{M}$.

Going over to quantum mechanics, we must take account of the fact that p_x, p_y, p_z are operators which do not commute with x, y, z, hence not with the components of $\boldsymbol{A}$, which are functions of x, y, z. Therefore, in the original expression for H, we have

$$\left(p_x - \frac{e}{c}A_x\right)\cdot\left(p_x - \frac{e}{c}A_x\right) = p_x{}^2 - \frac{e}{c}(A_x p_x + p_x A_x) + \ldots,$$

where the two terms in the bracket are not identical.

The operator representing the magnetic Hamiltonian is therefore

$$H_m = - \frac{e}{2mc} (A \cdot p + p \cdot A),$$

and the result of our considerations can now be formulated in this way: The modified Hamiltonian is obtained by adding H_m to the ordinary expression given on p. 137, and the modified Schrödinger equation differs from the ordinary one (p. 137) by adding the term

$$H_m \psi = - \frac{e}{2mc} (A \cdot p + p \cdot A)\psi$$

$$= - \frac{e}{2mc} \left(A_x \frac{h}{2\pi i} \frac{\partial}{\partial x} + \frac{h}{2\pi i} \frac{\partial}{\partial x} A_x \right)\psi + \dots$$

$$= - \frac{e}{2mc} \frac{h}{2\pi i} \left(2A_x \frac{\partial \psi}{\partial x} + \psi \frac{\partial A_x}{\partial x} \right) + \dots .$$

The next step consists in considering how the conservation law of particle numbers, derived in Appendix XX (p. 406), is affected by the additional term $H_m \psi$. Remembering our procedure there, we have to multiply the Schrödinger equation by ψ^* and its conjugate by ψ, and take the difference. If we do this with the new terms, we have

$$\psi^* H_m \psi - \psi H_m^* \psi^* = - \frac{e}{mc} \frac{h}{2\pi i} \left\{ A_x \left(\psi^* \frac{\partial \psi}{\partial x} + \psi \frac{\partial \psi^*}{\partial x} \right) + \psi \psi^* \frac{\partial A_x}{\partial x} + \dots \right\}$$

$$= - \frac{e}{mc} \frac{h}{2\pi i} \left(\frac{\partial}{\partial x} (\psi \psi^* A_x) + \dots \right) = - \frac{h}{2\pi i} \operatorname{div} s_m,$$

where

$$s_m = \frac{e}{mc} \psi \psi^* A$$

is the contribution to the total current due to the action of the magnetic field, which has to be added to the expression for s given in Appendix XX (p. 406).

It is true that the wave function ψ here is not the same as that without an external field, and that therefore the action of the field will modify the unperturbed expression s; but it can be shown that this is a negligible effect of higher order. Hence the above expression s_m represents the total particle current due to the field, if for ψ we take the unperturbed wave function.

The scattering can now be calculated by taking for A_x an incident plane wave propagating in the z-direction, hence proportional to e^{ikz}, while $A_y = 0$, $A_z = 0$. Thus the electric current in the atom has only an x-component, namely,

$$es_m = \frac{e^2}{mc} \mid \psi \mid^2 A e^{ikz},$$

and this produces a secondary field, where A has only a component Ψ parallel to x, satisfying the wave equation

$$\nabla^2 \Psi + k^2 \Psi = - \frac{4\pi}{c} es_m = -4\pi \frac{e^2}{mc^2} \mid \psi \mid^2 A e^{ikz}.$$

This has exactly the same form as that used in Appendix XX for the calculation of the α-ray scattering (p. 407), where $F(x, y, z)$ is replaced by $-4\pi \frac{e^2}{mc^2} \mid \psi \mid^2 A$. Hence the solution can be obtained in the same way as there (with the difference that the incident Ψ-wave has to be omitted); the asymptotic expression of the solution is

$$\Psi = \frac{e^{ikr}}{r} f(\theta),$$

where
$$f(\theta) = 4\pi \frac{e^2}{mc^2} A \int_0^\infty r'^2 \mid \psi \mid^2 \frac{\sin Kr'}{Kr'} \, dr'.$$

with
$$K = 2k \sin \tfrac{1}{2}\theta,$$

θ being the scattering angle, $k = 2\pi/\lambda$ the wave number. Now we have to go over from the vector potential Ψ ($=A_x$), to the field vectors E and H, and calculate the Poynting radiation vector. This is exactly the same procedure as given in Appendix VIII (p. 376) for the classical case of a free electron, and introduces the factor $(1 - \tfrac{1}{2}\sin^2\theta)$, which combines with $e^2/(mc^2)$ to give the Thomson scattering formula for free electrons $I_f(\theta) = I_0 \dfrac{1 - \tfrac{1}{2}\sin^2\theta}{r^2} \left(\dfrac{e^2}{mc^2} \right)^2$, I_0 being the incident intensity. The actual scattered intensity is then

$$I(\theta) = \mid F(\theta) \mid^2 I_f(\theta),$$

where
$$F(\theta) = 4\pi \int_0^\infty r'^2 \mid \psi \mid^2 \frac{\sin Kr'}{Kr'} \, dr'$$

is the so-called *atomic form factor* which represents the influence of the number density $\mid \psi \mid^2$ of electrons on the scattering.

The same factor determines the influence of the electronic cloud on the scattering of electrons by an atom. To see this we have to go back to the formulæ of Appendix XX, p. 407, which show that $f(\theta)$ can be written in the following form:

$$f(\theta) = \frac{1}{4\pi} \int\int\int e^{ikr' \cdot (n_o - n)} F(r') \, dx' \, dy' \, dz'.$$

For the case of an electron in interaction with a nucleus of charge Ze surrounded by a cloud of electrons with charge density $e|\psi|^2 = \rho(x, y, z)$, we have

$$F(r) = \frac{8\pi^2 m}{h^2} V(r) = \frac{8\pi^2 m}{h^2} \left(-\frac{Ze^2}{r} + e^2 \int\int\int \frac{\rho(x'', y'', z'')}{|r - r''|} \, dx'' \, dy'' \, dz'' \right).$$

If this is substituted in the previous formula, we can evaluate the integrals over $dx' \, dy' \, dz'$, by using the identity

$$\int\int\int \frac{e^{in \cdot r'}}{|r - r'|} \, dx' \, dy' \, dz' = \frac{4\pi}{|n|^2} e^{i(r \cdot n)},$$

and after some calculation we find

$$f(\theta) = \frac{8\pi^2 m e^2}{h^2} \frac{1}{K^2} \{Z - F(\theta)\};$$

substituting, according to de Broglie, $k = \dfrac{2\pi}{h} mv$,

$$f(\theta) = \frac{e^2}{2mv^2} \frac{1}{\sin^2 \frac{1}{2}\theta} \{Z - F(\theta)\}.$$

If here the term $F(\theta)$ is omitted, the square of $f(\theta)$ reduces to the well-known Rutherford formula for particles with electronic charge ($E = e$, not $E = 2e$ as for a-particles). The form factor $F(\theta)$ determines the screening effect of the electrons, by diminishing the actual nuclear charge number Z.

Both scattering formulæ are only approximations, valid under certain conditions. The optical one is restricted to velocities of the scattering electrons small compared with the velocity of light. The modified Rutherford formula holds only if the energy of the incident electrons is large compared with the interaction, so that the latter can be treated as a small perturbation. A closer examination shows that the approximation is fairly accurate for electrons with an energy

> 75 eV in hydrogen, > 100 eV in helium, > 1000 eV in neon and argon. Below these limits other methods of calculation have to be used.

We derive now the expression of the form factor obtained from the Thomas-Fermi model of the atom. Since $|\psi|^2$ represents the probability density, the charge density for (negative) electrons is $-e|\psi|^2 = \rho$. Introducing here the expression for ρ given in the text (p. 201) in terms of $\phi(r)$, and then replacing $\phi(r)$ by $\Phi(x)$, where $r = ax$ and a is related to the first Bohr radius $a_1 = h^2/(4\pi^2 me^2)$ by

$$a = a_1 \left(\frac{9\pi^2}{128Z} \right)^{1/3},$$

we find after elementary calculation the simple expression

$$4\pi|\psi|^2 = -\frac{4\pi\rho}{e} = \frac{Z}{a^3}\left[\frac{\Phi(x)}{x} \right]^{3/2}.$$

If this is substituted in the formula for $F(\theta)$ (p. 423), and x introduced as integration variable instead of r, we obtain at once

$$F(\theta) = Z \int_0^\infty x^{1/2} \left[\Phi(x) \right]^{3/2} \frac{\sin \sigma x}{\sigma x}\, dx, \quad \sigma = \frac{4\pi a}{\lambda} \sin \tfrac{1}{2}\theta,$$

which is the formula given in the text (p. 202).

In order to obtain the differential cross-section for electrons, we can either substitute this value of $F(\theta)$ into the formula for $f(\theta)$ given above, or go back to the original expression of $f(\theta)$ in terms of the potential $V(r)$, and replace this by the Fermi-Thomas function $V(r) = -e\phi(r) = -\dfrac{Ze^2}{ax}\Phi(x)$. Substituting this in the formula (Appendix XX, p. 408)

$$f(\theta) = \frac{8\pi^2 m}{h^2} \int_0^\infty r^2 V(r) \frac{\sin Kr}{Kr}\, dr,$$

and replacing r by $x = r/a$, we obtain

$$f(\theta) = -C\Psi(\sigma), \quad \Psi(\sigma) = \frac{1}{\sigma} \int_0^\infty \Phi(x) \sin \sigma x\, dx;$$

here

$$C = \frac{2Za^2}{a_1} = 2Za_1 \left(\frac{9\pi^2}{128Z} \right)^{2/3} = Z^{1/3} a_1 \pi^{1/2} C_0,$$

C_0 being a numerical constant, and, as before, $\sigma = \dfrac{4\pi a}{\lambda} \sin\frac{1}{2}\theta$; λ is the de Broglie wave-length of the electrons, $\lambda = h/(mv)$. Hence σ has the form

$$\sigma = \sigma_0 Z^{-1/3} v \sin\tfrac{1}{2}\theta,$$

σ_0 being a constant.

The differential cross-section is

$$Q(\theta) = |f(\theta)|^2 = C^2 |\Psi'(\sigma)|^2 = Z^{2/3} a_1{}^2 \pi C_0{}^2 |\Psi'(\sigma)|^2.$$

The total cross-section is obtained by multiplying this by $4\pi \sin\theta\, d\theta$, and integrating over θ, which is contained only in σ. The result can be written in the form

$$\frac{Q}{\pi a_1{}^2 Z^{2/3}} = F\left(\frac{v}{Z^{1/3}}\right),$$

where F is a universal function which can be determined by numerical integration. If finally the velocity v is replaced by $\sqrt{V}$, where V is the accelerating potential (according to the energy equation $\frac{1}{2}mv^2 = eV$), we have the relation represented by fig. 20 (p. 202) of the text.

XXV. The Formalism of Quantum Mechanics (pp. 96, 149, 409).

A complete account of quantum mechanics is beyond the scope of this book. Yet a short outline of the general formalism can be given which suffices to understand the statistical interpretation and the equivalence of matrix mechanics with wave mechanics. The connecting link is the conception of a linear operator.

In wave mechanics, there corresponds to every physical magnitude a (linear) *operator* A, which acts upon a wave function ψ of the co-ordinates q (q for brevity representing $q_1, q_2, \ldots$), and transforms it into another function $\phi(q)$:

$$\phi = A\psi.$$

To the co-ordinate q itself, considered as an operator, corresponds the operation " multiplication by q ", $\phi = q\psi$; to the momentum p corresponds the operator $\dfrac{h}{2\pi i}\dfrac{\partial}{\partial q}$, so that $\phi = \dfrac{h}{2\pi i}\dfrac{\partial\psi}{\partial q}$. Generally, we may think of A as any function of q and p; e.g. $p^2 + q^2$, which changes ψ into the function

$$\phi = (p^2 + q^2)\psi = \left(\frac{h}{2\pi i}\right)^2 \frac{\partial^2\psi}{\partial q^2} + q^2\psi.$$

Two operations A, B are in general not commutative; their *commutator* $AB - BA$ does not vanish; e.g. $pq - qp = h/2\pi i$. If an operator A reproduces a function ψ except for a factor, $A\psi = a\psi$, then ψ is called a *proper function* of A, and a a *proper value* of A. An example is given by the equation $p\psi = a\psi$, or $\dfrac{h}{2\pi i}\dfrac{\partial \psi}{\partial q} = a\psi$, with the solution $\psi = e^{(2\pi i/h)aq}$, which shows that the proper functions in general are complex numbers. The proper value a is here any number.

Generally, the following definition is found useful: to every operator A there corresponds an *adjoint* A^+ defined by the condition that for any two functions ϕ, ψ we are to have

$$\int \phi^* . (A\psi)\,dq = \int (A^+\phi)^* . \psi\,dq.$$

If $A = A^+$, A is said to be *self-adjoint* or real. That is the case, for example, for q and p. With q, this is trivial; with p, we have, by integration by parts,

$$\int \phi^* . (p\psi)\,dq = \int \phi^*\left(\frac{h}{2\pi i}\frac{\partial \psi}{\partial q}\right)dq = -\int \psi\,\frac{h}{2\pi i}\frac{\partial \phi^*}{\partial q}\,dq$$

$$= \int \psi . \left(\frac{h}{2\pi i}\frac{\partial \phi}{\partial q}\right)^* dq = \int \psi . (p\phi)^*\,dq,$$

where we assume that the integrated terms vanish. Generally, the proper values of real operators are real; for it follows from $A = A^+$, for $\phi = \psi$, that

$$\int \psi^*(A\psi)\,dq = \int (A\psi)^* . \psi\,dq = \left(\int (A\psi) . \psi^*\,dq\right)^*,$$

and is therefore real. If further $A\psi = a\psi$, then this integral is equal to $a\int \psi^*\psi\,dq$.

Let A be a real operator, and a_1, a_2 two of its proper values, ψ_1, ψ_2 the corresponding proper functions, so that we have

$$A\psi_1 = a_1\psi_1, \quad A\psi_2 = a_2\psi_2.$$

Multiply the first equation by ψ_2^*, the conjugate of the second by ψ_1; then integrate over the q-space and subtract: thus

$$\int [\psi_2^*(A\psi_1) - \psi_1(A\psi_2)^*]\,dq = (a_1 - a_2)\int \psi_1\psi_2^*\,dq.$$

The left side vanishes, since $A = A^+$. Hence, for any two unequal proper values $(a_1 \neq a_2)$,

$$\int \psi_1 \psi_2^* \, dq = 0;$$

the proper functions are said to be *orthogonal*.

Any proper function can be multiplied by an arbitrary factor; this we determine by the normalizing condition

$$\int \psi_1 \psi_1^* \, dq = \int |\psi_1|^2 \, dq = 1;$$

thus, for any two proper values a_m, a_n (assumed to be unequal)

$$\int \psi_m \psi_n^* \, dq = \delta_{mn},$$

where δ_{mn} is equal to 1 or 0, according as $m = n$, or $m \neq n$. A system of such functions is called *orthonormal*. Examples have already been given (oscillator, p. 396 ; hydrogen atom, p. 399).

A real operator has in general an infinite number of proper values $a_1, a_2, \ldots$, which we here assume for simplicity to be discrete and all different; and a set of orthonormal proper functions $\psi_1, \psi_2, \ldots$, which form a *complete* system. This means that we can expand any function ϕ in a series of such proper functions (generalized Fourier series):

$$\phi = \sum_n c_n \psi_n.$$

On multiplication by ψ_m^* this gives, in virtue of the orthonormality relations:

$$\int \phi \psi_m^* \, dq = c_m.$$

Further, we have

$$\int |\phi|^2 \, dq = \int \phi \cdot \phi^* \, dq = \sum_n c_n \int \psi_n \phi^* \, dq$$
$$= \sum_n c_n c_n^* = \sum_n |c_n|^2.$$

We can take for ϕ the function $B\psi_m$ where ψ_m is one of the proper functions of the operator A, and B another operator; the coefficients (c_n) depend now on both indices m and n:

$$B\psi_m = \sum_n B_{nm} \psi_n,$$
$$B_{nm} = \int \psi_n^* (B\psi_m) \, dq.$$

These B_{nm} can be considered as the elements of a *matrix*, representing the operator B in the system of the proper functions $\psi_1, \psi_2, \ldots$ of the operator A; for it can easily be shown that the B_{nm} have the properties of matrix coefficients. From the definition of the adjoint operator it follows that

$$B_{nm} = \int \psi_n{}^*(B\psi_m)\,dq = \int (B^+\psi_n)^* \cdot \psi_m\,dq;$$

interchanging n and m and taking the conjugate we get

$$B_{mn}{}^* = \int (B^+\psi_m) \cdot \psi_n{}^*\,dq = (B^+)_{nm}.$$

This shows that to find the matrix element of the adjoint B^+ one has to interchange lines and columns ($m \gtrless n$) and to replace each element by the conjugate value.

We consider now the matrix element of the composite operator BC:

$$(BC)_{nm} = \int \psi_n{}^*(BC\psi_m)\,dq = \int \psi_n{}^*B(C\psi_m)\,dq;$$

using the definition of the adjoint operator B^+ we get

$$(BC)_{nm} = \int (B^+\psi_n)^*(C\psi_m)\,dq.$$

Here we substitute the expansions

$$B^+\psi_n = \sum_k (B^+)_{kn}\psi_k, \quad C\psi_m = \sum_l C_{lm}\psi_l$$

and find

$$\begin{aligned}
(BC)_{nm} &= \int \sum_k (B^+)_{kn}{}^*\psi_k{}^* \sum_l C_{lm}\psi_l\,dq \\
&= \sum_{kl} B_{nk}C_{lm} \int \psi_k{}^*\psi_l\,dq \\
&= \sum_k B_{nk}C_{km}.
\end{aligned}$$

This formula shows that the matrix element of the operator product BC is equal to the element of the matrix product of the matrices belonging to B and C. In other words: The operator calculus and the matrix calculus are equivalent representations of the same mathematical structure.

It follows at once that wave mechanics and matrix mechanics of stationary states are equivalent. For the Schrödinger equation

$$\{H(p, q) - E_m\}\psi_m = 0$$

can be written as a matrix equation

$$\sum_n H_{nm}\psi_n - E_m\psi_m = 0,$$

and, by multiplying this from the left with $\psi_k{}^*$ and integrating over all q, we obtain

$$H_{km} = E_k \delta_{km};$$

or, in other words, the matrix of the energy operator calculated for its set of proper functions is diagonal. Hence the problem of wave mechanics, to find a set of matrices $p = (p_{nm})$, $q = (q_{nm})$, satisfying the commutation relation $pq - qp = h/2\pi i$, for which H is diagonal, is equivalent to determining the proper functions of the wave equation with the corresponding operator H.

It remains to show for the non-stationary case the equivalence of the time-dependent Schrödinger equation

$$\left(H + \frac{h}{2\pi i}\frac{\partial}{\partial t}\right)\psi = 0$$

with the canonical equations of mechanics

$$\frac{dq_a}{dt} = \frac{\partial H}{\partial p_a}, \quad \frac{dp_a}{dt} = -\frac{\partial H}{\partial q_a}.$$

The meaning of the partial derivatives herein is rather obvious: if H is given as a polynomial or a convergent power series in p_a, q_a, we can apply the ordinary rules of differentiation, with the one proviso, that the order of factors must be preserved, e.g. $\dfrac{\partial p^2 q}{\partial p} = 2pq$, $\dfrac{\partial pqp}{\partial p} = qp + qp$. But the meaning of the time-derivative of an operator, acting on functions of the q, is not obvious; it must in fact be suitably defined. This can only be done by referring to the physical meaning of the formalism. As explained in the text, quantum mechanics has to be interpreted in a statistical way. The following assumptions are made:

To each physical quantity or " observable " belongs a real operator A. The proper functions ψ_1, ψ_2, . . . correspond to the quantised or pure states, for which the operator takes the value a_1 or a_2 or a_3 . . . ; any function ϕ represents a state which can be considered as a mixture of pure states. The coefficients c_n of the expansion determine the strength with which the quantum state n occurs in the general state ϕ. The probability of then finding the proper value a_n in a measurement is given by

$$w_n = |c_n|^2.$$

If we assume that $\int |\phi|^2 dq = 1$, we have

$$\sum_n w_n = \sum_n |c_n|^2 = 1.$$

The *mean value*, or *expectation value*, of the quantity represented by A in the state ϕ is:

$$\bar{A} = \int \phi^* . (A\phi) \, dq = \int \phi^* \Sigma_n c_n A \psi_n . dq$$

$$= \underset{n}{\Sigma} \mid c_n \mid^2 a_n = \Sigma w_n a_n.$$

We can now define the total time rate of change dA/dt of an operator A as that operator the average of which is, for any solution of the time-dependent Schrödinger equation, identical with the time-derivative of the average of A for the same function,

$$\frac{\overline{dA}}{dt} = \frac{d\bar{A}}{dt}.$$

Now we have

$$\frac{d\bar{A}}{dt} = \frac{d}{dt} \int \psi^* (A\psi) \, dq$$

$$= \int \left\{ \frac{\partial \psi^*}{\partial t} (A\psi) + \psi^* \left(\frac{\partial A}{\partial t} \psi \right) + \psi^* \left(A \frac{\partial \psi}{\partial t} \right) \right\} dq.$$

Here it is assumed that A, apart from being a function of the q_a and $p_a = \dfrac{h}{2\pi i} \dfrac{\partial}{\partial q_a}$, may depend explicitly on time, and that $\partial A/\partial t$ is the corresponding partial derivative. Since ψ satisfies the wave equation, and ψ^* the conjugate equation, we have

$$\frac{\partial \psi}{\partial t} = - \frac{2\pi i}{h} H\psi, \quad \frac{\partial \psi^*}{\partial t} = \frac{2\pi i}{h} (H\psi)^*,$$

hence

$$\frac{d\bar{A}}{dt} = \int \left\{ \frac{2\pi i}{h} (H\psi)^* (A\psi) + \psi^* \left(\frac{\partial A}{\partial t} \psi \right) - \frac{2\pi i}{h} \psi^* A (H\psi) \right\} dq.$$

As H is self-adjoint (real), the first term under the integral can be written $\dfrac{2\pi i}{h} (H^+\psi)^* (A\psi)$ and, according to the definition of the adjoint, replaced by $\dfrac{2\pi i}{h} \psi^* H (A\psi)$. The brackets can be omitted if it is understood that all operators act to the right. Then we obtain

$$\frac{d\bar{A}}{dt} = \int \psi^* \left\{ \frac{\partial A}{\partial t} + \frac{2\pi i}{h} (HA - AH) \right\} \psi \, dq.$$

432

Equating this to

$$\overline{\frac{dA}{dt}} = \int \psi^* \frac{dA}{dt} \psi\, dq,$$

we find

$$\frac{dA}{dt} = \frac{\partial A}{\partial t} + \frac{2\pi i}{h}(HA - AH).$$

If A does not explicitly depend on time, we have

$$\frac{dA}{dt} = \frac{2\pi i}{h}(HA - AH).$$

This holds in particular when A is one of the p_a or q_a. Now it can be shown that for any function $F(p, q)$, which may also depend on t, we have

$$Fq_a - q_a F = \frac{h}{2\pi i}\frac{\partial F}{\partial p_a}, \quad Fp_a - p_a F = -\frac{h}{2\pi i}\frac{\partial F}{\partial q_a}.$$

This is obviously right if F is one of the p_a or q_a; further, suppose it holds for the functions F_1 and F_2 of p, q, then it can be directly verified that it holds also for $F_1 + F_2$ and for $F_1 F_2$. Hence it is correct for all polynomials, and all functions which can be represented by infinite series of polynomials, i.e. practically all functions.

Combining the last results, taking $F = H$, and A equal to one of the q_a and p_a, we get

$$\frac{dq_a}{dt} = \frac{2\pi i}{h}(Hq_a - q_a H) = \frac{\partial H}{\partial p_a},$$

$$\frac{dp_a}{dt} = \frac{2\pi i}{h}(Hp_a - p_a H) = -\frac{\partial H}{\partial q_a}.$$

Hence the canonical equations are a consequence of the time-dependent wave equation, and the inverse is also true, as can be seen by following the argument backwards.

In order to see what these equations mean in the matrix language, we observe first that we can interchange time-differentiation and formation of a matrix element.

To see this, take $\psi = a\psi_n + b\psi_m$, where ψ_n and ψ_m are two functions of an orthonormal system, with a and b real or complex constants satisfying $|a|^2 + |b|^2 = 1$, so that $\int |\psi|^2 dq = 1$. Then

$$\overline{A} = \int (a\psi_n + b\psi_m)^* A(a\psi_n + b\psi_m)\, dq$$
$$= |a|^2 A_{nn} + ab^* A_{mn} + a^* b A_{nm} + |b|^2 A_{mm},$$

and in the same way

$$\overline{\frac{dA}{dt}} = |a|^2 \left(\frac{dA}{dt}\right)_{nn} + ab^* \left(\frac{dA}{dt}\right)_{mn} + a^*b \left(\frac{dA}{dt}\right)_{nm} + |b|^2 \left(\frac{dA}{dt}\right)_{mm}.$$

According to the definition of dA/dt, the time-derivative of the first expression is equal to the second; but the same holds for the first and last term of the two sums respectively which are the averages for the states ψ_n and ψ_m. Hence we have

$$\frac{dA_{nm}}{dt} = \left(\frac{dA}{dt}\right)_{nm}$$

and its conjugate equation.

On the other hand, if H is explicitly independent of t, and E_n denotes one of its proper values, H has a diagonal matrix representation, $H_{nm} = E_n \delta_{nm}$; hence

$$(HA - AH)_{nm} = \sum_k (E_n \delta_{nk} A_{km} - A_{nk} E_k \delta_{km})$$

$$= (E_n - E_m) A_{nm} = h\nu_{nm} A_{nm}.$$

Combining the last two results, we find

$$\frac{dA_{nm}}{dt} = \frac{2\pi i}{h} (HA - AH)_{nm} = 2\pi i \nu_{nm} A_{nm}.$$

This is a differential equation for the matrix elements, with the solution

$$A_{nm} = a_{nm} e^{2\pi i \nu_{nm} t}, \quad h\nu_{nm} = E_n - E_m,$$

where the a_{nm} are constants. Thus we obtain the original formulation of matrix mechanics (Heisenberg, Born and Jordan, 1925), where the matrix elements were assumed to be periodical in time with periods satisfying the combination principle of Ritz (Chap. V, § 3, p. 128).

XXVI. General Proof of the Uncertainty Relation (pp. 151, 384).

The derivation of the uncertainty relation given on p. 98, with the help of diffraction phenomena and other processes capable of being visualized, gives only a result specifying the order of magnitude. To obtain an exact inequality, we must call upon the general formalism of quantum mechanics, as exposed in Appendix XXV (p. 426).

For any operator A the mean value of the product AA^+ is never negative; for, by the definition of A^+,

$$\overline{AA^+} = \int \phi^* \cdot (AA^+\phi)\,dq = \int \phi^* \cdot A(A^+\phi)\,dq$$
$$= \int (A^+\phi)^* \cdot (A^+\phi)\,dq = \int |A^+\phi|^2\,dq \geqq 0.$$

We can now deduce inequalities referring to the mean values of two real operators A, B—inequalities which lead to Heisenberg's uncertainty relation.

From the definition of A^+, it follows on multiplication by i that

$$\int \phi^* \cdot (iA\psi)\,dq = -\int (iA^+\phi)^* \cdot \psi\,dq,$$

that is to say, $(iA)^+ = -iA^+$. Somewhat more generally, we have also

$$(A + iB)^+ = A^+ - iB^+.$$

If A, B are real, and λ is a real number, then

$$\overline{(A + i\lambda B)(A + i\lambda B)^+} \geqq 0,$$

or

$$\overline{(A + i\lambda B)(A - i\lambda B)} = \overline{A^2} + \overline{B^2}\lambda^2 - i(\overline{AB - BA})\lambda \geqq 0.$$

From this it follows that the average of the commutator $AB - BA$ is purely imaginary. The minimum of the last expresssion occurs when

$$\lambda = \frac{i}{2}\frac{\overline{AB - BA}}{\overline{B^2}},$$

and is equal to

$$\overline{A^2} + \tfrac{1}{4}\frac{(\overline{AB - BA})^2}{\overline{B^2}} \geqq 0.$$

Hence

$$\overline{A^2} \cdot \overline{B^2} \geqq -\tfrac{1}{4}(\overline{AB - BA})^2.$$

Now replace A and B by $\delta A = A - \bar{A}$ and $\delta B = B - \bar{B}$; then

$$\delta A\,\delta B - \delta B\,\delta A = AB - A\bar{B} - \bar{A}B + \bar{A}\bar{B}$$
$$- BA + \bar{B}A + B\bar{A} - \bar{B}\bar{A}$$
$$= AB - BA,$$

and the preceding equation gives

$$\overline{(\delta A)^2} \cdot \overline{(\delta B)^2} \geqq -\tfrac{1}{4}(\overline{AB - BA})^2.$$

If p and q are conjugate operators (momentum and co-ordinate), we have

$$pq - qp = \frac{h}{2\pi i};$$

therefore

$$\overline{(\delta p)^2} \cdot \overline{(\delta q)^2} \geqq \frac{h^2}{16\pi^2}.$$

For the root-mean-square deviations

$$\Delta p = \sqrt{(\overline{(\delta p)^2})}, \quad \Delta q = \sqrt{(\overline{(\delta q)^2})}$$

we therefore have

$$\Delta p \Delta q \geqq \frac{h}{4\pi}.$$

The sign of equality holds only for one definite distribution (proper function), viz. the Gaussian error function, which occurs in the case of the linear oscillator (see Appendix XII, p. 383; Appendix XVI, p. 396; and Appendix XL, p. 471).

XXVII. Transition Probabilities (p. 151).

In quantum mechanics a closed system has definite stationary states in which it remains indefinitely. A physical process, consisting in transitions between these states, can only be initiated by coupling the system to another one; even the possibility of observing a system depends on such an interaction, say, with the surrounding electromagnetic field which transmits light waves or photons to the observer. The coupling must be weak, so that we can describe the effect of the coupling between the two systems as a small perturbation expressed in terms of quantities referring to the unperturbed systems. Let us consider such a perturbation by coupling with the help of the Schrödinger equation. We denote the unperturbed Hamiltonian of the uncoupled systems by H^0 (generally the sum of two or more Hamiltonians each belonging to one of the systems), and the coupling energy by H'.

Let E_n, $\psi_n(q)$ be the energy and normalized eigenfunctions of the unperturbed system; then we have

$$H^0\psi_n = E_n\psi_n.$$

The time-dependent Schrödinger equation of the perturbed (coupled) system is (Chap. V, § 4, p. 134)

$$\left(H^0 + H' + \frac{h}{2\pi i}\frac{\partial}{\partial t}\right)\psi = 0,$$

and we propose to solve it in terms of the unperturbed functions ψ_n. For this purpose we use the method of the variations of constants, expanding

$$\psi(q, t) = \Sigma_n a_n(t)e^{-\frac{2\pi i}{h}E_n t}\psi_n(q),$$

and try to determine the coefficients $a_n(t)$ in such a way that this superposition of the ψ_n satisfies the Schrödinger equation for ψ.

Now we see at once that the operators H_0 and $\dfrac{h}{2\pi i}\dfrac{\partial}{\partial t}$ applied to the product $\psi_n(q)\exp\left(-\dfrac{2\pi i}{h}E_n t\right)$ give opposite equal results; hence the equation reduces to

$$\Sigma_n e^{-\frac{2\pi i}{h}E_n t}\left(a_n H'\psi_n + \frac{h}{2\pi i}\frac{da_n}{dt}\psi_n\right) = 0.$$

If we multiply this by the conjugate function $\psi_m{}^*$ and integrate over the q, we obtain because of the orthogonality relations

$$\Sigma_n e^{-\frac{2\pi i}{h}E_n t}H'_{mn}a_n + \frac{h}{2\pi i}e^{-\frac{2\pi i}{h}E_m t}\frac{da_m}{dt} = 0,$$

where

$$H'_{mn} = \int \psi_m{}^* H'\psi_n dq$$

are the matrix elements of the operator H' (Appendix XXV, p. 426). We multiply by $\exp\left(\dfrac{2\pi i}{h}E_m t\right)$ and introduce the transition frequencies

$$\nu_{mn} = (E_m - E_n)/h;$$

then we obtain

$$\Sigma_n e^{-2\pi i\nu_{mn}t}H'_{mn}a_n + \frac{h}{2\pi i}\frac{da_m}{dt} = 0,$$

a set of linear differential equations for the $a_n(t)$ with given time-dependent coefficients.

Let us now assume that initially the system was in the unperturbed state E_k, ψ_k, so that the expression for $\psi(q, t)$ must reduce to $\psi_k(q)$ for $t = 0$. Hence all $a_n(0)$ must vanish except $a_k(0)$ which is unity, or

$$a_n(0) = \delta_{nk} = \begin{cases} 1 \text{ for } n = k, \\ 0 \text{ for } n \neq k. \end{cases}$$

As we assume H' to be a small perturbation, we can obtain an approximate solution by substituting these initial values into the sum of the differential equation; then

$$\frac{da_m}{dt} = -\frac{2\pi i}{h} H'_{mk} e^{-2\pi i \nu_{mk} t},$$

and this can be directly integrated so that the initial conditions remain satisfied:

$$a_m = \frac{2\pi i}{h} H'_{mk} \frac{1 - e^{2\pi i \nu_{mk} t}}{2\pi i \nu_{mk}}.$$

According to Appendix XXV, p. 430, the square of the modulus of this expression is the probability of finding the system in its mth unperturbed state at the time t, if it is known to have been in the state k at $t = 0$; this is the transition probability $p_{mk}(t)$,

$$p_{mk}(t) = \left(\frac{2\pi}{h}\right)^2 |H'_{mk}|^2 \left(\frac{\sin \pi \nu_{mk} t}{\pi \nu_{mk}}\right)^2,$$

which is symmetric in the two states m, k as H' is a real operator, $H'^+ = H$; hence $|H'_{mk}|^2 = H'_{mk} H'^*_{mk} = H'_{mk} H'_{km}$. This probability is periodic in time, therefore hardly observable for atomic systems (high frequencies ν_{mk}).

But the situation becomes quite different if the unperturbed system has energy levels which are so close together that they can be regarded as a continuous distribution. Let us consider as an example the case where a system with discrete energy states is coupled to one with a continuous range, and in particular the transition from a state k, where only the first system is excited, the second not, to a state m, where the first system has a lower energy, $E_m < E_k$, but some energy E is transferred to the second system. Then we have to replace ν_{mk} by $\nu_{mk} + \nu = \nu - \nu_{km}$, where $\nu = E/h$ and $\nu_{km} > 0$.

The matrix element of H' depends now, apart from the states m and k, on the energy E transferred to the continuous range, or on $\nu = E/h$, and we have to write $H'_{mk}(\nu)$.

If we make these substitutions in $p_{mk}(t)$ and multiply by the density function $\rho(\nu)$ of the final states, we obtain the total transition probability to all the lower states as the integral over $dE = h\,d\nu$:

$$P_{mk}(t) = \frac{4\pi^2}{h^2} \int_0^\infty |H'_{mk}(\nu)|^2 \left(\frac{\sin \pi(\nu - \nu_{km})t}{\pi(\nu - \nu_{km})}\right)^2 \rho(\nu)\,d\nu.$$

Now we can assume t to be large compared with all the atomic periods $(\nu - \nu_{km})^{-1}$ and determine the limit of $P_{mk}(t)$ for $t \to \infty$, by using the well-known formula

$$\lim_{t\to\infty} \frac{1}{t} \int_a^b f(x) \left(\frac{\sin \pi x t}{\pi x}\right)^2 dx = f(0),$$

which holds if the interval (a, b) includes the origin $x = 0$. With the substitution $\nu - \nu_{km} = x$, we obtain for the *transition probability per unit time*

$$P_{mk} = \lim_{t\to\infty} \frac{1}{t} P_{mk}(t) = \lim_{t\to\infty} \frac{4\pi^2}{ht} \int_{-\nu_{km}}^\infty |H'_{mk}(x + \nu_{km})|^2 \rho(x + \nu_{km}) \left(\frac{\sin \pi x t}{\pi x}\right)^2 dx,$$

and as $\nu_{km} > 0$,

$$P_{mk} = \frac{4\pi^2}{h^2} |H'_{mk}(\nu_{km})|^2 \rho(\nu_{km}).$$

If now $\rho(E)$ is defined in such a way that

$$\rho(E)\,dE = \rho(E)h\,d\nu = \rho(\nu)\,d\nu,$$

hence

$$\rho(E) = \frac{1}{h} \rho(\nu),$$

we obtain

$$P_{mk} = \frac{4\pi^2}{h} |H'_{mk}|^2 \rho(E).$$

This is the formula quoted in the text (Chap. V, § 7, p. 151). It shows that the transition probability per unit time is, under the assumptions made, determined by the value of the product of the perturbation matrix element and the density taken for the transition frequency of the discontinuous spectrum.

The application to the case of emission of radiation by an atom will be treated in Appendix XXVIII (p. 439).

XXVIII. Quantum Theory of Emission of Radiation (pp. 149, 438).

The method used here for the treatment of emission of radiation by an atomic system is in fact very general, and can be applied not only to other cases of interaction between atoms and the electromagnetic field (absorption, scattering), but also to any interaction of particles with fields, for instance of a nucleon with the meson field of Yukawa. It consists of analysing the field into its normal modes, which are, for a sufficiently big volume, practically plane waves; the amplitudes of these can be shown to behave like harmonic oscillators and can be treated by quantum mechanics. The interaction of the atom with this set of oscillators representing the field can then be described with the help of the quantum-mechanical transition probability.

We consider first the radiation alone, without the atomic system. The exact mathematical analysis of the electromagnetic vibrations in a very large volume shows that the solution at some distance from the boundary is practically independent of the shape of the surface and can be approximately regarded as a superposition of plane standing waves; one of these may be written down explicitly, namely, that for which the vector potential A vibrates in the x-direction, while it is propagated in the z-direction,

$$A_x = q(t)\sqrt{2} \cdot \cos(kz + \delta), \quad A_y = 0, \quad A_z = 0; \quad kc = \omega = 2\pi\nu,$$

where the spatial factor is normalized in such a way that the integral of the square over the unit of space is unity, e.g.

$$\int_{(1)} 2\cos^2(kz + \delta)\,dV = 1.$$

The phase δ depends on the rather accidental situation of the origin relative to the distant surface. $q(t)$ satisfies the equation

$$\ddot{q} + \omega^2 q = 0,$$

and can therefore be regarded as oscillator amplitudes of frequency $\nu = \omega/2\pi$.

The electric and magnetic field are found from the equations

$$E = -\frac{1}{c}\dot{A}, \quad H = \operatorname{curl} A;$$

the only non-vanishing components are

$$E_x = -\frac{1}{c}\sqrt{2}\dot{q}\cos(kz+\delta),$$

$$H_y = k\sqrt{2}q\sin(kz+\delta).$$

The total electromagnetic energy in the volume V can be easily evaluated in terms of q_1 and q_2, namely,

$$U = \frac{1}{8\pi}\int (E_x{}^2 + H_y{}^2)dV = \frac{V}{4\pi c^2}\,\tfrac{1}{2}(\dot{q}^2 + \omega^2 q^2).$$

It appears as the energies of a harmonic oscillator with frequency $\nu = \omega/2\pi$ and mass

$$m = V/4\pi c^2.$$

Now this oscillator can be treated according to quantum mechanics; it has a series of stationary states $n = 0, 1, 2, \ldots$, with energies $h\nu(n+\tfrac{1}{2})$. It is obvious that n can be interpreted as the number of photons $h\nu$ associated with the wave; but even the state of no photons present, $n = 0$, corresponds to a wave with zero-point energy $h\nu/2$.

If we now consider similar waves in different directions and with different frequencies, it is clear that these do not interfere (no cross terms in the energy integral) and that their energies are simply additive (owing to the normalization of the space part of the wave function). Hence the electromagnetic field corresponds dynamically to a mixture of independent photons flying about with the velocity of light.

We now consider an atomic system situated at the origin in interaction with the electromagnetic field. Let v be the velocity of one electron (if there are several electrons, we have simply to take the sum over all), and let us assume that the linear dimensions of the atomic system are negligibly small compared with the wave-length $\lambda = 2\pi/k$. Then the interaction energy is (Appendix XXIV, p. 421)

$$H' = -\frac{e}{c}\,v\cdot A(0) = -\frac{e}{c}\,v_x A_x(0) = -\frac{e}{c}\,v_x q\sqrt{2}\cos\delta.$$

We now apply the theory of transition as explained in Appendix XXVII. Then we have to form the matrix element between the initial and final state,

$$H'_{if} = -\frac{e}{c}\sqrt{2}\,(v_x q)_{if}\cos\delta = -\frac{e}{c}\sqrt{2}\,v_{ab}q_{nn'}\cos\delta,$$

where a, b denote two states of the atomic system and n, n' two states of the radiation oscillator q. Now $v_{ab} = \dot{x}_{ab} = \omega_{ab}x_{ab}$, where $h\omega_{ab} = 2\pi(E_a - E_b)$ and x_{ab} is the matrix element of the co-ordinate of the electron. Hence

$$H'_{if} = -\frac{e}{c}\sqrt{2}\,\omega_{ab}x_{ab}q_{nn'}\cos\delta.$$

We apply this to the case of emission by assuming that in the initial state i the atom is in the excited state a, and no photon present (all $n = 0$), whereas in the final state f the atom is in the lower state and some photons present; or what is the same, one of the radiation oscillators is excited. But as the matrix elements $q_{nn'}$ of an oscillator are always zero except if $n' = n \pm 1$, we have in our case only the matrix element q_{01} to consider, corresponding to the emission of one photon. We have found in Appendix XV, p. 395, that

$$|\,q_{01}\,|^2 = \frac{h}{4\pi m\omega} = \frac{hc^2}{V\omega},$$

where the value of m obtained above is used.

Before substituting this in the expression for H'_{if}, we remember that for any direction of propagation there are two directions of polarization normal to it; in our case where the wave moves along, we have an x- and y-vibration. Hence we obtain

$$|\,H'_{if}\,|^2 = \frac{2e^2}{c^2}\,\omega_{ab}(|\,x_{ab}\,|^2 + |\,y_{ab}\,|^2)\frac{hc^2}{V}\cos^2\delta.$$

To obtain the transition probability per unit time, we have to multiply this by $4\pi^2\rho(\nu_{ab})/h^2$, where $\rho(\nu)$ is the density of vibrations in the ν-scale (p. 218).

The transition probability per unit time is now

$$P_{if} = \frac{4\pi^2}{h^2}\,|\,H'_{if}\,|^2\,\rho = \frac{8\pi e^2}{hc^3}\,\omega_{ab}{}^3(|\,x_{ab}\,|^2 + |\,y_{ab}\,|^2)\cos^2\delta.$$

As the position of the atom (the origin) is arbitrary, we have to average over all phases δ; then the factor $\cos^2\delta$ becomes $\frac{1}{2}$. Further we have to average over all orientations of the atom and therefore to replace $|\,x_{ab}\,|^2 + |\,y_{ab}\,|^2$ by $\frac{2}{3}|\,r_{ab}\,|^2$, where r is the radius vector of the electron. Then the radiation emitted per second is obtained by multi-plying with $h\nu_{ab} = h\omega_{ab}/2\pi$:

$$I = \frac{4}{3}\frac{e^2}{c^3}\omega_{ab}{}^4 \mid r_{ab} \mid^2,$$

which agrees with the formula derived by a correspondence considera-tion (Chap. V, § 7, p. 149).

Suppose that in the initial state there is a radiation present, con-taining n photons of a special kind (direction, frequency), then the emission is different; it involves the matrix element $q_{n,\,n+1}$. Now the formulæ of Appendix XV, p. 395, show that

$$\mid q_{n,\,n+1} \mid^2 = (n+1) \mid q_{01} \mid^2;$$

hence the emission is multiplied by the factor $(1+n)$, or, in other words, there is, apart from the spontaneous emission calculated above, an induced emission proportional to n, i.e. to the strength of the radia-tion present.

The case of absorption is obtained by regarding the initial state i as that where the atom is in the lower energy level b and n quanta present, whereas in the final state f the atom is in the higher level a, and only $(n-1)$ quanta present. The matrix element involved is therefore $q_{n,\,n-1} = q^*_{n-1,\,n}$, and we have

$$\mid q_{n,\,n-1} \mid^2 = n \mid q_{01} \mid^2.$$

Substituting this in the transition probability instead of $\mid q_{01} \mid^2$, we obtain at once the coefficient of absorption. We shall not give its explicit expression, but consider the case of *thermal equilibrium* be-tween radiation and atoms, where in the average the same number of emissions and absorptions take place. Let $g = \rho\,d\nu$ be the number of waves in a small frequency interval $d\nu$, and $\bar{n} = \Sigma n$ the total number of photons in $d\nu$, then the ratio of the total number of emissions and absorptions in $d\nu$ is, according to the results just obtained, $(\bar{n}+g):\bar{n}$. If N_a and N_b are the number of atoms in the states a and b respec-tively, the condition for statistical equilibrium is obviously

$$N_a(\bar{n}+g) = N_b\bar{n},$$

hence

$$\bar{n} = \frac{g}{N_a/N_b - 1}.$$

We use now the general law of statistical mechanics, called Boltzmann's theorem (Chap. I, § 6, p. 14), which states that in statistical equilibrium

$$\frac{N_a}{N_b} = \frac{e^{-E_a/kT}}{e^{-E_b/kT}} = e^{-h\nu_{ab}/kT},$$

where T denotes the absolute temperature, and k Boltzmann's constant. Therefore the average number of light quanta $\bar{n}$ of a specified frequency ν, which can be emitted and absorbed by an atom at temperature T, is

$$\bar{n} = \frac{g}{e^{h\nu/kT} - 1}.$$

If we assume that there are atomic systems with so many energy levels present that practically all frequencies ν are involved, the formula represents the average distribution of photons in thermal equilibrium with matter. The radiation energy per frequency interval $d\nu$, $u_\nu d\nu$, is obtained by multiplying $\bar{n}$ by $h\nu$, and putting $g = \rho(\nu)d\nu = 8\pi\nu^2 V d\nu/c^3$ (corresponding to the existence of two polarized waves for each direction):

$$u_\nu = \frac{8\pi h\nu^3}{c^3} \frac{1}{e^{h\nu/kT} - 1}.$$

This is Planck's celebrated formula, derived by another method in Chap. VII, p. 210, which makes no use of the transition probabilities between states deviating from statistical equilibrium, but is based solely on the properties of the equilibrium distribution.

A historical remark must be added. The method used here is in fact older than quantum mechanics. Einstein (1917) in a deep analysis of the meaning of Planck's formula introduced the conception of transition probabilities for spontaneous emission, absorption, and induced emission, and showed that in order to obtain the correct equation we have to assume that they were in the ratios $g : \bar{n} : \bar{n}$. When multiplied by $h\nu$ these quantities are proportional to the Einstein A and B coefficients A_{nm}, B_{nm} and B_{mn} respectively. Thus from the expression for g above it follows that $B = c^3 A/(8\pi h\nu^3)$. The equality of the coefficients for absorption and induced emission means that a radiation field produces in an atom the same probability for the transitions $b \to a$ and $a \to b$. This symmetry was later one of the clues which led to matrix mechanics (where it follows from the symmetry of $|\, \mathbf{r}_{ab} \,|^2 = \mathbf{r}_{ab} \cdot \mathbf{r}_{ab}^* = \mathbf{r}_{ab} \cdot \mathbf{r}_{ba}$).

XXIX. **The Electrostatic Energy of Nuclei** (p. 308).

The electrostatic self-energy of a nucleus is given by the well-known formula

$$E = \tfrac{1}{2} \int \rho\phi\, dx\, dy\, dz,$$

where ρ is the density produced by all the Z protons, and ϕ the potential at a point in the interior due to all but one of the protons.

Assuming the shape of the nucleus to be a sphere of radius R, we have therefore

$$\rho = Ze\, \frac{3}{4\pi R^3},$$

$$\phi = (Z - 1)e\, \frac{3}{2R}\left(1 - \tfrac{1}{3}\frac{r^2}{R^2}\right);$$

hence

$$E(Z) = \tfrac{9}{4}\frac{Z(Z-1)e^2}{R^4}\int_0^R \left(1 - \tfrac{1}{3}\frac{r^2}{R^2}\right) r^2\, dr,$$

or

$$E(Z) = \tfrac{3}{5}Z(Z-1)\,\frac{e^2}{R}.$$

The difference in electrostatic energy between a nucleus $(Z + 1,\, A)$ and its "mirror-nucleus" $(Z,\, A)$ is therefore

$$W = E(Z + 1) - E(Z) = \tfrac{6}{5}Z\frac{e^2}{R}.$$

XXX. **Theory of α-Disintegration** (p. 311).

Let the potential for an α-particle, which has been emitted from a nucleus of atomic number Z and is therefore in the field of the residual nucleus $Z - 2$, be $V(r)$. For great distances this is the Coulomb potential, i.e.

$$V(r) = \frac{2(Z - 2)e^2}{r}, \text{ for } r > r_0;$$

for $r < r_0$ the form $V(r)$ is unknown, but special assumptions about it are unnecessary, apart from this, that it must have the crater-like character represented in fig. 1, p. 309.

According to Laue, the frequency λ of the emission can be split up into two factors. We think of the α-particle as oscillating to and fro in the crater, so that it strikes the wall n times per second. At each collision there is a certain probability p that it passes the wall. Hence

$$\lambda = np.$$

The order of magnitude of n we can take to be $v/2r_0$, where v is the velocity of the a-particle; the latter in its turn can be determined by putting the wave-length h/mv of the associated de Broglie wave equal to $2r_0$. We thus find

$$n = \frac{h}{4mr_0{}^2}.$$

To calculate p, we have to find a suitable solution of Schrödinger's equation

$$\frac{h^2}{8\pi^2 m} \frac{d^2\psi}{dr^2} + (E - V(r))\psi = 0.$$

For $r < r_0$, the function ψ will oscillate in some way or other; for large values of r it will be a progressive wave. If the energy of the a-particle is E, the latter state will be attained at about the distance where

$$E = \frac{2(Z - 2)e^2}{r_1}, \quad r_1 = \frac{2(Z - 2)e^2}{E}.$$

In the intermediate zone from r_0 to r_1, ψ will fall off exponentially. Thus, clearly we have approximately

$$p = \left| \frac{\psi(r_1)}{\psi(r_0)} \right|^2.$$

But Coulomb's law still holds in this intermediate zone; we have therefore to solve the equation (compare p. 137)

$$\frac{h^2}{8\pi^2 m} \psi'' + \left(E - \frac{2(Z - 2)e^2}{r} \right) \psi = 0.$$

Since Z is a large number, the following approximate method gives what is wanted. Put $\psi = e^{(2\pi/h)y(r)}$; we then obtain the equation

$$\frac{h}{2\pi} y'' + y'^2 - F(r) = 0,$$

where

$$F(r) = 2m \left(-E + \frac{2(Z - 2)e^2}{r} \right).$$

Neglecting the term multiplied by h, we get

$$y' = \sqrt{F(r)}, \quad y = \int_a^r \sqrt{F(r)}\, dr,$$

and therefore

$$\frac{\psi_1}{\psi_0} = e^{(2\pi/h)[y(r_1) - y(r_0)]}$$

$$= e^{(2\pi/h)\int_{r_0}^{r_1} \sqrt{F(r)}\, dr}.$$

If we substitute the expression for $F(r)$, the integration can be carried out, and we find

$$p = e^{-(2n_0 - \sin 2n_0)(8\pi e^2/h)(Z-2)/v}$$

where n_0 is given by the equation

$$\cos^2 n_0 = \frac{r_0 E}{2(Z-2)e^2}.$$

If we expand the exponent in powers of the quantity on the right, which for small α-energies and high potential barriers is a small quantity, we obtain finally the approximation

$$p = e^{-8\pi^2 e^2 (Z-2)/(hv) + (16\pi e \sqrt{m}/h)(\sqrt{(Z-2)r_0})}.$$

The disintegration constant λ is therefore given by

$$\log \lambda = \log \frac{h}{4mr_0^2} - \frac{8\pi^2 e^2 (Z-2)}{hv} + \frac{16\pi e}{h}\sqrt{m(Z-2)r_0},$$

or, numerically,

$$\log_{10}\lambda = 20.46 - 1.191 \times 10^9 \frac{Z-2}{v} + 4.084 \times 10^6 \sqrt{(Z-2)r_0};$$

here λ is measured in sec.$^{-1}$, and is connected with the half-value period by the equation

$$T = \frac{\log_e 2}{\lambda} = \frac{0.6931}{\lambda}.$$

The law thus obtained differs from the empirical formula of Geiger and Nuttall in giving a linear dependence of $\log \lambda$ on $1/v$, apart from a dependence on Z and r_0; however, since the variation of v is confined to comparatively narrow limits, from $v = 1.4 \times 10^9$ cm./sec. to $v = 2.0 \times 10^9$ cm./sec., and the variation of Z and r_0 is negligible for the elements of the radioactive series, the difference is slight. In consequence of the large factor in the second term—in which v appears— the range of values of the disintegration constant is extremely wide.

XXXI. **The Ground State of the Deuteron** (p. 318).

The wave equation for the relative motion of the two nucleons in the deuteron is

$$\nabla^2 \psi + \frac{4\pi^2}{h^2} M[E - V(r)]\psi = 0,$$

where $V(r)$ is the central potential, M the average mass of the nucleons, and $W = -E$ the binding energy. As suggested in the text, a box form of $V(r)$ is assumed:

$$V(r) = -V_0, \text{ for } r < a,$$
$$= 0, \quad \text{ for } r > a.$$

The range a is assumed to be of the order of e^2/mc^2, as nucleon scattering experiments indicate. Since $V(r)$ is spherically symmetrical, the ground state will be an S-state. By making the substitution $\psi(r) = u(r)/r$, the wave equations for the two regions of r are

$$\frac{d^2}{dr^2} u_1 + \frac{4\pi^2 M}{h^2} (V_0 - W)u_1 = 0, \text{ for } r < a;$$

$$\frac{d^2}{dr^2} u_2 - \frac{4\pi^2 M}{h^2} W u_2 = 0, \text{ for } r > a.$$

The boundary conditions for u_1, u_2 are $u_1 \to 0$ as $r \to 0$ and $u_2/r \to 0$ as $r \to \infty$, together with continuity conditions to be satisfied at $r = a$. The required solutions are

$$u_1 = A \sin kr,$$
$$u_2 = Be^{-ar},$$

where

$$k = \frac{2\pi}{h} \sqrt{M(V_0 - W)},$$

$$a = \frac{2\pi}{h} \sqrt{(MW)}.$$

A and B are arbitrary constants. A simple relation between the two parameters V_0 and a can be obtained by using the continuity condition

$$\left(\frac{1}{u_1}\frac{du_1}{dr}\right)_{r=a} = \left(\frac{1}{u_2}\frac{du_2}{dr}\right)_{r=a},$$

namely,

$$k \cot ka = -a.$$

The value of W is known from experiments to be 2·19 MeV. Substituting the value of $a = e^2/mc^2$, we get

$$V_0 = 41mc^2 = 21 \text{ MeV.}$$

It can be shown that the relation $k \cot ka = -a$ obtained above implies that no bound excited S-states can exist. For the value of

$$\cot ka = -\frac{a}{k} = -\left(\frac{W}{V_0 - W}\right)^{1/2}$$

is a small negative number already for the ground state, and will be smaller for excited states if they exist. The value of ka must be slightly greater than $\pi/2$ (ground state), or slightly greater than $3\pi/2$, &c. (excited states). Now from the definition of k, we have

$$W = V_0 - \frac{h^2}{(2\pi a)^2} \cdot \frac{(ak)^2}{M} = 21 - 5\cdot2\,(ak)^2,$$

where the energy is expressed in millions of electron volts. No value of $(ak)^2$ corresponding to the excited states is seen to be admissible, as W has to be positive for bound states.

XXXII. Meson Theory (p. 49).

Yukawa's theory of nuclear forces, which led him to the prediction of a new type of particles, later called " mesons ", is based on the conviction that all forces must be transmitted through space from point to point, to be consistent with the principle of relativity. That means mathematically that there must be a field of force satisfying a set of relativistically covariant differential equations, similar to Maxwell's equations of electrodynamics. The latter are equivalent to the wave equation valid for the scalar potential Φ, and for each component of the vector potential A,

$$\nabla^2 \Phi - \frac{1}{c^2}\frac{\partial^2 \phi}{\partial t^2} = 0,$$

where $\Delta^2 = \dfrac{\partial^2}{\partial x^2} + \dfrac{\partial^2}{\partial y^2} + \dfrac{\partial^2}{\partial z^2}$ is the Laplace operator, and c the velocity of light. The simplest possible generalization of this is the equation (Proca, 1936)

$$\Delta^2 V - \frac{1}{c^2}\frac{\partial^2 V}{\partial t^2} - \mu^2 V = 0,$$

where μ is a constant of the dimension of a reciprocal length. The elementary static solution of the electromagnetic equation is the Coulomb potential e/r. The corresponding static solution of Yukawa's equation is

$$V = f \frac{e^{-\mu r}}{r},$$

as can be easily verified by substitution. As explained in the text, this potential leads to short-range forces which vanish at a distance of the order μ^{-1}.

On the other hand, in order that a harmonic wave proportional to $\exp[2\pi i(\kappa x - \nu t)]$ is a solution of the electromagnetic equation, the wave number κ and the frequency ν must satisfy $\kappa^2 - (\nu/c)^2 = 0$, whereas in the case of Yukawa's equation we have

$$\kappa^2 - (\nu/c)^2 + (\mu/2\pi)^2 = 0.$$

If we now introduce de Broglie's relations $h\nu = E$, $h\kappa = p$, where E and p are the energy and momentum of a particle associated with the wave, we have in the electromagnetic case $p = E/c$ (apart from the sign), which is the characteristic relation for photons, i.e. particles moving with the velocity of light; in Yukawa's case we obtain

$$p^2 - (E/c)^2 + (\hbar\mu)^2 = 0,$$

which is the relativistic relation between energy and momentum for a particle with rest-mass

$$m_0 = \frac{\hbar\mu}{c}.$$

Hence Yukawa could predict the existence and mass of new types of particles, the mesons, from the existence and range of the nuclear forces.

The close analogy between the theory of light quanta and that of mesons can be extended to the interaction of these fields with their sources: the electrons (and charged particles) in the case of light quanta, and the nucleons in the case of mesons. Just as a light quantum may be created or annihilated by a change of state of its source particles (this change according to Dirac's theory can include negative energy levels), a meson may be created or annihilated by a change of state of the nuclear particles, the nucleons. A nucleon may make a transition from a neutron state to a proton state, and emit a negative meson or absorb a positive meson. In general, processes of this kind are not possible for free nucleons, since energy and momentum cannot be conserved. This is analogous to the inability of a free electron to emit light quanta. However, in contact with nuclear matter, a fast proton may emit a positive meson and be turned into a neutron, a process which leads to the production of cosmic-ray mesons in the earth's atmosphere. The instability of the mesons excludes them as primary particles of cosmic radiation. The coupling between the nucleons and the mesons is established by an interaction energy which has the form

$$H' = f\phi\rho,$$

where ϕ represents the meson and ρ the nucleon density. This is analogous to the electrostatic interaction energy $e\Phi\rho$ between a charge density ρ and the electrostatic field characterized by the potential Φ. In more elaborate versions of the meson theory the interaction H' is modified to account for the fact that there are both charged and neutral π-mesons responsible for nuclear forces. f is an interaction parameter playing a role similar to e, and ρ is a transition density of the form $\psi_N^*\psi_P$, where ψ_P and ψ_N are the eigenfunctions of the proton (which disappears) and the neutron (which is created when a positive meson is produced), analogous to the expression $\rho = \psi_f^*\psi_i$ of electrodynamics, where ψ_f is the wave function of the state f reached by the electron in a radiative transition $i \to f$. A measure of the strength of the forces between nucleons due to their mutual interaction with the meson field is the dimensionless quantity $f^2/\hbar c$, analogous to the fine-structure constant $e^2/\hbar c = 1/137$, which is a measure of the electric interaction between charged particles. We know from the theory of the deuteron (Chap. X, § 3, p. 317) that the potential well between two nucleons at an average distance of $2\cdot8 \times 10^{-13}$ cm. is approximately 20 MeV deep. On the other hand, the electrostatic interaction is only of the order $0\cdot5$ MeV at the same distance. We can therefore estimate f:

$$f^2/\hbar c \gtrsim \frac{20}{0\cdot5} \times \frac{1}{137} \sim 0\cdot3,$$ where the $>$ sign takes account of the exponential decrease of the Yukawa forces. The smallness of $1/137$ is the reason for the success of perturbation methods in electrodynamics, in which the solutions of the equations describing electrons and electromagnetic fields may be expanded into series in ascending powers of $1/137$. The fact that the corresponding factor $f^2/\hbar c$ is so much larger in Yukawa's theory makes most of the quantitative predictions of this theory rather unreliable. The significance of the number $1/137$ is illustrated by considering the system consisting of an electron in a radiation field. If the electron had no charge, and therefore would not

interact with radiation, it would forever remain in a given state. Its charge, however, causes it to exchange energy with the radiation field by rapid transitions from one state to another, so that, owing to the interaction with the electromagnetic field, it will spend 1/137 of its time in a different state, and a quantum will be there to balance momentum. In the case of mesons this means that a neutron spends a considerable fraction of its time as a proton and a negative meson. This dissociated neutron has electromagnetic properties, since the charges of the negative meson and of the positive proton will in the average be separated by an amount μ^{-1}. We should therefore expect the neutron to scatter charged particles, and the inverse of this effect (the scattering of slow neutrons by electrons) has indeed been observed (Havens, Rabi, Rainwater, 1947). This phenomenon also accounts in principle for the fact that although the neutron has no charge it nevertheless possesses a substantial magnetic moment.

XXXIII. The Stefan-Boltzmann Law and Wien's Displacement Law (p. 206).

The thermodynamical proof of the Stefan-Boltzmann law rests on the existence of radiation pressure. We imagine an enclosure shut off by a movable piston with a reflecting surface. The radiation field exerts a pressure on the piston; its magnitude is a function of u, the energy density of the radiation in the enclosure. In fact, both Maxwell's theory and the quantum (corpuscular) theory of light give the formula

$$p = \tfrac{1}{3}u.$$

This radiation pressure is due to the momentum which the radiation carries with it. In the case of the quantum theory this is clear, for according to this theory each light quantum of energy $h\nu$ possesses momentum $h\nu/c$. Maxwell's theory also ascribes to every radiation field, with energy density u, the "momentum density" $| g | = \dfrac{u}{c} = \dfrac{1}{c^2} | S |$,

where S is the Poynting vector; this is proved, e.g., by imagining a plane light wave to fall on a metal, where it is absorbed, and calculating the resulting force on the metal from Maxwell's equations. On both theories, then, the momentum contained in a definite volume of the radiation field is equal to the radiation energy contained in this volume divided by c. The rest of the calculation of the radiation pressure is now just like that of the mechanical pressure in the kinetic theory of gases. In the interval of time dt radiation energy $\dfrac{uc}{4\pi} \, dt \, d\omega \cos\theta$ falls on one square centimetre of the boundary from the solid angle $d\omega$; on reflection at the boundary it transfers to the latter a momentum equal to twice the component of the radiation momentum normal to the wall, i.e the momentum $2\,\dfrac{u}{c}\,\dfrac{c\,dt}{4\pi}\,d\omega\cos^2\theta$. If we integrate this expression with respect to θ and ϕ, where $d\omega = \sin\theta\,d\theta\,d\phi$, and θ goes from 0 to $\tfrac{1}{2}\pi$, ϕ from 0 to 2π, we obtain the pressure $p = \tfrac{1}{3}u$ as the momentum transferred to the boundary per unit time as a result of the reflection of the radiation.

We now regard the radiation as a thermodynamic engine, and apply the fundamental equation of thermodynamics, which includes both the first law and the second. If W is the total energy, S the entropy, T the absolute temperature, and V the volume, then, as we know,

$$T\,dS = dW + p\,dV.$$

Now $W = Vu$, where the energy density u is a function of T alone; hence

$$T\,dS = u\,dV + V\frac{du}{dT}\,dT + \tfrac{1}{3}u\,dV = V\frac{du}{dT}\,dT + \tfrac{4}{3}u\,dV.$$

From this we may conclude that

$$\frac{\partial S}{\partial V} = \frac{4}{3}\frac{u}{T}, \quad \frac{\partial S}{\partial T} = \frac{V}{T}\frac{du}{dT},$$

and hence

$$\frac{\partial}{\partial V}\left(\frac{V}{T}\frac{du}{dT}\right) = \frac{\partial^2 S}{\partial T\,\partial V} = \frac{4}{3}\frac{d(u/T)}{dT},$$

or

$$\frac{1}{T}\frac{du}{dT} = \frac{4}{3}\left(\frac{1}{T}\frac{du}{dT} - \frac{u}{T^2}\right).$$

Accordingly

$$\frac{du}{dT} = 4\frac{u}{T} \quad \text{or} \quad u = aT^4,$$

which is the Stefan-Boltzmann radiation law. It is also easy to verify that the entropy S is given by $S = \tfrac{4}{3}VT^3$.

Wien's displacement law depends on the existence of the Doppler effect. As is well known, to an observer at rest a wave motion whose source is moving appears to have its frequency altered. In actual fact, it is only the component of the velocity in the line of observation that is effective, and we have the formula $\dfrac{\Delta \nu}{\nu} = \dfrac{v}{c} \cos \theta$, where ν is the frequency, $\Delta \nu$ the change of frequency, v the velocity of the source, and θ the angle between the direction in which the source is moving and the line of observation.

Now if a mirror at which a wave of frequency ν is reflected moves with velocity v in the direction in which the light is propagated, we may think of the incident wave as coming from a source of light at rest. Then the reflected wave must behave as if it came from the mirror image of this imaginary source of light. Owing to the motion of the mirror, however, this mirror image moves in the direction of the normal to the mirror with velocity $2v$. Hence the change of frequency due to reflection at the moving mirror is $-2\dfrac{v}{c}\nu$, and for oblique incidence at the angle θ we have

$$\nu' = \nu \left(1 - \frac{2v}{c} \cos \theta \right).$$

Further, we can easily see that the intensity of radiation due to a train of waves falling on the moving mirror is altered by reflection in the same ratio; for the energy of radiation falling on the whole mirror of area A in time dt is $IA\,dt$, and that reflected is $I'A\,dt$. The difference must be equal to the work done by radiation pressure P in moving the mirror, i.e. $PAv\,dt$; but, as shown above, $P = \dfrac{2I}{c}\cos\theta$. We thus obtain

$$I' = I \left(1 - \frac{2v}{c} \cos \theta \right).$$

We shall now apply thermodynamics not to the total radiation but to a definite narrow range of wave-length. Here we must take note of the fact that whenever work is done in connexion with a motion of the reflecting piston the radiation will be displaced into another region of frequency according to the Doppler formula. This displacement does not by any means vanish when we make the mirror move infinitely slowly. To convince ourselves of this, we imagine the enclosure entirely surrounded by reflecting walls, so that a particular beam of light will go on moving in a zigzag path in the enclosure, meeting the mirrors again and again. If the velocity of the piston is halved, the number of reflections at it for a given total displacement of the piston will be just twice as great, so that the total Doppler displacement tends to a finite limit as $v \to 0$.

We shall now calculate $Vu_\nu d\nu$, the total change of energy of the radiation in an enclosure corresponding to a definite range of frequency $(\nu, \nu + d\nu)$ during time dt, arising from reflection at the mirror moving with velocity v. In the first place, all components of the radiation of this region of the spectrum which fall on the mirror are removed from the region in question owing to the Doppler effect. On the other hand, all components of the radiation which before reflection lay in the interval between ν' and $\nu' + d\nu'$ will now be brought into the region if $\nu' = \nu\left(1 + 2\dfrac{v}{c}\cos\theta\right)$. The quantity of energy from the range of frequency $(\nu, \nu + d\nu)$ falling from solid angle $d\omega$ on A, the area of the mirror, per second is

$$\Delta E = u_\nu \frac{c}{4\pi} A \cos\theta \, dt d\omega d\nu.$$

From this we obtain the energy thrown into the frequency range $(\nu, d\nu)$ on reflection by (1) multiplying by the energy factor $\left(1 - \dfrac{2v}{c}\cos\theta\right)$, (2) replacing $d\nu$ by $d\nu' = d\nu\left(1 + \dfrac{2v}{c}\cos\theta\right)$, and (3) introducing the expansion

$$u_{\nu'} = u_{\nu(1+2\nu/c\cos\theta)} = u_\nu + \nu \frac{\partial u_\nu}{\partial \nu} 2 \frac{v}{c}\cos\theta + \ldots .$$

The product of the first two factors, however, differs from $d\nu$ merely by a term of the second order in v/c, which can be omitted. For the energy diverted into the range $(\nu, d\nu)$ by reflection we accordingly have

$$\left(u_\nu + \nu \frac{\partial u_\nu}{\partial \nu} 2 \frac{v}{c}\cos\theta\right) \frac{c}{4\pi} A \cos\theta \, dt \, d\omega \, d\nu;$$

the increase in the energy of the beam as a result of reflection is therefore

$$\nu \frac{\partial u_\nu}{\partial \nu} \frac{1}{2\pi} dV d\nu \cos^2\theta \, d\omega,$$

since $Avdt = dV$. Integrating over the hemisphere (for which

$\int \cos^2 \theta \, d\omega = 2\pi/3$), we obtain the total increase of energy due to reflection, $d(u_\nu V) \, d\nu$. Hence

$$d(u_\nu V) = \tfrac{1}{3} \nu \frac{\partial u_\nu}{\partial \nu} \, dV.$$

This is a differential equation for u_ν as a function of ν and V:

$$\frac{\partial (V u_\nu)}{\partial V} = \tfrac{1}{3} \frac{\partial u_\nu}{\partial \nu} \nu \quad \text{or} \quad V \frac{\partial u_\nu}{\partial V} = \tfrac{1}{3} \nu \frac{\partial u_\nu}{\partial \nu} - u_\nu.$$

We easily see that this equation is satisfied by the relation

$$u_\nu = \nu^3 \phi(\nu^3 V),$$

where ϕ is an arbitrary function.

We now take a step further. We imagine that the change of volume is adiabatic, i.e. takes place without heat being supplied. This means that the entropy of the radiation in the enclosure remains constant during the compression. Now we saw above in deducing the Stefan-Boltzmann law that this entropy is proportional to the product of the volume V and the cube of the temperature, so that the constancy of the entropy implies that

$$VT^3 = \text{const.}$$

If in order to make the law of radiation independent of the size and shape of the enclosure we introduce the temperature instead of V by means of this relation, we have

$$u_\nu = \nu^3 F\left(\frac{\nu}{T}\right),$$

which is Wien's displacement law, given in the text.

XXXIV. Absorption by an Oscillator (p. 208).

We shall now add the proof of the formula

$$\delta W = \frac{1}{3} \frac{\pi e^2}{m} u_\nu$$

given in the text for the work done per second by a radiation field on an oscillator. We define the radiation field by stating the relationship between the electric vector E and the time. For reasons of convergence we shall assume that the radiation field exists only between the instants $t = 0$ and $t = T$; it is easy to pass to the limiting case

$T \to \infty$ subsequently. We now express one component, e.g. E_x, by its spectrum (Fourier integral),

$$E_x(t) = \int_{-\infty}^{+\infty} f(\nu)e^{2\pi i\nu t}\,d\nu,$$

where the amplitudes $f(\nu)$ are determined by

$$f(\nu) = \int_0^T E_x(t)e^{-2\pi i\nu t}\,dt,$$

$E_x(t)$ being zero outside the interval $t = 0$ to $t = T$; as E_x is real, the conjugate complex quantity satisfies the equation

$$f^*(\nu) = f(-\nu).$$

Analogous relationships hold for the two other components. According to the laws of electrodynamics, the total energy-density of the radiation field is then given by

$$u = \frac{1}{8\pi}(\overline{E^2 + H^2}) = \frac{1}{4\pi}\overline{E^2} = \frac{3}{4\pi}\overline{E_x^2};$$

the last of these relations follows from symmetry. The bars denote time-averages. Further,

$$\overline{E_x^2} = \frac{1}{T}\int_0^T E_x^2\,dt = \frac{1}{T}\int_0^T E_x\,dt \int_{-\infty}^{+\infty} f(\nu)e^{2\pi i\nu t}\,d\nu,$$

if we replace one factor by the above Fourier expression. If we now change the order of integration, we immediately obtain, as the value of the new integral with respect to t has been shown to be $f^*(\nu)$,

$$\overline{E_x^2} = \frac{1}{T}\int_{-\infty}^{+\infty} f(\nu)\,d\nu \int_0^T E_x e^{2\pi i\nu t}\,dt$$

$$= \frac{1}{T}\int_{-\infty}^{+\infty} f(\nu)f^*(\nu)\,d\nu = \frac{2}{T}\int_0^\infty |f(\nu)|^2\,d\nu,$$

as

$$|f(\nu)|^2 = f(\nu)f(-\nu) = |f(-\nu)|^2.$$

Hence for the total density of radiation we obtain the expression

$$u = \int_0^\infty u_\nu\,d\nu = \frac{3}{2\pi T}\int_0^\infty |f(\nu)|^2\,d\nu,$$

and for the distribution over the spectrum we therefore have

$$u_\nu = \frac{3}{2\pi T}|f(\nu)|^2.$$

After these preliminary remarks on the radiation field we now pass on to the equation giving the vibrations of the linear harmonic oscillator. If the oscillator is capable of vibrating only in the x-direction, this equation is

$$m\ddot{x} + ax = eE_x(t),$$

and the proper frequency is accordingly given by

$$m(2\pi\nu_0)^2 = a; \quad \nu_0 = \frac{1}{2\pi}\sqrt{\frac{a}{m}}.$$

Now, as we know, the most general solution of a non-homogeneous differential equation is obtained by adding one solution of the non-homogeneous equation to the general solution of the homogeneous equation. This last may be written in the form

$$x(t) = x_0 \sin(2\pi\nu_0 t + \phi),$$

where x_0 and ϕ are two arbitrary constants. The expression

$$x(t) = \frac{e}{2\pi\nu_0 m}\int_0^t E_x(t')\sin 2\pi\nu_0(t-t')\,dt'$$

is a solution of the non-homogeneous equation and satisfies the initial conditions $x(0) = 0$ and $\dot{x}(0) = 0$. $x(0) = 0$ is evident; to prove the other statements we carry out the differentiation:

$$\dot{x}(t) = \frac{e}{2\pi\nu_0 m}[E_x(t')\sin 2\pi\nu_0(t-t')]_{t'=t}$$

$$+ \frac{e}{m}\int_0^t E_x(t')\cos 2\pi\nu_0(t-t')\,dt';$$

here the first term vanishes, and we see that $\dot{x}(0) = 0$. Then

$$\ddot{x}(t) = \frac{e}{m}[E_x(t')\cos 2\pi\nu_0(t-t')]_{t'=t}$$

$$- \frac{2\pi\nu_0 e}{m}\int_0^t E_x(t')\sin 2\pi\nu_0(t-t')\,dt' = \frac{e}{m}E_x(t) - \frac{a}{m}x(t);$$

that is, the expression given above is actually a solution of the non-homogeneous differential equation.

We now proceed to investigate the work done by the field on the oscillator. From the differential equation of the vibrations we readily see (multiplication by $\dot{x}$ and integration with respect to the time leads to the energy theorem) that the work done per unit time is given by

$$\delta W = \frac{e}{T}\int_0^T \dot{x}(t)E_x(t)dt.$$

Now it is obvious that the part of the work done which arises from the free vibration (the solution of the homogeneous equation) vanishes. Hence the work done per second is obtained by evaluating the integral of the remaining part only:

$$\delta W = \frac{e}{T}\frac{e}{m}\int_0^T E_x(t)dt\int_0^t E_x(t')\,\cos2\pi\nu_0(t-t')dt'.$$

The integrand is obviously symmetrical in t and t'; hence, the expression for δW may be transformed in the following way. We see immediately that

$$\delta W = \frac{e^2}{mT}\int_0^T E_x(t')dt'\int_{t'}^T E_x(t)\,\cos2\pi\nu_0(t-t')dt.$$

For in the first instance we have to integrate with respect to t' between the limits 0 and t and subsequently with respect to t between the limits 0 and T; but of course we obtain the same result by integrating first with respect to t from t' to T and then with respect to t' from 0 to T. If we now merely interchange the letters t and t' δW may also be written in the form

$$\delta W = \tfrac{1}{2}\frac{e^2}{mT}\int_0^T E_x(t)dt\left\{\int_0^t+\int_t^T\right\}E_x(t')\,\cos2\pi\nu_0(t-t')dt',$$

or, if we replace $\cos 2\pi\nu_0(t-t')$ by

$$\tfrac{1}{2}\{e^{2\pi i\nu_0(t-t')}+e^{-2\pi i\nu_0(t-t')}\},$$

$$\delta W = \tfrac{1}{4}\frac{e^2}{mT}\left\{\int_0^T E_x(t)e^{2\pi i\nu_0 t}dt\int_0^T E_x(t')e^{-2\pi i\nu_0 t'}dt'\right.$$
$$\left.+\int_0^T E_x(t)e^{-2\pi i\nu_0 t}dt\int_0^T E_x(t')e^{2\pi i\nu_0 t'}dt'\right\}$$
$$= \frac{e^2}{2mT}\,|f(\nu_0)|^2.$$

Thus the work done by the field per second on the linear oscillator is

$$\delta W = \frac{e^2}{2mT}\frac{2\pi T}{3}u_\nu = \frac{\pi e^2}{3m}u_\nu,$$

if we use the formula deduced above for the density of radiation; we thus obtain the expression given in the text.

XXXV. Temperature and Entropy in Quantum Statistics (p. 230).

The proof of the fact that the quantity β occurring in statistics is inversely proportional to the absolute temperature can be presented in the same form for all three types of statistics, the Boltzmann (B.), the Bose-Einstein (B.E.), and the Fermi-Dirac (F.D.). In all three cases we have

$$\log W = \sum_s f(n_s),$$

where (see p. 12 for B.; p. 229 for B.E.; p. 234 for F.D.)

$$f(n_s) = \begin{cases} n_s \log g_s - n_s \log n_s & \text{for B.} \\ (g_s + n_s) \log(g_s + n_s) - g_s \log g_s - n_s \log n_s & \text{for B.E.} \\ -(g_s - n_s) \log(g_s - n_s) + g_s \log g_s - n_s \log n_s & \text{for F.D.} \end{cases}$$

The maximum of $\log W$, subject to the subsidiary conditions

$$\sum_s n_s = N, \qquad \sum_s n_s \epsilon_s = E,$$

is attained for

$$\alpha + \beta \epsilon_s = \frac{\partial f}{\partial n_s},$$

hence

$$\delta \log W = \sum_s \frac{\partial f}{\partial n_s} \delta n_s = \sum_s (\alpha + \beta \epsilon_s) \delta n_s$$

$$= \alpha \sum_s \delta n_s + \beta \sum_s \epsilon_s \delta n_s.$$

In equilibrium when all the ϵ_s as well as N and E are constant, the two sums disappear, as a consequence of the subsidiary conditions. But the formula can also be applied to the case of " quasi-static " processes where the external conditions are changed so slowly that the system can be considered as being at any moment in equilibrium. If the number of atoms is kept constant the two conditions lead to

$$\sum_s \delta n_s = 0, \quad \sum_s n_s \delta \epsilon_s + \sum_s \epsilon_s \delta n_s = \delta E.$$

The two terms in δE have a physical meaning. The first one represents the work done by expansion. If the volume changes by dV, we have

$$\sum_s n_s \delta \epsilon_s = \sum_s n_s \frac{\partial \epsilon_s}{\partial V} \delta V = - p \, \delta V,$$

where

$$p = -\sum_s n_s \frac{\partial \epsilon_s}{\partial V}$$

denotes the pressure as the sum of the forces $-\partial \epsilon_s / \partial V$ of the atoms in the different states.

The second term $\Sigma_s \epsilon_s \delta n_s$ represents the change of internal energy due to a rearrangement of the atoms with respect to the different states, produced by quantum jumps; this is the "heat applied" to the system,

$$\delta Q = \Sigma_s \epsilon_s \delta n_s,$$

and we have

$$\delta E = -p\,\delta V + \delta Q,$$

the first law of thermodynamics.

On the other hand

$$\delta \log W = \beta \Sigma_s \epsilon_s \delta n_s = \beta\,\delta Q.$$

This shows that $\beta \delta Q$ is a total differential, and β the "integrating factor" of the heat supplied. In this way the statistical theory leads automatically also to the second law of thermodynamics, which states that δQ has an integrating factor, namely, the reciprocal absolute temperature:

$$\frac{\delta Q}{T} = \delta S,$$

where S is the entropy. Hence β is proportional to $1/T$; putting

$$\beta = \frac{1}{kT},$$

we obtain by comparison

$$S = k \log W,$$

which is Boltzmann's celebrated formula.

XXXVI. Thermionic Emission of Electrons (p. 241).

Here we shall prove the two formulæ which we gave in the text for the thermionic emission of electrons (the Richardson effect), firstly on the basis of classical statistics and secondly on the basis of the Fermi-Dirac statistics. For this we require to calculate the number of electrons striking one square centimetre of the boundary in the metal per second, such that the kinetic energy of their motion normal to the boundary is sufficient to carry the electron over the energy barrier of height ϵ_i which represents the boundary. We have therefore to determine, on the basis of the distribution law, the number of electrons for which, e.g.,

$$\tfrac{1}{2}mv_x^2 \geqq \epsilon_i.$$

We begin with classical statistics. On this basis the number of electrons whose velocity lies between v and $v + dv$ is given by

$$dN = 4\pi n V \left(\frac{m}{2\pi kT}\right)^{\frac{3}{2}} e^{-\frac{1}{2}mv^2/kT} v^2\, dv$$

(p. 15); similarly (replacing $4\pi v^2 dv$ by $dv_x dv_y dv_z$ and integrating) the number of electrons with a velocity component between v_x and $v_x + dv_x$ is

$$dN_x = nV\left(\frac{m}{2\pi kT}\right)^{\frac{3}{2}} dv_x \int_{-\infty}^{+\infty}\int_{-\infty}^{+\infty} e^{-\frac{1}{2}m(v_x^2 + v_y^2 + v_z^2)/kT}\, dv_y\, dv_z$$

$$= nV\sqrt{\frac{m}{2\pi kT}}\, e^{-\frac{1}{2}mv_x^2/kT}\, dv_x.$$

To find the number of electrons falling on unit area of the boundary per second we have first to divide the above number by V, to get the density of the electrons, and then multiply by v_x, as in unit time there impinge on the boundary all the molecules with the component v_x which were contained in the layer of breadth v_x in front of the boundary (§ 3, p. 5). We thus obtain the emission current by evaluating the integral

$$i = en\sqrt{\frac{m}{2\pi kT}}\int_{\sqrt{(2\epsilon_i/m)}}^{\infty} v_x e^{-\frac{1}{2}mv_x^2/kT}\, dv_x.$$

This integral may be evaluated exactly, giving

$$i = en\sqrt{\frac{kT}{2\pi m}}\, e^{-\epsilon_i/kT},$$

which is the expression stated in the text.

The calculation takes a similar course in the case of the Fermi distribution. Here we start from the distribution function (p. 236)

$$dN = \frac{8\pi V}{h^3}\frac{\sqrt{2m^3}\sqrt{\epsilon}\, d\epsilon}{e^{a+\epsilon/kT}+1} \qquad (\epsilon = \tfrac{1}{2}mv^2),$$

where a, the degeneracy parameter, is determined from the subsidiary condition $\int dN = N$. If, however, we confine our attention to relatively low temperatures (hot cathodes), we can use the approximate formula given on p. 239,

$$dN = \frac{8\pi V}{h^3} \frac{\sqrt{2m^3}\sqrt{\epsilon}\,d\epsilon}{e^{(\epsilon-\epsilon_0)/kT}+1} = \frac{8\pi V m^3}{h^3} \frac{v^2 dv}{e^{(\epsilon-\epsilon_0)/kT}+1},$$

where

$$\epsilon_0 = \frac{h^2}{2m}\left(\frac{3n}{8\pi}\right)^{\frac{2}{3}}$$

represents the zero-point energy. Here, just as before, we obtain the Richardson current by evaluating the integral

$$i = \frac{2m^3 e}{h^3}\int_{-\infty}^{+\infty}\int_{-\infty}^{+\infty} dv_y dv_z \int_{\sqrt{(2\epsilon_i/m)}}^{\infty} \frac{v_x}{e^{(\epsilon-\epsilon_0)/kT}+1}\,dv_x.$$

Now in the heated metal of a cathode $(\epsilon_i - \epsilon_0)$ is always very large compared with kT, $(\epsilon_i - \epsilon_0)$ amounting to several electron-volts (p. 242), while kT at $300°$ K. corresponds to an energy of about 0·03 electron-volt. Hence in the integrand we always have $e^{(\epsilon-\epsilon_0)/kT} \gg 1$, so that we may neglect the 1 in the denominator, thus obtaining the integral

$$i = \frac{2m^3 e}{h^3}\, e^{\epsilon_0/kT}\int_{-\infty}^{+\infty}\int_{-\infty}^{+\infty} dv_y dv_z \int_{\sqrt{(2\epsilon_i/m)}}^{\infty} v_x e^{-\frac{1}{2}m(v_x^2+v_y^2+v_z^2)/kT}dv_x.$$

The integrations with respect to v_y and v_z are equivalent to evaluations of Gaussian error integrals (see Appendix I, p. 359); the integration with respect to v_x can, as before, be carried out by elementary means, giving

$$i = \frac{4\pi e m}{h^3}\,(kT)^2 e^{-(\epsilon_i-\epsilon_0)/kT},$$

the law stated in the text.

XXXVII. Temperature Variation of Paramagnetism (pp. 184, 251).

In order to understand the variation of paramagnetism with temperature we shall consider a simplified model of a paramagnetic substance. We think of the substance as consisting of a large number of particles, all of which have the same magnetic moment M. We shall also in the first instance ignore azimuthal quantisation, i.e. we assume that this moment can be inclined at *any* angle θ to the direction of the field, and not merely at certain specified angles.

So long as a particle is moving undisturbed in the magnetic field, the magnetic moment, which is always associated with an angular momentum about the same axis (cf. the top), will execute a precessional motion about the direction of the field, the angle θ at which it is inclined to the field remaining constant. The magnetic energy for this orientation is given by

$$E = -MH \cos\theta.$$

As a result of interaction with the other particles (collisions), however, this state of equilibrium is disturbed, the angle of inclination after collision being different from that before collision. From the kinetic theory we already know that the collisions due to thermal motions have the effect that in course of time the particles become uniformly distributed among all possible states (here possible angles of inclination to the field). The magnetic field, however, acts against this equalization; accordingly, settings of the moment in its direction are favoured as regards energy compared with those in the opposite direction. A state of equilibrium will be established, and can be determined by statistical methods.

In our discussion of the kinetic theory of gases we have already shown that a statistical argument gives Boltzmann's distribution law for the probability of a definite energy state, according to which a definite state with energy E has the probability

$$W \sim e^{-E/kT}.$$

The form in which the temperature T appears in this expression is deduced from thermodynamical considerations; k is Boltzmann's constant. In our case, therefore, the probability that the magnetic moment will set itself at a definite angle θ to the direction of the field is given by

$$W \sim e^{MH \cos\theta/kT} = e^{\beta \cos\theta} \qquad (\beta = MH/kT).$$

We now investigate the mean magnetic moment in the direction of the field. If the field were equal to zero or if the thermal motion (in the case of very high temperatures) dominated the directional effect of the magnetic field, so that the distribution of directions became almost uniform, the mean magnetic moment in a specified direction, in the direction of the field in particular, would be approximately or accurately equal to zero. At low temperatures, or with magnetic fields so strong that the magnetic energy MH is of the same order as the thermal energy kT, there will be a marked preference for the direction of the field, giving rise to a finite mean magnetic moment in this direction. If we adopt the classical ideas, according to which all directions of the moment are equally possible, the calculation is easily carried out. By definition

$$\overline{M \cos \theta} = M \frac{\int_0^\pi \cos \theta \, e^{\beta \cos \theta} \sin \theta \, d\theta}{\int_0^\pi e^{\beta \cos \theta} \sin \theta \, d\theta}.$$

Evaluating the integrals, we obtain

$$\overline{M \cos \theta} = M \frac{d}{d\beta} \log \int_0^\pi e^{\beta \cos \theta} \sin \theta \, d\theta = M \frac{d}{d\beta} \log \frac{e^\beta - e^{-\beta}}{\beta}$$

$$= M \left(\frac{e^\beta + e^{-\beta}}{e^\beta - e^{-\beta}} - \frac{1}{\beta} \right)$$

$$= M \left(\coth \beta - \frac{1}{\beta} \right).$$

In the limiting case $\beta \ll 1$, i.e. for weak fields or high temperatures, we obtain the expression

$$\overline{M \cos \theta} = M(\tfrac{1}{3}\beta + \ldots) = \frac{M^2 H}{3kT}$$

by expanding in powers of β.

In this limiting case, however, we could have obtained the result more easily by introducing a power series in β into the formula defining $\overline{M \cos \theta}$. We then have

$$\overline{M \cos \theta} = M \frac{\int_0^\pi \cos \theta (1 + \beta \cos \theta + \ldots) \sin \theta \, d\theta}{\int_0^\pi (1 + \beta \cos \theta + \ldots) \sin \theta \, d\theta}.$$

The first term in the numerator vanishes on integration, while the second gives $2\beta/3$. In the denominator the first term differs from zero, its value being 2. The quotient gives the expression stated above.

Now it is important to know that the same formula results from calculations on the basis of the quantum theory, i.e. if we take into account only a finite number of positions of the moment. We assume that β has small values and the resultant angular momentum large values. The fact that in this case we obtain the classical result for $\overline{M \cos \theta}$ is understandable if we consider the correspondence principle (the limiting case of large quantum numbers). It is perhaps advisable, however, for the reader to convince himself that this statement is correct by direct evaluation of the sums. If the resultant angular momentum is j, there are $2j + 1$ possible settings of the moment relative to the direction of the field; in fact, the projection of j on this can have the values $-j$, $-j + 1$, . . . , $+j$. In the previous calculation we therefore have merely to replace $\cos \theta$ by m/j and replace the integrals by sums:

$$\overline{M \cos \theta} = M \frac{\sum\limits_{-j}^{+j} \frac{m}{j} e^{\beta m l j}}{\sum\limits_{-j}^{+j} e^{\beta m l j}}.$$

Expanding in powers of β and making use of the formula

$$\sum_{0}^{n} \nu^2 = \frac{n(n+1)(2n+1)}{6},$$

we have

$$\overline{M \cos \theta} = M \frac{\sum\limits_{-j}^{+j} \frac{m}{j}\left(1 + \frac{\beta m}{j}\right)}{\sum\limits_{-j}^{+j}\left(1 + \frac{\beta m}{j}\right)} = M\beta \frac{\frac{1(j+1)(2j+1)}{3} \frac{}{j}}{2j+1}$$

$$= \frac{M\beta}{3}\frac{j+1}{j},$$

or, for large values of j, $\overline{M \cos \theta} = \dfrac{M\beta}{3}$, as we previously obtained by classical methods.

The susceptibility χ is defined as the magnetic moment per mole of the substance in question per unit field intensity H

$$\chi = \frac{LM^2}{3kT} = \frac{(LM)^2}{3RT}.$$

This is known as Curie's law; the magnetic susceptibility is inversely proportional to the temperature, i.e. decreases with increasing temperature.

We conclude with a few brief remarks on the simplifications made above. The account which we have just given certainly suffices for a rough survey of the variation of paramagnetism with temperature; but if, in particular, we wish to draw conclusions about the magnitude of the atomic magnetic moments from the measured values of the susceptibility, more detailed arguments become necessary. In the first place, as we have already indicated, the existence of azimuthal quantisation must be taken into account; this has the effect of introducing the factor $(j + 1)/j$ into the expression given above for the mean magnetic moment in the direction of the field. Further, we must take account of the fact that in general all the atoms of a substance do not have the same magnetic moment. When discussing the anomalous Zeeman effect we saw that the resultant magnetic moment of an atom is equal to the Bohr magneton multiplied by the resultant angular momentum j and the so-called Landé factor g, which depends on the three quantum numbers j, s, and l. We would therefore have to take the average for all possible combinations of the quantum numbers.

Historically it must be remarked that the first claim of an atomistic structure of magnetism was made by P. Weiss (1924); the Weiss magneton was about a fifth of the Bohr magneton. Modern research (van Vleck, 1932, &c.) has, however, disproved the existence of this smaller unit and established the existence of the Bohr magneton.

XXXVIII. Theory of Co-valency Binding (p. 271).

The fundamental idea of the Heitler-London theory of valency binding is as follows. As a model of the hydrogen molecule we imagine two nuclei a and b on the x-axis at a distance R apart, and two electrons 1 and 2 revolving about the nuclei. To the state of two widely separated neutral atoms there corresponds a large value of R and a motion of the electrons such that each one revolves round one of the two nuclei. Let the two atoms be in the ground state and have the same energy $E_0^1 = E_0^2 = E_0$. The motion of the electrons is described by proper functions u, which, relative to the corresponding nuclei, are identical; that is, one is obtained from the other by substituting $x + R$ for x; we shall briefly write them in the form

$$\psi_a^{(1)} = u(x_1, y_1, z_1),$$
$$\psi_b^{(2)} = u(x_2 + R, y_2, z_2).$$

The functions u are the same as the proper functions of atomic hydrogen [Appendix XVIII (p. 399)]. Hence the two Schrödinger equations

$$H_a^0 \psi_a^{(1)} = E_0 \psi_a^{(1)},$$
$$H_b^0 \psi_b^{(2)} = E_0 \psi_b^{(2)},$$

where H^0 denotes the energy operator of the hydrogen atom, are satisfied identically and the suffixes a and b indicate that in the one case the co-ordinates of the electron are relative to the nucleus a, in the other to the nucleus b (see above).

The energy operator of the molecule which arises when the atoms approach one another (when R diminishes) differs from the sum $H_a{}^0 + H_b{}^0$ by the interaction energy of the two atoms,

$$V = e^2\left(\frac{1}{r_{ab}} + \frac{1}{r_{12}} - \frac{1}{r_{a2}} - \frac{1}{r_{b1}}\right),$$

where r_{ab} denotes the distance between the two nuclei (R), r_{12} the distance between the two electrons, and r_{a2}, r_{b1} the distances between either electron and the nucleus of the other atom. For the molecule we accordingly have the Schrödinger equation

$$(H_a{}^0 + H_b{}^0 + V)\psi^{(1,\,2)} = E\psi^{(1,\,2)}.$$

We now seek to find an approximate solution for this equation by assuming that (to a first approximation) the function $\psi^{(1,\,2)}$ of the co-ordinates of the two electrons is a product of some proper function $\psi_a{}^{(1)}$ of one electron and some proper function $\psi_b{}^{(2)}$ of the other electron. Here, however, we have to bear in mind that the state of the system is a degenerate one. The total energy of the two separated atoms,

$$E = E_0{}^1 + E_0{}^2 = 2E_0,$$

corresponds not only to the product $\psi_a{}^{(1)}\psi_b{}^{(2)}$ but also to $\psi_a{}^{(2)}\psi_b{}^{(1)}$ and every possible combination of the two expressions. These two vibrational forms will interact with each other, owing to the coupling as the two atoms approach one another. (The interaction with vibrations corresponding to other energy levels will be small provided that the interaction energy V is, on the average, small compared to the spacing of atomic energy levels.) They accordingly suffice as a basis for a rough approximation, i.e. we attempt to represent the function $\psi^{(1,\,2)}$ approximately by a linear combination of the two functions $\psi_a{}^{(1)}\psi_b{}^{(2)}$ and $\psi_a{}^{(2)}\psi_b{}^{(1)}$. Instead of these we may also start from the symmetric and antisymmetric combinations

$$\psi_{\text{sym.}} = \psi_a{}^{(1)}\psi_b{}^{(2)} + \psi_a{}^{(2)}\psi_b{}^{(1)},$$
$$\psi_{\text{antis.}} = \psi_a{}^{(1)}\psi_b{}^{(2)} - \psi_a{}^{(2)}\psi_b{}^{(1)},$$

a course which has two advantages: (1) closer investigation shows that to a first approximation the symmetric function and the antisymmetric function are not coupled with one another by the Schrödinger equation, i.e. that each function by itself represents a separate state of the molecule; (2) they may readily be distinguished by means of the spin; for according to the exclusion principle the proper functions of a system must be antisymmetric in all the co-ordinates of the two electrons (of course taking the spin into account; cf. p. 227). If we were to give the electrons spin variables, as we did in the case of atomic spectra (§ 8, p. 187), the corresponding spin function would have to be antisymmetric for $\psi_{sym.}$ and symmetric for $\psi_{antis.}$ in order to satisfy the exclusion principle. This means that in the case of $\psi_{sym.}$ the spins are antiparallel and balance one another, whereas in the case of $\psi_{antis.}$ they are parallel and additive.

Now the perturbation calculation shows that when the two atoms are brought closer together (when the coupling is increased) the proper value of the uncoupled system $(2E_0)$ is split up into two values

$$E_1 = 2E_0 - W_1, \quad E_2 = 2E_0 + W_2$$

where the functions $W(R)$ have the following meanings:

$$W_1 = \frac{H_1 S + H_2}{1 + S}, \quad W_2 = \frac{H_1 S - H_2}{1 - S};$$

$$H_1 = \int\int (\psi_a{}^{(1)})^2 (\psi_b{}^{(2)})^2 V \, d\tau_1 \, d\tau_2,$$

$$H_2 = \int\int \psi_a{}^{(1)} \psi_a{}^{(2)} \psi_b{}^{(1)} \psi_b{}^{(2)} V \, d\tau_1 \, d\tau_2,$$

$$S = \int\int \psi_a{}^{(1)} \psi_a{}^{(2)} \psi_b{}^{(1)} \psi_b{}^{(2)} \, d\tau_1 \, d\tau_2,$$

the integrations being taken over the co-ordinates of the two electrons. As ψ^2, apart from a factor, represents the density of the charge-cloud of the electron, the first integral represents the Coulomb force due to the mutual actions of the charges distributed over each atom. The second integral is characteristic of the quantum theory; it does not contain the squares of the proper functions, so that an interpretation based on charge-densities is not possible. This integral is known as the " exchange integral ".

The evaluation of the integral as a function of R, which is very troublesome, gives the curves shown in fig. 7, p. 261, for the energy as a function of the distance between the two atoms. The function $\psi_{sym.}$, which is symmetrical in the co-ordinates of the electrons, gives the lower curve, according to which the energy for a given intramolecular distance has a minimum, while the antisymmetric function $\psi_{antis.}$ gives the monotonically ascending branch of the curve, which corresponds to a permanent repulsion between the two atoms. It turns out, therefore, that the state that leads to binding is that for which the electronic spins balance one another. Thus it comes about that we may regard the spin as a physical substitute for the chemical valency. But, as was mentioned in the text, attempts in this direction (Heitler, Rumer, Weyl, Born) were not very successful and have been replaced by semi-empirical theories.

XXXIX. Time-independent Perturbation Theory for Non-degenerate States.

We consider the Schrödinger equation for a system in which the Hamiltonian H can be divided into two parts H^0 and $\lambda H'$. H^0 is the unperturbed Hamiltonian, and we suppose that E_n and $\phi_n(q)$ are the energy and normalized eigenfunctions of H^0, thus

$$H^0\phi_n = E_n\phi_n.$$

$\lambda H'$ is some perturbation which is taken to be weak, i.e. $\lambda H' \ll H^0$ (the parameter λ is included so that orders of magnitude of small quantities can be identified easily—λ can be placed equal to unity at the end of the calculation). The problem then is to find approximate solutions for the perturbed Schrödinger equation

$$(H^0 + \lambda H')\psi_n = \epsilon_n\psi_n$$

where it is supposed that in the limit $\lambda H' \to 0$, $\psi_n \to \phi_n$ and $\epsilon_n \to E_n$. Here it is to be noted that such a procedure is only valid if the state ϕ_n is non-degenerate. If it is degenerate then a more complex procedure must be adopted. In the following we shall obtain solutions to the perturbed equation correct to first order in λ.

Let
$$\psi_n = \phi_n + \lambda\psi_n'$$
$$\epsilon_n = E_n + \lambda E_n'$$

Substituting into the perturbed equation gives

$$(H^0 + \lambda H')(\phi_n + \lambda\psi_n') = (E_n + \lambda E_n')(\phi_n + \lambda\psi_n').$$

Since λ is an arbitrary parameter, we can now equate terms corresponding to different orders in λ.

Zeroth order: $H^0\phi_n = E_n\phi_n$.

First order: $H'\phi_n + H_0\psi_n' = E_n\psi_n' + E_n'\phi_n$.

The first equation gives nothing new and is just the unperturbed Schrödinger equation. We now turn to the second equation and expand ψ_n' in terms of the complete set of eigenfunctions ϕ_m of H_0; thus

$$\psi_n' = \Sigma_m a_m \phi_m.$$

The problem now is to determine the coefficients a_m. Substituting for ψ_n' in the second equation gives

$$H'\phi_n + H_0(\Sigma_m a_m \phi_m) = E_n(\Sigma_m a_m \phi_m) + E_n'\phi_n$$

or

$$H'\phi_n + \Sigma_m a_m E_m \phi_m = E_n \Sigma_m a_m \phi_m + E_n'\phi_n$$

by using the unperturbed Schrödinger equation.

Now multiply on the left by ϕ_n^* and integrate over dq. Because of the orthogonality of the ϕ_m and their normalization this procedure leads at once to

$$\int \phi_n^* H' \phi_n dq = E_n'.$$

This means that the first-order change in the energy of the system is simply the expectation value of H' taken with respect to the unperturbed state ϕ_n.

Next multiply on the left of the same equation by $\phi_p^* (p \neq n)$ and integrate over dq. This gives

$$\int \phi_p^* H' \phi_n dq + a_p E_p = E_n a_p$$

or

$$a_p = \frac{H_{pn}'}{E_n - E_p}$$

where

$$H_{pn}' = \int \phi_p^* H' \phi_n dq.$$

We therefore have an expression for all the a_p *except* a_n which is not determined by this procedure. To determine a_n it is necessary to consider the normalization properties of ψ_n. We have

$$\psi_n = \phi_n(1 + \lambda a_n) + \lambda \Sigma_{m \neq n} a_m \phi_m$$

and to the first order in λ,

$$\int \psi_n^* \psi_n dq = 1 + \lambda(a_n + a_n^*).$$

Thus if ψ_n is to be normalized to the first order in λ, then $a_n + a_n^* = 0$ or $a_n = i\gamma_n$. This means that to the first order in λ we can write

$$\psi_n = e^{i\lambda\gamma_n}\Big[\phi_n + \lambda\sum_{m \neq n} a_m\phi_m\Big]$$

and γ_n therefore plays the role of an overall phase factor with no physical significance. For this reason γ_n can be put equal to zero. Finally then, putting $\lambda = 1$, we have to the first order in the perturbation H'

$$\psi_n = \phi_n + \sum_{m \neq n}\frac{H'_{mn}}{E_n - E_m}\phi_m.$$

Carrying out the above procedure to the second order in λ gives for the second-order perturbation in the energy

$$E''_n = \sum_{m \neq n}\frac{|H'_{mn}|^2}{E_n - E_m}.$$

It is important to note that this second-order shift in energy is always negative for the ground state ($E_o < E_m$) of a system.

XL. **Theory of the van der Waals Forces** (p. 275).

According to London, the theory of the van der Waals forces rests on a fact which is a distinctive feature of quantum theory, namely, the existence of a finite zero-point energy (cf. the case of the harmonic oscillator [Appendices XV, p. 392, XVI, p. 396]). According to the classical theory, the state of least energy of an oscillator is that of zero energy; this is the state of rest in the position of equilibrium. According to wave mechanics, however, the ground state has a finite energy $E_0 = \frac{1}{2}h\nu_0$, and the corresponding proper function is the Gaussian error function $\psi_0 = ae^{-\frac{1}{2}aq^2}$, where $a = \dfrac{4\pi^2 m\nu_0}{h}$ and ν_0 is the proper frequency of the oscillator. This zero-point energy can be explained by Heisenberg's uncertainty principle, according to which if the energy of a particle is prescribed accurately its position cannot be accurately determined. The proper function, which is a Gaussian error function, expresses the uncertainty directly. For the error curve immediately gives the mean square deviation of the co-ordinate (as $\bar{q} = 0$):

$$\overline{\delta q^2} = \overline{q^2} = \frac{\int_{-\infty}^{+\infty} q^2 \psi_0^2 \, dq}{\int_{-\infty}^{+\infty} \psi_0^2 \, dq} = \frac{1}{2a} = \frac{h}{8\pi^2 m \nu_0},$$

On the other hand, from the energy equation we have

$$E = \tfrac{1}{2} p^2/m + \tfrac{1}{2} m (2\pi\nu_0)^2 q^2;$$

if the energy is accurately determined, the mean square deviation of the momentum is

$$\overline{\delta p^2} = m^2 (2\pi\nu_0)^2 \overline{\delta q^2} = \tfrac{1}{2} h \nu_0 m.$$

Hence
$$\overline{\delta q^2} \, \overline{\delta p^2} = \frac{h^2}{16\pi^2} = \tfrac{1}{4}\hbar^2$$

or, with $(\Delta q)^2 = \overline{\delta q^2}$, $(\Delta p)^2 = \overline{\delta p^2}$:

$$\Delta q \Delta p = \frac{h}{4\pi} = \tfrac{1}{2}\hbar.$$

This is the exact form of Heisenberg's uncertainty principle (see Appendix XII, p. 384, and Appendix XXVI, p. 435).

After this digression on the zero-point energy and the theoretical uncertainties of the position and the momentum when the energy is determined accurately, we now return to London's explanation of the occurrence of the van der Waals forces. As a simple model we consider two linear oscillators at a distance R apart, vibrating in the direction of the line joining them (the x-axis). We think of these oscillators as vibrating electrical dipoles in which the positive charges e are held fast in the position of equilibrium while the negative charges $-e$ vibrate about these equilibrium positions, their displacements being x_1 and x_2. We express the restoring forces on the oscillators in the form $-\dfrac{e^2}{a} x_1$ and $-\dfrac{e^2}{a} x_2$; their potential energies are then $\tfrac{1}{2} \dfrac{e^2}{a} x_1^2$ and $\tfrac{1}{2} \dfrac{e^2}{a} x_2^2$. In addition there is the coupling force acting between the two oscillators, for which we assume Coulomb's law of force. The potential energy of this interaction is

$$\frac{e^2}{R} + \frac{e^2}{R + x_1 + x_2} - \frac{e^2}{R + x_1} - \frac{e^2}{R + x_2},$$

or, if we assume that R is very large compared with x_1 and x_2, and expand, we have

$$\frac{2e^2 x_1 x_2}{R^3}.$$

Hence the energy equation for the two oscillators is of the form

$$W = \frac{1}{2m}(p_1{}^2 + p_2{}^2) + \frac{e^2}{2a}(x_1{}^2 + x_2{}^2) + \frac{2e^2 x_1 x_2}{R^3}.$$

In the absence of the coupling term each of the resonators would vibrate with the frequency

$$\nu_0 = \frac{1}{2\pi}\sqrt{\frac{e^2}{am}}.$$

If the coupling is taken into account the frequency, as we have repeatedly stated before, is split up. Here we shall carry out the actual calculation of the splitting. For this purpose it is most convenient to refer the quadratic expression for the potential energy to " principal axes "; this does not affect the form of the kinetic energy. This is done by means of the transformation

$$x_s = \frac{1}{\sqrt{2}}(x_1 + x_2), \quad x_a = \frac{1}{\sqrt{2}}(x_1 - x_2);$$

as $p = m\dot{x}$, we also have

$$p_s = \frac{1}{\sqrt{2}}(p_1 + p_2), \quad p_a = \frac{1}{\sqrt{2}}(p_1 - p_2).$$

If we substitute these new co-ordinates and momenta in the expression for the energy, we obtain

$$W = \frac{1}{2m}(p_s{}^2 + p_a{}^2) + \frac{e^2}{2a}(x_s{}^2 + x_a{}^2) + \frac{e^2}{R^3}(x_s{}^2 - x_a{}^2),$$

or, in another form,

$$W = \left\{ \frac{1}{2m}p_s{}^2 + \left(\frac{e^2}{2a} + \frac{e^2}{R^3} \right) x_s{}^2 \right\} + \left\{ \frac{1}{2m}p_a{}^2 + \left(\frac{e^2}{2a} - \frac{e^2}{R^3} \right) x_a{}^2 \right\}.$$

This, however, is the energy equation for two non-coupled oscillators vibrating with the two frequencies

$$\nu_s = \frac{1}{2\pi}\sqrt{\frac{e^2}{m}\left(\frac{1}{a} + \frac{2}{R^3} \right)}, \quad \nu_a = \frac{1}{2\pi}\sqrt{\frac{e^2}{m}\left(\frac{1}{a} - \frac{2}{R^3} \right)}.$$

Hence the quantised energy of the system is

$$E_{n_s n_a} = h\nu_s(n_s + \tfrac{1}{2}) + h\nu_a(n_a + \tfrac{1}{2}),$$

which depends on R, as the new frequencies are functions of the distance between the oscillators. For the ground state we obtain the zero-point energy of the two vibrations of the oscillator,

$$E_{00} = \tfrac{1}{2}h(\nu_s + \nu_a),$$

that is, if we again expand,

$$E_{00} = \frac{h}{4\pi} \left\{ \sqrt{\frac{e^2}{m}\left(\frac{1}{a} + \frac{2}{R^3}\right)} + \sqrt{\frac{e^2}{m}\left(\frac{1}{a} - \frac{2}{R^3}\right)} \right\}$$

$$= \frac{h}{2\pi} \sqrt{\frac{e^2}{ma}} \left(1 - \frac{a^2}{2R^6} + \dots\right) = h\nu_0\left(1 - \frac{a^2}{2R^6} + \dots\right).$$

The additional energy is therefore negative and inversely proportional to the sixth power of the distance between the oscillators; the oscillators will accordingly attract one another with a force varying as the inverse seventh power of the distance between them. The magnitude of the attraction also depends on ν_0 and on the square of the constant a, which is obviously a measure of the deformability of the oscillators.

Exactly similar considerations apply to any atomic system in which there is interaction between the constituents, and invariably lead to the result that between two systems in the ground state there will be a force of attraction whose potential energy is inversely proportional to the sixth power of the distance and whose magnitude is proportional to the product of the deformabilities of the two atoms.

XLI. The Modes of Vibration of a Linear Monatomic Chain.

We shall consider here the simple case of an infinite linear chain of atoms of mass M and equilibrium spacing a, with restoring forces which are linear in the displacements u_n of the atoms. In general the restoring force on a given atom will depend on the relative displacements of several neighbouring atoms, but for simplicity we assume that each atom interacts only with its nearest neighbours. The equation of motion of the nth atom is then

$$M\ddot{u}_n = B(u_{n+1} - u_n) - B(u_n - u_{n-1})$$

where B is a stiffness constant which measures the strength of the restoring force.

In a normal mode of vibration the atoms have a common time dependence, $u_n(t) = u_n e^{\pm 2\pi i \nu t}$. Substituting this form into the equations of motion, the amplitudes u_n are seen to satisfy the equations

$$- M(2\pi\nu)^2 u_n = B(u_{n+1} - 2u_n + u_{n-1}),$$

that is

$$- M(2\pi\nu)^2 = B\left(\frac{u_{n+1}}{u_n} - 2 + \frac{u_{n-1}}{u_n}\right).$$

The left-hand side is independent of n, and in a normal mode the displacements of successive atoms have a constant ratio $u_{n+1}/u_n = e^{i\varphi}$. The corresponding frequency is given by

$$- M(2\pi\nu)^2 = 2B(\cos\phi - 1)$$

that is

$$\nu = \frac{1}{\pi}\sqrt{\frac{B}{M}} \mid \sin\left(\frac{\phi}{2}\right) \mid.$$

All physically distinct modes, that is all distinct ratios u_{n+1}/u_n, can be obtained by taking the angle ϕ in the range $-\pi < \phi < \pi$. The wave number is defined by writing the change in phase from atom to atom as $\phi = 2\pi\kappa a$. Then the ratio of the displacements of successive atoms is

$$\frac{u_{n+1}}{u_n} = e^{2\pi i \kappa}.$$

and the corresponding frequency is

$$\nu(\kappa) = \frac{1}{\pi}\sqrt{\frac{B}{M}} \mid \sin(\pi\kappa a) \mid,$$

with all distinct modes defined by κ in the interval $-\frac{1}{2a} < \kappa < \frac{1}{2a}$. The mode with $\kappa = \frac{1}{2a}$ has $\frac{u_{n+1}}{u_n} = -1$, and at this point $v_g = \frac{\partial\nu}{\partial\kappa} = 0$. This mode is a standing wave.

This explicit treatment of a simple model is seen to confirm the general arguments of Chapter IX, § 2.

BIBLIOGRAPHY

Although some of the books contained in this Bibliography are now rather dated it is felt, for historical and interest reasons, that none of them should be removed from the list.

GENERAL

N. Bohr, *Atomic Theory and the Description of Nature*, Cambridge University Press, 1934.

A. Sommerfeld, *Atomic Structure and Spectral Lines*, Methuen, 1930.

A. E. Ruark and H. C. Urey, *Atoms, Molecules and Quanta*, New York, McGraw-Hill, 1930.

E. N. da C. Andrade, *The Structure of the Atom*, Bell, 1934.

C. G. Darwin, *The New Conceptions of the Atom*, Bell, 1931.

F. K. Richtmyer, E. H. Kennard and T. Lauritsen, *Introduction to Modern Physics*, McGraw-Hill, 1955.

A. P. French, *Principles of Modern Physics*, J. Wiley and Sons, 1958.

R. B. Leighton, *Principles of Modern Physics*, McGraw-Hill, 1959.

R. E. Peierls, *The Laws of Nature*, George Allen, 1955.

R. P. Feynman and others, *The Feynman Lectures on Physics*, Addison-Wesley, 1963–5.

CHAPTER I

J. H. Jeans, *The Dynamical Theory of Gases*, Cambridge University Press, 1925.

J. H. Jeans, *Introduction to the Kinetic Theory of Gases*, Cambridge University Press, 1940.

E. Bloch, *The Kinetic Theory of Gases*, Methuen, 1924.

R. G. J. Fraser, *Molecular Rays*, Cambridge University Press, 1931.

R. C. Tolman, *The Principles of Statistical Mechanics*, Oxford University Press, 1938.

J. H. and M. Mayer, *Statistical Mechanics*, J. Wiley & Sons, 1940.

E. H. Kennard, *Kinetic Theory of Gases*, McGraw-Hill, 1938.

D. ter Haar, *Elements of Thermostatistics* (second edition), Holt, Rinehart and Winston, 1966.

A. Einstein, *Investigations on the Theory of the Brownian Movement*, Dover Publications, 1956.

S. G. Brush, *Kinetic Theory (Selected Readings in Physics)*, Vols. 1 and 2, Pergamon, 1965–6.

CHAPTERS II, III

Sir J. J. Thomson and G. P. Thomson, *Conduction of Electricity through Gases* (third edition), Cambridge University Press, 1928.

K. K. Darrow, *Electrical Phenomena in Gases*, Baillière, Tindall & Cox, London, 1932.

M. Born, *Einstein's Theory of Relativity*, Methuen, 1924.

A. S. Eddington, *Space, Time and Gravitation*, Cambridge University Press, 1920.

F. W. Aston, *Mass-Spectra and Isotopes* (second edition), Arnold, 1942.

Sir E. Rutherford, J. Chadwick and C. D. Ellis, *Radiations from Radioactive Substances*, Cambridge University Press, 1930.

A. Einstein, *The Meaning of Relativity*, Princeton University Press, 1956.

P. G. Bergman, *Introduction to the Theory of Relativity*, Prentice-Hall, 1942.

W. H. McCrea, *Relativity Physics*, Methuen, 1947.

R. A. Millikan, *Electrons (+ and —), Protons, Photons, Neutrons, and Cosmic Rays*, Cambridge University Press, 1935.

H. Dingle, *The Special Theory of Relativity*, Methuen, 1950.

J. D. Jackson, *The Physics of Elementary Particles*, Princeton University Press, 1958.

B. Rossi, *High Energy Particles*, Prentice-Hall, Inc., 1952.

L. D. Landau and G. B. Rumer, *What is Relativity?*, Oliver and Boyd, 1960.

C. Møller, *The Theory of Relativity*, Oxford, 1955.

C. E. Swartz, *The Fundamental Particles*, Addison-Wesley, 1965.

D. H. Frisch and A. M. Thorndike, *Elementary Particles*, Van Nostrand, 1964.

CHAPTERS IV, V

N. Bohr, *The Theory of Spectra and Atomic Constitution*, Cambridge University Press, 1922.

M. Born, *The Mechanics of the Atom*, Bell, 1927.

A. H. Compton, *X-Rays and Electrons*, Macmillan, 1927.

L. de Broglie and L. Brillouin, *Selected Papers on Wave Mechanics*, Blackie, 1928.

P. A. M. Dirac, *The Principles of Quantum Mechanics* (fourth edition), Clarendon Press, 1958.

F. Frenkel, *Wave Mechanics, Elementary Theory*, Clarendon Press, 1932.

W. Heisenberg, *The Principles of the Quantum Theory*, University of Chicago Press, 1930.

N. F. Mott, *An Outline of Wave Mechanics*, Cambridge University Press, 1930.

E. Schrödinger, *Collected Papers on Wave Mechanics*, Blackie, 1928.

V. Rojansky, *Introductory Quantum Mechanics*, Blackie, 1939.

L. Pauling and E. Bright Wilson, *Introduction to Quantum Mechanics*, McGraw-Hill, 1935.

A. Landé, *Principles of Quantum Mechanics*, Cambridge University Press, 1937.

M. R. Siddiqi, *Lectures on Quantum Mechanics*, Osmania University Press, 1938.

W. Heitler, *Elementary Wave Mechanics*, Oxford University Press, 1945.

N. F. Mott and F. N. Sneddon, *Wave Mechanics and its Applications*, Oxford University Press, 1948.

L. I. Schiff, *Quantum Mechanics*, McGraw-Hill, 1949.
D. Bohm, *Quantum Theory*, Prentice-Hall, 1951
F. Mandl, *Quantum Mechanics*, Butterworth, 1957.
H. A. Kramers, *The Foundations of Quantum Theory*, North-Holland, 1957.
L. D. Landau and E. M. Lifshitz, *Quantum Mechanics*, Pergamon, 1958.
R. H. Dicke and J. P. Wittke, *Introduction to Quantum Mechanics*, Addison-Wesley, 1960.
P. T Matthews, *Introduction to Quantum Mechanics*, McGraw-Hill, 1963.
E. Merzbacher, *Quantum Mechanics*, Wiley, 1961.
A. Messiah, *Quantum Mechanics*, Vols. I and II, North-Holland, 1962.
J. L. Powell and B. Crasemann, *Quantum Mechanics*, Addison-Wesley, 1961.
K. Gottfried, *Quantum Mechanics*, Vol. I : Fundamentals, Benjamin, 1966.
H. A. Bethe, *Intermediate Quantum Mechanics*, Benjamin, 1964.
H. A. Bethe and E. E. Salpeter, *Quantum Mechanics of One- and Two-Electron Atoms*, Springer, 1957.
A. S. Davydov, *Quantum Mechanics*, Pergamon, 1965.
B. L. van der Waerden, *Sources of Quantum Mechanics*, North-Holland, 1967.

CHAPTER VI

L. Pauling and S. Goudsmit, *The Structure of Line Spectra*, McGraw-Hill, 1930.
R. de L. Kronig, *Band Spectra and Molecular Structure*, Cambridge University Press, 1930.
Faraday Society, *Molecular Spectra and Molecular Structure*, 1930.
R. F. Bacher and S. Goudsmit, *Atomic Energy States*, McGraw-Hill, 1932.
H. E. White, *Introduction to Atomic Spectra*, McGraw-Hill, 1937.
E. U. Condon and G. H. Shortley, *The Theory of Atomic Spectra*, Cambridge University Press, 1935.
G. Herzberg, *Atomic Spectra and Atomic Structure*, Dover, 1944.
H. G. Kuhn, *Atomic Spectra*, Longmans, 1962.
G. W. Series, *Spectrum of Atomic Hydrogen*, Oxford, 1957.
R. M. Eisberg, *Fundamentals of Modern Physics*, Wiley, 1961.

CHAPTER VII

R. H. Fowler, *Statistical Mechanics*, Cambridge University Press, 1929.
F. Seitz, *The Modern Theory of Solids*, McGraw-Hill, 1940.
N. F. Mott and H. Jones, *The Theory of the Properties of Metals and Alloys*, Oxford University Press, 1936.
A. H. Wilson, *The Theory of Metals*, Cambridge University Press, 1936.
J. C. Slater, *Chemical Physics*, McGraw-Hill, 1939.
G. S. Rushbrooke, *Introduction to Statistical Mechanics*, Oxford University Press, 1949.
M. Born and Kun Huang, *Dynamical Theory of Crystal Lattices*, Oxford, Clarendon Press, 1954.
C. Kittel, *Introduction to Solid Physics*, Wiley, 1953.
R. E. Peierls, *Quantum Theory of Solids*, Oxford, Clarendon Press, 1955.
L. D. Landau and E. M. Lifshitz, *Statistical Physics*, Pergamon, 1958.

R. K. Chisholm and A. H. de Borde, *Introduction to Statistical Mechanics*, Pergamon, 1958.
D. ter Haar, *Elements of Thermostatistics* (second edition), Holt, Rinehart and Winston, 1966.
F. Reif, *Fundamentals of Statistical and Thermal Physics*, McGraw-Hill, 1965.
R. Reif, *Statistical Physics* (Berkeley Physics Course, Volume 5), McGraw-Hill, 1965.

CHAPTER VIII

P. Debye, *Polar Molecules*, Chemical Catalogue Company, New York, 1929.
G. B. B. M. Sutherland, *Infra-red and Raman Spectra*, Methuen, 1935.
R. de L. Kronig, *The Optical Basis of the Theory of Valency*, Cambridge University Press, 1935.
L. Pauling, *The Nature of the Chemical Bond*, Cornell University Press, 1939.
G. Herzberg, *Molecular Spectra and Molecular Structure*:
    Vol. I, *Diatomic Molecules*, Prentice-Hall, 1939.
    Vol. II, *Polyatomic Molecules*, D. van Nostrand, 1945.
N. F. Ramsey, *Molecular Beams*, Oxford, 1956.
C. A. Coulson, *Valency* (second edition), Oxford, 1961.
J. N. Murrell and others, *Valence Theory*, Wiley, 1965.
H. B. Gray, *Electrons and Chemical Bonding*, Benjamin, 1964.

CHAPTER IX

F. Seitz, *The Modern Theory of Solids*, McGraw-Hill, 1940.
N. F. Mott and H. Jones, *The Theory of the Properties of Metals and Alloys*, Oxford, 1936.
M. Born and K. Huang, *Dynamical Theory of Crystal Lattices*, Oxford, 1954.
R. E. Peierls, *Quantum Theory of Solids*, Oxford, 1955.
A. J. Dekker, *Solid State Physics*, Macmillan, 1958.
C. Kittel, *Introduction to Solid State Physics* (third edition), Wiley, 1966.
C. Kittel, *Elementary Solid State Physics*, Wiley, 1962.
C. Kittel, *Quantum Theory of Solids*, Wiley, 1963.
J. C. Slater, *Quantum Theory of Molecules and Solids*, Vols. I–III, McGraw-Hill 1963–7.
J. M. Ziman, *Electrons and Phonons*, Oxford, 1962.
J. M. Ziman, *Principles of the Theory of Solids*, Cambridge, 1965.
J. M. Ziman, *Electrons in Metals: a short guide to the Fermi surface*, Taylor, 1963.
A. B. Pippard, *The Dynamics of Conduction Electrons*, Blackie, 1965.
R. A. Smith, *Semiconductors*, Cambridge, 1964.
D. A. Wright, *Semiconductors*, Methuen, 1966.
C. A. Wert and R. M. Thomson, *Physics of Solids*, McGraw-Hill, 1964.
F. C. Brown, *The Physics of Solids*, Pergamon, 1967.

CHAPTER X

H. A. Bethe and P. Morrison, *Elementary Nuclear Theory*, John Wiley & Sons, 1956.

E. Rasetti, *Elements of Nuclear Physics*, Prentice-Hall, 1936.

L. Rosenfeld, *Nuclear Forces*, Vols. I and II, Interscience, 1948.

J. Mattauch and S. Flügge, *Nuclear Physics Tables*, Interscience, 1946.

J. M. Blatt and V. F. Weisskopf, *Theoretical Nuclear Physics*, John Wiley & Sons, 1952.

R. G. Sachs, *Nuclear Theory*, Addison-Wesley, 1953.

R. D. Evans, *The Atomic Nucleus*, McGraw-Hill, 1955.

S. Glasstone, *Source Book on Atomic Energy*, D. Van Nostrand Co., 1958.

A. E. S. Green, *Nuclear Physics*, McGraw-Hill, 1955.

D. Halliday, *Introductory Nuclear Physics*, John Wiley & Sons, 1955.

I. Kaplan, *Nuclear Physics*, Addison-Wesley, 1955.

L. R. B. Elton, *Introductory Nuclear Theory*, Pitman, 1965.

R. M. Eisberg, *Fundamentals of Modern Physics*, Wiley, 1961.

M. A. Preston, *Physics of the Nucleus*, Addison-Wesley, 1962.

W. E. Burcham, *Nuclear Physics*, Longmans, 1963.

S. de Benedetti, *Nuclear Interactions*, Wiley, 1964.

INDEX

Abegg, 267.
Abov, 341.
Abraham, 56.
absorption, by oscillator, 208, 455.
absorptive power, 205.
actinides, 183–4.
action variables, 384.
adiabatic invariants, 113, 386.
adjoint operator, 427, 429.
affinity, electronic, 268, 270.
alkalies, 123, 153, 165–6, 169, 185, 188, 270.
Allen, 68.
alpha disintegration, 40, 309, 444.
— rays, 30, 32, 62, 70, 377.
Alvarez, 314.
amplifier, 35.
Anderson, 47, 51, 194.
angle variables, 385.
Ångström unit, 248.
angular momentum in wave mechanics, 141.
— — nuclear, 66, 263, 311, 316, 327.
— — of electron, 144, 154.
— — orbital and spin (coupling), 155, 171, 327.
— — quantization of 111–12, 119–21, 131, 142, 154, 262, 404.
Angus, 335.
anharmonic forces, 280, 284.
anode rays, 28.
anti-neutron, 195.
anti-proton, 195.
astatine, 39, 43.
Aston, 29, 41, 65, 69, 70, 185, 349.
— mass-spectograph, 29, 41.
atom, electron shells in, 179.
atomic binding, 267.
— bomb, 353–5.
— energy, 355.
atomic form factor, 197, 198, 202, 409, 419.

atomic form factor number, 38, 63, 179.
— pile. 354.
— radius, 202.
— structure, 59, 103.
— theory, beginnings of, 62.
— — Bohr's.   See *Bohr* — —.
atomic theory in chemistry, 1.
— — many-electron, 198, 203.
— weight, 2, 39, 42.
Avogadro, 2.
— number (N_0), 3, 19, 361.

Bainbridge, 41, 73.
Balmer, 105.
— series, 105.
— terms, 106, 116, 119–20, 139, 166, 168, 256, 391.
band spectrum, 112, 256–66, 313.
Bardeen, 297, 305.
Barkla, 125.
Bartlett, 321, 330.
beats, 90, 148.
Becker, 44.
Becquerel, 30.
benzene, 267, 277.
Bernoulli, 5.
Berzelius, 267.
Bessel function, 398.
beta decay, 40, 67, 334.
— — *ft*-value, 339.
— rays, 30, 31, 40, 334.
betatron, 75–6.
Bethe, 52, 68, 95, 195, 333–4, 409.
Bhabha, 48.
Bhaghavantam, 220.
Bhimasenachar, 220.
binding, atomic, 267, 271, 275.
binding cohesion (van der Waals), 267, 275, 471.
— covalency (homopolar), 267, 271, 466.
— energies, nuclear, 64, 71, 75, 308, 318, 348, 350.

A CATALOG OF SELECTED

DOVER BOOKS
IN SCIENCE AND MATHEMATICS

A CATALOG OF SELECTED
DOVER BOOKS
IN SCIENCE AND MATHEMATICS

QUALITATIVE THEORY OF DIFFERENTIAL EQUATIONS, V.V. Nemytskii and V.V. Stepanov. Classic graduate-level text by two prominent Soviet mathematicians covers classical differential equations as well as topological dynamics and ergodic theory. Bibliographies. 523pp. 5⅜ × 8½. 65954-2 Pa. $10.95

MATRICES AND LINEAR ALGEBRA, Hans Schneider and George Phillip Barker. Basic textbook covers theory of matrices and its applications to systems of linear equations and related topics such as determinants, eigenvalues and differential equations. Numerous exercises. 432pp. 5⅜ × 8½. 66014-1 Pa. $9.95

QUANTUM THEORY, David Bohm. This advanced undergraduate-level text presents the quantum theory in terms of qualitative and imaginative concepts, followed by specific applications worked out in mathematical detail. Preface. Index. 655pp. 5⅜ × 8½. 65969-0 Pa. $13.95

ATOMIC PHYSICS (8th edition), Max Born. Nobel laureate's lucid treatment of kinetic theory of gases, elementary particles, nuclear atom, wave-corpuscles, atomic structure and spectral lines, much more. Over 40 appendices, bibliography. 495pp. 5⅜ × 8½. 65984-4 Pa. $11.95

ELECTRONIC STRUCTURE AND THE PROPERTIES OF SOLIDS: The Physics of the Chemical Bond, Walter A. Harrison. Innovative text offers basic understanding of the electronic structure of covalent and ionic solids, simple metals, transition metals and their compounds. Problems. 1980 edition. 582pp. 6⅛ × 9¼. 66021-4 Pa. $14.95

BOUNDARY VALUE PROBLEMS OF HEAT CONDUCTION, M. Necati Özisik. Systematic, comprehensive treatment of modern mathematical methods of solving problems in heat conduction and diffusion. Numerous examples and problems. Selected references. Appendices. 505pp. 5⅜ × 8½. 65990-9 Pa. $11.95

A SHORT HISTORY OF CHEMISTRY (3rd edition), J.R. Partington. Classic exposition explores origins of chemistry, alchemy, early medical chemistry, nature of atmosphere, theory of valency, laws and structure of atomic theory, much more. 428pp. 5⅜ × 8½. (Available in U.S. only) 65977-1 Pa. $10.95

A HISTORY OF ASTRONOMY, A. Pannekoek. Well-balanced, carefully reasoned study covers such topics as Ptolemaic theory, work of Copernicus, Kepler, Newton, Eddington's work on stars, much more. Illustrated. References. 521pp. 5⅜ × 8½. 65994-1 Pa. $11.95

PRINCIPLES OF METEOROLOGICAL ANALYSIS, Walter J. Saucier. Highly respected, abundantly illustrated classic reviews atmospheric variables, hydrostatics, static stability, various analyses (scalar, cross-section, isobaric, isentropic, more). For intermediate meteorology students. 454pp. 6⅛ × 9¼. 65979-8 Pa. $12.95

SPECIAL FUNCTIONS, N.N. Lebedev. Translated by Richard Silverman. Famous Russian work treating more important special functions, with applications to specific problems of physics and engineering. 38 figures. 308pp. 5⅜ × 8½.
60624-4 Pa. $7.95

OBSERVATIONAL ASTRONOMY FOR AMATEURS, J.B. Sidgwick. Mine of useful data for observation of sun, moon, planets, asteroids, aurorae, meteors, comets, variables, binaries, etc. 39 illustrations. 384pp. 5⅜ × 8¼. (Available in U.S. only)
24033-9 Pa. $8.95

INTEGRAL EQUATIONS, F.G. Tricomi. Authoritative, well-written treatment of extremely useful mathematical tool with wide applications. Volterra Equations, Fredholm Equations, much more. Advanced undergraduate to graduate level. Exercises. Bibliography. 238pp. 5⅜ × 8½.
64828-1 Pa. $6.95

CELESTIAL OBJECTS FOR COMMON TELESCOPES, T.W. Webb. Inestimable aid for locating and identifying nearly 4,000 celestial objects. 77 illustrations. 645pp. 5⅜ × 8½.
20917-2, 20918-0 Pa., Two-vol. set $12.00

MODERN NONLINEAR EQUATIONS, Thomas L. Saaty. Emphasizes practical solution of problems; covers seven types of equations. ". . . a welcome contribution to the existing literature. . . ."—*Math Reviews*. 490pp. 5⅜ × 8½. 64232-1 Pa. $9.95

FUNDAMENTALS OF ASTRODYNAMICS, Roger Bate et al. Modern approach developed by U.S. Air Force Academy. Designed as a first course. Problems, exercises. Numerous illustrations. 455pp. 5⅜ × 8½.
60061-0 Pa. $8.95

INTRODUCTION TO LINEAR ALGEBRA AND DIFFERENTIAL EQUATIONS, John W. Dettman. Excellent text covers complex numbers, determinants, orthonormal bases, Laplace transforms, much more. Exercises with solutions. Undergraduate level. 416pp. 5⅜ × 8½.
65191-6 Pa. $9.95

INCOMPRESSIBLE AERODYNAMICS, edited by Bryan Thwaites. Covers theoretical and experimental treatment of the uniform flow of air and viscous fluids past two-dimensional aerofoils and three-dimensional wings; many other topics. 654pp. 5⅜ × 8½.
65465-6 Pa. $16.95

INTRODUCTION TO DIFFERENCE EQUATIONS, Samuel Goldberg. Exceptionally clear exposition of important discipline with applications to sociology, psychology, economics. Many illustrative examples; over 250 problems. 260pp. 5⅜ × 8½.
65084-7 Pa. $7.95

LAMINAR BOUNDARY LAYERS, edited by L. Rosenhead. Engineering classic covers steady boundary layers in two- and three-dimensional flow, unsteady boundary layers, stability, observational techniques, much more. 708pp. 5⅜ × 8½.
65646-2 Pa. $15.95

LECTURES ON CLASSICAL DIFFERENTIAL GEOMETRY, Second Edition, Dirk J. Struik. Excellent brief introduction covers curves, theory of surfaces, fundamental equations, geometry on a surface, conformal mapping, other topics. Problems. 240pp. 5⅜ × 8½.
65609-8 Pa. $6.95

ROTARY-WING AERODYNAMICS, W.Z. Stepniewski. Clear, concise text covers aerodynamic phenomena of the rotor and offers guidelines for helicopter performance evaluation. Originally prepared for NASA. 537 figures. 640pp. 6⅛ × 9¼.
64647-5 Pa. $14.95

DIFFERENTIAL GEOMETRY, Heinrich W. Guggenheimer. Local differential geometry as an application of advanced calculus and linear algebra. Curvature, transformation groups, surfaces, more. Exercises. 62 figures. 378pp. 5⅜ × 8½.
63433-7 Pa. $7.95

INTRODUCTION TO SPACE DYNAMICS, William Tyrrell Thomson. Comprehensive, classic introduction to space-flight engineering for advanced undergraduate and graduate students. Includes vector algebra, kinematics, transformation of coordinates. Bibliography. Index. 352pp. 5⅜ × 8½.   65113-4 Pa. $8.95

A SURVEY OF MINIMAL SURFACES, Robert Osserman. Up-to-date, in-depth discussion of the field for advanced students. Corrected and enlarged edition covers new developments. Includes numerous problems. 192pp. 5⅜ × 8½.
64998-9 Pa. $8.95

ANALYTICAL MECHANICS OF GEARS, Earle Buckingham. Indispensable reference for modern gear manufacture covers conjugate gear-tooth action, gear-tooth profiles of various gears, many other topics. 263 figures. 102 tables. 546pp. 5⅜ × 8½.   65712-4 Pa. $11.95

SET THEORY AND LOGIC, Robert R. Stoll. Lucid introduction to unified theory of mathematical concepts. Set theory and logic seen as tools for conceptual understanding of real number system. 496pp. 5⅜ × 8¼.   63829-4 Pa. $10.95

A HISTORY OF MECHANICS, René Dugas. Monumental study of mechanical principles from antiquity to quantum mechanics. Contributions of ancient Greeks, Galileo, Leonardo, Kepler, Lagrange, many others. 671pp. 5⅜ × 8½.
65632-2 Pa. $14.95

FAMOUS PROBLEMS OF GEOMETRY AND HOW TO SOLVE THEM, Benjamin Bold. Squaring the circle, trisecting the angle, duplicating the cube: learn their history, why they are impossible to solve, then solve them yourself. 128pp. 5⅜ × 8½.   24297-8 Pa. $3.95

MECHANICAL VIBRATIONS, J.P. Den Hartog. Classic textbook offers lucid explanations and illustrative models, applying theories of vibrations to a variety of practical industrial engineering problems. Numerous figures. 233 problems, solutions. Appendix. Index. Preface. 436pp. 5⅜ × 8½.   64785-4 Pa. $9.95

CURVATURE AND HOMOLOGY, Samuel I. Goldberg. Thorough treatment of specialized branch of differential geometry. Covers Riemannian manifolds, topology of differentiable manifolds, compact Lie groups, other topics. Exercises. 315pp. 5⅜ × 8½.   64314-X Pa. $8.95

HISTORY OF STRENGTH OF MATERIALS, Stephen P. Timoshenko. Excellent historical survey of the strength of materials with many references to the theories of elasticity and structure. 245 figures. 452pp. 5⅜ × 8½. 61187-6 Pa. $10.95

CATALOG OF DOVER BOOKS

GEOMETRY OF COMPLEX NUMBERS, Hans Schwerdtfeger. Illuminating, widely praised book on analytic geometry of circles, the Moebius transformation, and two-dimensional non-Euclidean geometries. 200pp. 5⅜ × 8¼.
63830-8 Pa. $6.95

MECHANICS, J.P. Den Hartog. A classic introductory text or refresher. Hundreds of applications and design problems illuminate fundamentals of trusses, loaded beams and cables, etc. 334 answered problems. 462pp. 5⅜ × 8½. 60754-2 Pa. $8.95

TOPOLOGY, John G. Hocking and Gail S. Young. Superb one-year course in classical topology. Topological spaces and functions, point-set topology, much more. Examples and problems. Bibliography. Index. 384pp. 5⅜ × 8¼.
65676-4 Pa. $8.95

STRENGTH OF MATERIALS, J.P. Den Hartog. Full, clear treatment of basic material (tension, torsion, bending, etc.) plus advanced material on engineering methods, applications. 350 answered problems. 323pp. 5⅜ × 8½. 60755-0 Pa. $7.50

ELEMENTARY CONCEPTS OF TOPOLOGY, Paul Alexandroff. Elegant, intuitive approach to topology from set-theoretic topology to Betti groups; how concepts of topology are useful in math and physics. 25 figures. 57pp. 5⅜ × 8½.
60747-X Pa. $2.95

ADVANCED STRENGTH OF MATERIALS, J.P. Den Hartog. Superbly written advanced text covers torsion, rotating disks, membrane stresses in shells, much more. Many problems and answers. 388pp. 5⅜ × 8½. 65407-9 Pa. $9.95

COMPUTABILITY AND UNSOLVABILITY, Martin Davis. Classic graduate-level introduction to theory of computability, usually referred to as theory of recurrent functions. New preface and appendix. 288pp. 5⅜ × 8½. 61471-9 Pa. $6.95

GENERAL CHEMISTRY, Linus Pauling. Revised 3rd edition of classic first-year text by Nobel laureate. Atomic and molecular structure, quantum mechanics, statistical mechanics, thermodynamics correlated with descriptive chemistry. Problems. 992pp. 5⅜ × 8½. 65622-5 Pa. $19.95

AN INTRODUCTION TO MATRICES, SETS AND GROUPS FOR SCIENCE STUDENTS, G. Stephenson. Concise, readable text introduces sets, groups, and most importantly, matrices to undergraduate students of physics, chemistry, and engineering. Problems. 164pp. 5⅜ × 8½. 65077-4 Pa. $6.95

THE HISTORICAL BACKGROUND OF CHEMISTRY, Henry M. Leicester. Evolution of ideas, not individual biography. Concentrates on formulation of a coherent set of chemical laws. 260pp. 5⅜ × 8½. 61053-5 Pa. $6.95

THE PHILOSOPHY OF MATHEMATICS: An Introductory Essay, Stephan Körner. Surveys the views of Plato, Aristotle, Leibniz & Kant concerning propositions and theories of applied and pure mathematics. Introduction. Two appendices. Index. 198pp. 5⅜ × 8½. 25048-2 Pa. $6.95

THE DEVELOPMENT OF MODERN CHEMISTRY, Aaron J. Ihde. Authoritative history of chemistry from ancient Greek theory to 20th-century innovation. Covers major chemists and their discoveries. 209 illustrations. 14 tables. Bibliographies. Indices. Appendices. 851pp. 5⅜ × 8½. 64235-6 Pa. $17.95

DE RE METALLICA, Georgius Agricola. The famous Hoover translation of greatest treatise on technological chemistry, engineering, geology, mining of early modern times (1556). All 289 original woodcuts. 638pp. 6¾ × 11.
60006-8 Pa. $17.95

SOME THEORY OF SAMPLING, William Edwards Deming. Analysis of the problems, theory and design of sampling techniques for social scientists, industrial managers and others who find statistics increasingly important in their work. 61 tables. 90 figures. xvii + 602pp. 5⅜ × 8½.
64684-X Pa. $15.95

THE VARIOUS AND INGENIOUS MACHINES OF AGOSTINO RAMELLI: A Classic Sixteenth-Century Illustrated Treatise on Technology, Agostino Ramelli. One of the most widely known and copied works on machinery in the 16th century. 194 detailed plates of water pumps, grain mills, cranes, more. 608pp. 9 × 12. (EBE)
25497-6 Clothbd. $34.95

LINEAR PROGRAMMING AND ECONOMIC ANALYSIS, Robert Dorfman, Paul A. Samuelson and Robert M. Solow. First comprehensive treatment of linear programming in standard economic analysis. Game theory, modern welfare economics, Leontief input-output, more. 525pp. 5⅜ × 8½.
65491-5 Pa. $13.95

ELEMENTARY DECISION THEORY, Herman Chernoff and Lincoln E. Moses. Clear introduction to statistics and statistical theory covers data processing, probability and random variables, testing hypotheses, much more. Exercises. 364pp. 5⅜ × 8½.
65218-1 Pa. $9.95

THE COMPLEAT STRATEGYST: Being a Primer on the Theory of Games of Strategy, J.D. Williams. Highly entertaining classic describes, with many illus-trated examples, how to select best strategies in conflict situations. Prefaces. Appendices. 268pp. 5⅜ × 8½.
25101-2 Pa. $6.95

MATHEMATICAL METHODS OF OPERATIONS RESEARCH, Thomas L. Saaty. Classic graduate-level text covers historical background, classical methods of forming models, optimization, game theory, probability, queueing theory, much more. Exercises. Bibliography. 448pp. 5⅜ × 8¼.
65703-5 Pa. $12.95

CONSTRUCTIONS AND COMBINATORIAL PROBLEMS IN DESIGN OF EXPERIMENTS, Damaraju Raghavarao. In-depth reference work examines orthogonal Latin squares, incomplete block designs, tactical configuration, partial geometry, much more. Abundant explanations, examples. 416pp. 5⅜ × 8¼.
65685-3 Pa. $10.95

THE ABSOLUTE DIFFERENTIAL CALCULUS (CALCULUS OF TENSORS), Tullio Levi-Civita. Great 20th-century mathematician's classic work on material necessary for mathematical grasp of theory of relativity. 452pp. 5⅜ × 8½.
63401-9 Pa. $9.95

VECTOR AND TENSOR ANALYSIS WITH APPLICATIONS, A.I. Borisenko and I.E. Tarapov. Concise introduction. Worked-out problems, solutions, exer-cises. 257pp. 5⅜ × 8¼.
63833-2 Pa. $6.95

THE FOUR-COLOR PROBLEM: Assaults and Conquest, Thomas L. Saaty and Paul G. Kainen. Engrossing, comprehensive account of the century-old combinatorial topological problem, its history and solution. Bibliographies. Index. 110 figures. 228pp. 5⅜ × 8½.   65092-8 Pa. $6.95

CATALYSIS IN CHEMISTRY AND ENZYMOLOGY, William P. Jencks. Exceptionally clear coverage of mechanisms for catalysis, forces in aqueous solution, carbonyl- and acyl-group reactions, practical kinetics, more. 864pp. 5⅜ × 8½.   65460-5 Pa. $19.95

PROBABILITY: An Introduction, Samuel Goldberg. Excellent basic text covers set theory, probability theory for finite sample spaces, binomial theorem, much more. 360 problems. Bibliographies. 322pp. 5⅜ × 8½.   65252-1 Pa. $8.95

LIGHTNING, Martin A. Uman. Revised, updated edition of classic work on the physics of lightning. Phenomena, terminology, measurement, photography, spectroscopy, thunder, more. Reviews recent research. Bibliography. Indices. 320pp. 5⅜ × 8¼.   64575-4 Pa. $8.95

PROBABILITY THEORY: A Concise Course, Y.A. Rozanov. Highly readable, self-contained introduction covers combination of events, dependent events, Bernoulli trials, etc. Translation by Richard Silverman. 148pp. 5⅜ × 8¼.
65232-7 Pa. $5.95
63544-9 Pa. $5.95

THE CEASELESS WIND: An Introduction to the Theory of Atmospheric Motion, John A. Dutton. Acclaimed text integrates disciplines of mathematics and physics for full understanding of dynamics of atmospheric motion. Over 400 problems. Index. 97 illustrations. 640pp. 6 × 9.   65096-4 Pa. $17.95

STATISTICS MANUAL, Edwin L. Crow, et al. Comprehensive, practical collection of classical and modern methods prepared by U.S. Naval Ordnance Test Station. Stress on use. Basics of statistics assumed. 288pp. 5⅜ × 8½.
60599-X Pa. $6.95

DICTIONARY/OUTLINE OF BASIC STATISTICS, John E. Freund and Frank J. Williams. A clear concise dictionary of over 1,000 statistical terms and an outline of statistical formulas covering probability, nonparametric tests, much more. 208pp. 5⅜ × 8½.   66796-0 Pa. $6.95

STATISTICAL METHOD FROM THE VIEWPOINT OF QUALITY CONTROL, Walter A. Shewhart. Important text explains regulation of variables, uses of statistical control to achieve quality control in industry, agriculture, other areas. 192pp. 5⅜ × 8½.   65232-7 Pa. $6.95

THE INTERPRETATION OF GEOLOGICAL PHASE DIAGRAMS, Ernest G. Ehlers. Clear, concise text emphasizes diagrams of systems under fluid or containing pressure; also coverage of complex binary systems, hydrothermal melting, more. 288pp. 6½ × 9¼.   65389-7 Pa. $10.95

STATISTICAL ADJUSTMENT OF DATA, W. Edwards Deming. Introduction to basic concepts of statistics, curve fitting, least squares solution, conditions without parameter, conditions containing parameters. 26 exercises worked out. 271pp. 5⅜ × 8½.   64685-8 Pa. $7.95

TENSOR CALCULUS, J.L. Synge and A. Schild. Widely used introductory text covers spaces and tensors, basic operations in Riemannian space, non-Riemannian spaces, etc. 324pp. 5⅜ × 8¼. 63612-7 Pa. $7.95

A CONCISE HISTORY OF MATHEMATICS, Dirk J. Struik. The best brief history of mathematics. Stresses origins and covers every major figure from ancient Near East to 19th century. 41 illustrations. 195pp. 5⅜ × 8½. 60255-9 Pa. $7.95

A SHORT ACCOUNT OF THE HISTORY OF MATHEMATICS, W.W. Rouse Ball. One of clearest, most authoritative surveys from the Egyptians and Phoenicians through 19th-century figures such as Grassman, Galois, Riemann. Fourth edition. 522pp. 5⅜ × 8½. 20630-0 Pa. $10.95

HISTORY OF MATHEMATICS, David E. Smith. Nontechnical survey from ancient Greece and Orient to late 19th century; evolution of arithmetic, geometry, trigonometry, calculating devices, algebra, the calculus. 362 illustrations. 1,355pp. 5⅜ × 8½. 20429-4, 20430-8 Pa., Two-vol. set $23.90

THE GEOMETRY OF RENÉ DESCARTES, René Descartes. The great work founded analytical geometry. Original French text, Descartes' own diagrams, together with definitive Smith-Latham translation. 244pp. 5⅜ × 8½. 60068-8 Pa. $6.95

THE ORIGINS OF THE INFINITESIMAL CALCULUS, Margaret E. Baron. Only fully detailed and documented account of crucial discipline: origins; development by Galileo, Kepler, Cavalieri; contributions of Newton, Leibniz, more. 304pp. 5⅜ × 8½. (Available in U.S. and Canada only) 65371-4 Pa. $9.95

THE HISTORY OF THE CALCULUS AND ITS CONCEPTUAL DEVELOPMENT, Carl B. Boyer. Origins in antiquity, medieval contributions, work of Newton, Leibniz, rigorous formulation. Treatment is verbal. 346pp. 5⅜ × 8½. 60509-4 Pa. $7.95

THE THIRTEEN BOOKS OF EUCLID'S ELEMENTS, translated with introduction and commentary by Sir Thomas L. Heath. Definitive edition. Textual and linguistic notes, mathematical analysis. 2,500 years of critical commentary. Not abridged. 1,414pp. 5⅜ × 8½. 60088-2, 60089-0, 60090-4 Pa., Three-vol. set $29.85

GAMES AND DECISIONS: Introduction and Critical Survey, R. Duncan Luce and Howard Raiffa. Superb nontechnical introduction to game theory, primarily applied to social sciences. Utility theory, zero-sum games, n-person games, decision-making, much more. Bibliography. 509pp. 5⅜ × 8½. 65943-7 Pa. $11.95

THE HISTORICAL ROOTS OF ELEMENTARY MATHEMATICS, Lucas N.H. Bunt, Phillip S. Jones, and Jack D. Bedient. Fundamental underpinnings of modern arithmetic, algebra, geometry and number systems derived from ancient civilizations. 320pp. 5⅜ × 8½. 25563-8 Pa. $8.95

CALCULUS REFRESHER FOR TECHNICAL PEOPLE, A. Albert Klaf. Covers important aspects of integral and differential calculus via 756 questions. 566 problems, most answered. 431pp. 5⅜ × 8½. 20370-0 Pa. $8.95

CHALLENGING MATHEMATICAL PROBLEMS WITH ELEMENTARY SOLUTIONS, A.M. Yaglom and I.M. Yaglom. Over 170 challenging problems on probability theory, combinatorial analysis, points and lines, topology, convex polygons, many other topics. Solutions. Total of 445pp. 5⅜ × 8½. Two-vol. set.
Vol. I 65536-9 Pa. $6.95
Vol. II 65537-7 Pa. $6.95

FIFTY CHALLENGING PROBLEMS IN PROBABILITY WITH SOLUTIONS, Frederick Mosteller. Remarkable puzzlers, graded in difficulty, illustrate elementary and advanced aspects of probability. Detailed solutions. 88pp. 5⅜ × 8½.
65355-2 Pa. $3.95

EXPERIMENTS IN TOPOLOGY, Stephen Barr. Classic, lively explanation of one of the byways of mathematics. Klein bottles, Moebius strips, projective planes, map coloring, problem of the Koenigsberg bridges, much more, described with clarity and wit. 43 figures. 210pp. 5⅜ × 8½.   25933-1 Pa. $5.95

RELATIVITY IN ILLUSTRATIONS, Jacob T. Schwartz. Clear nontechnical treatment makes relativity more accessible than ever before. Over 60 drawings illustrate concepts more clearly than text alone. Only high school geometry needed. Bibliography. 128pp. 6⅛ × 9¼.   25965-X Pa. $5.95

AN INTRODUCTION TO ORDINARY DIFFERENTIAL EQUATIONS, Earl A. Coddington. A thorough and systematic first course in elementary differential equations for undergraduates in mathematics and science, with many exercises and problems (with answers). Index. 304pp. 5⅜ × 8½.   65942-9 Pa. $7.95

FOURIER SERIES AND ORTHOGONAL FUNCTIONS, Harry F. Davis. An incisive text combining theory and practical example to introduce Fourier series, orthogonal functions and applications of the Fourier method to boundary-value problems. 570 exercises. Answers and notes. 416pp. 5⅜ × 8½.   65973-9 Pa. $9.95

THE THEORY OF BRANCHING PROCESSES, Theodore E. Harris. First systematic, comprehensive treatment of branching (i.e. multiplicative) processes and their applications. Galton-Watson model, Markov branching processes, electron-photon cascade, many other topics. Rigorous proofs. Bibliography. 240pp. 5⅜ × 8½.   65952-6 Pa. $6.95

AN INTRODUCTION TO ALGEBRAIC STRUCTURES, Joseph Landin. Superb self-contained text covers "abstract algebra": sets and numbers, theory of groups, theory of rings, much more. Numerous well-chosen examples, exercises. 247pp. 5⅜ × 8½.   65940-2 Pa. $6.95

Prices subject to change without notice.
Available at your book dealer or write for free Mathematics and Science Catalog to Dept. GI, Dover Publications, Inc., 31 East 2nd St., Mineola, N.Y. 11501. Dover publishes more than 175 books each year on science, elementary and advanced mathematics, biology, music, art, literature, history, social sciences and other areas.